AKiL

NEUROSECRETION AND BRAIN PEPTIDES

Implications for Brain Functions and Neurological Disease

Advances in Biochemical Psychopharmacology
Volume 28

Advances in Biochemical Psychopharmacology

Series Editors

E. Costa, M.D.
Chief, Laboratory of Preclinical Pharmacology
National Institute of Mental Health
Washington, D.C.

Paul Greengard, Ph.D.
Professor of Pharmacology
Yale University School of Medicine
New Haven, Connecticut

Neurosecretion and Brain Peptides

*Implications for Brain Functions
and Neurological Disease*

---

*Advances in Biochemical
Psychopharmacology
Volume 28*

Volume Editors

Joseph B. Martin, M.D., Ph.D.
*Bullard Professor of Neurology
Harvard Medical School
and
Chief, Neurology Service
Massachusetts General Hospital
Boston, Massachusetts*

Seymour Reichlin, M.D., Ph.D.
*Chief, Endocrine Division
Department of Medicine
Tufts University School of Medicine
New England Medical Center Hospital
Boston, Massachusetts*

Katherine L. Bick, Ph.D.
*Deputy Director, Neurological Disorders Program
National Institute of Neurological and Communicative Disorders and Stroke
National Institutes of Health
Bethesda, Maryland*

Raven Press ■ New York

Raven Press, 1140 Avenue of the Americas, New York, New York 10036

Made in the United States of America

Great care has been taken to maintain the accuracy of the information contained in the volume. However, Raven Press cannot be held responsible for errors or for any consequences arising from the use of the information contained herein.

Materials appearing in this book prepared by individuals as part of their official duties as U.S. Government employees are not covered by the above-mentioned copyright.

Library of Congress Cataloging in Publication Data

Main entry under title:

Neurosecretion and brain peptides.

(Advances in biochemical psychopharmacology ; v. 28)
Includes index.
1. Brain chemistry. 2. Peptides—Metabolism.
3. Neurosecretion. 4. Nervous system—Diseases.
I. Martin, Joseph B. II. Reichlin, Seymour.
III. Bick, Katherine L. IV. Title: Brain peptides.
V. Series. [DNLM: 1. Brain—Physiology.
2. Neurosecretion. 3. Peptides—Secretion.
4. Nervous system diseases—Pathology. W1 AD437
v. 28 / WL 102 N5063]
RM315.A4   vol. 28   [QP376]   615′.78s   80-5375
ISBN 0-89004-535-6   [612′.8042]   AACR1

Preface

Investigations of the neuronal localization and secretion of structurally defined polypeptides have now occupied three decades of scientific endeavor. The historic elucidation of the amino acid composition of vasopressin and oxytocin by du Vigneuad and co-workers provided the first framework for biochemical approaches to the field. The characterization of specific hypothalamic releasing factors essential for regualtion of anterior pituitary hormone secretion by Guillemin and Schally and their respective colleagues catapulted peptides into neurobiology. The elucidation of the amino acid sequences of Substance P, neurotensin, and the opioid peptides drew attention to a broader spectrum of neuropeptides important for neurobiology, endocrinology, neurology, and psychiatry.

There have been many publications on the neuropeptides and neuroendocrinology in the past few years. One might ask, "Why another?" The consideration that led to the publication of this volume was the recognition that sufficient information was now available about the synthesis, anatomical distribution, and function of neuropeptides to justify a critical evaluation of their potential importance for the understanding of the pathogenesis and manifestations of neurological disease.

To provide coherence to the volume, individual chapters are organized by topic areas, each of which is summarized by a brief introduction that highlights the topics covered. In the first section, the biology of the neurosecretory neuron, general principles of peptide hormone biosynthesis, and specific examples of peptide processing and degradation are considered. The anatomical distribution of selected neuropeptides including oxytocin, vasopressin, the opioid peptides, Substance P, and neurotensin are reviewed in Sections II and III. Physiological control of neurosecretion and the mechanisms by which neuropeptides regulate neuronal functions are examined in a variety of systems. The topics covered in Section III range from a consideration of general principles of peptide–receptor interaction and characterization of opiate receptors in brain to evaluation of the function of Substance P and endorphins in pain perception.

The role of peptides in regulating neuronal growth and differentiation and their function in visceral homeostasis in pituitary control and regulation of brain volume are considered in Sections IV through VII. Other topics covered in the volume include the function of peptides in shock, circadian rhythms, the significance of cerebrospinal fluid as a pathway of neuroendocrine and nervous system control and the anatomy of the circumventricular organs.

The usefulness of pituitary function studies as an indicator of hypothalmic funciton, of the practical aspects of evaluation of peptide secretion in man, and of the use of peptide analogues as pharmacological modulators of brain function are evaluated critically in the last two sections. The possible role of peptides in the pathogenesis of genetic disorders of brain development is reviewed. Speculations about the potential implication of neuropeptides in specific neurological disease are offered.

This volume will be of interest to clinical neurologists, neurosurgeons, and psychiatrists and to basic scientists working in fundamental studies on the neuropeptides.

Joseph B. Martin
Seymour Reichlin
Katherine L. Bick

Acknowledgments

We are grateful to the many individuals who were responsible for the completion of this effort. Dr. Donald B. Tower, Director of the National Institute of Neurological and Communicative Disorders and Stroke (NINCDS), recognizing the relevance of neuroendocrine and neuropeptide research for problems of neurological disease, inititated the planning for the conference on which this volume is based and has provided continuing encouragement and advice. In addition to serving as coeditor, Dr. Katherine L. Bick, Deputy Director, Neurological Disorders Program, NINCDS, and her staff provided invaluable administrative support for the conference and the preparation of this volume. Many individuals helped in the final assembly of the papers for publication, in editorial review, and in referencing. Special thanks are due to Martha Conant, Rebecca Frost, Therese Kendall, Kathryn Phillips, and Kathy Sullivan.

The conference was supported by USPHS Grants NS15743 and AM27129 from the National Institute of Neurological and Communicative Disorders and Stroke and the National Institute of Arthritis, Metabolism and Digestive Diseases.

Contents

Contributors

Huda Akil
Department of Psychiatry
Mental Health Research Institute
University of Michigan
Ann Arbor, Michigan 48109

Y.-A. Barde
Department of Neurochemistry
Max-Planck Institute for Psychiatry
D-8033 Martinsried, Federal Republic
   of Germany

M. Benuck
Center for Neurochemistry
Rockland Research Institute
Ward's Island, New York 10035

Katherine L. Bick
Neurological Disorders Program
National Institutes of Health
Bethesda, Maryland 20205

Edward D. Bird
McLean Hospital
Belmont, Massachusetts 02178

Ira B. Black
Division of Developmental Neurology
Department of Neurology
Cornell University Medical College
New York, New York 10021

C.B. Bræstrup
Department of Physchopharmacology
Sct. Hans Hospital
DK-400, Roskilde, Denmark

Leon D. Braun
Department of Neurology
University of California School of
   Medicine
Los Angeles, California 90073

Marvin R. Brown
Peptide Biology Laboratory
The Salk Institute
San Diego, California 92138

Michael J. Brownstein
Unit on Neuroendocrinology
Laboratory of Clinical Science
National Institutes of Health
Bethesda, Maryland 20205

M. Bunge
Department of Anatomy and
   Neurobiology
Washington University School of
   Medicine
St. Louis, Missouri 63110

Richard P. Bunge
Department of Anatomy and
   Neurobiology
Washington University School of
   Medicine
St. Louis, Missouri 63110

S. Caldecott-Hazard
Department of Psychology
Brain Research Institute
University of California
Los Angeles, California 90024

J. T. Cannon
Department of Psychology
Brain Research Institute
University of California
Los Angeles, California 90024

T. Caraceni
Istituto Neurologic C. Besta
Milan, Italy

Marie S. Carmichael
Department of Psychology
University of California
Berkeley, California 94720

Verne S. Caviness, Jr.
Neurology Service
Massachusetts General Hospital
Boston, Massachusetts 02114 and
The Eunice Kennedy Shriver Center for
  Mental Retardation, Inc.
Waltham, Massachusetts 02154

D. Cocchi
Department of Pharmacology
University of Milan
Milan, Italy

Eain M. Cornford
Department of Neurology
University of California School of
  Medicine
Los Angeles, California 90073

Paul D. Crane
Department of Neurology
University of California School of
  Medicine
Los Angeles, California 90073

Charles A. Czeisler
Laboratory of Human Chronophysiology
Department of Neurology
Montefiore Hospital
Albert Einstein College of Medicine
Bronx, New York 10467

John R. Delfs
Department of Neurosciences
Children's Hospital Medical Center
Boston, Massachusetts 02115

Marc A. Dichter
Department of Neurosciences
Children's Hospital Medical Center
Boston, Massachusetts 02115

Jane Dodd
Department of Neurobiology
Harvard Medical School
Boston, Massachusetts 02115

Kathleen Dunlap
Department of Pharmacology
Harvard Medical School
Boston, Massachusetts 02115

D. Edgar
Department of Neurochemistry
Max-Planck Institute for Psychiatry
D-8033 Martinsried, Federal Republic
  of Germany

P. D. Edminson
Pediatric Research Institute
Rikshospitalet
Oslo 1, Norway

Betty A. Eipper
Department of Physiology
University of Colorado Health Sciences
  Center
Denver, Colorado 80262

Alan N. Epstein
Leidy Laboratory of Biology
University of Pennsylvania
Philadephia, Pennsylvania 19104

Alan I. Faden
Division of Neuropsychiatry
Department of Neurosciences
Walter Reed Army Institute of Research
Washington, D.C. 20012

Howard L. Fields
Departments of Neurology and
  Physiology
University of California
San Francisco, California 94153

Gerald D. Fischbach
Department of Pharmacology
Harvard Medical School
Boston, Massachusetts 02115

Jeffrey S. Flier
Department of Medicine
Beth Israel Hospital
Harvard Medical School
Boston, Massachusetts 02215

Harrison J. L. Frank
Division of Endocrinology and
Metabolism
Department of Medicine
University of California School of
Medicine
Los Angeles, California 90073

E. J. Furshpan
Department of Neurobiology
Harvard Medical School
Boston, Massachusetts 02115

Harold Gainer
Section on Functional Neurochemistry
Laboratory of Developmental
Neurobiology
National Institutes of Health
Bethesda, Maryland 20205

Detlev Ganten
Department of Pharmacology
University of Heidelberg
6900 Heidelberg, Federal Republic
of Germany

J. Gibbs
Department of Psychiatry
The Edward W. Bourne Behavioral
Research Laboratory
New York Hospital–Cornell Medical
Center, Westchester Division
White Plains, New York 10605

P. Giovannini
Istituto Neurologica C. Besta
Milan, Italy

Denis Gospodarowicz
Cancer Research Institute
University of California Medical Center
San Francisco, California 94143

Joel F. Habener
Laboratory of Molecular Endocrinology
Massachusetts General Hospital
Boston, Massachusetts 02114

A. Hamberger
Department of Neurobiology
University of Gothenberg
S-40053, Gothenberg, Sweden

John W. Holaday
Division of Neuropsychiatry
Department of Medical Neurosciences
Walter Reed Army Institute of Research
Washington, D.C. 20012

K. Hole
Department of Physiology
University of Bergen
N-5000 Bergen, Norway

Hiroo Imura
Department of Medicine
Kyoto University
Sakyo-Ku
Kyoto 606, Japan

Ivor M. D. Jackson
Tufts University School of Medicine
New England Medical Center
Boston, Massachusetts 02111

Thomas M. Jessell
Department of Pharmacology
St. George's Hospital Medical School
London SW17 0RE, England

H. Katakami
Department of Medicine
Kyoto University
Sakyo-Ku
Kyoto 606, Japan

Y. Kato
Department of Medicine
Kyoto University
Sakyo-Ku
Kyoto 606, Japan

John S. Kelly
Department of Pharmacology
St. George's Hospital Medical School
London SW17 0RE, England

John A. Kessler
Division of Developmental Neurology
Department of Neurology
Cornell University Medical College
New York, New York 10021

Dorothy T. Krieger
Division of Endocrinology
Department of Medicine
Mt. Sinai School of Medicine
New York, New York 10029

Dennis M. D. Landis
Department of Neurology
Massachusetts General Hospital
Harvard Medical School
Boston, Massachusetts 02115

S. C. Landis
Department of Neurobiology
Harvard Medical School
Boston, Massachusetts 02115

P. Ledaal
Pediatric Research Institute
Rikshospitalet
Oslo 1, Norway

Susan E. Leeman
Department of Physiology
Harvard Medical School
Boston, Massachusetts 02115

J. W. Lewis
Department of Psychology
Brain Research Institute
University of California
Los Angeles, California 90024

John C. Liebeskind
Department of Psychology
Brain Research Institute
University of California
Los Angeles, California 90024

O. Linjærde
University Psychiatric Hospital
Asgard Hospital
N-9010 Tromso, Norway

Anthony S. Liotta
Division of Endocrinology
Department of Medicine
Mt. Sinai School of Medicine
New York, New York 10029

Ivan S. Login
Departments of Internal Medicine and
* Neurology*
University of Virginia School of
* Medicine*
Charlottesville, Virginia 22908

Robert L. Macdonald
Department of Neurology
University of Michigan
Ann Arbor, Michigan 48109

Richard E. Mains
Department of Physiology
University of Colorado Health Sciences
* Center*
Denver, Colorado 80262

Neville Marks
Center for Neurochemistry
Rockland Research Institute
Ward's Island, New York 10035

Joseph B. Martin
Department of Neurology
Massachusetts General Hospital
Boston, Massachusetts 02114

A. Martinez-Campos
Department of Pharmacology
University of Milan
Milan, Italy

N. Matsushita
Department of Medicine
Kyoto University
Sakyo-Ku
Kyoto 606, Japan

Robert Y. Moore
Departments of Neurology and
Neurobiology
State University of New York at Stony
Brook
Stony Brook, New York 11794

Martin C. Moore-Ede
Department of Physiology
Harvard Medical School
Boston, Massachusetts 02115

F. Moya
Departments of Anatomy and
Neurobiology
Washington University School of
Medicine
St. Louis, Missouri 63110

Anne Mudge
Department of Pharmacology
Harvard Medical School
Boston, Massachusetts 02115

Eugenio E. Muller
Department of Pharmacology
University of Milan
Milan, Italy

James A. Nathanson
Departments of Neurology and
Pharmacology
Massachusetts General Hospital
Harvard Medical School
Boston, Massachusetts 02114

Linda M. Nowak
Neurosciences Program
Unversity of Michigan
Ann Arbor, Michigan 48109

William H. Oldendorf
Research Service
Veterans Administration Medical
Center, Brentwood
Los Angeles, California 90024

H. Orbeck
Pediatric Research Insitute
Rikshospitalet
Oslo 1, Norway

E. A. Parati
Istituto Neurologica C. Besta
Milan, Italy

William M. Pardridge
Division of Endocrinology and
Metabolism
Department of Medicine
University of California School of
Medicine
Los Angeles, California 90073

Marilyn Perrin
Peptide Biology Laboratory
The Salk Institute
San Diego, California 92138

Candace B. Pert
Section of Biochemistry and
Pharmacology
Biological Psychiatry Branch
National Institute of Mental Health
Bethesda, Maryland 20205

Quentin Pittman
Arthur Vining Davis Center for
Behavioral Neurobiology
The Salk Institute
San Diego, California 92138

David D. Potter
Department of Neurobiology
Harvard Medical School
Boston, Massachusetts 02115

Marcus E. Raichle
Department of Neurology and
Neurological Surgery
Mallinckrodt Institute of Radiology
Washington University School of
Medicine
St. Louis, Missouri 63110

Leo P. Renaud
Division of Neurology
Montreal General Hospital
Montreal, Quebec H3A 2B4, Canada

Karl L. Reichelt
Pediatric Research Institute
Rikshospitalet
Oslo 1, Norway

Seymour Reichlin
Endocrine Division
Department of Medicine
Tufts University School of Medicine
New England Medical Center Hospital
Boston, Massachusetts 02111

Donald J. Reis
Laboratory of Neurobiology
Department of Neurology
Cornell University Medical College
New York, New York 10021

Catherine Rivier
Peptide Biology Laboratory
The Salk Institute
San Diego, California 92138

Jean Rivier
Peptide Biology Laboratory
The Salk Institute
San Diego, California 92138

G. Sælid
Pediatric Research Institute
Rikshospitalet
Oslo 1, Norway

P. Schelling
Department of Pharmacology
University of Heidelberg
6900 Heidelberg, Federal Republic of
Germany

Gerard P. Smith
The Edward W. Bourne Behavioral
Research Laboratory
New York Hospital–Cornell Medical
Center, Westchester Division
White Plains, New York 10605

Mark Smith
Peptide Biology Laboratory
The Salk Institute
San Diego, California 92138

Solomon H. Snyder
Departments of Pharmacology and
Experimental Therapeutics and
Psychiatry and Behavioral Sciences
Johns Hopkins University School of
Medicine
Baltimore, Maryland 21202

M. V. Sofroniew
Department of Anatomy
Ludwig Maximilians University
8000 Munich, Federal Republic of
Germany

G. Speck
Department of Pharmacology
University of Heidelberg
6900 Heidelberg, Federal Republic of
Germany

A. Suhar
Jozef Stefan Institute
Ljubljana, Yugoslavia

Yvette Tache
Peptide Biology Laboratory
The Salk Institute
San Diego, California 92138

Hans Thoenen
Department of Neurochemistry
Max-Planck Institute for Psychiatry
D-8033 Martinsried, Federal Republic
of Germany

Michael O. Thorner
Departments of Internal Medicine and
Neurology
University of Virginia School of
Medicine
Charlottesville, Virginia 22908

Donald B. Tower
*National Institute of Neurological and
    Communicative Disorders and
    Stroke
National Institutes of Health
Bethesda, Maryland 20205*

George R. Uhl
*Department of Medicine
Stanford University School of Medicine
Stanford, California 94301*

Th. Unger
*Department of Pharmacology
University of Heidelberg
6900 Heidelberg, Federal Republic of
    Germany*

Wylie W. Vale
*Peptide Biology Laboratory
The Salk Institute
San Diego, California 92138*

Stanley J. Watson
*Department of Psychiatry
Mental Health Research Institute
University of Michigan
Ann Arbor, Michigan 48109*

Adolf Weindl
*Department of Neurology
Technical University
8000 Munich, Federal Republic of
    Germany*

Elliot D. Weitzman
*Laboratory of Human
    Chronophysiology
Department of Neurology
Montefiore Hospital
Albert Einstein College of Medicine
Bronx, New York 10467*

Hajime Yamaguchi
*Division of Endocrinology
Department of Medicine
Mt. Sinai School of Medicine
New York, New York 10029*

P. Zanardi
*Istituto Neurologica C. Besta
Milan, Italy*

Earl A. Zimmerman
*Department of Neurology
College of Physicians and Surgeons
Columbia University
New York, New York 10038*

Janet C. Zimmerman
*Laboratory of Human
    Chronophysiology
Department of Neurology
Montefiore Hospital
Albert Einstein College of Medicine
Bronx, New York 10467*

Irving Zucker
*Department of Psychology
University of California
Berkeley, California 94720*

Neurosecretion and Brain Peptides,
edited by J. B. Martin, S. Reichlin, and K. L. Bick.
Raven Press, New York © 1981.

Introduction to Section I:
Biosynthesis, Processing, and Release of Polypeptides

Seymour Reichlin

*Endocrine Division, Department of Medicine, Tufts University School of Medicine,
New England Medical Center, Boston, Massachusetts 02111*

By any reckoning, the most dramatic advance in neurobiology of this decade has been the recognition of the importance of peptides as neurotransmitters and neuromodulators, both within the central nervous system and in peripheral and autonomic nerves. The impact of these discoveries is at least as important as the discovery of the chemotransmitter function of acetylcholine, of neurotransmitter amino acids, and of the central bioaminergic neuronal systems. All of these advances have in common the recognition that cell-to-cell communication within the brain is mediated by chemical substances that are secretory products of neurons and that cell-to-cell communication within the nervous system is coded chemically as well as anatomically. What makes the discovery of neuroregulatory effects of peptides particularly important is the quantum increase in the number of known neuroregulators that can now be taken into account in understanding brain function (or dysfunction), and the enormous strides that have also been made recently in the molecular biology of peptide synthesis and of peptide receptor mechanisms. Thus, within the span of a few years, it has become potentially possible to define specific brain functions in the most fundamental molecular terms. It is obvious that scientists have not as yet had time to digest these insights and to integrate them fully into the corpus of pre-existent knowledge of brain function and disease. The overall function of this volume is to further this objective. The first section endeavors to summarize some of the fundamental aspects of neuropeptide function in brain.

In the first presentation, Dr. Gainer describes the current state of knowledge of the neurohypophysial system, defined in molecular biological terms. That this paper was chosen to be the first presentation is a reflection of the significance of studies of the neurohypophysis in the overall development of neuropeptide biology and neuroendocrinology. Outside of some early work on neurosecretory cells in the fish and skate, it was the study of the neurohypophysial system of vertebrates that had the greatest influence on the development of the concept of the neurosecretory neuron by Scharrer and his colleagues. Initially, the concept of neurosecretion was restricted to the histochemically identifiable proteinaceous secretory granules of the neurohypophysial neurons that terminated in the neural lobe, were released into the systemic circulation and were, in fact, neurons acting as glandular cells of internal secretion. But this concept has been progressively modified so that now many neurobiologists recognize neurosecretion to include any secretory product of a neuron, even if secreted in close apposition to another neuron within the substance of the brain itself. Coming full circle, recent studies outlined by Dr. Gainer have shown that the paraventricular nuclei, whose main projection is the neural lobe, unequivocally contribute neuron trajectories to mid-brain and spinal structures. Dr. Gainer describes current knowledge of the anatomy of the system for vasopressin and oxytocin formation and the biochemistry of its synthesis defined in modern molecular biological terms. More is known about this group of neurons than any other peptidergic system, and workers in the field have assumed for many years that the fundamental plan of the neuropeptide neuron would be analogous to this group of cells. For anatomical reasons, the synthetic, transport and release mechanisms of the neurohypophysis are particularly easy to study separately, and the physiological regulatory inputs (plasma osmolarity and suckling) well described. Indubitably, the supraoptico- and paraventriculo-hypophysial neuron are the archetype peptidergic neuron, and conceptually the basis for much of this conference volume.

Although studies of neurohypophysial biosynthesis of vasopressin by Sachs and his collaborators were among the earliest studies of protein synthesis in general and among the first to show that smaller peptides were synthesized as part of a larger precursor molecule, the greatest advances in knowledge of peptide synthesis have come from experiments using endocrine tissue. Dr. Habener reviews these advances, including the nature of genetic control of the secretory process, the mode by which newly synthesized hormone precursor forms are vectorially translocated within the cell itself, and the mechanism of enzymatic processing of larger precursor molecules to form the biologically active molecules, whether utilized for transport outside the cell as a secretion, or within the cell as a receptor molecule or a structural protein. He also deals with the aspect of control of these various molecular events within the cell. As the commonality of function of all cells including neurons has been increasingly emphasized, each of the topics considered, though based largely on non-neural cells, will prove to have profound relevance to the understanding of neural function.

In his paper, Dr. Mains reviews the work of his group and that of others on the detailed knowledge of structure and synthesis of one particular peptide system, that for formation of ACTH. This molecule is of great interest to the neurobiologist because it is associated in its precursor form with β-endorphin, a potent endogenous opiate, is present in a population of neurons and hence is a neuropeptide as well as a pituitary tropic hormone, and is reported to have actions on the brain independent of the pituitary. ACTH and the related peptides are of interest to the molecular biologist because the complete molecular sequence of the prohormone has been established recently using the most up-to-date genetic cloning techniques and base-sequence analysis, and because much is known about the steps in intracellular breakdown and modification of the prohormone (post-translational processing). The formation of ACTH is a very good example of the convergence of the concept of peptidergic neuron with modern molecular biology. The topic of ACTH-related peptides is also discussed later in this conference by Dr. Krieger who deals with the question of the form of ACTH and related peptides found in the blood. This is important because different ACTH-related peptides have different biological effects, as illustrated, also in this volume, by the work of Dr. Faden and Holaday on the role of endorphins in shock.

The final paper in this section by Dr. Marks emphasizes the processing of neuropeptides within the brain. This topic is relevant for two main reasons. The first has already been alluded to in relation to intracellular post-translational processing of secretion products. What is the chemical basis of enzyme action? How is it regulated? The second point of interest in brain peptide enzymology is the problem of extracellular degradation of secretory product after it has left the cell and come into contact with the receptor. This is a classical question of synaptology and has been asked repeatedly in considering the nature of control of neurotransmitter action. For example, acetylcholine is degraded by acetylcholinesterase and the biogenic amines by monoamine oxidase. Drugs that modify the breakdown of these neurotransmitters have striking effects on brain function and, in many cases, therapeutic benefits. Knowledge of peptide degradation, sites of localization of control, and of drug modification (through analogues) are in a primitive state. Dr. Marks has attempted to clarify these questions by providing several examples of enzyme degradation of peptides. The work gives an indication of the scope of potential developments in this area.

Neurosecretion and Brain Peptides,
edited by J. B. Martin, S. Reichlin, and K. L. Bick.
Raven Press, New York © 1981.

The Biology of Neurosecretory Neurons

H. Gainer

*Section on Functional Neurochemistry, Laboratory of Developmental Neurobiology,
NICHHD, National Institutes of Health, Bethesda, Maryland 20205; and Department of
Zoology, Tel Aviv University, Ramat Aviv, Tel Aviv, Israel*

It is now over fifty years since Speidel (88) and Ernst Scharrer (79) first described the existence of neurons with the morphological properties of glandular cells. The original concept of neurosecretion, that neurons secreted hormones directly into the bloodstream, was considered so revolutionary that it took a considerable number of experimental efforts and debates (6,24,81,-82) to overcome the resistance of the skeptics. Although the concept of neurosecretion is now unequivocally accepted (8,59,60,99), efforts to provide a specific definition for the neurosecretory cell has met with less success (see ref. 48 for a history of the semantic debates).

In a recent paper (80), B. Scharrer writes, "neurosecretory neurons can be defined as nerve cells that engage in secretory activity to a degree which greatly surpasses that of conventional neurons, and which is comparable to that of gland cells". It is interesting that this definition does not restrict the term to the classical case, i.e., where the neuron terminates in a neuro-hemal organ and releases its content directly into the bloodstream (this form of neurosecretory neuron has been referred to elsewhere as an "endocrine neuron", see ref. 22). That the classical definition was untenable was apparent from the fact that neurosecretory cells occur in the nervous systems of virtually all metazoans from the primitive nerve nets of coelenterates to the complex brains of humans (31,53,59,80). Some of these organisms (e.g., coelenterates) do not even possess circulatory systems, and in these species the secreted substances must diffuse through the extracellular space to reach their receptors. What is now abundantly clear is that the neurosecretory cell, as the nervous system itself, has been shaped by evolutionary forces into a multitude of morphological and functional types. The specific features of these neurons are influenced by such factors as whether the organism has a circulatory system, a blood-brain barrier (hence the need for a neurohemal organ), endocrine cells interposed between the nervous system and the rest of the organism, or appropriate rceptors in the central nervous system to be used in paracrine or synaptic mechanisms. Analysis of various living forms provides some insights into the evolutionary history of the neurosecretory cell (59,80). In the more primitive, non-arthropod invertebrates, neurohormones are used to regulate relatively long-term processes such as morphogenesis, growth, development, gonad maturation, and regeneration. In more advanced invertebrates, neurohormones are involved in the control of more rapid processes such as egg laying, ion and water regulation, heart rate, and color changes, with speeds comparable to that found in vertebrates (59). Although these phylogenetic correlations suggest certain evolutionary trends, one must be chastened by the fact that the most "advanced" use of neuropeptides, in synaptic contacts, appears to occur both in invertebrates (89) and vertebrates (44).

The significance of the neurosecretory cell today extends far beyond its original conception. With the emergence of the "peptidergic neuron" as a major cellular component in the nervous system (33,39), the question can be raised whether there is such a cell-type as the "conventional neuron" (le neurones banales). The fact that neurons share so many morphological, biochemical, and electrophysiological properties with other cell types has led to various evolutionary speculations about peptide hormone and brain peptide inter-relations (27,91), as well as to such notions as the APUD theory (68) and the paraneurone concept (30). It appears that the only really unique feature of the neuron which remains is its axon, i.e., the long process devoid of ribosomes which is specialized for axonal transport and rapid conduction of action potentials. Thus, the peptidergic neurosecretory cell is no longer a unique phenomenon. It simply represents one of the diverse forms in which peptidergic neurons are found. As one author has noted (91), "the field of biologically active peptides is a study in structural similarity and functional diversity". It is in this context that the classical neurosecretory system, the hypothalamo-neurohypophysial system, has become more than a subject of historical interest. Because of the background of information that has been accumulated about this system and its peptides, and its unique morphological features, it is an especially valuable and exploitable model for the study of various aspects of peptidergic neuron function.

THE HYPOTHALAMO-NEUROHYPOPHYSIAL SYSTEM REVISITED

The existence of the hypothalamo-neurohypophysial complex was established over 25 years ago by the classical Gomori histochemical technique (6,82). Since that time this system has been the object of an enormous number of studies and review articles (for the most recent reviews see refs. 21, 35, 59,62, 69). The cellular units of the system are the magnocellular neurons which are located in bilateral nuclei in the hypothalamus (as a diffuse nucleus in the preoptic area of lower vertebrates; but as two separate nuclei, i.e., the supraoptic and paraventricular nuclei, in mammals), and which project to the posterior pituitary where they secrete their neurohypophysial peptides and proteins (neurophysins) into the circulatory system. It has been calculated that in the rat, the 18,000 neurosecretory cells found in the hypothalamus give rise to about 40 million axon endings in the posterior pituitary (i.e., each neuron generates about 2000 endings in the neurohemal area).

The Neurohypophysial Secretory Products

The peptide hormones.
The neurohypophysial hormones have been studied in 40 species belonging to about 7 vertebrate classes (1). These hormones, irrespective of vertebrate class, are all nonapeptides with a hemicystinyl residue at the amino-terminus linked by a disulfide bond to a cysteine at position 6, and a glycinamide at the carboxyl terminus. Variations occur only in residues 3, 4, and 8 producing the different neurohypophysial hormones. From the comparative studies, it can be seen that each species has two distinct neurohypophysial peptides (with the exception of the cyclostomes which appear to contain only arginine vasotocin). The peptides appear to be characteristic in a given vertebrate class. Mammals contain oxytocin and lysine or arginine vasopressin. The non-mammalian tetrapods have mesotocin (Ile -oxytocin) and arginine vasotocin (Arg^8-oxytocin). The bony fishes contain arginine vasotocin and isotocin (Ser^4-Ile^8-oxytocin), whereas the cartilagenous fishes appear to contain arginine vasotocin and a more heterogeneous group of oxytocin-like peptides (1).

Considerations of the comparative biochemical data in the light of available paleontological data has led Acher (1) to hypothesize two evolutionary lines for neurohypophysial peptides, i.e., an oxytocin line with 3 steps (isotocin, mesotocin and oxytocin) and a vasopressin line with 2 steps (vasotocin and vasopressin). Acher further suggests that the abrupt change in both peptides from non-mammalian tetrapods (birds, reptiles, amphibians and lung fishes) to mammals, may be correlated with the emergence of new functions in the latter, i.e., lactation and antidiuresis.

The neurophysins.
Each of the neurohypophysial hormones is associated with a specific, low molecular weight (about 10,000 daltons), acidic protein, with which they are synthesized, packaged in granules, intraxonally transported, and secreted, co-ordinately. These proteins are rich in disulfide bonds (containing 12-14 cysteines per 92-95 amino acid residues), as well as glycine and proline residues, and are known collectively as the neurophysins (1,2,11,21, 70).

To date there are complete or partial sequences available for 9 distinct mammalian neurophysins (11,21,70). There appears to be considerable sequence homology between them, and generally the central part of the molecule (residues 10-74) shows the least variation. Segment 38-57, which is virtually invariant, represents a polar region of the molecule believed to contain the binding site (residues 43-49) for the neurohypophysial peptides (11,21). The maximum binding constant is of the order of 10^{-5} to $10^{-6} M^{-1}$, and the nature of the interactions between these "carrier" proteins and the hormones has been extensively studied (11,21).

In vivo, chemically distinct neurophysins have been found to be associated separately with oxytocin and vasopressin. However, this biological compartmentation does not imply a physiochemical selectivity, since the oxytocin associated neurophysin can bind vasopressin equally as well as oxytocin. Chauvet et al. (20) have proposed that the neurophysins be classified into two major groups dependent upon the amino acids in residues 2, 3, 6 and 7. In this scheme the vasopressin associated neurophysins, called MSGL-neurophysins, have Met, Ser, Glu, and Leu in these positions, while the VLAV neurophysins (Val, Leu, Asp, Val) are associated with oxytocin, in situ. Because of the imperfect fit of rat and other neurophysins to this classification scheme, it has failed to gain general acceptance in the field (11,70).

An interesting structural feature of the neurophysins is that there appears to be sequence duplication in the single polypeptide chain (11,21,70). For example, residues 12-31 have a greater than 50% homology with residues 60-77, whereas the 32-59 positions in the chain are not duplicated (note that the latter positions contain the presumptive binding site). This observation has led to the proposal that the origin of neurophysin was via an ancestral gene coding for a 50-60 amino acid peptide which underwent gene duplication and partial fusion to yield the ancestral neurophysin gene (11,70). It would be interesting, in this regard, to obtain the sequences of arginine vasotocin associated neurophysins, since presumably their structures would be closer to that of the ancestral neurophysin molecule. Although aldehyde-fuchsin stainable cells have been reported in various invertebrate ganglia (31,53,59), and sulfur-rich proteins are found in certain insect neurosecretory granules (90); no systematic effort has been made to identify neurohypophysial peptide-like and neurophysin-like molecules in invertebrate species. Recently, some immunohisto-chemical studies have succeeded in showing immunoreactivity to antibodies directed against vasopressin and neurophysin in insect neurosecretory cells (74,75). Furthermore, vasopressin-like molecules and receptors have been reported in molluscan nervous systems (7,43). In view of the above hypothesis regarding the evolution of neurophysin, it would be especially interesting to do radioimmunoassays of invertebrate systems using antibodies directed specifically against the duplicated regions (residues 12-31 or 60-77) of neurophysins.

Separate Hormones in Separate Neurons

In contrast to earlier beliefs that oxytocin and vasopressin were present in the supraoptic and paraventricular nuclei, respectively, recent microbiochemical and immunocytochemical data show that these peptides are about equally distributed in both nuclei (102). It is now generally accepted, on the basis of immunocyto-chemical work, that the one cell-one hormone concept is correct as it regards oxytocin and vasopressin in the the magnocellular neur-ons (26,103), and that these cell-types show distinct regional separation within the nuclei (102,103).

Hormone secretion studies also support the idea that the pep-tide hormones are released from separate cells. Legros and coworkers (54,55) have shown that bovine neurophysin II (vaso-pressin associated neurophysin) is selectively released during hemorrhage, whereas bovine neurophysin I (oxytocin associated neurophysin) is selectively released during milking and suckling. Similarly, in the human, the vasopressin and oxytocin associated neurophysins are selectively secreted into the blood in response to stimulation by nicotine and estrogen, respectively (85). Elec-trophysiological studies in the rat are also consistent with this concept. Two distinct cell-types in the supraoptic (SON) and paraventricular (PVN) nuclei can be distinguished physiologically; the oxytocin cell which responds with a rapid (2-4 sec) burst of impulses causing a selective release of oxytocin into the blood during the milk ejection reflex (59,96), and the vasopressin cell which responds relatively selectively to hemorrhage as a stimulus (71,97).

Input-Output Relations

In 1964, Kandel introduced the technique of recording antidro-mic spikes in the hypothalamus in response to stimulation of the posterior pituitary as an electrophysiological identification criterion for the magnocellular neurons (46). Using this criter-ion, Cross and his coworkers (22) have made detailed electrophys-iological analyses of these neurons. In addition to the demon-stration of independent inputs to the oxytocin and vasopressin cells (59,71,96,97), both cell-types respond to increases in blood osmotic pressure with vigorous increases in activity (12). Although both cells are osmoresponsive, i.e., they increase in firing frequency in response to the osmotic stimulus, only the vasopressin cell produces a profound phasic (or bursting) pattern of firing. This characteristic pattern of firing has been used as an identifying criterion for the vasopressin cell in the rat (12,22). However, a recent study suggests that this criterion may not be relevant in the cat and dog hypothalamus (100). Bursting (phasic) patterns of firing are characteristic of many neurosecre-tory cells (34), and this property is maintained in cultured neurosecretory cells (32). It has been suggested that this pat-tern of firing is particularly efficacious for electro-secretory coupling (34), and recent studies on the posterior pituitary in vitro lend support to this supposition (29).

Considerable information is available regarding the different pathways to the magnocellular neurons (41), and about the nature of the transmitters found in the SON and PVN (13). Although

extensive iontophoretic pharmacology has been done on the magnocellular neurons (13), there has been little effort to identify
the peptidergic nature of the cells which were studied. In view
of the heterogeneity of the peptides found in neurons in the hypothalamic nuclei which project to the posterior pituitary (and
hence would be identified by antidromic firing),it behooves future
investigators in this area to employ, if possible, the elegant
identification criteria recently described by Reaves and Hayward
(72,73). The [^{14}C] deoxyglucose technique of Sokoloff has also
recently been applied to the hypothalamo-neurohypophysial system
with success (84), and this technique, particularly in its higher
resolution form, may be very useful in elucidating the input-
output relations in this system under physiological conditions.

A CELL BIOLOGICAL MODEL FOR PEPTIDERGIC NEURONS

The magnocellular neurons of the hypothalamo-neurohypophysial
system are typical of neurons in general, in that ribosomally
directed protein and peptide biosythesis occurs exclusively in the
perikaryon, and these newly synthesized molecules are then transported intraxonally at different rates (50). The case that these
neurons are particularly good models for the study of the secretory process, because of the topographical segregation of the
synthesis, translocation, and the release processes in the cell
body, axon, and terminals, respectively, has been made more extensively elsewhere (35,69).

Biosynthesis and Axonal Transport of Proteins

Evidence that the neurosecretory proteins and peptides are synthesized in the magnocellular neuron perikarya and the intraxonally transported to their terminals in the posterior pituitary has
been obtained by a variety of autoradiographic and biochemical
techiques (for reviews of these experiments see refs. 14,35,62,-
69). Application of the mitotic inhibitor, colchicine, prevents
the rapid axonal transport of the newly synthesized neurosecretory
material, and produces a pile up of neurosecretory granules in the
cell bodies. A particularly elegant, quantitative autoradiographic study at the electron microscopic level was performed by
Kent and Williams (47), who showed that ^{3}H-cysteine labeled
proteins synthesized in the cell bodies were transported to the
neural lobe within neurosecretory granules, and that this labeled
protein (neurophysin) was lost from the neural lobe after the induction of secretion.
As is the case for neurons in general (51,58), the magnocellular neurons exhibit more than one rate of axonal transport
with different sets of proteins characteristic of each transport
wave. Using ^{35}S-methionine in a standard axonal transport paradigm, we have found that in addition to the fast transported
component (containing primarily the neurosecretory proteins and
peptides), the magnocellular neurons exhibit at least 3 other dis-

tinct waves, at slower rates (1-24 mm/day), in a manner similar to non-neurosecretory neurons (in preparation). The slower waves contain primarily cytoskeletal proteins, and actin appears to be exclusively transported in these slower components. Neurophysin appears restricted to the fast component. While the axonal transport mechanisms in the magnocellular neurons appear to be similar to that of ordinary neurons, there is one important difference. The neurosecretory neurons, because of their high level of peptide production and transport, show a larger investment in the fast component of transport. Several cytoskeletal-like proteins appear to be co-transported together with the neurosecretory granules (and their contents), and further elucidation of the relationship between these proteins and the granules may provide a unique opportunity for understanding the molecular biology of the axonal transport mechanism.

Biosynthetic Evidence for Prohormones

The idea that the nonapeptides, vasopressin and oxytocin, were synthesized as common precursors together with their respective neurophysins derives from the work of Sachs and his colleagues (76,77). Recently, evidence for higher molecular weight immunoreactive forms of vasopressin (18) and neurophysin (52) has been reported. In biosynthetic experiments, conducted in vivo in the rat, two common precursors of about 20,000 molecular weight have been identified, one for vasopressin and its associated neurophysin and another for oxytocin and its associated neurophysin (for a review of these biosynthetic studies, see ref. 14). The vasopressin prohormone appears to be a glycoprotein, whereas the oxytocin prohormone is not. Consistent with these observations is the fact that the Brattleboro rat, a genetic strain which does not synthesize the vasopressin precursor, does not contain the glycoprotein product of this precursor in its posterior pituitary, whereas the normal rat does (14). The existence of higher molecular weight, biosynthetic forms of neurophysin (17-25,000 molecular weight) have been confirmed in cell-free translation experiments using hypothalamic mRNA (38,56,83). Furthermore, Schmale et al. (83) in co-translation experiments with microsomal membranes from dog pancreas, could show various post-translational processing steps including the glycosylation of one form of pro-neurophysin.

The Neurosecretory Granule as a Site of Post-Translational Processing

It has been known for some time that neurosecretory granules appear to be larger in the magnocellular neuron perikarya than in their terminals in the posterior pituitary (62). A more direct morphological demonstration of granule maturation is provided by

Morris and Cannata (19,61). These authors showed that the electron density in the neurosecretory granules observed by electron microscopy was a function both of the fixation procedure and the site of the granules in the hypothalamo-neurophypophysial system. When the tissues are prefixed for 2 hours in a triple aldehyde mixture buffered at pH 5.0, all the granules appear electron dense, irrespective of site. However, when the tissues are prefixed for 2 hours at pH 8.0, all the granules in the neural lobe become swollen and electron lucent, in contrast to the granule cores in the Golgi elements in the perikarya which remain electron dense (19). Comparisons of the changes in granule density throughout the system from Golgi zone, perikaryon, axon, to axon terminal in the neural lobe suggested that there was a morphological maturation during axonal transport.

Conversion of the putative prohormone to its peptide products during axonal transport was suggested earlier by the work of Sachs (77). The demonstration that the prohormone was indeed the form in which the neurosecretory material was transported out of the perikaryon into the axon (73), and that the conversion occurs intraxonally (36,37) in neurosecretory granules (47), provided compelling evidence that the granule is the site of the post-translational processing events. Hence, the neurosecretory granule would seem to be an ideal place to search for post-translational processing enzymes. Unfortunately, only one study has addressed this question to date (66). In this work, the authors describe a chymotryptic-like enzyme involved in the degradation of neurophysins. It is unlikely that this enzyme is involved in prohormone processing, since the latter process appears to involve tryptic-like and carboxypeptidase-like cleavages (14).

The pH dependent morphological change in granule electron density described above (19,61), may be explained by the fact that the neurohormone-neurophysin complex within the mature granule has an optimum binding at pH 5.5-5.7. We have recently found, using methylamine distribution methods, that the intragranule pH is around 5.5 (unpublished data). Exposure of the granule to pH 8.0, under fixation conditions, would therefore inhibit the binding of neurohormone to neurophysin and lead to osmotic lysis of the granule contents. The immature granule (e.g., in the Golgi) would presumably contain only prohormone, and this lytic process would not occur at pH 8.0.

The life-cycle of neurosecretory granules may involve not only rmaturation, but also senescence. In an autoradiographic study, Heap et al. (42)) demonstrated that granules entering the neural lobe, containing newly synthesized hormones, are first transported to the axonal endings (release sites), and then, are subsequently transported to axonal swellings (Herring bodies) for storage. The granules are quite stable in the swellings (half-life greater than 2 weeks), and can be recruited for release upon demand. In a recent study, Nordmann et al. (65) compared the granule (vesicle) diameters in the axon endings and swellings in the rat neural lobe, and found the former to have an average diameter of 170 nm and the latter a diameter of 188 nm. When granules from the neural lobe were separated by sub-cellular fractionation using an isosmotic sucrose-metrizamide gradient, two granule populations (containing both oxytocin and vasopressin) were distinguished. The lighter population (band 3) had vesicle diameters under isosmotic conditions equal to the vesicles in the swellings, and the denser population had vesicle diameters comparable to those in the endings. What was particularly interesting was that the band 4 vesicles were osmotically inactive, whereas the band 3 vesicles behaved as perfect osmometers. The authors hypothesized that this difference is osmotic properties of the two vesicles reflects the aging process, and that the osmotic lability of the older granules is a prerequisite for subsequent turnover. This hypothesis can be easily tested, in the future, by combining this sub-cellular fractionation procedure with the axonal transport paradigm described earlier.

Cellular Correlations with Functional Activity

Long-term release of the neurohypophysial hormones induced by dehydration of the animal is accompanied by a 5-fold increase in biosynthesis and axonal transport of the hormones (37,62), and a 90% depletion of the hormone stores in the neural lobe (45). In addition to an increase in biosynthesis rate, the prohormones are post-translationally processed at about a 3-fold increase in rate (unpublished data). As in other neuronal systems, the rate of axonal transport is unaffected by functional activity; only the amount of neurosecretory material transported is increased. The increase in rate of processing of the prohormone is consistent, in a teleogical sense, with the augmented secretion of newly synthesized hormone during dehydration. In this way, the newly synthesized hormone being transported to the axon endings (42) would be completely processed before release. The mechanisms underlying this increase in processing rate are not yet understood. Bearing on this issue, are the recent findings of Alonso and Assenmacher (3-5) that there is a substantial increase in smooth endoplasmic reticulum (SER) in the axons during dehydration. These authors suggest that the neurosecretory material is transported in the SER during dehydration (as opposed to granules in the normal animal), and is released from microvesicles budding from the SER. Immunocytochemical studies (28) in dehydrated rats failed to confirm this hypothesis (i.e., immunoreactivity was found only in neurosecretory granules). However, a more rigorous study employing quantitative electron microscopic autoradiography (and ^{3}H-cysteine), along the lines done for the normal rat (49), is necessary before any final conclusions about this hypothesis can be made.

An alternate role for the abundant axonal SER during functional activation may be as a source of lysosomal and other proteolytic enzymes to dispose of the excessive membranes generated during augmented secretion (see below). When rats are subjected to prolonged dehydration followed by rehydration, there is a proliferation of tubular and SER like elements in the axon (28). Under these circumstances, considerable granulolysis takes place in the axons and terminals (10), usually by autophagy (i.e., the sequestration of the secretory granules in autophagic vacuoles). Some of the SER elements in the axon may be providing the enzymes necessary for this process. In contrast, the granulolytic process in the perikarya and the pituicytes (in the neural lobe) appear to occur by crinophagy (i.e., fusion of the granules with primary lysosomes, see refs. 9,92). The intracellular signals to initiate granulolysis must be very specific during rehydration, since overloading the perikaryon (by inhibition of axonal transport) with granules, does not induce granulolysis (10).

Several possibilities exist to explain how the information about the state of release at the nerve terminal (in the neural lobe) is signalled to the cell perikaryon (in the hypothalamus) in order to regulate the biosynthetic level of the latter. These include: 1) the action potential itself, which at the terminal is coupled to secretion, in the perikaryon may be coupled to biosynthesis, 2) information about release from the terminal may be returned to the perikaryon via retrograde axonal transport (possibly in the form of recovered vesicle membranes after exocytosis), or 3) a direct action of the synaptic transmitters (conceivably involving second messengers) from the afferent input, could be the signal for increased hormone biosynthesis. In order to dissect which of these possibilities, if any, can account for the relevant regulatory signal, it may be first necessary to have an in vitro model in which it will be possible to control these parameters independently.

The Secretory Process

The neural lobe represents an unusually high concentration of relatively homogeneous peptidergic endings (the neurosecretory axons, endings, and swellings occupy about 42% of its total volume, (64). No other neural elements are present in this region, and it is unlikely that a better model system for the study of mechanisms of peptidergic secretion will be found. It is now clearly established that the neurohormones are secreted by a calcium-dependent exocytosis mechanism (62). The ease with which purified neurosecretory granules (vesicles) and isolated axon terminal plasma membranes can be obtained from this experimental preparation, makes it ideal for the study of the molecular events involved in peptide secretion (40).

The issue of membrane retrieval following exocytosis is still unresolved. Calculations of the resting release of neurohormones in vivo assuming an exocytotic mechanism, suggests an increase of terminal membrane area of 16% per hour, and maximum stimulation in vitro would produce a two-fold change in membrane area (59). Clearly, the excess membrane due to exocytosis must be retrieved from the axolemmal surface. Substantial numbers of microvesicles are found near the neurosecretory axonal endings, and there are reports indicating that these microvesicles take up extracellular horseradish peroxidase (HRP) following stimulation (63). However, other HRP and morphometric studies indicate that large vacuoles, comparable in size to the neurosecretory granules, take up more HRP after stimulation than microvesicles by a factor of 100:1, and that the retrieved membrane is largely in the form of the large vacuoles (for a review see ref. 62). Unfortunately, the latter experiments do not entirely resolve this issue, since it is possible that the microvesicles are also involved in membrane uptake but can rapidly fuse to form the large vacuoles. In any case, the role of the microvesicles is still uncertain, as is the ultimate fate of the retrieved membrane. Since all biosynthesis occurs at the level of the parikaryon, it is unlikely that this membrane is recycled at the axonal level. HRP is retrogradely transported to the perikarya, but there have been no rigorous quantitative studies to show that this transport is increased with stimulation. In fact, in an elegant study, Swann and Pickering (94) showed that ^{3}H-choline-labeled granule membrane, in contrast to ^{35}S-cysteine labeled neurophysin, showed no decrease in the "granule" fraction of the neural lobe after 5 days of massive secretion induced by saline-drinking. It is possible that most of the retrieved membrane is destroyed within the neural lobe, possibly by the phagocytotic and lysosomal activity of the pituicytes which are activated during periods of enhanced secretion (9,62,92).

NEW ROLES FOR THE NEUROHYPOPHYSIAL PEPTIDES

Probably the most exciting new data in this area has come from recent immunocytochemical studies which indicate that nerve fibers containing oxytocin and vasopressin project to many brain regions other than the neural lobe. In addition to the demonstration of immunoreactive vasopressin fibers projecting from the PVN to the zona externa of the median eminence (23,25,86,107), a wide variety of extrahypothalamic projections have also been found. Previous anatomic connectivity studies, using HRP and autoradiographic techniques, indicated that there were projections from the PVN to the mesencephalic grey and the intermediolateral column in the spinal cord (78). More recent, retrograde HRP studies by Ono et al. (67) have shown that the projections to the intermediolateral column of the spinal cord come from magnocellular neurons in the dorsomedial part of the PVN, whereas the neural lobe projections derive from cells in the dorsolateral cap of the nucleus.

The first indication that these projections were indeed peptidergic, came from the immunohistochemical studies of Swanson (93), who showed PVN fibers containing immunoreactive neurophysin projecting to the Edinger-Westphal Nucleus, locus coeruleus, vagal complex, and spinal cord. Subsequent studies using vasopressin and oxytocin antisera confirmed and extended these findings, and reported more vasopressin than oxytocin fibers in diencephalic and mesencephalic subcortical brain areas, but the reverse situation in the medulla oblongata and spinal cord (15,17). The extrahypothalamic fibers also derive from parvicellular neurons in the suprachiasmatic nucleus, which contain only vasopressin, and project to the organum vasculosum of the lamina terminalis (OVLT), lateral septum, medial dorsal thalamus, lateral habenula, interpeduncular nucleus, and nucleus of the solitary tract (15, 87). The neuropeptide containing vesicles in the extrahypothalamic areas appear to be smaller (around 100 nm) than the 140-170 nm neurosecretory vesicles found in the neural lobe. The vesicles in the external zone of the median eminence, and in the suprachiasmatic nucleus also seem to be small (90-100 nm), suggesting that different cells in the PVN project to the neural lobe versus other brain areas (15,49,95).

The physiological significance of these new peptidergic pathways is not entirely clear at present. In view of the behavioral effects produced by oxytocin and vasopressin and their analogues (98), it is possible that these extra-hypothalamic pathways may be involved. The recent observation that immunoreactive vasopressin synapses are present in the lateral septum and habenular nucleus, and oxytocin and vasopressin synapses are located in the amygdala (16), raises the possibility that these peptides may play roles as neurotransmitters (or neuromodulators). Most of these observations are based solely on immunocytochemical data, and it is therefore essential that biochemical analyses of the peptides in these extra-hypothalamic areas be done. Recent data from our laboratory suggests that the system contains other immunologically similar but biochemically distinct peptides (unpublished data), and the biosynthesis-axonal transport paradigm, discussed earlier, should be of great value in resolving this issue. In conclusion, although attention is now focused on a wide variety of neuropeptides and peptidergic pathways, the hypothalamo-neurohypophysial system is still relevant and there remains much to be learned from it.

REFERENCES

1. Acher, R. (1978): In: Neurosecretion and Neuroendocrine Activity, Evolution, Structure, and Function, edited by W. Bargman, A. Oksche, A. Polenov, and B. Scharrer, pp 31-43 Springer-Verlag, Berlin.
2. Acher, R. (1979): Angew. Chem. Int. Ed. Engi., 18:846-860.
3. Alonso, G., and Assenmacher, I. (1978): Biol. Cellulaire, 32: 203-206.
4. Alonso, G., Boudier, J.L., and Tasso, F. (1974): C.R.Acad. Sci. Paris, Serie D, 279:1471-1474.
5. Alonso, G., Rambourg, A., and Assenmacher, I. (1974): J. Physiol. Paris, 72:6B-7B.
6. Bargmann, W. (1949): Z. Zellforsch. Mikrosk. Anat., 34: 610-634.
7. Barker, J.L., Ifsbin, M., and Gainer, H. (1975): Brain Res., 84:501-513.
8. Berlind, A., (1977): Int. Rev. of Cytology, 49:171-251.

9. Boer, G.J., Notten, J.W.L., Koenders, Y., and Rheenen-Verberg, C.F. (1976): Brain Res, 114:257--277.
10. Boudier, J.A., and Picard, D. (1976): Cell Tiss. Res., 172: 39-58.
11. Breslow, E. (1979): Ann. Rev. Biochem. 48:251-274.
12. Brimble, M.J. and Dyball, R.E.J. (1977): J. Physiol., Lond., 271:367-384.
13. Brownstein, M.J. (1979): In: Brain Peptides: A New Endocrinology, edited by A.M. Gotto, E.J. Peck, and A.E. Boyd, pp. 27-40. Elsevier/North Holland, Amsterdam.
14. Brownstein, M.J., Russell, J.T., and Gainer, H. (1980): Science, 207:373-378.
15. Buijs, R.M. (1978): Cell Tiss. Res., 192:423-435.
16. Buijs, R.M. and Swaab, D.F. (1979): Cell Tiss. Res., 204: 355-365.
17. Buijs, R.M., Swaab, D.F., Dogterom, J., and Van Leeuwen, F.W. (1978): Cell Tiss. Res., 186:423-433.
18. Camier, M., Lauber, M., Mohring, J., and Cohen, P. (1979): FEBS Lett., 108:369-373.
19. Cannata, M.A., and Morris, J.F. (1973): J. Endocrinol., 57: 531-538.
20. Chauvet, M.T., Chauvet, J., Acher, R. (1976): Eur. J. Biochem. 69:475-485.
21. Cohen, P., Nicholas, P., and Camier, M. (1979): Current Topics in Cellular Regulation, 15:263-318.
22. Cross, B.A., Dyball, R.E.J., Dyer, R.G., Jones, C.W., Lincoln, D.W., Morris, J.F., and Pickering, B.T., (1975): Res. Progr. Horm. Res., 31:243-294.
23. DeMey, J., Dierickx, K., and Vandesande, F. (1975): Cell Tiss. Res., 157:517-519.
24. De Robertis, E., (1962): In: Neurosecretion: Memoirs of the Soc. Endocrinology, Vol. 12, edited by H. Heller and R.B. Clark, pp. 3-20. Academic Press, New. York.
25. Dierikx, K., Vandesande, F., and DeMey, J. (1976): Cell Tiss. Res., 168:141-151.
26. Dierickx, K., Vandesande, F., and Goossens, N. (1977): In: Cell Biology of Hypothalamic Neurosecretion, edited by J.D. Vincent and C. Kordon, pp. 391-398. CNRS, Paris.
27. Dockray, G.J., (1979): Fed. Proc., 38:2295-2301.
28. Dreifuss, F., Burlett, A., Chateau, M., and Czernichow, P. (1979): Biol. Cellulaire, 35:141-146.
29. Dutton, A. and Dyball, R.E.J. (1979): J. Physiol. Lond., 290: 433-440.
30. Fujita, T., and Kobayashi, S. (1979): Trends in Neurosci., 2:27-30.
31. Grabe, M., (1966): Neurosecretion. Pergamon Press, New. York.
32. Gahwiler, B.H., and Dreifuss, J.J. (1979): Brain Res., 177:95-103.
33. Gainer, H., (1977): Peptides in Neurobiology. Plenum Press, New York.

34. Gainer, H. (1978): In: Comparative Endocrinology, edited by P.J. Gaillard and H.H. Boer, pp. 293-304. Elsevier/North Holland Press, Amsterdam.
35. Gainer, H., Russell, J.T., and Fink, D. (1979): In: Brain Peptides: A New Endocrinology, edited by A.M. Gotto, E.J. Peck, and A.E. Boyd, pp. 139-159. Elsevier/North Holland Press, Amsterdam.
36. Gainer, H., Sarne, Y., and Brownstein, M.J. (1977): Science, 195:1354-1356.
37. Gainer, H., Sarne, Y., and Brownstein, M.J. (1977): J. Cell Biol., 73:366-381.
38. Giudice, L.C., and Chaiken, I.M. (1979): Proc. Natl. Acad. Sci., USA, 76:3800-3804.
39. Gotto, A.M., Peck, E.J., Boyd, A.E. (1979): Brain Peptides: A New Endocrinology. Elsevier/North-Holland Press, Amsterdam.
40. Gratzl, M., Dahl, G., Russell, J.T., and Thorn, N.A. (1977): Biochim. Biophys. Acta., 470:45-57.
41. Harris, M.C. (1977): In: Cell Biology of Hypothalamic Neurosecretion, edited by J.D. Vincent and C. Kordon, pp. 47-61. CNRS, Paris.
42. Heap, P.F., Jones, C.W., Morris, J.F., and Pickering, B.T. (1975): Cell Tiss. Res., 156:483-497.
43. Ifshin, M., Gainer, H., and Barker, J.L. (1975): Nature, Lond. 254:72-73.
44. Jan, Y.N., Jan, L.Y., and Kuffler, S.W., (1979): Proc. Natl. Acad. Sci., USA, 76:1501-1505.
45. Jones, C.W., and Pickering, B.T. (1969): J. Physiol., Lond., 203:499-558.
46. Kandel, E.R. (1964): J. Gen. Physiol., 47:691-717.
47. Kent, C., and Williams, M.A. (1974): J. Cell. Biol., 60:554-570.
48. Knowles, F., (1974): In: Neurosecretion - the Final Neuroendocrine Pathway, edited by F. Knowles and L. Vollrath, pp. 3-11. Springer-Verlag, Berlin.
49. Krisch, B. (1980): Prog. in Histochem. Cytochem., 13:1-163.
50. Lasek, R.J. (1970): Inst. Rev. Neurobiol., 13:289-324.
51. Lasek, R.J., and Hoffman, P.N. (1976): In: Cell Motility, edited by R. Goldman, T. Pollard, and J. Rosenbaum, pp. 1021-1049. Cold Spring Harbor Press, New York.
52. Lauber, M., Camier, M., and Cohen, P. (1979): FEBS Letters, 97:343-347.
53. Lentz, T.L., (1968): Primitive Nervous Systems. Yale University Press, New Haven.
54. Legros, J.J., Reynaert, R., and Peeters, G. (1975): J. Endocrinol., 67:297-302.
55. Legros, J.J., Reynart, R., and Peetrs, G. (1975): J. Endocrinol., 60:327-332.
56. Lin, C., Joseph-Bravo, P., Sherman, T., Chan, L., and McKelvey J.F. (1979): Biochem. Biophys. Res. Comm., 89:943-950.
57. Lincoln, D.W., and Wakerley, J.B. (1974): J. Physiol., Lond., 242:533-554.

58. Lorenz, T., and Willard, M. (1978): Proc. Natl. Acad. Sci. USA, 75:505-509.
59. Maddrell, S.H.P., and Nordmann, J.J., (1979): Neurosecretion. Blackie Press, Glasgow.
60. Mason, C.A., and Bern, H.A. (1977): In: Handbook of Phyysiology, Vol 1, The Nervous System, edited by E.R. Kandel, pp. 651-689. Amer. Physiol. Soc., Bethesda.
61. Morris, J.F., and Cannata, M.A. (1973): J. Endocrinol., 57: 517-530.
62. Morris, J.F., Nordmann, J.J., and Dyball, R.E.J. (1978): Int. Rev. of Exptl. Path., 18:1-95.
63. Nagasawa, J., Douglas, W.W., and Schultz, R.A. (1971): Nature Lond., 232:341-344.
64. Nordmann, J.J. (1977): J. Anat., 123:213-218.
65. Nordmann, J.J., Louis, F., and Morris, S.J. (1979): Neuro-Science, 4:1367-1379.
66. North, W.G., Vattin, H., Morris, J.F., an La Rochelle, F.T. (1977): Endocrinology, 101:110-118.
67. Ono, T., Nishino, H., Sasaka, K., Muramoto, K., Yano, I, and Simpson, A., (1978): Neurosci. Lett., 10:141-146.
68. Pearse, A.G.E. (1969): J. Histochem. Cytochem., 17:303-313.
69. Pickering, B.T. (1978): Essays in Biochemistry, 14:45-81.
70. Pickering, B.T., and Jones, C.W. (1978): Horm. Prot. and Pept., 5:103-158.
71. Poulain, D.A., Wakerley, J.B., and Dyball, R.E.J. (1977): Proc. R. Soc. Lond. B., 196:367-384.
72. Reaves, T.A. and Hayward, J.N. (1979): Cell Tiss. Res., 200: 147-151.
73. Reaves, T.A., and Hayward, J.N. (1979): 9th Ann. Meet. for Neurosciences, 5:537.
74. Remy, C., Girardie, J., and Dubois, M. (1977): C.R. Acad. Sci. Paris, 285:1495-1497.
75. Remy, C., Girardie, J., and Dubois, M.P. (1978): C.R. Acad. Sci. Paris, 286:651-653.
76. Sachs, H., and Takabatake, Y. (1964): Endocrinology, 75:943-948.
77. Sachs, H., Fawcett, P., Takabatake, Y., and Portanova, R. (1969): Rec. Prog. Horm. Res., 25:447-491.
78. Saper, C.B., Loewy, A.L., Swanson, L.W., and Cowan, W.M. (1976): Brain Res., 117:305-312.
79. Scharrer, E. (1928): Z. Vgl. Physiol., 7:1-38.
80. Scharrer, B., (1978): In: Neurosecretion and Neuroendocrine Activity. Evolution, Structure, and Function, edited by W. Bargmann, A. Oksche, A. Polenov, and B. Scharrer, pp. 9-14. Springer-Verlag, Berlin.
81. Scharrer, E., and Scharrer, B. (1945): Physiol. Rev., 25: 171-181.
82. Scharrer, E. and Scharrer, B. (1954): Rec. Prog. Horm. Res., 10:183-240.
83. Schmale, H., Leipold, B., and Richter, D. (1980): FEBS Lett., (in press).

84. Schwartz, W.J., Smith, C.B., Davidsen, L., Savaki, H., Sokoloff, L., Mata, M., Fink, D., and Gainer, H. (1979): Science, 205:723-725.
85. Seif, S.M., and Robinson, A.G. (1978): Ann. Rev. Physiol., 40: 345-376.
86. Silverman, A.J. (1976): J. Histochem. Cytochem., 24:816-827.
87. Sofronview, M. V., and Weindl, A. (1978): Amer. J. Anat., 153:391-430.
88. Speidel, C.G. (1919): Carnegie Inst. Wash. Publ., 13:1-31.
89. Starratt, A.N., and Brown, B.E. (1975): Life Sci., 1253-1258.
90. Steel, C.G.H., and Morris, G.P. (1975): Gen. Comp. Endocrinol., 26 517-524.
91. Stewart, J.M., and Channabasavaiah, K., (1979): Fed. Proc., 38:2302-2308.
92. Sunde, D., Osinchak, J., and Sachs, H. (1972): Brain Res., 47: 195-216.
93. Swanson, L.W. (1977): Brain Res., 128:346-353.
94. Swann, R.W. and Pickering, B.T. (1976): J. Endocrinol., 68: 95-108.
95. Van Leeuwen, F.W., Swaeb, D.F., and Raay, C. (1978): Cell Tiss. Res., 193:1-10.
96. Wakerley, J.B., and Lincoln, D.W. (1973): J. Endocrinol., 57: 477-493.
97. Wakerley, J.B., Poulain, D.B., Dyball, R.E.J., and Gross, B.S. (1975): Nature, Lond., 258:82-84.
98. Wied, D. de, and Bohus, B. (1979): In: Brain Mechanisms in Memory and Learning: From the Single Neuron to Man, edited by M.A.B. Brazier, pp. 139-149. Raven Press, New York.
99. Yagi, K., (1977): Int. Rev. of Cytology, 48:141-186.
100. Yamashita, H., Koizumi, K., and McC. Brooks, C. (1979): Proc. Natl. Acad. Sci. USA, 76:6684-6688.
101. Zimmerman, E.A. (1976): In: Frontiers in Neuroendocrinology, edited by A. Martinosi and W.F. Ganong, pp. 25-62. Raven Press, New York.
102. Zimmerman, E.A. and Defendini, R. (1977): In: Neurohypophysis, edited by A.M. Moses and L. Share, pp. 22-29. Karger, Basel.
103. Zimmerman, E.A., Stillman, M.A., Recht, L.D., Michaels, J. and Nilaver, G. (1977): In: Cell Biology of Hypothalamic Neurosecretion, edited by J.D. Vincent and C. Kordon, pp. 376-389. CNRS, Paris.

Neurosecretion and Brain Peptides,
edited by J. B. Martin, S. Reichlin, and K. L. Bick.
Raven Press, New York © 1981.

Principles of Peptide-Hormone Biosynthesis

Joel F. Habener

*Laboratory of Molecular Endocrinology, Massachusetts General Hospital, and
Howard Hughes Medical Institute Laboratories, Harvard Medical School,
Boston, Massachusetts 02114*

INTRODUCTION

During the past several decades, endocrinologists and cellular
biologists have become increasingly interested in the details of
the biosynthesis of proteins to learn how the control of the
cellular production of secreted protein hormones is geared to the
function of these hormones. A wealth of information has been
uncovered so far concerning the biochemical processes involved in
the synthesis of hormones. A combination of factors is responsible for this acquisition: a) the use of radioisotopes to label
biosynthetic precursors and intermediate products in the pathways
of biosynthesis, b) the development of more refined and sensitive
techniques for analyses of cellular constituents, c) improved
whole-cell and cell-free systems in which to carry out biosynthetic studies under in vitro conditions, and d) the discovery and
application of recombinant-DNA technology for the detailed investigation of gene structure and function, as well as for the rapid
determination of the sequences of large hormonal precursors. In
addition, there have been continued improvements in the resolving
power of the transmission electron microscope, coupled with the
development of sensitive immunochemical, cytochemical, and autoradiographic techniques for the subcellular localization of hormones and newly synthesized proteins, that have contributed to the
elucidation of the precise cellular locations where specific
synthetic reactions take place.

Studies of the protein hormones and other secreted proteins
over the past few years have revealed that synthesis involves
precursor forms, larger than the molecules ultimately secreted,
that are processed to the final cellular products by cleavages
during this transport within specialized subcellular organelles of
the secretory cells. The purpose of this chapter is to review
briefly the current understanding of the cellular pathways of
polypeptide-hormone biosynthesis, transport, and release; to
provide evidence for a role of the precursors and their co- and
post-translational processing in intracellular signalling and

segregation of proteins; and to point to directions in which this understanding may be applied to investigations of the processes involved in the neuronal synthesis of the polypeptide-hormone-like substances.

GENERAL ASPECTS OF PROTEIN BIOSYNTHESIS

The processes of protein biosynthesis encompass many complex reactions (Fig. 1). The expression of genetic information in terms of the formation of a biologically active polypeptide begins in the nucleus of the cell with the activation of transcriptional processes. Then follows a cascade of events that includes processing of the initial RNA transcripts by excision of intervening sequences (introns), rejoining of specific RNA segments (exons), and modifications of the 3' ends of the RNAs by polyadenylation and of the 5' ends by the addition of 7-methyl guanosine caps. The mature, fully-assembled mRNAs are transported into the cytoplasm where they serve as templates for the translation of genetic

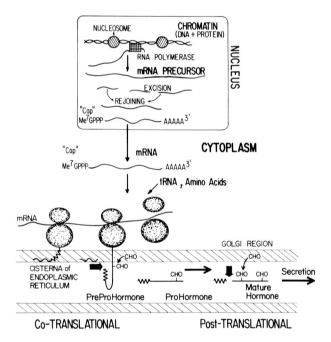

FIG. 1. Schema depicting the cellular events in the biosynthesis of polypeptide hormones. Nascent polypeptides are vectorially discharged across the membrane bilayer into the cisterna of the rough endoplasmic reticulum. Cleavages of NH_2-terminal leader sequences and glycosylation (CHO) of the nascent polypeptide are shown. Furthermore post-translational modifications of the polypeptide take place in the Golgi region.

information from polynucleotides into proteins by directing the assembly of amino acids into polypeptide chains. The initial polypeptides thus formed are precursors that undergo a series of modifications of their primary structures, leading to the formation of the final bioactive proteins.

CO- AND POST-TRANSLATIONAL PROCESSING OF HORMONAL PRECURSORS

Most, if not all, polypeptide hormones are synthesized by way of co- and post-translational processing of larger precursor molecules. These biosynthetic processes include, solely or in combination, proteolytic cleavages, glycosylation, phosphorylation, acetylation, and (or) other derivatizations of amino acids, and formation of intra- and inter-molecular disulfide linkages. Analyses of the products of cell-free translations of mRNAs coding for polypeptide hormones and of the products formed during pulse and pulse/chase labeling studies in intact cells have shown that cleavages, glycosylation, and assembly of polypeptides occur during their passage through the endoplasmic reticulum and transport to the Golgi complex. The modifications of the precursor polypeptides initially synthesized are often extensive, resulting in substantial alterations in the physico-chemical properties of the molecules. It is likely that these modifications alter the biologic activities of, and, in turn, may lend biologic specificities to, the polypeptides.

Studies of co- and post-translational processing of hormones have brought to light several functions that these precursors appear to fulfill in biologic systems. Two of these proposed functions, for which there is considerable supporting evidence, are a) intracellular signalling, by which the cell distinguishes among specific classes of proteins and directs them to their specific sites of action and b) the generation of multiple biologic activities from a common gene product by regulated or cell-specific variations in the co- or post-translational modifications. Examples of these two functions are discussed below.

Intracellular Signalling

The problem of how proteins are segregated within the cell after they are synthesized has occupied the attention of cellular biologists for many years. It has been estimated that a typical eukaryotic cell synthesizes about 20,000 different proteins at some time during its cycle (14). Present evidence suggests that these many different proteins produced by a cell are synthesized by a common pool of polyribosomes. Each of the different proteins synthesized is directed to a specific location where its particular biologic function is expressed. For example, specific groups of proteins are transported to the nucleus and to other subcellular organelles, whereas other groups of proteins are synthesized specifically for export from the cell, e.g., immunoglobulins, blood coagulation factors, serum albumin, and the protein and peptide hormones. It is clear that the forces involved in this process of directional transport of proteins must involve a highly sophisticated set of information signals. Inasmuch as the information for this transport can only reside either wholly or in part within the primary structure or conformational properties of the protein itself, post-translational modification (Fig. 1) may be important for specificity of protein function. Once the newly synthesized protein is released from the mRNA--ribosome--nascent-chain complex, a further regulatory role for the RNA seems unlikely.

Two lines of evidence point to a role of precursor sequences and post-translational modifications of polypeptides in the processes of intracellular signalling and directional transport. The first line of evidence is the discovery of NH_2-terminal "leader", "signal", or "pre" sequences on the products resulting from the cell-free translations of mRNAs coding for hormones and other secretory proteins (10), and the demonstration that these sequences are present on nascent polypeptides during their synthesis in intact cells. Such NH_2-terminal sequences probably serve as signals that recognize specific sites on the rough endoplasmic reticulum (RER), resulting in transport of the nascent polypeptides into the secretory pathway of the cell (2,3,5,16,32). The second line of evidence is the discovery that phosphorylation of carbohydrate groups attached to certain lysosomal enzymes determines whether these enzymes are packaged in lysosomes or are secreted from the cell.

Signalling in the rough endoplasmic reticulum.

Cell-free translation products containing NH_2-terminal leader sequences, termed pre-proteins (hormones), or pre-proproteins (hormones), in instances in which intermediate precursor forms exist, have been found for numerous protein hormones including parathyroid hormone, insulin, growth hormone, prolactin, thyroid-stimulating hormone, calcitonin, somatostatin, glucagon, and adrenocorticotropin, as well as the non-hormonal secreted proteins such as pancreatic enzymes, egg-white proteins, serum albumin, immunoglobulins, melittin (bee venom), and several membrane-associated bacterial proteins (see refs. 2,3,5, and 10 for reviews). A characteristic of the precursor-specific sequences is that they vary in length from 15 to 25 amino acids; hydrophobic regions of 10 to 12 amino acids are found within the central portion of the sequences. This high degree of hydrophobicity is found in proteins known to be specifically associated with membranes. One exception to the rule that the precursors of secreted proteins contain NH_2-terminal leader sequences is ovalbumin in which the hydrophobic sequence is located within the actual sequence of albumin itself (15).

Recent views favor the probability that precursors play a role in the post-translational processes that lead to the intracellular transport and compartmentalization of hormones in the secretory pathway (2,3,5,10). This hypothesis, known as the signal hypothesis, was introduced by Milstein et al. (7) and by Blobel and Sabatini (4) to explain, at a molecular level, the mechanisms by which proteins destined for secretion from cells are able to selectively obtain access to the membrane-enclosed subcellular organelles involved in the transport, packaging, and secretion of the proteins. The amino-terminal signal, or leader, sequence of nascent proteins recognizes putative attachment sites in the membrane of the endoplasmic reticulum. The polyribosome--nascent polypeptide complexes form junctions with the membrane, and the growing polypeptides are vectorially transported across the membrane into the cisterna of the RER. Co-translationally, during the transport of the nascent polypeptides, the leader sequences are removed, presumably by the action of specific peptidases located on the inner face of the membrane, and after completion of the polymerization of amino acids, the polypeptides are released into the cisterna.

Experimental support of the signal hypothesis has come from studies carried out both in cell-free systems and in intact cells. Translation of mRNAs in cell-free systems supplemented with microsomal membranes results in the translocation of the secretory proteins into the interior of the microsomal vesicles concomitant with the removal of the leader sequence (3,22) and, in certain circumstances, glycosylation (1,22) of specific sites on the polypeptide. In addition, "read-out" experiments conducted under cell-free conditions showed that these events occur co-translationally (3,22). Pulse labeling (for 30-60 sec) of nascent proteins in intact parathyroid gland slices using [^{35}S] methionine revealed the presence of labeled proparathyroid hormone before the appearance of pre-proparathyroid hormone, a finding indicative of co-translational rather than post-translational processing of the nascent polypeptide (9). From these observations it appears reasonable to postulate that the basic function of the leader sequence is to provide the means for transmembrane movement. Thus, as a result of the specialized nature of the precursor sequences, proteins destined for secretion are selected from a great many other cellular proteins for sequestration and subsequent transport and packaging in the secretory pathway of the cell.

Signalling in the Golgi complex.

Recent studies uncovered evidence suggesting that post-translational modifications of certain proteins may be involved in directional signalling at the level of the Golgi complex (12,28). Mucopolysaccharidosis I (I-cell disease), is an inherited disease (autosomal recessive) that is characterized by deficiency of lysosomal function. Enzymes normally stored in the lysosomes are secreted from the cells in large amounts. The defect in the disorder (28) appears to be the absence of an enzyme that phosphorylates mannose residues in carbohydrate moieties attached to the glycoprotein enzymes. The phosphoman- nose groups on the enzymes have been shown to serve as recognition markers sensed by receptors in the Golgi complex that target the enzymes for incorporation into lysosomes. In the absence of the phosphomannosyl-recognition marker, the enzymes are misdirected into the secretory rather than the lysosomal pathway and are therefore not retained in, but are exported from, the cell.

These two examples, cited, leader sequences and phosphomannose, may represent unique instances in which co- and post-translational processes serve important functions in intracellular signalling. On the other hand, it is likely that many more similar instances of such intracellular signalling may become apparent upon further studies of the biosynthetic pathways of specific cellular proteins.

Precursors as a Source of Diverse Biologic Activities

A function that appears to be fulfilled by certain biosynthetic precursors is that of a prohormone that provides a means to generate multiple biologic activities from a single gene product. Inasmuch as all of the many mRNAs thus far identified from eukaryotes are monocistronic, i.e., code for only a single protein, multiple but separate biologic activities appear to be generated at the level of post-translational cleavages of precursors. As many as seven different proteins are derived from a single precursor coded for by polio virus RNA (27). A prototypic hormonal precursor is the Mr-31000 protein that contains within its sequence (in addition to a leader sequence) the sequences of the hormones ACTH, LPH, β-MSH, α-MSH, the endorphins, and enkephalins, as well as a Mr-16000 "cryptic" sequence in which reside mutated, "vestigial" sequences corresponding to MSH and to calcitonin (18) (Fig. 2). The Mr-22000 common precursor of vasopressin and neurophysin is an example of two distinctly separate biologic activities residing in a single precursor (25) (Fig. 2). Both of these precursors, Mr-22000 vasopressinneurophysin (pro-pressophysin) and the Mr-31000 ACTH--LPH--endorphin (pro-opiocortin) are glycosylarted at specific sites in the molecules, and, furthermore, the Mr-31000 precursor appears to undergo acetylation under certain circumstances (see Mains and Eipper, this volume). One can apprecirate that different combinations of cleavages, glycosylation, and acetylation provides the possibility for the generation of a large number of chemically different peptides and, as a result, the potential for further diversification of biologic activities of the peptides so generated. An example of this type of diversification comes from the Mr-31000 pro-opiocortin precursor in which it was shown that the products of cleavages in the intermediate lobe of the pituitary differ from those in the anterior lobe (23). In the anterior lobe, the cleavages generate ACTH as a major product; in the intermediate lobe, however, they generate α-MSH and the "CLIP" peptide from the sequence of ACTH in the Mr-31000 precursor.

Processing by proteolytic cleavages of the prohormones of parathyroid hormone (proparathyroid hormone) and of insulin (proinsulin) was shown to occur in the Golgi complex (7,13). The prohormones are synthesized in the RER, after co-translational cleavages of leader sequences, and are transported to the Golgi

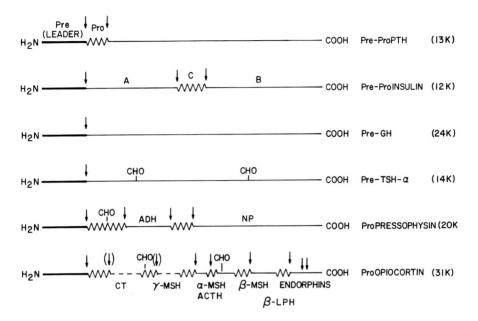

FIG. 2.   Schematic representations of some polypeptide-hormone precursors. Arrows indicate sites of co- and post-translational cleavages. Zig-zag lines denote intervening "cryptic" sequences. Heavy lines represent NH₂-terminal leader, or signal, sequences. CHO=carbohydrate moieties.

complex from the RER via the smooth endoplasmic reticulum. Unlike the situation with pre-hormones, in which the amino acids at the site of cleavage between the leader sequence and the remainder of the molecule (hormone or prohormone) varies from one pre-hormone to the next (5,10), the cleavage site(s) of the prohormone intermediates uniformly consists of the basic amino acids lysine or arginine, or both, usually two or three together. This substrate is readily and preferentially attacked by endopeptidase cleavage; the basic residues are susceptible to selective removal by exopeptidases with activity resembling that of carboxypeptidase B.

It is quite likely that all proproteins are cleaved by a common enzymic process within the Golgi complex of cells of diverse origins. The significance of a general cleavage process for prohormones found in the secretory pathway, however, remains unknown, as does the explanation of the existence of proprotein intermediates in some but not all secreted proteins. It seems likely, however, that many more examples, other than pro-pressophysin and pro-opiocortin, will be found in which a large precursor molecule codes for multiple biologically active peptides.

Other Apparent Functions of Precursors

Biosynthetic precursors may express functions other than the two just described. One such function appears to be in the formation of secondary structure in proteins and in the facilitation of the assembly of individual protein chains into higher-ordered structures. The C-peptide of proinsulin, the protein segment that connects the A and B chains of insulin, appears to be requisite for the accurate juxtaposition of the A and B chains and for the formation of the intrachain disulfide bonds that link the two chains together in the mature insulin molecule (29). Another function may be to serve as inactive, or masked, stores of biologically important proteins that are activated, or unmasked, when needed, by a single post-translational modification. As examples of this particular function one may cite the proteins involved in blood clotting (fibrinogen, prothrombin) and the pancreatic enzymes (trypsinogen, procarboxypeptidase).

REGULATION OF PEPTIDE-HORMONE BIOSYNTHESIS

A major concern of investigators working in the area of polypeptide-hormone biosynthesis has been the determination of how the biosynthetic and secretory processes are regulated. Studies of regulatory processes have been directed principally toward two areas: a) investigations of the nature of the cellular mechanisms involved in the coupling of extracellular regulatory stimuli to intracellular events that lead to changes in the formation and release of hormones, and b) attempts to define the level in the various steps of protein synthesis at which regulation takes place, i.e., which of the transcriptional, post-transcriptional (and pre-translational), translational, or co- and post-translational processes is involved.

The molecular processes involved in the coupling of extracellular stimuli to the secretion and, ultimately, to the biosynthesis of hormone required to replace that which is secreted are not known in final detail. These processes, when they are understood, may differ from one protein-hormone-secreting cell to another. There is evidence, however, that a common coupling mechanism may be involved in the secretion of many of the hormones. This mechanism utilizes cAMP as a second messenger (Fig. 3).

In this model, the stimulatory factor, instead of entering the cell, interacts with a receptor located within the plasma membrane. In some manner, the stimulatory factor, when bound to the plasma-membrane receptor, activates adenylate cyclase, resulting in the generation of 3',5'-AMP, which, in turn, converts an inactive form of a phosphorylating enzyme, protein kinase, to an active form by way of dissociation of a regulatory subunit from the active catalytic subunit. The protein kinase (active subunit) catalyzes the phosphorylation of certain intracellular proteins, and the phosphoproteins thus formed are believed to function in the processes of hormone transport and secretion, perhaps via the activation of microtubules and (or) fusion of secretory-granule membranes with the plasma membrane. These events are highly speculative at present because of a lack of experimental evidence to support them. In fact, very little is known about the nature of the processes and the forces that are involved in the active form by way of dissociation of a regulatory subunit from intracellular movement of proteins.

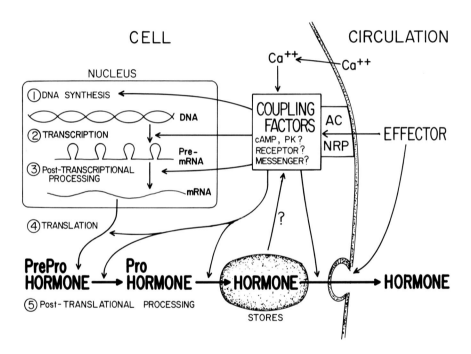

FIG. 3. Diagram of putative regulatory steps in the biosynthesis of polypeptide hormones. AC = adenylate cyclase, NRP = nucleotide regulatory protein, PK = protein kinase. Hypothetical coupling factors may include cytosolic protein receptors. Many effectors are believed to recognize specific protein receptors in the plasma membrane.

It should be noted that calcium appears to have an important function in the secretory process (24). Fluxes of calcium from the extracellular fluid into the cell, as well as fluxes from intracellular organelles (e.g., mitochondria) into the cytosol, are closely coupled to the event of secretion. When extracellular fluid levels of calcium are insufficient, secretion is severely

inhibited. It is possible that the influx of calcium into the
cytosol is linked, in some as-yet-undefined manner, to the
activation of adenylate cyclase.

The secretion of protein hormones by a number of endocrine
organs is also influenced by catecholamines through actions on
adrenergic receptors that in turn, are coupled to the receptors
that respond to the primary stimuli. It is believed that these
adrenergic effects serve to modulate secretory activity, which is
principally under the influence of the primary stimulating factor.

Biosynthetic processes must be coupled in some manner to the
secretion processes. Synthesis of new hormone is required to
replace that which is released, and, conversely, when secretory
demands decrease, synthesis of new hormone must therefore also
decrease to prevent the overloading of the cell with hormone.
Little is known about the cellular mechanisms that couple secretory events to biosynthetic events---that is, whether the extracellular stimulatory factors that regulate rates of secretion also
directly affect rates of hormone biosynthesis or whether the
process of secretion itself provides regulatory signals that are
transmitted to the steps in biosynthesis. The closeness of the
coupling of secretory activity to biosynthetic activity in a
particular endocrine gland may depend to a large degree on the
relative magnitude of the amount of hormone that is in stored
form. Of course, a gland that has a relatively large store of
hormone can meet secretory demands for a longer time than can a
gland with a lesser store. All endocrine cells store hormone to
some extent, as evidenced by the major morphologic hallmark of the
glandular cells, the secreted granules. It is likely that such a
storage system has evolved to provide endocrine secretory cells
with a "buffer," or reservoir, of hormone that can be called upon
to meet secretory demands over a very short time without the
necessity for the occurrence of abrupt changes in rates of hormone
biosynthesis.

The specific level (or levels) in the pathway of hormone
biosynthesis at which regulation takes place is not well
understood at present. In concept, regulation could take place at
one or more of several levels (Fig. 3): a) DNA synthesis (cell
division), b) transcription, c) post-translational processing, d)
translation, and e) post-translational processing. Evidence was
presented in support of regulatory processes at the level of
transcription (Step 2, Fig. 3). For example, the regulation, by
glucocorticoids or by triiodothyronine (T_3), of growth-hormone
synthesis in pituitary cells maintained in culture appears to take
place via the formation of complexes consisting of cytosolic
receptors and T_3 or glucocorticoids taken up from the extracellular fluid by the receptors (26). The complexes are transported to the nucleus where they stimulate the transcription of
mRNA coding for growth hormone. This regulatory process is
similar to that described for the regulation of the synthesis of
egg-white proteins mediated by estrogens (19,20,31). Such
regulation at the level of transcription appears to occur in

tissues that are induced to produce high levels of a specific protein. When stimulated, these tissues contain as many as 100,000 copies of the particular mRNA per cell (11).

A second level of regulation for which some supporting evidence exists is that of translational events (Step 4, Fig. 3). Glucose appears to act at this level (indirectly) in the regulation of proinsulin synthesis within minutes, inasmuch as changes in extracellular levels of glucose alter amounts of proinsulin synthesized (21). The exact mechanisms involved in this regulatory step are not known but presumably involve changes in rates of initiation of proinsulin synthesis utilizing pre-formed mRNA.

Chronic stimulation of secretory cells leads to cell division (hyperplasia) and, conversely, suppression of glandular activity results in glandular atrophy or inhibition of cell growth. Thus, in a sense, stimulation or suppression of the secretory activity of cells is ultimately reflected in changes in DNA synthesis (Step 1, Fig. 3).

The few studies that have been undertaken to examine regulation at the level of post-translational events have yielded largely negative results (Step 5, Fig. 3). For example, changes in concentrations of calcium in the extracellular fluid do not appear to influence the rates of conversion of proparathyroid hormone to parathyroid hormone (8). One post-translational process, however, that under certain circumstances may serve as a regulated step is that of the intracellular degradation, or turnover, of hormone. Under conditions in which secretion is generally suppressed, the parathyroid gland degrades approximately 40% of the hormone synthesized. Stimulation of the gland by lowering of the calcium in the extracellular fluids results in a sharp curtailment in the rate of hormone degradation, thus providing increased amounts of hormone to meet secretory demands (8).

Little is known about the possible regulatory influences on hormone biosynthesis at the level of post-transcriptional processing (step 3, Fig. 3) or, for that matter, at the level of the availability for translation of mRNA in the cytoplasm. Davidson and Britten (6) however, proposed that the synthesis of certain proteins produced in scant or only moderate amounts may be under nuclear control at the level of RNA-to-RNA interactions involving the processing of gene transcripts. The hypothesis includes the concept that "integrative" RNAs transcribed from repetitive DNA sequences in a regulated manner interact with "constitutive" mRNA precursors transcribed from genes coding for specific proteins. The interaction influences the processing of the RNA transcript and, as a result, regulates the amounts of mature mRNA produced for transport into the cytoplasm.

An additional point of potential regulation is at the level of the availability of cytoplasmic mRNAs for translation into protein. Inhibition of new RNA synthesis by treatment of cells with actinomycin D was shown to stimulate the synthesis of specific proteins (30). This paradoxical effect of actinomycin D is believed to result from inhibition of the synthesis of a relatively labile mRNA coding for cytoplasmic "repressor" proteins with resultant prevention of the translation of mRNAs coding for the specific proteins. Whether the repressor is regulated is unknown, but the potential for regulation at this level appears to exist.

BIOSYNTHESIS OF PEPTIDE HORMONES IN THE NERVOUS SYSTEM

The discovery in the nervous system of peptide-hormone-like substances immunologically similar to the endocrine hormones suggests that similar gene products may serve functions both as apocrine neurotransmitters (cybernins) and as endocrine hormones. Approximately 30 different polypeptide-hormone-like activities have now been detected in various regions of the central nervous system by either immunocytochemical localizations or radioimmuno-assays for the hormones. It is of great interest to ascertain the specific functions of these cybernins: how they are synthesized, how their synthesis and release are regulated, and how they are related to the endocrine hormones. It is tempting to speculate that, in the course of evolution, these peptides served as neuro-transmitters before the appearance of the hormonal functions that evolved with the growing complexity of and the requirement of the organism for a nervous system. Later, as a result of genetic diversification and the need for more highly regulated metabolic systems related to growth and development, similar gene products were utilized as hormones.

As discussed earlier, genetic diversification may arise by at least two mechanisms: a) gene duplication (or multiplication) and mutation or b) variability in the co- and post-translational processing of a given gene product. Therefore, one might predict that the structures of the cybernins, when determined,will differ from those of the hormones by either genetic variability in the amino acid sequences or the nature of their processing from precursors.

At present, little is known about the processes involved in the biosynthesis of the cybernins. The information available is largely derived from radioimmunoassay measurements of hormone-like activities in extracts prepared from nerve tissue. The relatively low amounts of these hormone-like substances in nerve tissue com-pared with the concentrations of the hormones in endocrine organs render difficult studies of the biosynthesis of these substances by techniques of pulse-chase analyses using radiolabelled amino acids. Likewise, the paucity of mRNAs coding for the substances in nerve tissue has precluded identification of specific cell-free translation products---an approach that has been highly successful in the identification of products (precursors of hormones) using messenger RNAs prepared from endocrine glands. An alternative approach, however, that should prove successful is the use of recombinant-DNA technology to identify and amplify mRNA sequences coding for the cybernins. cDNAs complementary in sequence to mRNAs coding for specifc hormones can be prepared using the large number of mRNAs obtained from endocrine glands. The cDNAs may then be greatly increased in number by the introduction and replication of plasmids containing recombinant DNAs (cDNA plus plasmid DNA) within bacterial hosts; these more numerous cDNAs may then be used to select, by hybridization techniques, the scarce mRNAs in nerve tissue that are genetically related to the mRNAs coding for the hormones. cDNAs complementary in sequence to the neural mRNAs can, in turn, be amplified by molecular cloning of recombinant molecules and their respective nucleic acid sequences can be analyzed, thus determining the amino acid sequences. By this approach it may be possible to determine the primary structures of the hormone-like cybernins and thereby draw certain conclusions regarding the genetic relatedness among the endocrine hormones and cybernins.

SUMMARY

Specialized and specific biologic functions can arise as a result of genetic diversification, as well as by way of bio-chemical modifications of a given gene product. Additional, more detailed investigations of the structure and organization of the genome and the processes involved in the co- and post-translational modifications of polypeptide hormones may provide clues to the functions of the hormone-related peptides (cybernins) produced by peptidergic neurons.

REFERENCES

1. Bielinska, M., and Boime, I. (1978): Proc. Natl. Acad. Sci. USA, 75:1768-1772.
2. Blobel, G. (1980): Proc. Natl. Acad. Sci. USA, 77:1496-1500.
3. Blobel, G., and Dobberstein, B. (1975): J. Cell Biol., 67:835-851.
4. Blobel, G., and Sabatini, D.D. (1971): Biomembranes, 2:193-195.
5. Chan, S.J., Patzelt, C., Dugid, J.R., Quinn, P., Labrecque, A., Noyes, B., Keim, P., Heinrikson, R.L., and Steiner, D.F. (1979): In: From Gene to Protein: Information Transfer in Normal and Abnormal Cells (Proceedings of the 11th Miami Winter Symposium), edited by T.R. Russell, K. Brew, H. Faber, and J. Schultz, Vol. 16, pp. 361-378. Academic Press, New York.
6. Davidson, E.H., and Britten, R.J., (1979): Science, 204:1052-1059.
7. Habener, J.F., Amherdt, M., Ravazzola, M., and Orci, L. (1979): J. Cell Biol., 80:715-731.
8. Habener, J.F., Kemper, B., and Potts, J.T., Jr. (1975): Endocrinology, 97:431-441.
9. Habener, J.F., Maunus, R., Dee, P.C., and Potts, J.T., Jr. (1980): J. Cell Biol., 85:292-298.
10. Habener, J.F., and Potts, J.T., Jr. (1978): N. Engl. J. Med., 299:580-585.

11. Harris, S.E., Rosen, J.M., Means, A.R., and O'Malley, B.W. (1975): Biochemistry, 14:2072.
12. Hickman, S., and Neufeld, E.F. (1972): Biochem. Biophys. Res., 49:992-999.
13. Kemmler, W., Steiner, D.F., and Borg, J. (1973): J. Biol. Chem., 248:4544-4551.
14. Lehninger, A.L. (1975): Biochemistry: The Molecular Basis of Cell Structure. Worth Publishers, New York.
15. Lingappa, V.R., Lingapppa, J.R., and Blobel, G. (1979): Nature, 281:117-121.
16. Lingappa, V.R., Shields, D., Woo, S.L.C., and Blobel, G. (1978): J. Cell. Biol., 79:567-572.
17. Milstein, C., Brownlee, G.G., Harrison,T.M., and Mathews, M.B. (1972): Nature (New Biol.), 239:193-195.
18. Nakanishi, S., Inoue, A., Kita, T., Nakamura, M., Chang, A.C.Y., Cohen, S.N., and Numa, S. (1979): Nature, 278:423-427.
19. O'Malley, B.W., McGuire, W.L., Kohler, P.O., and Korenman, S.G. (1969): Recent Prog. Horm. Res., 25:105-160.
20. Palmiter, R.O., Moore, P.B., Mulvihill, E.R., and Emtage, S. (1976): Cell, 8:557-572.
21. Permutt, M.A., and Kipnis, D.M. (1975): Fed Proc., 34:1549-1555.
22. Rothman, J.E., and Lodish, H.F. (1977): Nature, 269:775-780.
23. Roberts, J.L., Phillips, M.A., Rosa, P.A, and Herbert, E. (1978): Biochemistry, 17:3609-3617.
24. Rubin, R.P. (1970): Pharmacol. Rev., 22:389-428.
25. Russell, J.T., Brownstein, M.J., and Gainer, H. (1979): Proc. Natl. Acad. Sci. USA, 76:6086-6090.
26. Shapiro, L.E., Samuels, H.H., and Yaffe, B.M. (1978): Proc. Natl. Acad. Sci. USA, 75:45-49.
27. Shih, D.S., Shih, C.T., Kew, O., Pallansch, M., Rueckert, R., and Kaesberg, P. (1978): Proc. Natl. Acad. Sci. USA, 75:5807-5811.
28. Sly, W.S., and Stahl, P. (1978): In: Transport of Macromolecules in Cellular Systems, edited by S.C. Silverstein, pp. 229-244. Dahlem Konferenzen, Berlin.
29. Steiner, D.F., and Clark, J.L. (1968): Proc. Natl. Acad. Sci. USA, 60:622-629.
30. Tompkins, G.M., Levinson, B.B., Baxter, J.D., and Dethlefsen, L. (1972): Nature (New Biol.), 239:9-14.
31. Towle, H.G., Tsai, M.J., Tsai, S.Y., Schwartz, P.J., Parker, M.G., and O'Malley, B.W. (1976): In: The Molecular Biology of Hormone Action, edited by J. Papaconstantinou, pp. 107-136. Academic Press, New York.
32. Walter, P., Jackson, R.C., Marcus, M.M., Lingappa, V.R., and Blobel, G. (1979): Proc. Natl. Acad. Sci. USA, 76:1795-1799.

Neurosecretion and Brain Peptides,
edited by J. B. Martin, S. Reichlin, and K. L. Bick.
Raven Press, New York © 1981.

Synthesis and Secretion of ACTH, β-Endorphin, and Related Peptides

Richard E. Mains and Betty A. Eipper

Department of Physiology, University of Colorado Health Sciences Center, Denver, Colorado 80262

In this article we present a brief review of the biosynthesis and secretion of pituitary ACTH, β-endorphin and related peptides; a more comprehensive review of this field can be found in ref. 14 and 33. We also present some preliminary data on release of these peptides in vivo.

STRUCTURE OF THE COMMON PRECURSOR, PRO-ACTH/ENDORPHIN

The structure of the common precursor to ACTH, β-endorphin and related peptides is diagrammed in Fig. 1 (9,14,53). The complete amino acid sequence of a bovine intermediate pituitary common precursor has been predicted (53); structural studies on mouse pro-ACTH/endorphin produced a similar picture for the structure of the precursor molecule (9). ACTH is situated in the middle of the precursor molecule; β-lipotropin (βLPH) accounts for the carboxyl terminal third of the precursor; 16K fragment forms the amino terminal third of the precursor. A conventional hydrophobic signal is cleaved from the amino terminus of the initial translation product (23).

As previously observed for proinsulin and proparathyroid hormone, pairs of basic amino acids mark the major proteolytic cleavage sites in pro-ACTH/endorphin (5,20). In the bovine intermediate pituitary precursor there are ten pairs of basic amino acids (Fig. 1). The sequence Lys-Arg separates 16K fragment and ACTH, ACTH and βLPH, γLPH and β-endorphin and also occurs within 16K fragment and within ACTH. Other pairs of basic amino acids (Arg-Lys, Arg-Arg, Lys-Lys) also occur within 16K fragment, ACTH, γLPH and β-endorphin.

Similar precursor molecules are thought to occur in both the anterior and intermediate lobes of the pituitary. The post-translational proteolytic processing of the precursor molecules differs in the two lobes of the pituitary (Fig. 2) (10,24,49,64). However in both lobes of the pituitary the first two proteolytic cleavage

35

steps almost always occur in the same order: the first proteolytic cleavage step separates βLPH from the carboxyl terminus of ACTH; soon thereafter 16K fragment is cleaved from the amino terminus of ACTH. The cleavage of βLPH to γMSH plus β-endorphin and of ACTH to αMSH plus CLIP (in intermediate pituitary) occurs on a slower time scale (47-50); CLIP is ACTH (18-39) and αMSH is N-acetyl-ACTH (1-13)NH$_2$ (64). Although it would be possible, based on the structure of pro-ACTH/endorphin to produce β-endorphin with a single proteolytic cleavage of the precursor, in pituitary tissue βLPH

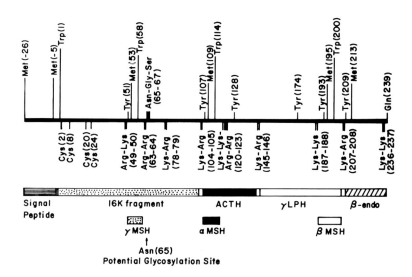

FIG. 1. Structure of the common precursor to ACTH and β-endorphin. Data from studies on the bovine intermediate precursor (53) and the mouse pituitary precursor (9,24) have been combined because the precursors appear to be quite similar (except in the γLPH region (11)). Pairs of basic amino acids that occur in the predicted bovine amino acid sequence are indicated. Studies in mouse and rat pituitary tissue indicate that the paired basic amino acids do serve as proteolytic processing sites (48-50). The numbering of the bovine amino acid sequence has been adjusted to reflect cleavage of the signal peptide (23,28); the Trp at the amino terminus of pro-ACTH/endorphin is thus numbered Trp[1] (from ref. 14).

serves a biosynthetic intermediate in the production of β-endorphin (48-50). Methionine enkephalin, which could be derived from
pro-ACTH/endorphin, is not a product in pituitary tissue and
appears to have its own separate precursor (27,35,49,60,73).

Structural studies of pro-ACTH/endorphin and its product peptides from different species have demonstrated that most of the
structure of pro-ACTH/endorphin is highly conserved among species.
Both ACTH and β-endorphin are known to be highly conserved among
species (36,55). Recent data on 16K fragment from mouse, hog, and
bovine tissues indicate that the amino terminal region of 16K
fragment is also highly conserved (21,28,53). The middle region
of bovine 16K fragment contains a third MSH-like sequence called
γMSH that is flanked by pairs of basic amino acids (53). A similar region occurs in mouse 16K fragment (unpublished observations). The high degree of conservation of certain regions of 16K
fragment suggests that additional biologically important regions
of pro-ACTH/endorphin may remain to be identified and studied.
Trypsin treatment of purified mouse 16K fragment generates a
smaller peptide capable of activating adrenal cortical cholesterol
ester hydrolase and potentiating the steroidogenic action of ACTH
(56). The γLPH and CLIP regions of the precursor show the most
substantial species specific variation (11,36,44,55).

POST-TRANSLATIONAL PROCESSING OF THE COMMON PRECURSOR MOLECULE

Proteolytic cleavage, N-acetylation, amidation and glycosylation are the post-translational modifications already known to be
of importance in the pro-ACTH/endorphin system. As indicated in
Fig. 2, intermediate pituitary tissue in the rat and mouse processes the common precursor molecule to a collection of product
peptides which are smaller than those seen in the anterior pituitary. The vast majority of the ACTH produced in the rat intermediate pituitary is cleaved to form peptides similar to αMSH and
CLIP; αMSH and CLIP are not major anterior pituitary products
(10,16,32,49,64). In the rat intermediate pituitary almost all of
the βLPH produced is converted to β-endorphin-like products; in
the anterior pituitary both βLPH and β-endorphin are final pro-
rducts (3,10,15,24,26,39,40,57,65,74). It has been suggested that
a β-endorphin is found in the anterior pituitary only as a result
of proteolytic degradatin during extraction (38,61). The fact
that pulse-chase protocols carried out on pituitary tissue in
several different labs demonstrate a time dependent conversion of
labeled βLPH to β-endorphin indicates that the extraction methods
used do block proteolytic degradation fully, and that a β-endorphin-like peptide is a genuine anterior pituitary product (7,23,
24,41,42,48,49, 58). The glycoprotein (16K fragment) accounting
for the non-ACTH, non-β LPH region of pro-ACTH/endorphin was first
identified in mouse tumor cells (10,48) and in rat pituitary
(10,49), and was subsequently shown, in beef, to contain an MSH-
like segment (γMSH) (53). The processing of 16K fragments differs
in anterior and intermediate pituitary (Fig. 3, a and b), but the

structure of the 16K fragment-derived products has not been completely elucidated.

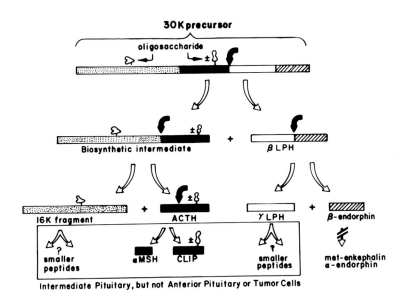

FIG. 2.   Post-translational processing of pituitary pro-ACTH/β-endorphin.   The processing scheme shown was developed from studies on rat pituitary cells and mouse anterior pituitary tumor cells in vitro (47-50) and shows only the major proteolytic processing events.   In rat and mouse, glycosylated forms of ACTH and CLIP are found (11,49); in sheep, beef, pig and human pituitary an oligosaccharide chain is not attached within ACTH itself (53,55,65). 16K fragment is a glycoprotein, but its oligosaccharide composition has not yet been determined (9,59).   Microheterogeneity of the oligosaccharide chains creates many different forms.   Other proteolytic cleavages (such as that converting β-endorphin(1-31) to β-endorphin (1-27) can also occur (67,68,75).   Acetylation (e.g.,of ACTH, αMSH and β-endorphin) occurs to varying degrees in different tissues (13,50,68,75) (from ref. 50).

Fig. 2 summarizes only the major proteolytic cleavages known to occur in pituitary pro-ACTH/endorphin.   The pair of basic amino acids near the carboxyl terminus of β-endorphin can also be used as a cleavage site (67,68).   When cleavage occurs at this site, the opiate active peptide β-endorphin(1-31) is converted to the less potent peptide β-endorphin(1-27) [also known as C'-fragment]

(66). The relative levels of β-endorphin(1-31) and β-endorphin-(1-27) differ substantially among tissues (75).

Several of the peptides produced in the intermediate lobe of the pituitary are α-N-acetylated. Roughly three quarters of the αMSH-like material produced by rat intermediate pituitary cells in vitro is α-N-acetylated; about half of the small amount of intact ACTH(1-39) found in the intermediate lobe is α-N-acetylated (49,50). Diacetyl αMSH has been detected in pituitary tissue

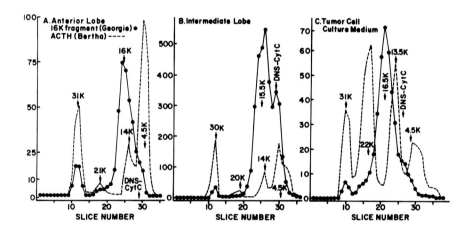

FIG. 3. Immunoactive 16K fragment in rat anterior and intermediate pituitary. Pituitaries from two male rats (500 g) were separated into anterior (A) and intermediate/posterior (B) lobes and extracted (47). Extracts were fractionated on SDS polyacrylamide gels. Spent AtT-20 mouse pituitary tumor cell culture medium (C) was concentrated on CG-50 (11) and fractionated on an SDS polyacrylamide gel. Gels were sliced, eluted and immunoassayed for ACTH (middle ACTH antiserum,....) and for 16K fragment (antiserum Georgie, ●). Apparent molecular weights are indicated. (See ref. 10 for further details.)

(62). α-N-acetylation enhances the melanocyte stimulating potency of αMSH but greatly decreases the steroidogenic potency of ACTH-(1-39) (54). In the intermediate lobe of the pituitary, over half of the β-endorphin-like material is α-N-acetylated and thus inactivated as far as most conventional opiate bioassays are concerned (13,75).

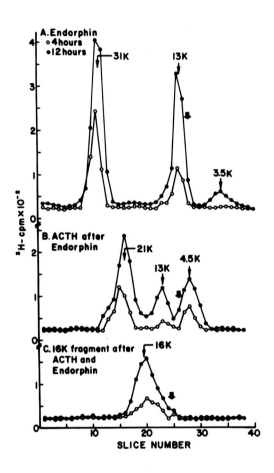

FIG. 4. Secretion of ACTH, β-endorphin and 16K fragment from rat anterior pituitary cells in vitro. Dissociated rat anterior pituitary cells were incubated in complete medium plus [³H]-phenylalanine. Medium was harvested at the times indicated and immunoprecipitated sequentially with antiserum to β-endorphin (A), followed by antiserum to the middle region of ACTH (B), and finally by antiserum to 16K fragment (C). Dissociated immunoprecipitates were fractionated by SDS polyacrylamide gel electrophoresis; apparent molecular weights are (See ref. 12 for further details.)

Amidation of the valine residue at the carboxyl terminus of αMSH is required for the production of authentic αMSH. Many of the small, biologically active neural and endocrine peptides currently under investigation have modified amino and/or carboxyl termini. The pro-ACTH/endorphin-derived peptides are presently the only ones with a clearly defined biosynthetic pathway.

Glycosylation of pro-ACTH/endorphin has been directly demonstrated in only a few species. Mouse, rat, and frog pituitary

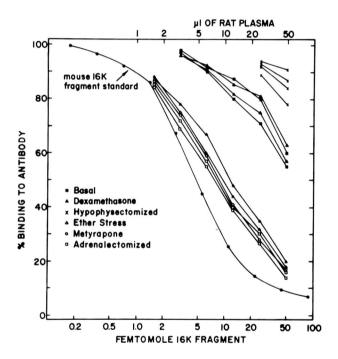

FIG. 5. Immunoassay of 16K fragment in rat plasma. Rat blood was collected as described in Table 1. Immunoassays for 16K fragment were performed as described (10,49) with one modification; samples plus antiserum Georgie (1: 40,000) were incubated at 4° for 30 h and ^{125}I-mouse 16K fragment was added for the final 16 h. The difference between samples levels were altered without antibody (usually 3% of input radioactivity) and maximal binding in the absence of unlabeled protein (usually about 23% of input) was adjusted to 100%.

cells in tissue culture have been shown to incorporate radiolabeled sugars into pro-ACTH/endorphin (8,9,42,49,51,59). Purified
mouse and hog 16K fragment contain hexosamines by amino acid
analysis (unpublished observations, 21). The amino acid sequence
predicted for bovine 16K fragment contains a single site appropriate for attachment of a N-glycosidically linked oligosaccharide
(Fig. 1) (53). Attachment of an oligosaccharide chain with the
carboxyl terminal region of ACTH(1-39) has been shown by radio-
labeling of mouse and rat molecules and by analysis of acid
hydrolysates of "13K ACTH" purified from mouse tumor cell culture
medium (8,9,49,59). It is clear that the suggestion that glycosylation of pro-ACTH/endorphin is restricted to tumor cells in
tissue culture was incorrect (29). The occurrence of glycosylated
ACTH(1-39) appears to be species specific; bovine, human, ovine,
and porcine ACTH(1-39) do not contain an amino acid sequence
appropriate for the addition of a complex asparagine linked
oligosaccharide chain (53,55,65).

SECRETION OF PEPTIDES DERIVED FROM PRO-ACTH/ENDORPHIN

It is now quite clear from work with both normal anterior pituitary cells and pituitary tumor cells in tissue culture that,
under basal conditions, all of the smaller peptide products derived from pro-ACTH/endorphin are secreted together in a coordinate, equimolar fashion (Fig. 3C; Fig. 4) (2,10,12,48,63,71). For
the tumor cells, stimulated release still consists of equimolar
amounts of ACTH (glycosylated and not), 16K fragment and β LPH/
β-endorphin.
The clinical observation of coordinate release of immunoreactive ACTH and immunoactive βMSH (now known to represent β LPH and
γLPH) was one of the key pieces of data supporting the common
precursor hypothesis (1,6,17,25,45,69). Therefore it was anticipated that release of 16K fragment-related material would accompany release of ACTH and β-endorphin in vivo. In order to test
this hypothesis, standard experimental protocols known to alter
release of rat anterior pituitary ACTH (34) were carried out;
plasma levels of immunoreactive 16K fragment (10,49), immunoactive
β-endorphin (48) and glucocorticoids (as a measure of ACTH secretion) were determined. Table 1 shows that the experimental protocols produced the expected changes in plasma corticosterone.
Likewise, plasma immunoactive β-endorphin levels were altered in
the expected manner. Unextracted rat plasma was analyzed using
the 16K fragment immunoassay originally developed for mouse tumor
cell 16K fragment (Fig. 5). Plasma from basal and dexamethasone-
treated rats had immunoactive 16K fragment concentrations that
were close to equimolar to the concentration of immunoactive β-
endorphin (Table 1). The three treatments known to increase
anterior pituitary ACTH secretion (ether stress, adrenalectomy,
administration of metyrapone) also substantially increased plasma
levels of immunoactive 16K fragment and immunoactive β-endorphin.
Although knowledge of the clearance rates for each of the
different product peptides is needed, coordinate and roughly equi-

molar secretion of pro-ACTH/endorphin derived peptides appears to occur in vivo as well as in vitro. Further analyses will be necessary to determine the molecular form of the immunoactive 16K fragment-related material in plasma.

TABLE 1.   Release of immunoactive 16K fragments in rats

Treatment Group (number of animals)	Corticosterone (ng/ml)	Immunoreactive β-endorphin (pM)	Immunoreactive 16K fragment (pM)
Basal (3)	14 + 4	75 + 14	85 + 11
Hypophysectomized (8)	3 + 1	47 + 4	39 + 8
Dexamethasone (3)	17 + 4	75 + 12	79 + 10
Ether Stress (3)	273 + 28	623 + 106	403 + 37
Adrenalectomized (6)	3 + 2	820 + 190	562 + 73
Metyrapone (1)	990[a]	520	465

Trunk blood was collected into heparin plus aprotinin to give a final concentration of 33 U/ml heparin and 0.40 TIU/ml aprotinin; phenylmethylsulfonyl fluoride and iodoacetamide were added to the plasmas to give a final concentration of each of 0.3 mg/ml and plasmas were stored frozen until assayed. Basal rats (250 g males) were killed at 0700 h. Dexamethasone-treated rats (250 g males) received 1 mg dexamethasone phosphate by intraperitoneal injection twice daily for a total of four injections, and were bled at 0700 h (9 h after the last injection). Ether stress was carried out on 250 g male rats at 0700 h for 15 min. Adrenalectomized rats (4 days) and hypophysectomized rats (4 days) were 100-120 g males and were purchased from Charles River; completeness of surgical removal of adrenals or pituitary was confirmed by visual examination of each animal. A single 317 g male rat was given drinking water containing 0.45 mg/ml metyrapone, 0.9% NaCl for 7 d; water intake averaged 40 ml/d and the rat gained 40 g by the end of the treatment week. Corticosterone was determined in unextracted or methylene chloride extracted plasma with the New England Nuclear cortisol kit, using corticosterone as the standard. The immunoassays for β-endorphin and 16K fragment have been described (10,48,49). Data are mean + standard deviation. [a]Expressed as deoxycorticosterone, which reacts fully but at potency with respect to corticosterone.

NONPITUITARY ACTH AND β-ENDORPHIN

There are a large number of examples of multiple genes coding for molecules with similar functions in different tissues: actin, myosin, cytochrome c, amylase, lactate dehydrogenase, the hemoglobins (4,18,22,30,52,72). There are also many examples of different post-translational processing of the same or very similar molecules in different tissues: pro-ACTH/endorphin, ribonuclease, conalbumin/transferrin, cytochrome b_5, β-galactosidase, and βglucuronidase (19,30,31,46,70). Thus it seems important to be cautious when making comparisons between the known structures and processing pathways of pituitary pro-ACTH/endorphin (Figs. 1 and 2) and related molecules and pathways in the hypothalamus and other nonpituitary tissues.

ACKNOWLEDGEMENTS

Supported by National Institutes of Health Grants AM 19859 and AM 18929.

REFERENCES

1. Abe, K., Nicholson, W.E., Liddle, G.W., Orth, D.N., and Island, D.P. (1969): J. Clin. Invest., 48:1580-1585.
2. Allen, R.G., Herbert, E., Hinman, M., Shibuya, H., and Pert, C.B. (1978): Proc. Natl. Acad. Sci. USA, 75:4972-4976.
3. Baizman, E.R., Cox, B.M., Osman, O.H., and Goldstein, A. (1979): Neuroendocrinology, 28:402-414.
4. Bunn, H.F., Forgot, B.G, and Ranney, H.M. (1979): Human Hemoglobins. W.B. Saunders Co, Philadelphia.
5. Chan, S.J., and Steiner, D.F. (1977): Top. Biochem. Sci., 2: 254-256.
6. Coscia, M., Brown, R.D., Miller, M., Tanaka, K., Nicholson, W.E., Parks, K.R., and Orth, D.N. (1977): Am. J. Med., 62: 303-307.
7. Crine, P., Gianoulakis, C., Seidah, N.G., Gossard, F., Pezalla, P.D., Lis, M., and Chretien, M. (1978): Proc. Natl. Acad. Sci. USA, 75:4719-4723.
8. Eipper, B.A., and Mains, R.E. (1977): J. Biol. Chem., 252: 8821-8832.
9. Eipper, B.A., and Mains, R.E. (1978): J. Biol. Chem., 253: 5732-5744.
10. Eipper, B.A., and Mains, R.E. (1978): J. Supramol. Struct., 8 247-262.
11. Eipper, B.A., and Mains, R.E. (1979): J. Biol. Chem., 254: 10190-10199.
12. Eipper, B.A., and Mains, R.E. (1979): Acta Endocrinol., (in press).
13. Eipper, B.A., and Mains, R.E. (1980): Excerpta Medica Int. Conq. Ser., (in press).
14. Eipper, B.A., and Mains, R.E. (1980): Endocrine Reviews, 1: 1-27.
15. Fukata, J., Nakai, Y., and Imura, H. (1979): Life Sci., 25: 541-546.
16. Gianoulakis, C., Seidah, N.G., Routhier, R., and Chretien, M. (1979): J. Biol. Chem., 254:11903-11906.
17. Gilkes, J.J.H., Bloomfield, G.A., Scott, A.P., Lowry, P.J., Ratcliffe, J.G., Landon, J., and Rees, L.H. (1975): J. Clin. Endocrinol. Metab., 40:450-457.
18. Goldberg, E., Sberna, D., Wheat, T.E., Urbanski, G.J., and Margoliash, E. (1977): Science, 196:1010-1012.

19. Graham, I., and Williams, J. (1975): Biochem. J., 145:263-279.
20. Habener, J.F., and Potts, J.T. Jr. (1978): N. Engl. J. Med. 299:580-586, 635-644.
21. Hakanson, R, Ekman, R., Sundler, F., and Nilsson, R. (1980): Nature, 283:789-792.
22. Harding, J.D., and Rutter, W.J. (1978): J. Biol. Chem., 253: 8736-8740.
23. Herbert, E., Budarf, M., Phillips, M., Rosa, P., Policastro, P., Oates, E., Roberts, J.L., Seidah, N.G., and Chretien, M. (1980): Ann. N.Y. Acad. Sci., (in press).
24. Herbert, E., Roberts, J.L., Phillips, M., Rosa, P.A., Budarf, M., Allen, R.G., Policastro, P.F., Pacquette, T.L., and Hinman, M. (1979): In: Endorphins in Mental Health Research, edited by E. Usdin, W.E. Bunney, Jr., and N.S. Kline, pp. 159-180. Macmillan Press, New York.
25. Hirata, Y., Matsukura, S., Imura, H., Nakamura,M., and Tanaka, A. (1976): J. Clin. Endocrinol. Metab., 42:33-38.
26. Hong, J.S., Yang, H.Y.T., and Costa, E. (1977): Neuropharmacology, 16:451-453.
27. Hughes, J. (1978): In: Endorphins '78, edited by L. Graf, M. Palkovits, and A.Z. Ronai, pp. 157-175. Akademiai Kiado, Budapest.
28. Keutmann, H.R., Eipper, B.A., and Mains, R.E. (1979): J. Biol. Chem., 254:9204-9208.
29. Kimura, S., Lewis, R.V., Gerber, L.D., Brink, L., Rubenstein, M., Stein, S., and Udenfriend, S. (1979): Proc. Natl. Acad. Sci. USA, 76:1756-1759.
30. Kindle, K.L., and Firtel, R.A. (1978): Cell, 15:763-778.
31. Kornfeld, R., and Kornfeld, S. (1976): Ann. Rev. Biochem., 45:217-237.
32. Kraicer, J. (1977): Front. Hormone Res., 4:200-207.
33. Kreiger, D.T., Liotta, A.S., Brownstein, M.J., and Zimmerman, E.A. (1980): Rec. Prog. Horm. Res., 36:(in press).
34. Krieger, D.T., Liotta, A.S., Hauser, H., and Brownstein, M.J. (1979): Endocrinology, 105:737-742.
35. Lewis, R.V., Stein, S., Gerber, L.D., Rubinstein, M., and Udenfriend, S. (1978): Proc. Natl. Acad. Sci. USA, 75:4021-4023.
36. Li, C.H., and Chung, D. (1976): Nature, 260:622-624.
37. Li, C.H., Chung, D., Oelofsen, W., and Naude, R.J. (1978): Biochem. Biophys. Res. Commun., 81:900-906.
38. Liotta, A.S., Suda, T., and Krieger, D.T. (1978): Proc. Natl. Acad. Sci. USA, 75:2950-2954.
39. Lissitsky, J.C., Morin, O., Dupont, A., Labrie, F., Seidah, N.G., Chretien, M., Lis, M., and Coy, D.H. (1978): Life Sci., 22:1715-1722.
40. Loeber, J.G., Verhoef, J., Burbach, J.P.H., and Witter, A. (1979): Biochem. Biophys.. Res. Commun., 86:1288-1295.
41. Loh, Y.P. (1979): Proc. Natl. Acad. Sci. USA, 76:796-800.
42. Loh, Y.P., and Gainer, H. (1979): Endocrinology, 105:474-487.

43. Lowney, L.I., Gentleman, S.B., and Goldstein, A. (1979): Life Sci., 24:2377-2384.
44. Lowry, P.J., Bennett, H.P.J., and McMartin, C. (1979): Life Sci,. 141:427-437.
45. Lowry, P.J., Rees, L.H., Tomlin, S., Gillies, G., and Landon, J. (1976): J. Clin. Endocrinol. Metab., 43:831-835.
46. Lusis, A.J., Breen, G.A.M., and Paigen, K. (1977): J. Biol. Chem., 252:831-835.
47. Mains, R.E., and Eipper, B.A. (1976): J. Biol. Chem., 251: 4115-4120.
48. Mains, R.E., and Eipper, B.A. (1978): J. Biol. Chem., 253: 651-655.
49. Mains, R.E. and Eipper, B.A. (1979): J. Biol. Chem., 254:7885-7894.
50. Mains, R.E., and Eipper, B.A. (1979): Ann. New York Acad. Sci., (in press).
51. Mains, R.E., and Eipper, B.A. (1979): Soc. Exp. Biol. Med., 33:37-55.
52. Markert, C.L., Shaklee, J.B., and Whitt, G.S. (1975): Science, 189:102-114.
53. Nakanishi, S., Inoue, A., Kita, T., Nakamura, M., Chang, A.C.Y., Cohen, S.N., and Numa, S. (1979): Nature 278: 423-427.
54. Ney, R.L., Ogata, E., Shizume, N., Nicholson, W.E., and Liddle, G.W. (1964): Proc. Int. Cong. Endocrinol., 2: 1184-1191.
55. Orth, D.N., Tanaka, K., and Nicholson, W.E. (1976): In: Hormones in Human Blood, edited by H.N. Antoniades, pp. 423-448. Harvard University Press, Cambridge, MA.
56. Pedersen, R.C., and Brownie, A.C. (1980): Proc. Nat. Acad. Sci. USA., 77:2239-2243.
57. Przewlocki, R., Hollt, V., and Herz, A. (1978): Eur. J. Pharmacol., 5:179-183.
58. Przewlocki, R., Hollt, V., Osborne, H., and Herz, A. (1979): Acta Endocrinol. Suppl., 225:72.
59. Roberts, J.L., Phillips, M., Rosa, P.A, and Herbert, E. (1978): Biochemistry, 17:3609-3618.
60. Rossier, J., Vargo, T.M., Minick, S. Ling, N., Bloom, F.E., and Guillemin, R. (1977): Proc. Natl. Acad. Sci. USA, 74: 5162-5165.
61. Rubenstein, M., Stein, S., Gerber, L.D., and Udenfriend, S. (1977): Proc. Natl. Acad. Sci. USA, 74:3052-3055.
62. Rudman, D., Chawla, R.K., and Hollins, B.M. (1979): J. Biol. Chem., 254:10102-10108.
63. Sabol, S.L., Ling, A., and Daniels, M.P. (1979): In: Hormones and Cell Culture, Cold Spring Harbor Conf., Vol. 6:843.
64. Scott, A.P., Ratcliffe, J.G., Rees, L.H., Landon, J., Bennett, H.P.J., Lowry, P.J., and McMartin, C. (1973): Nature (New Biol.), 244:65-69.
65. Silman, R.E., Holland, D., Chard, T., Lowry, P.J., Hope, J., Robinson, J.S., and Thorburn, G.D. (1978): Nature, 276: 526-528.

66. Smyth, D.G., Massey, D.E., Zakarian, S, and Finnie, M.D.A. (1979): Nature, 279:252-254.
67. Smyth, D.G., Snell, C.R., and Massey, D.E. (1978): Biochem. J., 175:261-270.
68. Smyth, D.G., and Zakarian, S. (1979): In: Endorphins in Mental Health Research, edited by E. Usdin, W.E. Binney, Jr., and N.S. Kline, pp. 84-92. Macmillan Press, New York.
69. Tanaka, K., Nicholson, W.E., and Orth, D.N. (1978): J. Clin. Invest., 62:94-104.
70. Tulsiani, D.R.P., Six, H., and Touster, O. (1978): Proc. Natl. Acad. Sci. USA, 75:3080-3084.
71. Vale, W., Rivier, C., Yang, L., Minick, S., and Guillemin, R. (1978): Endocrinology, 103:1910-1915.
72. Vandekerckhove, J., and Weber, K. (1978): Proc. Natl. Acad. Sci. USA, 75:1106-1110.
73. Watson, S.J., Akil, H., Richard, C.W., III, and Barchas, J.D. (1978): Nature, 275:226-228.
74. Yoshimi, H., Matsukura, S., Sueoka, S., Fukase, M., Yokota, M., Hirata, Y., and Imura, H. (1978): Life Sci. 22: 2189-2196.
75 Zakarion, S., and Smyth, D. (1979): Proc. Natl. Acad. Sci. USA, 76:5972-5976.

Neurosecretion and Brain Peptides,
edited by J. B. Martin, S. Reichlin, and K. L. Bick.
Raven Press, New York © 1981.

Peptide Processing in the Central Nervous System

Neville Marks, *A. Suhar, and M. Benuck

*Center for Neurochemistry, Rockland Research Institute, Ward's Island, New York 10035;
and *Jozef Stefan Institute, Ljubljana, Yugoslavia*

There is evidence that most neuropeptides are formed by cleavage of precursor materials, yet little information exists on the nature of the enzymes involved in processing or the manner they are regulated in vivo. Processing encompasses a sequence of events that commences with the formation of mRNA (from a pro-mRNA form), translation by a ribosomal-membrane coupled system, the translocation of the growing peptide to within the lumen of the endoplasmic reticulum, post-translational modifications and transport to the active site, followed by release of the active moiety and its inactivation. A detailed account of some of these processes is provided elsewhere (9) and only highlights are selected below in order to illuminate a number of key features.

Most secreted proteins are synthesized as a larger precursor with an additional hydrophobic (N-terminal) signal sequence that is rapidly removed during translation by a membrane bound protease. Evidence for this was provided by studies on the in vitro translation of mRNA in cell-free systems lacking homologous membranes. In such cases, it was demonstrated that the gene products for placental hormones or immunoglobulins was about 3000 daltons larger than when membranes were present during translation (11,8). The enzyme releasing the presequence is rate-limiting and is operative only during translation and does not cleave the pre-prohormone once synthesis is completed. Recent studies suggest it is a metalloendopeptidase with chymotryptic-like properties capable of cleaving the presequence from a large number of prepro-hormones of diverse structure (19). The presence of a presequence and its composition can be predicted from studies on the cloning of the relevant mRNA and assigning amino acid residues to codons of the cDNA (12,14).

A second and critical step in processing are the action of converting enzymes (termed convertases) (9). Clues as to their nature are provided by the prevalence of pairs of basic residues adjacent to some of the vulnerable bonds indicating the presence in tissues of general purpose "tryptic-like" enzymes. Progress in characterizing such enzymes has been slow but it can be demonstrated in vitro that conversion of prohormones occurs by action of thiol proteinases (cathepsin B) and carboxypeptidase acting sequentially on substances such as proinsulin (9). In addition to "tryptic-like"enzymes, conversion of some prohormones can be mediated by specific enzymes such as renin and angiotensin converting enzyme.

The formation of opiate peptides from a precursor present in pituitaries (termed 31K) illustrates some of the features involved in neuropeptide processing. Of importance is the fact that ACTH and β-endorphin are derived from the same precursor and appear only after the obligatory appearance of inactive intermediates (15). It can be predicted that an absence of a key enzyme in this pathway will affect the levels of opiate peptides found with possible endocrinological and behavioral consequences. Another finding, unexpected at the time, is that one precursor can give rise to a host of fragments of widely differing activities. This leads to the expectation that similar principles apply to the formation of other neuropeptides as illustrated further in the case of neurophysin-neurophypophysial peptides (H. Gainer, see this volume). The development of suitable assay procedures can be seen as the key to the discovery of opiate peptides which were present as cleavage products in extracts of pituitary but were overlooked. Different levels of processing enzymes in neurosecretory regions could account for the higher content of α-MSH and CLIP in the intermediate as compared to anterior lobes of the pituitary. The formation of α-MSH is of particular interest since it illustrates the nature of post-translational modifications required to form an active fragment; cleavage of ACTH by a tryptic-like enzyme followed by chain shortening on the C-terminus by a carboxypeptidase, and introduction of N-acetyl and C-terminal amide groupings (9). Post-translational modifications (glycosylation, phosphorylation, methylation, introduction of N-acetyl, pyroglutamyl, or C-terminal amide) constitute an important area of research with respect to activation and in some cases inactivation of neuropeptides.

In this study, two enzymes have been purified from pituitary and from brain for studies on their potential role in processing of opiate peptides. The first enzyme represents a class of thiol proteinases implicated in conversion of prohormones in other tissues and which has not been studied in detail in the CNS. The second enzyme is a dipeptidyl carboxypeptidase that plays dual roles since in one case it can convert angiotensin-I to II, and in other cases it inactivates neuropeptides by removal of the C-terminal dipeptide moiety.

PITUITARY THIOL PROTEINASE

Thiol proteinases such as cathepsin B may represent a class of intracellular proteinases involved in conversion and inactivation processes. There is evidence that they can release albumin from proalbumin and cleave bonds adjacent to basic residues of proinsulin. Such enzymes may play a key role in intracellular protein turnover and are involved in pathological processes such as demyelination, muscular dystrophy and malignancy (20). Thiol proteinases are assayed with N-protected basic arylamides such as B_z-Arg-Nap or its cogeners at pH 6.5 in presence of -SH groups and EDTA. In crude homogenates assayed under these conditions a high concentration was found in the pituitary as compared to other brain regions (Table 1), and this area was used as a source of enzymes for purification utilizing the scheme of Suhar and Marks (20) (autolysis, extraction, ammonium salt precipitation, CM-cell-ex chromatography, gel filtration). This scheme when applied to human pituitaries gave an overall purification of about 400 fold.

TABLE 1. <u>Distribution of dipeptidyl carboxypeptidase and cathepsin B</u>

	relative activity	
Region	Met-enkephalin	B z-Arg-Nap[+]
Pituitary	100	100
Cerebellum	67	49
Corpus striatum	31	28
Midbrain	26	33
Pons-medulla	20	35
Hypothalamus	12	40
Cerebral cortex	7	44
Spinal cord	7	24

Membrane fractions or whole homogenates ([+]) of rat brain were incubated with ^{3}H-Tyr-Gly-Gly-Phe-Met at pH 8.0 or B z-Arg-Nap at pH 6.0. Under the assay conditions the values for pituitary were 6.1 and 0.63 nmoles of product released per min per mg protein respectively.

Elution of enzyme from CM-cellex showed the highest activity in fractions eluted with 0.03 -0.06 M NaCl (peak II, Fig. 1). Since this peak also showed activity towards Lys-β-naphthylamide it resembles in part a BANA hydrolase of lung with aminoendopeptidase functions recently described by Singh and Kalnisky (18).

The properties of purified pituitary enzyme indicated some resemblance to thiol proteinases of other tissues in terms of pH optima (6.5) and its absolute dependence on -SH compounds for activity. Among agents inhibiting activity were acetyl-L-Leu-Leu-arginal (or leupeptin): and a synthetic analog boc-D-Phe-Pro-arginal (Table 2). Unlike leupeptin, the latter acted noncompetitively and is of interest since it can block fibrinogen-thrombin Pconversion probably by interaction with the thrombin sensitive region Val-Arg----GlyPro of the fibrinogen Aa chain.

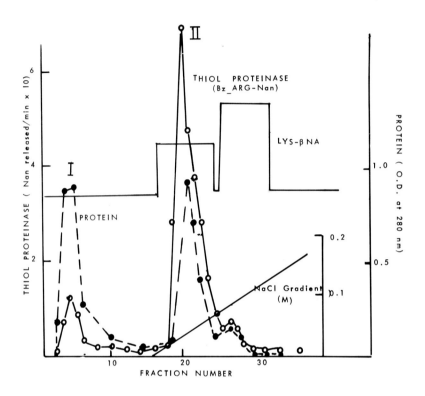

FIG. 1. Elution profile of human pituitary extracts on a CM-cellex column eluted with a salt gradient and tested with B_z-ARG-Nap (thiol proteinase) and lysyl β-naphthylamide.

Purified thiol proteinase of pituitary degraded a number of substrates ranging in size from histones, myelin basic protein, lipotropins to smaller peptides such as substance P (Table 3). In previous studies with a calf brain thiol proteinase showing properties comparable to that of the pituitary enzyme we demonstrated breakdown of ^{125}-I labelled β-LPH with release of an 8 kilodalton fragment suggesting cleavage at the Thr^{76}-Leu-Phe-Lys-Asn^{80} region (Fig. 2). It is known that this region has a conformation favoring action by several endopeptidases that include brain or pituitary cathepsin D (cleaving Leu-Phe), trypsin (Lys-Asn), and chymotrypsin (Phe-Lys) (4,9).

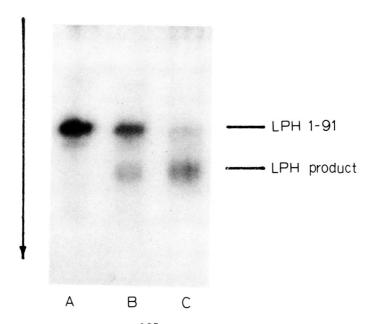

FIG. 2. Radioautogram of ^{125}-I-labeled LPH I-91 incubated for 3 h (B), 20 h (C) with a purified brain thiol proteinase as compared with the unincubated control (A). (Reproduced by courtesy of Eur. J. Biochem.) (20).

TABLE 2. Effect of inhibitors on thiol proteinase of human pituitary

Agent	Concentration (nM)	Activity (per cent)
HgCl$_2$	0.35	33
ZnCl$_2$	0.2	100
Leupeptin	0.0007	52
Boc-D-Phe-Pro-Arg-H	0.0007	52
Antipain	0.006	16
Chymostatin	0.006	0
Pepstatin	0.07	100
Bestatin	0.1	100
Chloroquine	0.07	35

The reaction volume contained 0.5 µg enzyme protein, 130 µmoles phosphate buffer, pH 6.2 plus additions as noted below and incubated with Bz-Arg-Nap (1 mg in 50 µl dimethyl sulfoxide). Results are expressed relative to enzyme activity in presence of 2.5 mM cysteine and 1 mM EDTA as 100% (56.8 nmol β-naphthylamine released per minute per mg protein).

TABLE 3.   Hydrolysis of proteins and polypeptides by purified
           human pituitary and calf brain thiol proteinase

Substrate	Pituitary	Brain
Histone (Lys-rich)	100	100
Glucagon	67	--
Basic protein (myelin)	40	91
Substance P	40	--
β-endorphin (human)	37	--
β-lipotropin(β-LPH) (porcine)	18	41
Neurophysin	26	--
Hemoglobin	11	0
Albumin	0	0
Insulin B chain	0	--

Reaction mixture contained 2.5 μg of enzyme, 50 nmoles substrate
and incubated at pH 6.2 in presence of EDTA and Clelands reagent
for the times indicated. Breakdown was evaluated on the basis of
the production of ninhydrin positive materials as determined by
an automated procedure and expressed relative to that for histone
as 100. The relative rates for a calf brain thiol proteinase are
included for comparison (20).

It appears unlikely that pituitary thiol proteinase is an enzyme activating release of β-endorphin which by itself can act as
a substrate (Table 2). The pituitary enzyme may play a role in
release of intermediate endorphins, some of which have opiate-
like effects or which may act as substrates for production of
enkephalin-like fragments. Cathepsin D can cleave β-endorphin to
yield γ-endorphin and in addition cleave human β-LPH to yield a
fragment containing an additional four amino acids on the N-terminal of the putative β-MSH sequence. This could account for the
22-β-MSH fragment observed in earlier studies.
Incubation of Substance P with pituitary thiol proteinase gave
a time-related release of cleavage products (Fig. 3). One of the
products migrated with Met.NH$_2$, Gly, Leu and was derived by cleavage of the Gln-Phe bond. An almost identical pattern was observed with a brain thiol proteinase (data not shown). One of
the cleavage products consisted of two components when submitted
to a two-dimensional TLC procedure (Fig. 3). Breakdown was decreased but not completely blocked by addition of leupeptin.
The data for Substance P together with the breakdown of glucagon could suggest a role of thiol proteinases in turnover of
peptides by an endo- or exopeptidase type action. It remains to

be determined if exopeptidase function is an inherent property of pituitary or brain thiol proteinase or is a membrane enzyme present as a contamination. Recent studies show that one of the thiol proteinases of lung has combined endoexopeptidase functions (18) supporting the view that the pituitary enzyme present in peak II (hydrolysing B_z-ARG-Nap and Lys-Nan) has similar actions. Other enzymes known to degrade Substance P are an aminopeptidase, a prolyl endopeptidase, and neutral and acidic endopeptidases(2,4, 7,13).

Additional information on specificity of pituitary thiol proteinases was gained by the use of synthetic substrates. Of these, the pituitary enzyme favored peptides with basic residues adjacent or penultimate to the chromophoric group. The best substrates based on Kcat/Km ratios were B_z-Arg-Nap followed by its 4-methoxy derivative, Lys-Lys-and Lys-Ala. Monoacylated substrates with free α-amino groups including Arg-Nap were poor substitutes (Table 4).

Although the specificity pattern with synthetic substrates argued well for conversion of β-LPH at the Arg-Tyr bond, studies with the brain enzyme did not indicate this type of conversion (Fig. 2). Since brain enzyme was comparable to that of pituitary in terms of specificity with the identical substrates (20), it is probable that pituitary thiol proteinase is not acting as a convertase for production of LPH 61-91. This illustrates the difficulties of making predictions based on the use of synthetic substrates since they do not mimic in all respects the conformation of native polypeptides nor the effects of allosteric binding sites on catalysis. Studies with synthetic substrates also would not have predicted the types of cleavages observed for Substance P.

ANGIOTENSIN CONVERTING ENZYME AND ENKEPHALINASES

For purification of membrane bound enkephalinases, a rat brain P_2 fraction was extracted by procedures outlined previously for angiotensin converting enzyme (ACE) (3). Briefly, enzymes were solubilized by use of Triton X-100 in Tris buffer followed by DEAE cellulose chromatography. Fractions were monitored with Met-enkephalin as substrate and assayed using an automated ninhydrin assay procedure capable of detecting degrading enzymes acting either on the N-terminus or internally. This revealed three peaks of activity when tested with enkephalin and other peptide substrates (Fig. 4, A-C).

In seeking to separate a specific enkephalinase from ACE, we adopted a strategy used previously which involved passage of the peaks over an IgG Sepharose column prepared with a specific antibody to purified lung ACE (kindly provided by Richard Soffer, Cornell Medical Center N.Y.) (3). A dipeptidyl carboxypeptidase was found in these peaks which did not bind to the IgG Sepharose but which cleaved enkephalin with release of the C-terminal Phe-Met as detected by TLC.

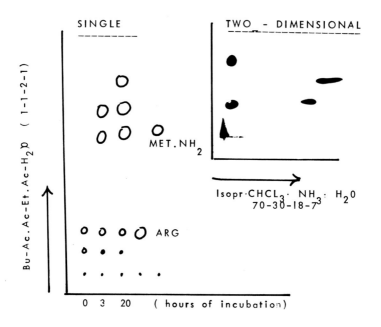

FIG. 3. Separation of cleavage products of Substance P (Arg-Pro-Lys-Pro-Gln-Gln-Phe-Phe-Gly-Leu-Met.NH$_2$) incubated with human pituitary for the periods shown. The inset represents a photograph (reduced in size) of a two-dimensional chromatogram and printed in reverse form. Spots were localized by fluorescamine. Arg present at zero time may be a contamination in the mixture or have resulted from release in trace amounts from Substance P.

Peak I activity (unadsorbed to the DEAE cellulose, Fig. 4) was chromatographed on CM-cellulose prior to the immunoaffinity step. The effluent from the Sepharose-IgG column degraded enkephalin, and its activity was unaffected by addition of Cl, divalent metals (Mn^{2+}, Ca^{2+}, Mg^{2+}) or SQ 20881. Analysis of products by TLC indicated release of Tyr, Gly-Gly-Phe-Met, Gly-Gly and Phe-Met indicating a combined action of an amino peptidase and a carboxy-peptidase. This fraction also degraded bradykinin with release of the C-terminal dipeptide Phe-Arg (identified by HPLC using a re-verse phase C-18 column) but not hippuryl-His-Leu (HHL) or angiotensin I (based on release of His-Leu).

Peak II, eluted with a low NaCl concentration, cleaved HHL, angiotensin I, bradykinin, enkephalin, as well as Leu-Gly-Gly. Passage through IgG-Sepharose led to the adsorption of ACE. The immobilized enzyme cleaved all of the above substrates except for Leu-Gly-Gly with release of the C-terminal dipeptide (identified by HPLC or TLC). The unadsorbed material also contained a

TABLE 4. <u>Substrate specificity of purified pituitary thiol proteinase</u>

Substrate – β–naphthylamide	K_{cat} min^{-1}	K_{cat}/K_m $mM^{-1}min^{-1}$	Leupeptin % inhibition
B_z–Arg–	1.1	4.78	70
Z–Arg–4–methoxy	0.77	3.08	41
Arg–	–	–	–
Leu–	–	–	–
Lys–	0.24	2.0	38*
Lys–Lys–	1.28	2.42	44
Lys–Ala–	0.95	0.76	20
Arg–Arg–	1.48	1.11	25
Z–Arg–Tyr–	–	–	–
Z–(Gly)$_2$–Arg–	0.68	2.35	45
Z–(Gly)$_2$–Leu–	–	–	–

*50 μM leupeptin

Kinetic constants for synthetic substrates hydrolysed by purified human pituitary thiol proteinase present in CM–cellex peak IIa (Fig. 1). Enzyme 0.3 μg was incubated with different concentrations of substrate and activity monitored by fluorometry. K_m values were obtained from double reciprocal plots, and K_{cat} calculated assuming an enzyme mol. wt. of 29,000 daltons. Inhibition by leupeptin 0.25 μm was carried out at substrate concentrations of 1.5 mM.

dipeptidyl carboxypeptidase which cleaved enkephalin releasing Phe–Met, and bradykinin, releasing Phe–Arg, but not HHL or angiotensin–1. This separate enkephalin was not inhibited by SQ 20881 or SQ 14225, specific inhibitors of ACE. The unadsorbed material did cleave Leu–Gly–Gly, but no release of N–terminal Tyr from Met enkephalin was observed. Thus, an aminopeptidase is present similar to one described in soluble supernatant by Sachs and Marks (16) which does not act on various polypeptides including enkephalin.

Peak III contained sizable arylamidase activity, as indicated by cleavage of Leu–β–naphthylamide. This peak also cleaved enkephalin, with release primarily of Tyr. Both reactions were inhibited by puromycin, suggesting a similarity to soluble amino-peptidases which degrade arylamides and which are sensitive to inhibition by puromycin, and act as potent enkephalinases (10,17).

These preliminary data indicate that brain contains one or more membrane bound enzymes acting as "enkephalinases", including an aminopeptidase releasing Tyr, and more than one dipeptidyl carboxypeptidase releasing Phe–Met or Phe–Leu. It is evident that

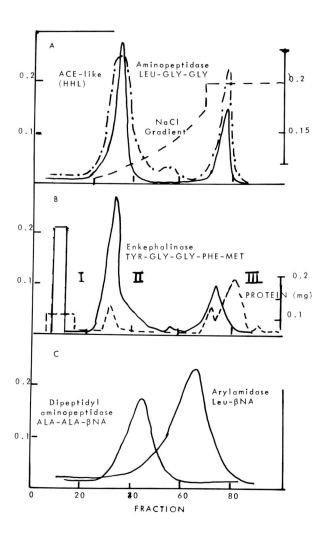

FIG. 4. Elution profile of membrane bound enkephalinases degrading Met-enkephalin (B) as compared to aminopeptidase (Leu-Gly-Gly)
and dipeptidyl carboxypeptidase (ACE-like) (A), and monoacyl or
dipeptidyl arylamide substrates (C). Note peak II (B) overlaps
aminopeptidase and ACE-like enzymes (A), and peak III (B) partially overlaps in addition an arylamidase (C). Dipeptidyl arylamidase (C) forms a separate peak between peaks II and III of A.
Peaks I, II, III showing enkephalinase activity were subjected to
IgG immunoaffinity chromatography as described in the text.

one of the latter enzymes is comparable to ACE of lung by virtue of the fact that it binds to pulmonary derived antibody, but that other dipeptidyl carboxypeptidases exist which may have unique properties. The C-terminal of enkephalin does not appear to fulfill all the structural requirements for binding to the obligatory and accessory subsites of ACE as postulated by Cheung et al. (5) and other enzymes with higher affinity for enkephalin may exist. In addition to the presence of a membrane bound dipeptidyl carboxypeptidase, consideration must be given also to the action of membrane bound aminopeptidases which rapidly inactivate enkephalin by removal of the N-terminal Tyr. Aminopeptidases are known to be present in membrane preparations of other tissues notably the brush border of kidney and intestinal mucosa cells and may play additional roles in the inactivation of opiate peptides. In soluble fractions of brain they represent the most rapidly acting enzymes degrading enkephalin (9). If enkephalin is involved in sensory mechanisms of pain then the development of inhibitors of the many categories of enzymes may have clinical applications. A choice must be made among the various enzymes involved in enkephalin breakdown as to which plays the most important role in metabolism. As this study has shown, these include ACE-like enzymes, a dipeptidyl carboxypeptidase, and aminopeptidases. It is conceivable that enkephalin itself represents a degradation product formed from other more active precursors. Consideration will have to be given to the role of those enzymes in converting the larger forms to liberate the pentapeptide.

SUMMARY

1. A thiol proteinase from human pituitaries was purified approximately 400 fold and shown to have different chromatographic properties from that of calf brain. Among substrates cleaved were myelin basic protein, histones, β-lipotropin, neurophysin, and Substance P.

2. The enzyme showed properties associated with a cathepsin-B like enzyme: dependence on -SH groups, pH optimum of 6.5, inhibition by leupeptin and a synthetic analog, Boc-D-Phe-Pro-arginal, and cleavage of dipeptidyl arylamides with basic residues adjacent to or penultimate to the chromatographic grouping.

3. Membranes present in the P2 fraction of rat brain contained three or more enkephalinases when submitted to DEAE-cellulose chromatography. Further purification on an IgG-Sepharose affinity column prepared with antibody to lung angiotensin converting enzyme indicated the presence of dipeptidyl carboxypeptidase(s) with properties distinct from those of ACE. In addition, the DEAE-cellulose fractions contained various aminopeptidase activities when tested with Leu-Gly-Gly, Leu-Nap, and Ala-Ala-Nap as substrates.

ACKNOWLEDGEMENTS

This work was supported in part by grants from the USPHS (NS 12578), and J. Stefan Institute, Ljubljana, Yugoslavia (A.S. and N.Y. State HRC 9-013 (M.B.).

REFERENCES

1. Barat, E., Patthy, A.,and Graf, L. (1979): Proc. Natl. Acad. Sci. USA, 76:6120-6123.
2. Benuck, M.,and Marks, N. (1975): Biochem. Biophys. Res. Commun., 65:153-160.
3. Benuck, M.,and Marks, N. (1979): Biochem. Biophys. Res. Commun., 88:215-221.
4. Benuck, M., Grynbaum, A., Cooper, T.B.,and Marks, N. (1978) Neurosci. Lett., 10:3-9.
5. Cheung, H.S., Wang, F.L., Ondetti, A., Sabo, E.F., and Cushman, D.W. (1980): J. Biol. Chem., 255:401-407.
6. Gorenstein, C., and Snyder, S.H. (1979): Life Sci., 25:2065-2070.
7. Hersh, L.B., and McKelvy, T.F. (1979): Brain Res., 168:553-564.
8. Lingappa, V.R., Thiery, A.D., and Blobel, G. (1977): Proc. Natl. Acad. Sci. U.S.A., 74:2432-2436.
9. Marks, N. (1978): In: Frontiers in Neuroendocrinology, edited by W.F. Ganong,and L. Martini, Vol. 5, pp. 329-377. Raven Press, New York.
10. Marks, N., Datta, R.K.,and Lajtha, A. (1968): J. Biol. Chem., 243:2882-2889.
11. Milstein, C., Brownlee, G.G., Harrison, T.M., and Mathews, M.B. (1972): Nature, New Biol., 239:117-120.
12. Nakanishi, S., Inoue, A., Kita, T., Nakamura, M., Chang, A.C.Y., Cohen, S.N.,and Numa, S. (1979): Nature, 278:423-427.
13. Orlowski, M., Wilk, E., Pearce, S.,and Wilk, S. (1979): J. Neurochem., 33:461-469.
14. Roberts, J.L.,and Herbert, E. (1979): Proc. Natl. Acad. Sci. USA, 74:4826-4830.
15. Roberts, J.L., Phillips, M.A., Rosa, P.A.,and Herbert, E. (1978): Biochemistry, 17:3609-3617.
16. Sachs, L.,and Marks, N. (1980): Proc. Soc. for Neurosci., (in press).
17. Schnebli, H.P., Phillipps, M.A.,and Barclay, R.K. (1979): Biochem. Biophys. Acta, 569:89-98.
18. Singh, H., and Kalnitsky, G. (1980): J. Biol. Chem., 255:369-373.
19. Strauss, A.W., Zimmerman, M., Boime, I., Ashe, B., Mumford, C.A., and Alberts, A.W. (1979): Proc. Natl. Acad. Sci. USA, 76:4225-4229.
20. Suhar, A.,and Marks, N. (1979): Eur. J. Biochem., 101:23-30.

Neurosecretion and Brain Peptides,
edited by J. B. Martin, S. Reichlin, and K. L. Bick.
Raven Press, New York © 1981.

Introduction to Section II:
Neuroanatomy of Peptidergic Distribution in Brain

Michael J. Brownstein

*Unit on Endocrinology, Laboratory of Clinical Science, National Institutes of Health,
Bethesda, Maryland 20205*

Neurological practitioners have a wealth of information about functional neuroanatomy at their disposal. They learn a great deal during the course of training and years of practice about the physiological consequences of damage to specific brain regions or neuronal populations. It now seems likely that the neurologist will soon gain clinical information from knowledge about the location of the peptidergic perikarya in the central nervous system and the nature of their projections. Such data can be integrated into concepts of brain function to permit formulation of hypotheses about the role of peptide containing neurons in health and disease.

Anatomical information is also essential for a variety of basic scientists -- biochemists, physiologists, pharmacologists, and behavioral psychologists. Immunocytochemistry and new orthograde and retrograde tract-tracing techniques have enabled neuroanatomists to describe peptide containing neuronal systems in detail at the light and electron microscopic levels. The results of these studies have changed the way that we view the brain. Ten years ago we thought of the magnocellular and parvicellular hypothalamic neurons as being analogous to lower motor neurons. That is, we felt that they were simply involved in secretion of their hormones in response to changes in neural input. Furthermore, it was believed that hypothalamic hormones were confined to neurons in the basal hypothalamus and to their projecting axons in the posterior lobe of the pituitary or the median eminence. It is now known that this concept of the neuroendocrine brain was much too simple. Hypothalamic neurons that are intimately involved in regulating pituitary functions also innervate other brain regions remote from the median eminence and pituitary. Moreover, neurons that contain "hypothalamic hormones" are found throughout the brain and in periphery tissues as well.

In addition to the hypothalamic hormones a number of other biologically active peptides have been found in the brain. Some of these, e.g., αMSH, ACTH, and β-endorphin, are made by both central neurons and pituitary cells. Others, e.g., cholecystokinin, vasoactive intestinal polypeptide, glucagon, and insulin, seem to be synthesized by both brain and gut cells. Hence, a single peptide can play three distinct roles: it can be released at synapses to act as a neurotransmitter, it can be released onto neighboring cells to exert a "paracine" action, or it can be secreted into the blood to act as a hormone on distant target cells.

Peptides are not limited to these roles alone. Recent studies in a number of laboratories have shown that peptides can be made in and released from the same cell that secretes other active substances, such as biogenic amines.

The chapters in this section by Zimmerman, Watson and Akil, and Uhl and Snyder discuss several biologically active peptides that have been shown to be present in specific pathways in the central nervous system. Such anatomical information has provided important clues about roles of these peptides that were not expected when they were discovered.

Neurosecretion and Brain Peptides,
edited by J. B. Martin, S. Reichlin, and K. L. Bick.
Raven Press, New York © 1981.

The Organization of Oxytocin and Vasopressin Pathways

Earl A. Zimmerman

*Department of Neurology, College of Physicians and Surgeons, Columbia University,
New York, New York 10032*

INTRODUCTION

The Magnocellular System

The magnocellular system of the hypothalamus has been an important model for the study of peptide-forming neurons for nearly half a century. We now know it was mainly the sulfur-rich neurophysin proteins (NPs) associated with oxytocin (OT) and vasopressin (VP) (39) which reacted with Gomori stains which were so useful in establishing the neurosecretory nature of supraoptic (SON) and paraventricular (PRN) neurons (33). Concentration of their axonal projections and hormone storage in their terminals in the posterior pituitary gland facilitated early purification and subsequent synthesis of the active hormones (11). The presence of relatively large granules within the neurons aided both transmission electron microscopy (16) and biochemical studies of granule content (8). In addition, the concentration of large neuronal perikarya into nuclei with an identified pathway from them to the posterior pituitary gland and systemic circulation encourage further studies using electrophysiological techniques (7,29) and biosynthetic methods using radioisotopes (3,30). As a result of these excellent studies, we now know much more about the biosynthesis and secretion of the "neurohypophysial peptides" than of any others in the brain (3).

Immunocytochemistry

When immunocytochemical methods (immunofluorescence 12,20; immunoperoxidase 54,55) came into use about ten years ago to study neuroendocrine pathways in the hypothalamus, the magnocellular system offered unique opportunities. Antisera to more than one peptide in a particular neuron became available (e.g., to VP and to its associated NPs). This provided some assurance that the neuron localized was actually synthesizing VP because its NPs was also present, since both arise from the same precursor molecule (3,30). Without the localization of NP, the VP reactivity might reflect visualization of shared sequences in some other unknown peptide such as in the case of some antisera to enkephalin which detect β-endorphin, or possibly visualization of VP uptake in a cell which does not produce it as has been reported for ACTH cells in the anterior pituitary (5). More recently, a similar problem concerning the synthesis or uptake of ACTH by the hypothalamus (23) was largely resolved by the availability of antisera to several different parts of its precursor and products, which visualized the in same arcuate neurons (17,24,51). Unfortunately, additional antigens (precursor parts) are not yet available to study other peptide systems such as those containing Substance P or LHRH. Finally, in terms of the magnocellular system, it was fairly certain from previous studies which neurons contained VP and OT, and, therefore, excellent morphological controls were available.

LOCALIZATION OF CELL BODIES

Magnocellular Neurons

Supraoptic and paraventricular nuclei: vasopressin and oxytocin.

Oxytocin and VP and NPs are concentrated in the large neurons of SON and the PVN in the hypothalamus of every mammal studied by immunocytochemistry, including man (Fig. 1) (10,12,26,46,49,54). In the ox (48), rat (41), man and monkey (1,9), a particular NP was shown to be associated with each hormone. The antiserum to rat NP contained antibodies to both NPs. This problem was solved by comparing (25) normal with homozygous Brattleboro rats with diabetes insipidus (HODI) which cannot produce VP, VP-NP, or its precursor (3,41), or by absorption of the antiserum with HODI pituitary extracts (25). NP-positive cells in HODI rats reacted with antiserum to OT, while the negative cells contained NP and VP in the normal rat in ventral SON and central PVN (41). There is general agreement that in mammals VP cells are more concentrated in ventral SON and OT cells more rostral and dorsal (9,41,46). Although this rostrocaudal difference is again seen in PVN (46), the two cell types are considered more admixed in this region (9). Using three-dimensional computer plotting techniques we recently found that most of the VP neurons in rat PVN are arranged as a continuous ellipsoidal group forming the ventrolateral wall (F. Langendorf, D. Margulies, C. Levinthal, E.A. Zimmerman, unpublished). Not all magnocellular neurons containing OT or VP lie withhin these major nuclei. Some are scattered between them, others are found lateral to PVN or rostral to it as far as the organum vasculosum and the anterior commissure, and a few are found far dorsal to the PVN between the capsule and the anterior thalamus (1,9,40,54).

SON and PVN, other peptides.

Recent data indicate that these nuclei are not the sole domain of OT and VP. Our own work (15) supports that of Phillips et al. (27) that antisera to angiotensin II (AII) react with many cell bodies. In addition, antisera to enkephalin (32), cholecystokinin (47) and Substance P (21) react with cells in or near them. It is not yet certain whether any of these other peptides are located in those containing OT or VP. So many cells react with AII antiserum in our studies, however, that it seems probable that further work will confirm its reactivity in VP- and/or OT-containing cells. Although absorptions confirm specificity for sequences of AII, one does not know if these cells actually produce it or if a similar sequence is being seen in an unrelated molecule.

FIG. 1. Low power photomicrograph of magnocellular NP-containing neurons in mouse PVN. Immunoperoxidase technique on a 75 µm coronal vibratome section using antiserum to bovine NP I. X160. (Courtesy of Ann-Judith Silverman, unpublished.)

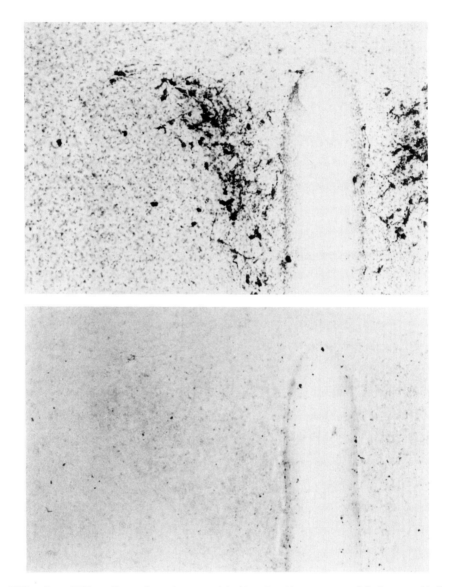

FIG. 2. <u>TOP</u>: Neurotensin reactivity in the parvocellular medial
portion of a rat PVN.   Immunoperoxidase technique on an 80 μm-
thick cryostat section, cresyl violet counterstain.
    <u>BOTTOM</u>:   Reactivity is abolished in an adjacent section in
which the antiserum was preincubated with synthetic neurotensin.
No counterstain. X 120. (Courtesy of David Kahn, Gary M. Abrams,
Donald Hoffman, A. Zimmerman, Robert Carraway and Susan E. Leeman,
unpublished).

Parvocellular Neurons

Neurotensin.

Recent studies in our laboratory using antisera to neurotensin reveal that reactive neurons occupy the paravocellular medial portion of the PVN (14) (Fig. 2). Some of the cells are scattered within the magnocellular portion as well. The SON does not contain reactivity. Many other hypothalamic neurons do react including perikarya in the periventricular, arcuate and preoptic areas. Fibers project to median eminence and posterior pituitary gland reminiscent of other neurosecretory pathways such as those containing somatostatin (12) and VP and OT.

Suprachiasmatic nucleus.

One of the surprises provided by immunoctyochemistry and not revealed earlier by Gomori stains was the presence of immunoreactive VP and VP-NP in dorsal and medial small cells of the suprachiasmatic nucleus (SCN) (49,54). No OT or OT-NP was found in this nucleus (46,54). The role of the hormone in this region and possible projections from these cells are not known. A participation of this VP system in adrenal diurnal rhythms is ruled out by persistence of the rhythm in HODI rats (18).

In the course of our recent immunocytochemical studies of vasoactive intestinal polypeptide (VIP) in rat and mouse, another group of SCN neurons were found to be reactive (38) (Fig. 3). The findings appear to be identical to those obtained independently by Loren et al. (22). The majority of VIP-positive cells are located in a different region of the nucleus from those containing VP. They occupy the basal regions of the nucleus with some embedded in the optic chiasm, and a dense tract can be traced from them passing dorsally to the PVN and then possibly onward to the thalamus. That the SCN may innervate the PVN is further suggested by our studies in progress of afferents to PVN using horseradish peroxidase (HRP) methods (35,36). Iontophoresis of HRP into PVN reveals labeling of perikarya within SCN which tend to lie primarily in its ventral part. Combined tracing and immunocytochemical studies of VIP are needed to establish this possibility fully.

PROJECTIONS OF THE MAGNOCELLULAR SYSTEM

Neurosecretory

Axonal contacts between VP and OT neurons are made with capillaries of the organum vasculosum, the posterior pituitary gland, and the median eminence (Fig. 4) (1,54). The immunocytochemical approach has proven very useful in establishing that VP and OT cells of the PVN project to the hypophysial portal system in the zona externa of the median eminence (1,20,49,54). This was previously considered unlikely since the large neurosecretory granules (120-180 nm dia.) seen by transmission electron microscopy (16) were difficult to find in zona externa nerve terminals where they are usually smaller (80-100 nm dia.). By immunoelect-

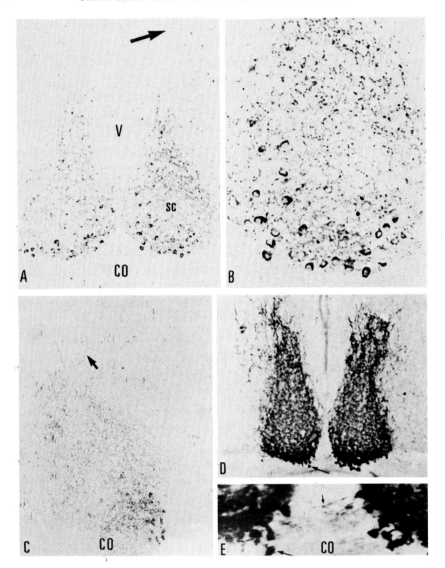

FIG. 3.   VIP in mouse SCN by immunoperoxidase techniques.   (A)
Coronal 6 μm section showing VIP in neuronal perikarya at the
basal portion of the nucleus and in fibers some of which pass
dorsally towards the PVN of the hypothalamus (arrow), X100.   At
higher magnification of A, note that some of the cell bodies are
also found in the medial aspect of the nucleus and appear to
project to terminal fields throughout the nucleus, X250. (C) A
sagittal 6 μm section showing that the cells tend to lie in the
caudal portion of the nucleus and that fibers travel dorsally from
its apex (arrow), X100. (D) Additional detail is provided by 50 μm
frozen sections.   Cell bodies extend down into the optic chiasm
(arrow) and fibers projecting from the nucleus in a dorsal direc-

FIG. 3 caption continued--

tion are seen to be very numerous, X100. (E) Higher magnification of D demonstrating a cell (lower arrow) and a fiber (upper arrow) in the optic chiasm (CO), X400. (Courtesy of K.B. Sims et al. (38),with permission).

---

ron microscopy smaller granules containing VP and NP were found in the zona externa of the median eminence than in posterior pituitary (37). Furthermore, adrenalectomy markedly enhanced the visualization of VP or VP-NP fibers over those containing OT or OT-NP (43). Replacement doses of corticosterone or dexamethasone were much more effective than desoxycorticosterone in suppression of this increase after adrenalectomy (A.J. Silverman and E.A. Zimmerman, unpublished). Bilateral lesions in PVN but not the SCN abolish the zona externa VP response to adrenalectomy (49). Lesions of one PVN in normal monkeys eliminate OT-VP- and NP-containing fibers in the ipsilateral zona externa (1). Although it is known that both PVN and SON project to posterior pituitary, the PVN also projects to the portal capillary system in the median eminence. Furthermore, the VP part of that system is selectively inhibited by glucocorticoid (43). Where and how that feedback acts is not yet known. We have recently shown by combined autoradiography for radiolabelled cytidine and immunocytochemistry for VP, that there is increased uptake of the radiolabel in VP cells in PVN, but not in SON or SCN in adrenalectomized rats, suggesting selective increases in DNA turnover (34).

Vasopressin in portal blood.
Very high concentrations of VP are found in individual portal vein blood plasma of both monkeys (55) and rats (26). Levels are 100-1000 times greater than peripheral blood levels, amounts which could release ACTH (55). The PVN-median eminence system is the probable source of VP in the portal system. Although a previous report by Oliver et al. (26) indicated that most of the VP in long portal veins arrives by retrograde flow from posterior pituitary since there was a 10 fold fall after removal of the gland, we have not been able to find a significant decrease (L.D. Recht, D.L. Hoffman, J. Haldar, A.J. Silverman and E.A. Zimmerman, unpublished). The data strongly suggest that the PVN-median eminence system participates in the pituitary-adrenal system possibly along with other CRFs (18). Although antiserum to VP can inhibit the ACTH-releasing activity of hypothalamic CRF extracts (13), ACTH regulation in HODI rats is reasonably intact, although not totally normal in response to all stresses (53). Our present concept is that VP may be one of several hypothalamic releasers of ACTH, the other(s) as yet to be discovered.

EXTRAHYPOTHALAMIC PROJECTIONS

Several years ago, Conrad and Pfaff (6) showed that magnocellular PVN neurons projected to many forebrain and brainstem
regions by radioautographic orthograde tracing methods. Retrograde HRP studies from spinal cord and brainstem also establish
labeling in PVN but not in SON (19,31) Subsequently, Swanson (44)
using antiserum to bovine OT-NP in immunocytochemical studies on
rat and ox showed that many of these projections contained NP. In
addition, NP and OT fibers were found in the spinal cord of the
rat and monkey (45). Weindl and Sofroniew (52) demonstrated fibers containing VP and NP in the amygdala of the guinea pig and
repeated the studies of Swanson (44) in the rat using antiserum to

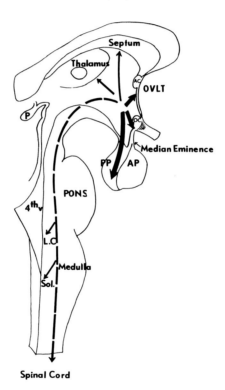

FIG. 4. Sagittal diagram of magnocellular pathways containing
vasopressin, oxytocin and neurophysins. Heavy arrows represent
neurosecretory pathways to posterior pituitary (PP), median
eminence, and the organum vasculosum of the lamina terminalis
(OVLT); smaller arrows indicate extrahypothalamic projections.
AC, anterior commissure; AP, anterior pituitary; LC, locus
coeruleus; OC, optic chiasm; P, pineal; SOL, solitary n.; 4th$_V$,
fourth ventricle.

porcine NP (40). We have confirmed these findings in rat using an antiserum which reacts with both NPs and compared the results in HODI normal rats after absorption with HODI rat pituitary extract (25). The OT-NP pathway appears to be much more prominent in most caudal projections which was also found by studies using antiserum to OT (4,24,45). Although these fibers react better with antiserum to OT than to VP, neither antiserum stains as well as antiserum to NP in our laboratory. This may represent other partly false negative results or the presence of some other peptides in this part of the magnocellular system derived from the OT and VP precursors. Buijs (4), however, demonstrated both OT and VP in most of the extrahypothalamic pathways including spinal cord and found that VP predominated over OT in the more rostral pathways, whereas OT predominated over VP in the more caudal pathways. These anatomical observations are summarized in Fig. 4.

Areas in which NP-reactive fibers appear to form innervation patterns by branching and infiltration of all or part of various nuclei include, notably, lateral septal nucleus, thalamus, and amygdala in the forebrain; central grey, parabrachial nucleus, and nuclei solitarius in brainstem; and the intermediolateral grey and substantia gelatinosa of the spinal cord (Fig. 5). Other areas of note include some fibers containing OT and VP in rat entorhinal

FIG. 5. Darkfield photomicrograph of a longitudinal section of thoracic spinal cord of a rat through the superficial dorsal horn. Left side is rostral and bottom is medial. Immunoreactive neurophysin fibers travel in a band of rostrocaudal fibers in the dorsal funiculus (top) and send perpendicular branches into the substantia gelatinosa (bottom). Immunoperoxidase technique, 100 μm section. X100. (Courtesy of G. Nilaver et al. ref. 25, with permission).

cerebral cortex, subiculum of the hippocampus, substantia nigra pars reticulata, locus coeruleus and nucleus ambiguus in the brainstem (4). In our material we have seen NP and OT but not VP-containing fibers in several areas of cerebral cortex and substantia nigra, pars reticulata. In addition, some NP fibers also appear to innervate the lateral reticular nucleus (G. Nilaver, D.L. Hoffman, and E.A. Zimmerman, unpublished).

By light microscopy the PVN magnocellular system appears to innervate a wide variety of brain regions in the CNS, although the total picture of the respective contributions of OT and VP remains to be fully appreciated. It is likely that immunoelectron microscopy will demonstrate some type of synaptic-like contacts with other neurons as has been shown for Substance P and enkephalin in the solitarius (28). The wide radiation of projections from a central nuclear concentration would suggest a system by which many different but related activities might be coordinated. As more is learned about the anatomy and biochemistry of the "PVN-Brain", physiological experiments can be designed to study its functions. Similar problems are raised by our present knowledge of the ACTH/ β-LPH pathways originating in the arcuate and nearby regions of the hypothalamus and projecting to many of the same areas as the NP system (17,51). A traditional concept at present for the role of the PVN system might center around simultaneous induction of stress-related activities, such as release of VP and ACTH by the neurosecretory component, alerting and memory activation by forebrain components (50), autonomic (e.g., regulation of blood pressure, 42), and pain modulation by brainstem or spinal cord (2). The system is much more complex, however, since many of the same regions which receive projections from PVN also send axons to the PVN as shown by retrograde tracing with HRP: for example, nucleus accumbens septi, subiculum of the hippocampus, substantia nigra, central grey, parabrachial nucleus, locus coeruleus, solitarius and lateral reticular nucleus (35,36). How influences from these many areas are integrated by the PVN, and what its peptides do in turn in these sites are expected to be the subject of many studies in the future.

ACKNOWLEDGEMENTS

The author thanks his colleagues for their contributions to this work and C. Topping for assistance with the manuscript. Supported by USPHS, NIH Grants AM 20337, HD 13147 and 05077, and HG 24105.

REFERENCES

1. Antunes, J.L., Carmel, P.W., and Zimmerman, E.A. (1977): Brain Res., 137:1-10.
2. Bodnar, R.J., Zimmerman, E.A., Nilaver, G., Mansour, A., Thomas, L.W., Kelly, D.D., and Glusman, M. (1980): Life Sci., (in press).

3. Brownstein, M.J., Russell, J.T., and Gainer, H. (1980): Science, 207:373-378.
4. Buijs, R.M. (1978): Cell Tiss. Res., 192:423-435.
5. Çastel, M., and Hochman, J. (1976): Sixth European Congress on Electron Microscopy: 606-607.
6. Conrad, L.C.A., and Pfaff, D. (1976): J. Comp. Neurol., 169: 185-220.
7. Cross, BA. (1973): In: Frontiers in Neuroendocrinology 1973, edited by W.F. Ganong and L. Martini, pp. 133-171. Oxford University Press, New York.
8. Dean, C.R., Hope, D.B., and Kazic, I. (1968): Br. J. Pharmacol., 24:192p-193p.
9. Defendini, R., and Zimmerman, E.A. (1978): In: The Hypothalamus, edited by S. Reichlin, R.J. Baldessarini and J.B. Martin, pp 137-154. Raven Press, New York.
10. Dogterom, J., Snijdewint, F.G.M., and Buijs, R.M. (1978): Neurosci. Lett., 9:341-346.
11. DuVigneaud, V. (1956): In: The Harvey Lectures 1954-55, pp. 1-26. Academic Press, New York.
12. Elde, R., and Hokfelt, T. (1978): In: Frontiers in Neuroendocrinology, edited by W.F. Ganong and L. Martini, pp 1-33. Raven Press, New York.
13. Gillies, G., and Lowry, P. (1979): Nature, 278:463-464.
14. Kahn, D., Abrams, G.M., Zimmerman, E.A., Carraway, R., and Leeman, S.E. (1980): Endocrinology, (in press).
15. Kilcoyne, M.M., and Zimmerman, E.A. (1980): Clinical Science. (In press.)
16. Knigge, K.M., and Scott, D.E. (1970): Am. J. Anat., 129: 223-244.
17. Krieger, D.T., Liotta, A.S., Brownstein, M.J., and Zimmerman, E.A. (1980): Recent Prog. Horm. Res., (in press).
18. Krieger, D.T., and Zimmerman, E.A. (1978). In: Clinical Neuroendocrinology, edited by M. Besser and L. Martini, pp. 363-391. Academic Press, New York.
19. Kuypers, H.G.J.M., and Maisky, V.A. (1975): Neurosci. Lett., 1:9-14.
20. Livett, B.G., Uteenthal, L.O., and Hope, D.B. (1971): Philos. Trans. R. Soc. Lond. (Biol. Sci.), 261:371-378.
21. Ljungdahl, A., Hokfelt, T., Nilsson, G. and Goldstein, M. (1978): Neuroscience, 3:945-976.
22. Loren, I., Emson, P.C., Fahrenkrug, J., Bjorklund, A., Alumets, J., Hakanson, R., and Sundler, F. (1979): Neuroscience, 4:1953-1976.
23. Moldow, R., and Yalow, R. (1978): Proc. Natl. Acad. Sci. USA, 75:994-998.
24. Nilaver, G., Zimmerman, E.A., Defendini, R., Liotta, A.S., Krieger, D.T., and Brownstein, M.J. (1979): J. Cell Biology, 81:50-58.
25. Nilaver, G., Zimmerman, E.A., Wilkins, J., Michaels, J., Hoffman, D., and Silverman, A.J. (1980): Neuroendocrinology, (in press).

26. Oliver, C., Mical, R.S., and Porter, J.C. (1977): Endocrinology, 101:598-604.
27. Phillips, M.I., Weyhenmeyer, J., Felix, D., Ganten, D., and Hoffman, W.E. (1979): Federation Proc., 28:2260-2266.
28. Pickel, W.M., Joh, T.H., Reis, D.J., Leeman, S.E., and Miller, R.J. (1979): Brain Res., 160:387-400.
29. Renaud, L.P., Pittman, Q.U., and Blume, H.W. (1979): In: Central Regulation of the Endocrine System, edited by K. Fuxe, T. Hokfelt, and R. Luft, pp 119-136. Plenum Press, New York.
30. Sachs, H., and Takabatake, Y. (1964): Endocrinology, 75:943-948.
31. Saper, C.B., Loewy, A.D., Swanson, L.W. and Cowan, W.M. (1976): Brain Res., 117:305-312.
32. Sar, M., Stumpf, W.E., Miller, R.J., Chang, K.J., and Cuatrecasas, P. (1978): J. Comp. Neurol., 182:17-38.
33. Scharrer, E., and Scharrer, B. (1954): Recent Prog. Horm. Res., 10:183-240.
34. Silverman, A.J., Gadde, C.A., and Zimmerman, E.A. (1980): Neuroendocrinology, (in press).
35. Silverman, A.J., Hoffman, D.L., and Zimmerman, E.A. (1979): Anat. Record, 193:685.
36. Silverman, A.J., Hoffman, D.L., and Zimmerman, E.A. (1980): Anat. Record, (in press).
37. Silverman, AJ., and Zimmerman, E.A. (1975): Cell Tiss. Res., 159:291-301.
38. Sims, K.B., Hoffman, D.L., Said, S.I., and Zimmerman, E.A. (1980): Brain Res., 186:165-183.
39. Sloper, J.C. (1966): In: The Pituitary Gland, Vol. 3, edited by G.W. Harris, and B.T. Donovan, pp. 131-239. University of California Press, Berkeley.
40. Sofroniew, M.V. and Weindl, A. (1978): Endocrinology, 102: 334-337.
41. Sokol, H.W., Zimmerman, E.A., Sawyer, W.H., and Robinson, A.G. (1976): Endocrinology, 98:1176-1188.
42. Sterba, G., Hoffman, E., Solecki, R., Naumann, W., Hoheisel, G., and Schober, E. (1979): Cell Tissue Res., 196:321-336.
43. Stillman, M.A., Recht, L.D., Rosario, S.L., Seif, S.M., Robinson, A.G., and Zimmerman, E.A. (1977): Endocrinology, 101:42-49.
44. Swanson, L.W. (1977): Brain Res., 128:346-353.
45. Swanson, L.W. and McKellar, S. (1979): J. Comp. Neurol., 188: 87-105.
46. Swaab, D.F., Pool, C.W., and Nijveldt, F. (1975): Cell Tiss. Res., 164:153-162.
47. Vanderhaeghen, J.J., and Lotstra, F. (1979): Soc. for Neurosci. Abstr., p. 543.
48. Vandesande, F., Dierickx, K., and DeMey, J. (1975): Cell Tiss. Res., 156:189-200.
49. Vandesande, F., Dierickx, K., and DeMey, J. (1977): Cell Tissue Res., 180:443-452.

50. VanWimersma Greidanus, T.J.B., and Versteeg, D.H.G. (1980): In: Behavioral Neuroendocrinology, edited by C.B., Nemeroff and A.J. Dunn. Spectrum Publishing Inc., Jamaica, New York. (In press.)
51. Watson, S.J., Akil, H., Richard III, C.W., and Barchas, J.D. (1978): Nature (Lond.), 275:226-228.
52. Weindl, A., and Sofroniew, M.V. (1976): Pharmakopsych, 9: 226-234.
53. Yates, F.E., Russell, S.M., Dallman, M.F., Hedge, G.A., McCann, S.M., and Dhariwal, A.P.S. (1971): Endocrinology, 88: 3-15.
54. Zimmerman, E.A. (1976): In: Frontiers in Neuroendocrinology, Vol. 4, edited by L. Martini and W.F. Ganong, pp. 25-62. Raven Press, New York.
55. Zimmerman, E.A., Carmel, P.W., Husain, M.K., Ferin, M., Tannenbaum, M., Frantz, A.G., and Robinson, A.G. (1973): Science, 182:925-927.

Neurosecretion and Brain Peptides,
edited by J. B. Martin, S. Reichlin, and K. L. Bick.
Raven Press, New York © 1981.

Opioid Peptides and Related Substances: Immunocytochemistry

Stanley J. Watson and Huda Akil

Mental Health Research Institute, Department of Psychiatry, University of Michigan, Ann Arbor, Michigan 48109

Shortly after the discovery of the pentapeptides methionine- and leucine-enkephalin (12), the longer opiate peptide β-endorphin (β-END) (6,10,16) was discovered by several investigators. It was soon realized that the entire structure of methionine-enkephalin and β-END were to be found in the C terminal portion of the 91 amino acid peptide known as β-lipotropin (β-LPH) (12). β -LPH had been sequenced by Li et al. (15) over a decade earlier, but no specific biological function had been assigned to this molecule. In 1973, Moon, Jennings and Li (18) were able to map the localization of β-LPH in the rat pituitary gland and found that it occurred in the corticotrophs of the anterior lobe and all of the cells of the intermediate lobe of pituitary. A few years later, Pelletier and co-workers (22) carried the study of β-LPH and ACTH to the electron microscopic level. They demonstrated that both β-LPH and ACTH antisera stained all of the granules in both corticotrophs and intermediate lobe cells. This work led to a physiological study, in which stress and other stimuli were shown to result in the co-ordinate release of β-END, β-LPH and ACTH from anterior pituitary.

A biochemical relationship between β-END, β-LPH and ACTH was demonstrated by Mains, Eipper and Ling (17), in a classic paper in which they used pulse chase and immunoprecipitation studies to show that β-LPH, β-END and ACTH were derived from a common precursor molecule (the 31K precursor, or pro-opiocortin). These studies were carried out in a mouse pituitary cell line (ATt-20-D 16-v) and were later confirmed in normal rat pituitary preparations. Nakanishi et al. (19) have elegantly confirmed and extended the protein work on the 31K precursor by preparing cDNA to the mRNA for that precursor. That DNA sequence allows a clearer understanding of the structure of the 16K piece at the N-terminal of the precursor. Thus, there were several lines of work leading to the conclusion of a strong relationship between β-END, β-LPH

and ACTH. Finally, Bloom and co-workers (5) were able to visualize β-END-like immunoreactivity in cells which also contained ACTH
(or α-MSH) and β-LPH (i.e., the corticotrophs and all intermediate
lobe cells).

It soon became clear (by radioimmunoassay and extraction techniques) that brain also contained   β-END and β-LPH size peptides
(23). Several groups (3,4,32,34,36) began the study of the brain
β-END/β-LPH system simultaneously, using immunocytochemical techniques. These anatomical studies located β-END and β-LPH in a
single major neuronal cell group in the basal arcuate nucleus of
hypothalamus (Fig. 1).   When C-terminal lipotropin antibodies
(anti-β-END antibodies) were used, the same cell group and fiber
system could be visualized (32). In contrast to the large number
of widely spread enkephalin cell groups and fibers, β-END and
β-LPH were contained within the single arcuate cell group but had
a large fiber system associated with those cells. The fibers
were distributed throughout the limbic system to the thalamus,
amygdala, nucleus accumbens, septum, periaqueductal central grey,
locus coeruleus and nuclei of the medulla oblongata. This system
was distinctly different from the enkephalin system, and tended to
reinforce the idea that enkephalins and β-END neuronal pools represent two distinct neuronal networks.

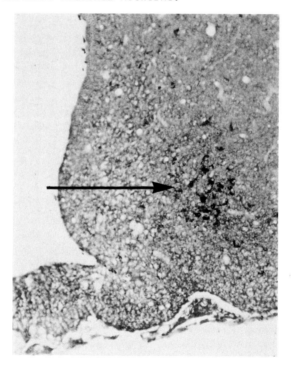

FIG. 1.   β-END stained cells in the rat arcuate nucleus (see
arrow).   Note the relationship to the third ventricle and the
median eminence. X=100.

Because of the intimate linkage in pituitary between β-END and ACTH, it seemed reasonable to study brain for ACTH-like immuno-reactivity as well. Using several different anti-ACTH antisera, it was possible to demonstrate a single major ACTH-containing cell group in the arcuate nucleus of hypothalamus, and a fiber system very much like that seen for β-END and β-LPH (35). The data were in agreement with Krieger et al. (14), who found ACTH immunoreac-tivity in brain extracts using RIA. The ACTH positive system was so much like the endorphin/lipotropin system that we carried out further studies investigating the co-existence of these peptides (33). This work demonstrated that all cells in the arcuate nuc-leus which contained ACTH immunoreactivity also contained β-END and β-LPH. These observations were replicated by others (20,25) and extended to man.

The presence of ACTH, β-END and N-terminal lipotropin (34) immunoreactivity tended to strongly support the hypothesis that brain contained a 31K molecule, similar to that seen in pituitary. However, there had been no studies on the N-terminal portion of 31K, i.e., the 16K or "silent" portions of that molecule. Eipper and Mains (7,8) were able to extract 16K from their pituitary tumor line and to affinity purify antisera directed against several portions of that molecule. They provided us with these affinity purified antibodies for immunohistochemical studies. Again, the 16K antisera revealed the same arcuate cell group and fiber pattern as seen with ACTH, β-LPH and β-END (unpublished data; Fig. 2). It has become evident that all of the major protein segments associated with the 31K precursor in pituitary are found within the same cells and fibers in brain.

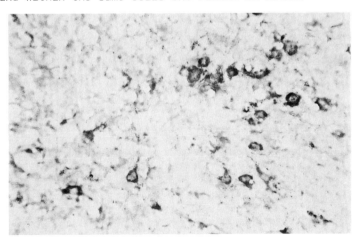

FIG. 2. 16K positive cells in the arcuate nucleus. X=210.

More recent evidence has raised questions about the exact nature of the processing of the 31K molecule in pituitary. Eipper and Mains (8) and Gianoulakis et al. (9) demonstrated that cells

of the anterior pituitary cleave the 31K precursor into ACTH 1-39,
16K and β-LPH with a somewhat slower conversion to β-END. This is
in contrast to the intermediate lobe cells which appear to process
ACTH and β-LPH one step further. These cells rapidly cleave β-LPH
into β-END and γ-LPH, and cleave ACTH into α-MSH (N-Ac-ACTH 1-13
amide) and CLIP (corticotropin-like intermediate lobe peptide).
Thus, there were two possible end points to which the brain could
cleave its 31K precursor. The several studies showing ACTH
immunoreactivity in brain (see above), including our own, had used
antibodies aimed at the ACTH peptide backbone, and would not
differentiate between ACTH and α-MSH and CLIP. In our own study
two antibodies were revealing in the information they gave about
the nature of brain ACTH immunoreactivity. One antiserum was
directed against ACTH 34-39. This antibody, provided by Mains and
Eipper (7), was incapable of recognizing the C-terminus of ACTH
when it was contained in the midst of the 31K precursor, but was
capable of recognizing it as the C-terminal end of the 20K product. In other words, this antibody could only bind to ACTH 34-39
when it was free at its extreme C-terminus (position 39). Our
brain immunohistochemical demonstrations with this serum seemed to
indicate that the ACTH immunoreactivity detected in brain was at
least partially free of its precursor molecule. The second
antibody was associated with the basic region of ACTH (between
positions 14 and 17). This particular antibody detects the 17-24
region in full ACTH (R. Mains, personal communication). For
example, it is capable of visualizing the corticotrophs in the
anterior lobe readily; however, it is a very poor antibody for
visualizing intermediate lobe cells. In the process of cleaving
ACTH to α-MSH and CLIP, the region between 14 and 17 is lost.
Therefore, since this antiserum in part was directed against that
region, it was very poor for detecting α-MSH and CLIP-containing
structures. Using this antibody, the brain 31K neurons and fibers
were very poorly stained, implying an absence of the 14-17 region.
This result suggested that brain ACTH was further processed to
α-MSH and CLIP. The immunocytochemcial evidence in brain suggested that brain ACTH-like immunoreactivity was at least in part due
to a free peptide (as opposed to a precursor molecule), and might
be carried out to the production of α-MSH (as opposed to stopping
at the ACTH step).

Radioimmunoassay studies (24,27) indicated that brain contains
α-MSH-like material of approximately the proper size. Four groups
(3,13,21,26) had demonstrated α-MSH immunoreactive neurons and
fibers in brain. These studies tended to agree with the distribution scheme of β-END, β-LPH and ACTH. Therefore, we studied
brain α-MSH and compared it to brain β-END. Using several different α-MSH antisera, we could visualize cells in the arcuate region
and fiber patterns similar to those seen with β-END (Fig. 3)
(29,30). Fibers never reported before for β-END/β-LPH could also
be detected with α-MSH antisera. For example, fornix, caudate,
hippocampus (Fig. 4) and cortex (Fig. 5) were noted to have
α-MSH-positive fibers as well as all of the areas reported previously for β-END antibodies. In a study using serial 4 μm

sections, it was possible to demonstrate that all of the cells of
the arcuate nucleus that contain β-END also contained α-MSH and
vice versa (29). We concluded that brain neurons contain all of
the pieces of the 31K precursor and, in fact, that brain carried
out the cleavage of that precursor in a fashion analagous to that
of intermediate lobe, i.e., brain tended to cleave ACTH and
lipotropin to smaller pieces, thereby producing β-END, β-LPH,
α-MSH and CLIP. It is unclear, as of this writing, what the fate
of 16K is in any one of the three cell-containing areas.

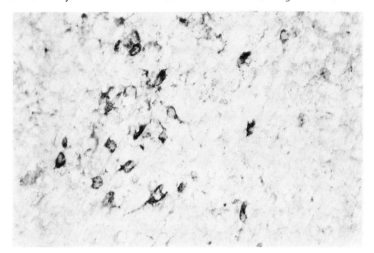

FIG. 3. α-MSH positive cells in the arcuate nucleus. X=210.

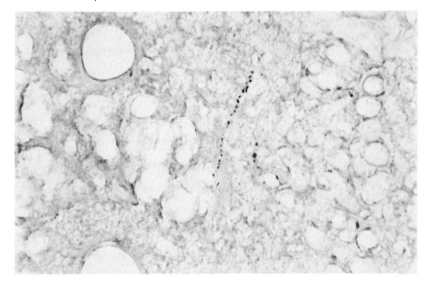

FIG. 4. α-MSH positive fiber in the hippocampus. X=400.

In the process of this work, we realized that brain not only contained α-MSH in arcuate neurons, but also contained α-MSH-like immunoreactivity (with all seven antibodies) in a more dorsal and laterally distributed cell group as well (Fig. 6) (30). This cell group appears to be quite large when compared to the arcuate system and extends from the dorsal, third ventricular region at mid-hypothalamus, in a caudal-lateral fashion, ending at the lateral hypothalamic sulcus at the extreme posterior portion of the

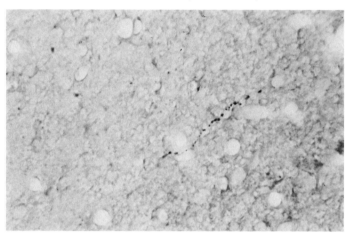

FIG. 5.   α-MSH positive fiber in the cortex. X=400.

hypothalamus.   This rather broad wing of cells contains α-MSH immunoreactivity which is blocked by α-MSH.   However, none of the other 31K antibodies stained these cells (β -END, β -LPH, C-terminal ACTH or 16K).   Thus, it would appear that this extra arcuate α-MSH group (termed α-2 cell group), while containing α-MSH (or a similar peptide), does not contain the remainder of the 31K precursor, even under colchicine conditions.   The biosynthesis in the α-2 cells is unclear.   It is possible that it does come from the 31K system, but only α-MSH is packaged, the remainder of the precursor being destroyed.   It is equally possible that the α-MSH-like material is derived from a different biosynthetic route.
    In an attempt to study the anatomical distribution of the α-2 cells and fibers more carefully, monosodium glutamate (MSG) studies were carried out in collaboration with Dr. John Olney of the Washington University School of Medicine.   Using monosodium glutamate lesioned rats, we could demonstrate an approximately 80% loss of arcuate 31K cells but no substantial loss of the α-2 neurons (Fig. 7).   In animals so treated, much of the 31K fiber system is lost in thalamus, hypothalamus, amygdala and midbrain.   However, none of the fibers associated with the α-2 system in caudate, cortex, hippocampus or fornix appeared reduced.   Although

this anatomical data is preliminary, it would appear that α-MSH has two distinctly different distributions, the first being associated with β-END which has been described in numerous papers

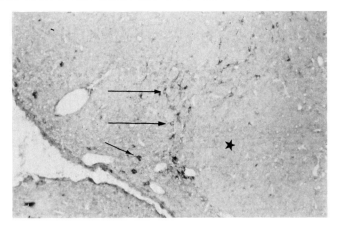

FIG. 6. α-MSH positive cell group (arrows) in the lateral hypothalamus (star is in the optic tract). X=100.

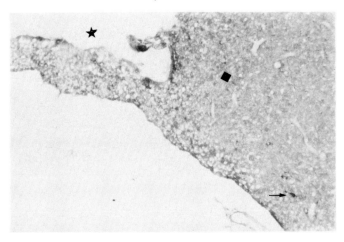

FIG. 7. Monosodium glutamate-treated rats show a loss of most arcuate β-END/α-MSH cells. The star is in the third ventricle. The square is the approximate position of 31K cells in a normal animal (see Fig. 1). Only one positive cell is visible (see small arrow). The more dorsal and lateral α-2 cells were untouched (see Fig. 6). X=100.

previously. The second distribution is related to the α-2 system recently described by us (29,30). Its main projections are to caudate, hippocampus and cortex. Clearly, a great deal of work remains to be done in determining finer anatomical points of both systems.

As indicated above, the β-END/β-MSH system is unusual in that it appears to contain at least two active neuronal substances, β-END and α-MSH. Both of these substances have been demonstrated to be biologically active materials, and in our hands both substances would appear to have CNS receptors (1 and Akil, unpublished observation). In recent studies, central microinjections of α-MSH and related peptides yielded significant behavioral changes, including analgesia (28). Thus, both α-MSH and β-END derive from the same neuron and can modulate pain responsiveness via opiate and non-opiate mediated mechanisms.

We are thus faced with the possibility that neuronally active substances such as β-END and α-MSH could be co-released from the same neuron (31). The prospect of "dual" neurotransmission would appear to introduce a major set of pharmacological and physiological concerns, such as questions of neuronal regulation, control of synthesis and the nature of receptor systems. It is unclear whether endorphin and α-MSH are always released in equal amounts or whether the neuron has the capacity to dampen the release and/or activity of one or the other of these compounds. Questions about the number, nature and interdependence of β-END and α-MSH receptor sites must be addressed. Such studies would shed light not only on 31K system physiology and regulation, but on the whole issue of dual neurotransmitter systems in brain.

ACKNOWLEDGEMENT

This paper was supported in part by NIDA Grant #02265-02 and Scottish Rite Schizophrenia Research Foundation. The authors wish to express their gratitude to Ms. Carol Criss for assistance in manuscript preparation and Rosalie Beer and Robert Thompson for technical assistance.

REFERENCES

1. Akil, L, Hewlett, W., Barchas, J.D., and Li, C.H.: (1980): Eur. J. Pharm., (in press).
2. Akil, H., Watson, S.J., Berger, P.A., and Barchas, J.D. (1978): In: The Endorphins: Advances in Biochemical Psychopharmacology, Vol. 18, edited by E. Costa and E.M. Trabucchi, pp. 125-139. Raven Press, New York.
3. Bloch, B., Bugnon, C., Fellman, D., and Lenys, D. (1978): Neurosci. Lett., 10:147-152.
4. Bloom, F.E., Battenberg, E., Rossier, J., Ling, N., and Gullemin, R., (1978): Proc. Natl. Acad. Sci. USA,75: 1591-1595.

5. Bloom, F.E., Battenberg, E., Rossier, J., Ling, N., Leppaluoto, J., Vargo, T.M., and Guillemin, R. (1977): Life Sci., 20:43-48.
6. Bradbury, A.F., Feldberg, W.F., Smyth, D.G., and Snell, C. (1976): In: Opiates and Endogenous Opioid Peptides, edited by H.W. Kosterlitz, pp. 9-17. Elsevier/North Holland, Amsterdam.
7. Eipper, B., and Mains, R. (1980): Endocrine Rev., (in press).
8. Eipper, B., and Mains, R. (1978): J. Supramol. Struct., 8:247-262.
9. Gianoulakis, C., Seidah, N.G., and Chretien, M. (1980): In: Endogenous and Exogenous Opiate Agonists and Antagonists, edited by E.L. Way, pp. 289-292. Pergamon Press, New York.
10. Guillemin, R., Ling, N., and Burgus, R. (1976): C.R. Hebd. Seances Acad. Sci. Ser. D., 282:783-785.
11. Guillemin, R., Vargo, T., Rossier, J., Minick, S., Ling, N., Rivier, C., Vale, W., and Bloom, F. (1977): Science, 197:1367-1369.
12. Hughes, J., Smith, T.W., Kosterlitz, H.W., Fothergill, L.A., Morgan, B.A., and Morris, H.R. (1975): Nature, 258:577-579.
13. Jacobowitz, D.M., and O'Donohue, T.L. (1978): Proc. Natl. Acad. Sci. USA, 75:6300-6304.
14. Krieger, D.T., Liotta, A., and Brownstein, M.J. (1977): Proc. Natl. Acad. Sci. USA, 74:648-652.
15. Li, C.H., Barnafi, L., Chretien, M., and Chung, D. (1965-66): Excerpta Medica, 3:111-112.
16. Li, C.H., and Chung, D. (1976): Proc. Natl. Acad. Sci. USA, 73:1145-1148.
17. Mains, R.E., Eipper, B.A., and Ling, N. (1977): Proc. Natl. Acad. Sci. USA, 74:3014-3018.
18. Moon, H.D., Li, C.H., and Jennings, B.M. (1973): Anat. Rec., 173:524-538.
19. Nakanishi, S., Inoue, A., Kita, T., Nakamura, M., Chang, A.C.Y., Cohen, S.N., and Numa, S. (1979): Nature, 278:423-427.
20. Nilaver, G., Zimmerman, E.A., Defendini, R., Liotta, A., Krieger, D.T., and Brownstein, M.J. (1979): J. Cell Biol., 81:50-58.
21. Pelletier, G., and Dube, D. (1977): Am. J. Anat., 150:201-206.
22. Pelletier, G., Leclerc, R., Labrie, F., Cote, J., Chretien, M., and Lis, M. (1977): Endocrinology, 100:770-776.
23. Rossier, J., Vargo, T.M., Minick, S., Ling, N., Bloom, F.E., and Guillemin, R. (1977): Proc. Natl. Acad. Sci. USA, 74:5162-5165.
24. Rudman, D., Scott, J.W., Del Rio, A.E., Houser, D.H., and Sheen, S. (1974): Am. J. Physiol., 226:682-686.
25. Sofroniew, M.V. (1979): Amer. J. Anat., 154:283-289.
26. VanLeeuwen, F.W., Swaab, D.F., deRaay, C., and Fisser, B. (1979): J. of Endocrinol., 80:59P-60P.

27. Vaudry, H., Tonon, M.C., Delarue, C., Vaillant, R., and Kraicer, J., (1978): Neuroendocrinology, 27:9-24.
28. Walker, J.M., Akil, H., and Watson, S., (in preparation).
29. Watson, S.J., and Akil, H. (1980): Brain Res., 182:217-223.
30. Watson, S.J., and Akil, H. (1979): Eur. J. Pharmacol., 58:101-1203.
31. Watson, S.J., and Akil, H. (1980): In: The Brain as an Endocrine Target Organ in Health and Disease, edited by D. De Wied, MTP Press, Lancaster. (In press).
32. Watson, S.J., Akil, H., and Barchas, J.D. (1979): In: Endorphins in Mental Health Research, edited by E. Usdin, W.E. Bunney, and N.S. Kline, pp. 30-44. MacMillan Press, New York.
33. Watson, S.J., Akil, H., Richard, C.W., and Barchas, J.D. (1978): Nature, 275:226-228.
34. Watson, S.J., Barchas, J.D., and Li, C.H. (1977): Proc. Natl. Acad. Sci. USA, 74:5155-5158.
35. Watson, S.J., Richard, C.W., and Barchas, J.D. (1978): Science, 200:1180-1182.
36. Zimmerman, E.A., Liotta, A., and Krieger, D.T. (1978): Cell Tiss. Res., 186:393-398.

Neurosecretion and Brain Peptides,
edited by J. B. Martin, S. Reichlin, and K. L. Bick.
Raven Press, New York © 1981.

Neurotensin

George R. Uhl and *Solomon H. Snyder

*Department of Medicine, Stanford University School of Medicine, Stanford,
California 94301; and *Departments of Pharmacology and Experimental Therapeutics and
Psychiatry and Behavioral Sciences, Johns Hopkins University School of Medicine,
Baltimore, Maryland 21202*

Neurotensin is a basic tridecapeptide that was purified by Leeman and co-workers from acid-acetone extracts of bovine hypothalamus (31). Carraway and Leeman isolated the material (6) and, in
1975 (7), published the sequence: pyroGlu-Leu-Tyr-Glu-Asn-Lys-Pro-
Arg-Arg-Pro-Tyr-Ile-Leu-OH. The peptide, closely related molecular species, and its receptor are present in the peripheral
tissues where neurotensin exerts several hormone-like activities.
It is also found in selected regions of the central nervous system
(CNS), suggesting that it is a neurotransmitter (or neuromodulator) in specific brain circuits.

We briefly review the peripheral localizations and activities
of neurotensin, discuss evidence that it is a CNS neurotransmitter, analyze in detail the brain distributions of the peptide and
its receptor, mention some of the possible implications of its
differential localization, and briefly discuss several potential
clinical implications of this basic knowledge.

PERIPHERAL NEUROTENSIN DISTRIBUTION AND ACTIVITIES

Distribution

The distribution of neurotensin is remarkable for its concentration in the CNS, intestines and stomach (8). It is clearly a
"brain-gut" peptide (50). The tridecapeptide has been isolated
from intestine (11). Immunohistochemical studies show that gut
neurotensin is largely concentrated in apparently secretory cells
in intestinal villi and crypts (45,61). Neurotensin in these

cells may be packaged in large granules (52). The stomach, however, appears to contain shorter molecular species (probably C-
terminal fragments) closely related to neurotensin (8). Neurotensin immunoreactivity is found in plasma (31,36,3), where food ingestion augments its levels (36,3). Release of immunoreactive
components into the gut lumen has also been demonstrated (37). It
appears likely that plasma immunoreactivity is biochemically heterogeneous and may differ from the neurotensin tridecapeptide.

Peripheral Activities

Neurotensin has several powerful effects on peripheral systems.
Although several of these activities are probably pharmacologic
and several exhibit species variability, some of the effects may
mimic physiologic roles for neurotensin systems.
    The peptide has hemodynamic effects. The peptide's hypotensive
activity was originally detected by its ability to produce a visible vasodilation in rat skin (6,56). It influences vascular tone
in cutaneous, adipose, intestinal, and coronary systems (53,56).
The material also exerts species-specific inotropic and chronotropic effects on isolated heart muscle preparations (54).
    Neurotensin modulates glucoregulatory systems, leading to rapid
hyperglycemia and hyperglucagonemia (31,9,4,40). These changes
are at least partially antagonized by histamine receptor blockers,
suggesting some histamine mediation of the effects (5,41). More
variable changes in insulin levels have been noted by different
groups; both elevations and depressions have been observed in
slightly differing experimental paradigms (4,40,73). In isolated
pancreatic islets, neurotensin rapdily stimulates insulin, glucagon, and somatostatin release when low ambient glucose concentrations are maintained (15). Release of these same substances is
inhibited, however, with longer exposures under conditions of high
glucose (15). These in vitro glucose- and time-dependencies of
neurotensin's insulin effects may explain the variable in vivo results noted above.
    The peptide transiently increases rat plasma cholesterol levels
(51). Several neurotensin effects on pituitary hormone release
have been described. In some experimental systems the peptide
significantly elevates plasma levels of gonadotropins, prolactin,
growth hormone and thyroid-stimulating hormone when administered
intravenously (74,33,34). When applied to cultured pituitaries,
neurotensin releases prolactin into the medium (74,34). Intracerebroventricular administration, however, leads to decreased re-
rlease of all of the above-mentioned pituitary hormones (74,33).
Peripheral neurotensin administration also causes increases in
plasma corticosteroid levels that are blocked by prior hypophysectomy and therefore probably mediated by ACTH release (31). Multiple neuroendocrine systems are thus significantly modulated by
neurotensin.

Several classical smooth muscle test systems are influenced by the peptide. It causes contraction of relaxed native guinea pig ileum (6) but appears to induce transient relaxation of contracted ileum before leading to renewed contractions (26). The rat duodenum (6) and contracted ileum (23) are both relaxed by neurotensin, while it contracts estrous rat uterus (6). Most of these effects seem to be mediated directly through neurotensin receptors in the smooth muscle. Blockers of neural transmission such as tetrodotoxin and anticholingergics alter the neurotensin effect on the guinea pig ileum, suggesting a neuronal or cholinergic mediation there, but they fail to affect the other gut organ activities (23,24). Further, in vitro binding studies show high affinity specific neurotension binding to membranes from guinea pig ileum (23) and estrous rat uterus (62). Some evidence has been adduced that activation of these receptors leads to changes in sodium and calcium ion conductance without alterations in cyclic nucleotide levels (23,25). Neurotensin alters gut secretory function, markedly reducing gastic acid secretion after it is given either peripherally or centrally (1,47). The peptide has also been reported to have kinin-like effects in causing leukocyte chemotaxis, producing pain and leading to increased vascular permeability after peripheral administration (31,5).

This confluence of evidence points toward involvement of peripheral neurotensin in at least three major physiological activities. Handling of food and energy metabolism could be biochemically affected by its effects on glucose, cholesterol, glucagon, insulin, growth hormone and thyroid-stimulating hormone, and mechanically affected by its activities on gastric secretion and gut motility. Perhaps some of this control could be exerted through neurotensin release from gut cells. A neurotensin effect on reproductive systems could be inferred from its influence on uterine smooth muscle. Finally, actions on vascular smooth muscle and cardiac muscle might indicate a hemodynamic role.

Structure-Activity Relationships

Studies of the biological effects of neurotensin and closely related peptides have shown definite structure-activity relationships. Although there are subtle differences from system to system, these analyses reveal that the C-terminus of neurotensin is essential for biological activity. Potency is only gradually reduced when residues are successively deleted from the N-terminus (10,32).

CNS NEUROTENSIN

Evidence for a Neurotransmitter Role

Neurotensin is now reasonably well established as a neurotransmitter or neuromodulator-candidate in the central nervous system as a consequence of data from anatomical, biochemical and electrophysiological studies (68,72).

The tridecapeptide is present in neural tissues. It was originally isolated and sequenced from hypothalamus (5,7). Physico-
chemical studies of brain neurotensin immunoreactivity suggest
that the bulk of this material co-migrates with authentic neurotensin (8,61). However, there is some evidence for physico-chemical heterogeneity of the material in brain extracts based on reactions with less selective neurotensin antisera (7; R. Carraway,
personal communication). This may reflect the presence in brain
of neurotensin-related antigens or fragments, by analogy to the
apparent fragments noted in the stomach (27). Nevertheless, the
predominant portion of the immunoreactivity determined by selective neurotensin antisera seems to be indistinguishable from
neurotensin itself (8,62).

As would be anticipated for a neurotransmitter candidate (63,
29), this immunoreactivity is concentrated in certain grey matter
areas and in synaptosome-enriched subcellular fractions. Immunohistochemical studies show neurotensin immunoreactivity in patterns resembling neuronal perikarya, nerve fibers and terminals
(64,65).

Calcium-dependent release of neurotensin from these tissue
stores is demonstrable by in vitro depolarization (21). Apparent
neurotensin receptors have been demonstrated in several studies.
First, high-affinity, saturable, specific, rapid and reversible
binding of neurotensin is found in brain membrane preparations
(66,67,28,30). This binding is concentrated in certain grey matter regions and displays structure-activity relationships parallel
to those noted for neurotensin's biological effects. These properties suggest the existence of physiologic neurotensin receptors.
With microiontophoretic application of the peptide, marked changes
in cell firing rate are obtained that also suggest mediation by
neurotensin receptors. The peptide decreases firing rates of
cells in the locus coeruleus and nucleus accumbens while increased
rates are recorded from cells in the dorsal horn of the spinal
cord (77,38,39). In vitro autoradiographic techniques (75) have
disclosed the presence of heterogeneously distributed neurotensin
receptors in tissue slices (76). Finally, intraventricular neurotensin exerts profound effects on thermoregulatory, hypophyseal,
hormonal, and nociceptive systems (see below) in a manner
consistent with receptor activation (42,12,43). Mechanisms for
inactivation of the peptide through presumably enzymatic degradation are also easily demonstrable in brain (17,67).

ANATOMIC DISTRIBUTION OF NEUROTENSIN-CONTAINING BRAIN SYSTEMS

Detailed studies of the localizations of neurotensin-containing
and neurotensin-receptive brain systems have shown localizations
of these elements to specific brain regions. These localizations,
in turn, have provided hints as to functional roles for the
peptide in the CNS.

Two approaches have been used to determine the differential
distribution of neurotensin in the CNS. A radioimmunoassay of the

neurotensin immunoreactivity extracted from brain subdivisions has provided a gross understanding of its preferential localizations. This approach allows some degree of biochemical characterization of the extracted immunoreactivity. As noted above, however, some neurotensin antisera appear to recognize heterogeneous materials separable by chromatographic methods while others are apparently more specific (8,62). Immunohistochemistry adds tremendously to the level of anatomic resolution achievable. Although antisera used in immunohistochemistry can (and should) be characterized in radioimmunoassay studies, fewer controls for the nature of the immunoreactivity can be applied. In practical terms, use of high-affinity, high titer, specific primary antisera and subtraction of structures postively stained in adjacent control sections in which the primary antiserum has been pre-adsorbed with peptide (Fig. 1) seems to yield results correlating well with those obtained by other techniques.

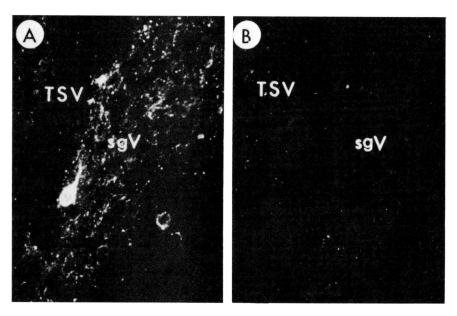

FIG. 1. Photomicrographs of neurotensin immunohistofluorescence in the substantia gelatinosa of the nucleus caudalis of the spinal tract of the trigeminal (sgV) and the spinal tract of the trige-minal (TSV) of the rat. The primary antibody in B was preadsorbed overnight at 4°C with 0.1 mM neurotensin to establish a control.

Receptors can also be studied by binding experiments in brain
homogenates and, recently, by autoradiographic techniques following binding in tissue slices. Here again, subtraction of binding
that is not displacable by unlabeled neurotensin provides a control. Techniques and pitfalls of binding studies in homogenates
have been extensively discussed elsewhere (59). Autoradiographic
receptor localization <u>in vitro</u> (75) allows association of radio-

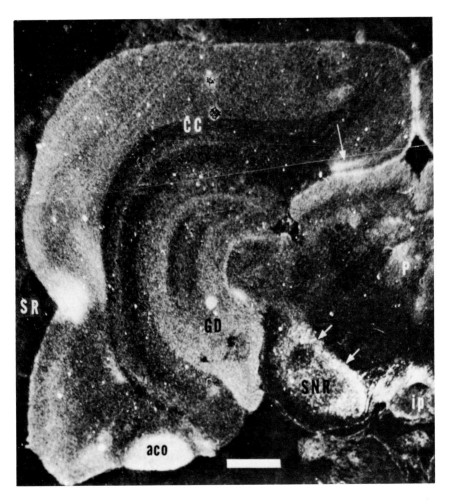

FIG. 2.   Darkfield photomicrograph of neurotensin receptor autoradiography in rat brain. ACO = cortical amygdaloid nucleus; CC =
corpus callosum; GD = dentate gyrus; ip = interpeduncular nucleus;
P = periaqueductal grey; SNR = substantia nigra, pars reticulata;
SR = rhinal sulcus. Bar = 1000 μ. (Adapted from ref. 76, with
permission.)

labeled neurotensin to receptors in cryostat slices of perfused rat brain with or without the presence of high levels of unlabeled peptide. After washing, autoradiograms are generated by apposing emulsion-coated coverslips. Following exposure, tissue staining and emulsion development, radiographic grains are seen overlying areas of elevated receptor density. This technique thus allows light-microscopic localization of peptide receptor densities (Fig. 2).

Spinal Cord

Neurotensin levels in segments of rat and calf spinal cord are relatively low (Table 1) (29,63). Immunohistofluorescent studies suggest that much of this immunoreactivity is concentrated in the "substantia gelatinosa" (64), corresponding to lamina II-III of Rexed (60). Following intrathecal injection of colchicine, neuro-tensin-immunoreactive fibers seen in the white matter lateral to the substantia gelatinosa could conceivably represent proprio-spinal connections linking different spinal cord segments (64).

Binding to whole calf spinal cord is also low, about one-twen-tieth of values obtained from the most receptor-dense regions of CNS in this species (67). Autoradiographically determined neuro-tensin receptors are also concentrated in the substantia gelatino-sa (76). A small spinal cord region, the substantia gelatinosa, thus appears to contain most of the spinal neurotensin systems.

This distribution corresponds closely with the localization of single units in cat spinal cord responsive to microiontophoresed peptide (39). More than half of the units located in lamina I-III are excited (albeit slowly) by the peptide, while none of the cells lying in laminae IV-VII respond.

Medulla Oblongata

Neurotensin binding to whole cat and rat medulla is low (30, 67). Immunoreactivity in these species is similarly low in radioimmunoassay studies (29,63). However, when immunohistochemi-cal and autoradiographic examinations of the medulla are under-taken discrete densities of apparent peptide and receptor are associated with several nuclei. The "substantia gelatinosa" zone of the spinal tract of the trigeminal possesses dense neuroten-sin-like fluorescence (Fig. 3) (65). Cell bodies found here may correspond to the "spiny" cells identified in Gogli preparations by Gobel and co-workers (20). This region contains a concentra-tion of autoradiographically determined neurotensin receptors (76).

TABLE 1. Immunoreactive neurotensin and ^{125}I-neurotensin binding
in calf brain regions

Region	Immunoreactive Neurotensin (pmol/g wet wt. ± SEM)	% Specific Binding (frontal pole = 100%)
Cerebral Cortex		
Parahippocampal gyri	6.76 + 0.67	180 + 15
Cingulate gyrus (mid)	3.84 + 0.24	132 + 13
Frontal pole	1.80 + 0.11	100
Occipital pole	2.53 + 0.30	126 + 5
Precentral gyrus	2.44 + 0.15	96 + 13
Parietal cortex	2.36 + 0.07	92 + 11
Hypothalamus		
Anterior	17.3 + 3.3	144 + 21
Medial	19.1 + 3.1	165 + 22
Mamillary bodies	11.1 + 0.88	132 + 16
Thalamus		
Dorsomedial	2.47 + 0.09	249 + 4
Ventral	1.27 + 0.22	136 + 38
Anterior	4.32 + 0.66	112 + 8
Pulvinar	1.42 + 0.03	99 + 12
Basal Ganglia		
Caudate	11.0 + 0.13	100 + 13
Globus pallidus	9.34 + 0.66	84 + 20
Limbic Regions		
Hippocampus	2.99 + 0.39	88 + 21
Amygdala	1.94 + 0.38	98 + 4
Brainstem		
Pons	1.25 + 0.23	59 + 5
Medulla oblongata	1.62 + 0.05	27 + 15
Colliculi	2.74 + 0.27	87 + 9
Others		
Cervical spinal cord	1.06 + 0.08	13 + 5
Cerebral white matter	0.74 + 0.07	27 + 7
Cerebellar cortex	≤ 0.32	30 + 3

Regions from three fresh calf brains were dissected on ice.
Each dissection involved regions from one calf except hypothalamic
subdivisions, thalamic subdivisions, amygdala, and globus pallidus, where tissues from three calf brains were pooled. Reported
values for each brain region are the mean ± SEM for three regional
dissections. (Adapted with permission from ref. 63 and 67.)

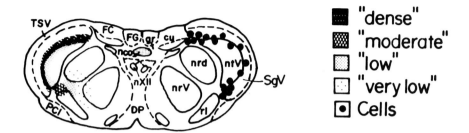

FIG. 3. Neurotensin immunofluorescence in the rat medulla. cu = cuneate nucleus; DP = decussation of the pyramids; FG = fasciculus gracilis; FC = fasciculus cuneatus; gr = gracile nucleus; nco = commissural nucleus; nrd = dorsal reticular nucleus; ntV = ventral reticular nucleus; nXII = hypoglossal nucleus; rl = lateral reticular nucleus; SgV = substantia gelatinosa of the trigeminal; TSV = spinal tract of the trigeminal. (Adapted with permission from ref. 65.)

The concentrations of apparent neurotensin systems in the substantia gelatinosa zones of both the spinal cord and the trigeminal nuclear complex suggested that neurotensin-containing systems might have a modulatory influence on nociceptive input, which is thought to be integrated in these two regions (22,70). Subsequently, centrally administered neurotensin has been found to potently and specifically inhibit responses to noxious stimuli, as though it were an analgesic. The peptide has activity in several behavioral models for analgesia, including hot plate, acetic-acid writhing, and tail-flick tests (43,12). In the latter test, the peptide is more potent than either met-enkephalin, leu-enkephalin or morphine (43). Importantly, the apparent analgesic effect of neurotensin is not reversed by the opiate antagonist naloxone (12). This finding suggests that neurotensin analgesia is not likely to be dependent on activation of enkephalin- or β-endorphin containing neurons. This independence is also supported by differences in the distributions of neurotensin and enkephalin immunoreactivities within the substantia gelatinosa zone when they are examined in adjacent sections (69,65).

The nucleus of the solitary tract possesses substantial densities of immunoreactive fibers and cell bodies while the adjacent

vagal nucleus contains a higher density of autoradiographic receptor grains and moderate amounts of neurotensin immunofluorescence
(65,76). Neurotensin in this region might conceivably play a part
in the integration of some of the "visceral reflexes" thought to

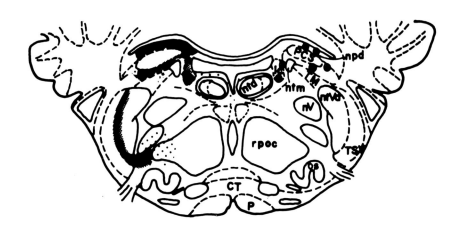

FIG. 4. Neurotensin immunofluorescence in the rat pons (for key
to fluorescence, see Fig. 3). CT = trapezoid body; lc = locus
coeruleus; ntm, nV, ntVd, TM, TSV = trigeminal tracts and nuclei;
ntd = dorsal tegmental nucleus of Gudden; npd = dorsal parabrachial nucleus; os = superior-olive; PCS = superior cerebellar peduncle; P = pyramid; rpoc = parvocellular vetricular nucleus.
(Adapted with permission from ref. 65.)

be mediated here. Several other medullary regions appear to possess lower densities of immunoreactive fibers, especially the nucleus of the spinal tract of the trigeminal, lateral areas of the
reticular formation, and the grey matter of the floor of the
fourth ventricle, where receptors also are concentrated (65,76).

Pons

Binding-site density in the calf pons is about twice that of
the medulla. Furthermore, the dorsal pons contains significantly
more apparent receptors than the ventral pons (67). Though immunoreactivity in radioimmunoassay studies of whole calf and rat pons
is low, immunohistochemical studies indicate that this immunoreactivity is concentrated in cell bodies, fibers and terminals in the
locus coeruleus and the parabrachial nuclei (Fig. 4) (65).

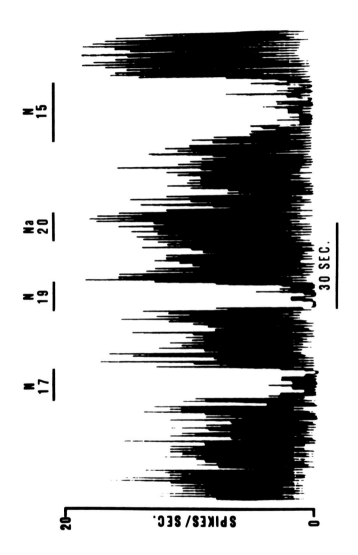

FIG. 5. Depression of single-unit response of locus coeruleus neuron to microiontophoresis of neurotensin (N) or carrier solution (Na). (Adapted with permission from ref. 77.)

These fourth ventricular floor areas also have modest concentrations of autoradiographically determined receptors (76). The
parabrachial region has been associated with several functional
systems, including gustatory sensibilities and respiratory control
(2,44); neurotensin in these regions might interact with these
functions. The moderate concentrations of neurotensin immunoreactivity and of neurotensin receptors in the locus coeruleus but not
in immediately dorsal or ventral areas afforded an opportunity for
the initial microiontophoretic studies of the peptide's action
(Fig. 5) (77). Microiontophoretic application of neurotensin
promptly depresses firing rates of half of the units found in the
locus coeruleus, though few cells in the neurotensin-poor regions
ventral to the locus are responsive (77).
    Pontine areas with lower amounts of neurotensin immunoreactiv-
ity include the lateral reticular formation and a dorsal midline
band (65).

Cerebellum

    The cerebellum is the grey-matter region with perhaps the low-
est density of neurotensin-related elements as indicated   by
receptor binding (67), radioimmunoassay (29,63), and immunohistochemistry (64,65). Cells here are not responsive to microiontophoresed peptide (38).

Mesencephalon

    Receptor binding to midbrain areas displays low to moderate
density (30,67). Radioimmunoassay yields similar results, demonstrating elevations of levels in the central mesencephalic grey
and midbrain tegmental areas (8,29,63).
    Immunohistochemistry reveals concentrations of fibers, termin-
als and perikarya in the ventral tegmental area of Tsai, the periaqueductal grey and the dorsal raphe nucleus (Fig. 6) (65).
Autoradiography shows corresponding receptor densities in the ventral tegmentum and periaqueductal grey (76). However, markedly
elevated receptor grain density in the substantia nigra, pars compacta is accompanied by only moderate densities of fibers and terminals in immunohistochemical studies (64,76). The dorsal raphe,
periaqueductal grey and ventral tegmentum have all been implicated
in multiple functions. Some of neurotensin's antinociceptive
activity, for example, could be attributable to interactions with
pain-controlling circuitry thought to lie in the dorsal raphe and
periaqueductal grey (18,55). Neurotensin has been shown to increase measures of turnover of serotonin, dopamine and norepinephrine (19). The dorsal raphe and ventral tegmental substantia
nigra areas contain dopaminergic neurons while the locus coerulus
localization could allow effects on the norepinephrine system
(13,48).

A band extending to the cuneiform region from the central grey, the lateral reticular formation, the dorsal midline and the compacta zone of the substantia nigra are other areas with more modest neurotensin fluorescence (65).

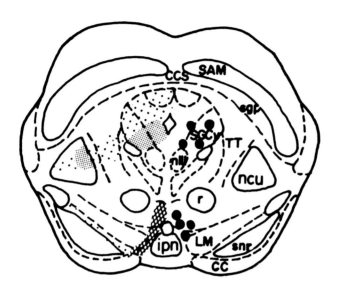

FIG. 6. Neurotensin immunofluorescence in the rat mesencephalon. CC = cerebral peduncle, ipn = interpeduncular nucleus, LM = medial lemniscus, ncu = cuneiform nucleus, nIII = oculomotor nucleus, r = red nucleus, TT = tectospinal tract. (Adapted with permission from ref. 65.)

Diencephalon

The hypothalamus contains in many of its nuclei large amounts of apparently neurotensin-containing and neurotensin-receiving elements. In general, anterior and basal nuclei seem to be most enriched. The high receptor binding and immunoreactivity in the entire bovine hypothalamus are most concentrated in anterior and medial regions (63,67). In the rat, high hypothalamic levels are concentrated in the preoptic area and in the basal hypothalamus/-median eminence (29). These findings correspond well with immunohistochemical observations of densities of fibers and terminals in basal and preoptic hypothalamic zones and in the median eminence (Fig. 7) (64). Preliminary results reveal cell

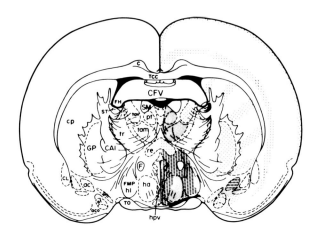

A 5780 μ

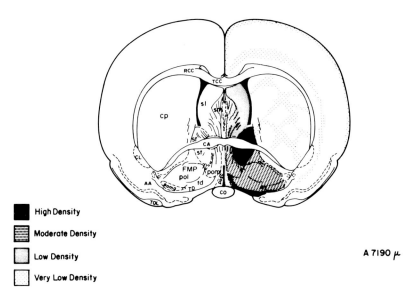

■ High Density

▦ Moderate Density

▒ Low Density

☐ Very Low Density

A 7190 μ

FIG. 7. Neurotensin immunofluorescence in the rat diencephalon
and telencephalon. Important abbreviations: ac = central amygdala; ca = anterior commissure; cp = caudate/putamen; TCC,RCC =
corpus callosum; FMP = median forebrain bundle; F = fornix; GP =
globus pallidus; ha = anterior hypothalamus; hl = lateral hypothalamus; poma, pol = preoptic areas; ST = stria terminalis; st =
interstitial nucleus of the stria terminalis; sl, sm = septal nuclei; tam, tr, tav = thalamic nuclei. (Adapted with permission
from ref. 65.)

bodies in the medial and lateral preoptic areas, in the peri-ven-tricular hypothalamus, and in the basal hypothalamus (62). These hypothalamic sites could be the loci for another potent effect of centrally administered neurotensin, hypothermia. Intracisternal injection of the peptide produces a decrease in body temperature, which is accentuated in a cold environment, and is greater than that produced by bombesin, melanocyte inhibitory factor, or opioid peptides (43,42).

Neurotensin releases somatostatin from incubated hypothalamic fragments (57) and increases acetylcholine turnover in the diencephalon when administered intracerebroventricularly (35). These activities may point to hypothalamic interactions between endogenous neurotensin systems and neurons containing these two neurotransmitter candidates.

As mentioned above, central neurotensin injection has several effects on pituitary hormones, decreasing release of gonadotro-pins, prolactin, growth hormone, and thyroid-stimulating hormone (see above, 33,74). Neurotensin fiber and terminal densities noted in the median eminence (65) could conceivably play a corres-ponding physiological role, perhaps by releasing the peptide into portal capillaries. Interestingly, evidence for pituitary to brain neurotensin transport has also been adduced. Intrapituitary peptide injections result in a hypothermic effect that is not eli-cited when neurotensin is injected next to the pituitary or intra-venously (16). This effect is blocked when the pituitary stalk is previously severed. Since neurotensin immunoreactivity is found in the plasma and in the pituitary, this transport could conceiv-ably have a physiologic role.

Thalamus/Epithalamus

In the calf and rat, the thalamus possesses moderate to high levels of specific binding, but more modest concentrations of immunoreactive neurotensin (29,63,30,67). Immunofluorescence results suggest that medial zones of the rat thalamus, especially the peri-ventricular nucleus but also the medial, paratenial and anteromedial nuclei, possess most of the thalamic neurotensin (64). The medial habenular nucleus, which possesses modest neurotensin immunofluorescence, appears to contain a density of receptors in autoradiographic studies (76).

Telecephalon

Hippocampus.
This structure possess modest to low receptor binding density, and relatively low levels of immunoreactivity in calf and rat (63,67,29). Immunofluorescent examination reveals only sparse fibers and terminals (64).

Striatum.
The caudate-putamen and globus pallidus of calf possess moder-ate levels of both neurotensin immunoreactivity and receptor bind-

ing (62,66). However, in the rat, both radioimmunoassay levels and immunofluorescence densities are low (29,64).

Amygdala/Stria terminalis.
Receptor binding and radioimmunoassay studies in calf and rat assign moderate values to neurotensin system densities in the whole amygdala (63,29,67). Most of the immunoreactivity is concentrated in fibers, terminals, and cell bodies in the central nucleus of the amygdala in immunofluorescence studies (Fig. 7) (64,71). Fiber-like fluorescent patterns of some density are visible running in the ventrolateral quadrant of the stria terminalis (64). Classical anatomic degeneration studies (14) suggest that this region of the stria carries fibers from the central and other amgydaloid nuclei that end in the interstitial nucleus of the stria terminalis. After lesions that destroy the central amygdala, fluorescence in the ventrolateral stria is dramatically reduced (71). Fluorescent material also accumulates caudal to and disappears rostral to knife cuts of the stria terminalis. Neurotensin-containing cell bodies in the central amygdala thus seem to give origin to rostral projecting fibers in the stria terminalis. An autoradiographic receptor density in the interstitial nucleus of the stria terminalis fits well with this scheme, perhaps corresponding to a zone of termination of some of these fibers (76).

Basal forebrain regions.
The interstitial nucleus of the stria terminalis possesses a dense network of immunoreactive fibers and cell bodies, as well as a high concentration of receptor grains, as noted above (71,76). The lateral portion of the nucleus accumbens similarly displays densities of fibers and terminals (64) where elevated concentrations of immunoreactivity are also found in radioimmunoassay studies (29). Other studies suggest that neurotensin has physiological activity in the accumbens. It is said to block certain amphetamine effects when microinjected there (43). When applied microiontophoretically, it rapidly and potently depresses the firing of the accumbens dopamine-sensitive cells (38). Although neurotensin systems appear at several brain sites associated with locomotor activity, accumbens neurotensin systems could be plausible sites for its activities in causing "muscle relaxation" (46) and decreased locomotion (42).

Cerebral cortex.
The cerebral cortex, like the thalamus, displays elevated levels of binding that are higher than its relative density of immunoreactivity. In the calf, the parahippocampal and cingulate cortices possess especial elevations of binding and also have the highest immunoreactivity found among cortical regions (63,67). Autoradiographic studies in rat reveal a corresponding pattern; cortical receptor grains are concentrated in the cingulate cortex and along the rhinal sulcus (76). In the rat, low levels of

immunoreactivity are seen in immunohistochemical studies to correspond to relatively sparse fiber and terminal patterns (29,65).

The selective association of increased neurotensin immunoreactivity and binding with limbically associated cortical and thalamic brain regions may hint at a preferential role in more "emotion related" regions (49).

Pituitary.

Neurotensin immunoreactivity has been extracted from the pituitary (8). However, the immunologic behavior of this material in radioimmunoassay suggests that it may differ from authentic neurotensin. Striking fluorescence seen in cells of the anterior pituitary, therefore, is probably not authentic neurotensin (64). The localization of neurotensin-like material to anterior pituicytes does suggest that the material might be secreted into the circulation. Neurotensin immunoreactivity detected in plasma could conceivably partially arise from the pituitary, though the gut seems a more likely source.

POTENTIAL CLINICAL SIGNIFICANCE

Studies of central and peripheral neurotensin have several implications for neurology and general medicine. It can be predicted that neurotensin-secreting tumors will be found, perhaps in patients displaying a clinical constellation similar to that seen in experimental animals injected with the peptide (58). These patients might exhibit hypotension, diarrhea, and/or hypoglycemia, possibly in association with intestinal malignancies or other "APUDomas".

Perhaps more importantly, the physiological activities of neurotensin could be exploited to produce therapeutic agents efficacious in hemodynamic, gastric secretory, thermoregulatory, gonadotropin-releasing, glucoregulatory, and pain-perceptive systems. Especially in the latter case, there seems reasonable hope that neurotensin analogs with the ability to cross the brain-blood barrier could offer potent, powerful, nonopiate and, conceivably, nonaddicting analgesia.

REFERENCES

1. Andersson, S., Chang, D., Folkers, K., and Rosell, S. (1976): Life Sci., 19:367-370.
2. Berger, A. Mitchell, R., and Severinghaus, J.W. (1977): N. Engl. J. Med., 297:138-143.
3. Besterman, H., Bloom, S.R., Sarson, D.L., Blackburn, A.M., Johnston, D.I., Patel, H.R., Modigliani, R., Guerin, S., and Millinson, C.N. (1978): Lancet, 1:758-788.

4. Brown, M., and Vale, W. (1976): Endocrinology, 98:819-822.
5. Brown, M., Villarreal, J., and Vale, W. (1976): Metabolism, 25:1459-1461.
6. Carraway, R, and Leeman, S. (1973): J. Biol. Chem., 248: 6854-6861.
7. Carraway, R., and Leeman, S. (1975a): J. Biol. Chem., 250: 1907-1911.
8. Carraway, R., and Leeman, S. (1976): J. Biol. Chem., 251: 7045-7052.
9. Carraway, R.E., Demers, L.M., and Leeman, S.E. (1976): Endocrinology, 99:1452-1462.
10. Carraway, R., and Leeman, S. (1975b): In: Peptides: Chemistry, Structure and Biology, edited by R. Walter, and J. Meienhofer, pp. 679-685. Science Publishers, Ann Arbor, Michigan.
11. Carraway, R., Kitobgi, P., and Leeman, S. (1978): J. Biol. Chem., 253:7996-7998.
12. Clineschmidt, B.V., and McGuffin, J.C. (1977): Eur. J. Pharm., 46:395-396.
13. Dahlstrom, A., and Fuxe, K. (1965): Acta. Physiol. Scand., 62: Suppl., 232:1-55.
14. deOlmos, J.S. (1972): In: The Neurobiology of the Amygdala, edited by B. Eleftheriou, pp. 145-204. Plenum Press, New York.
15. Dolais-Kitabgi, J., Kitabgi, P.. Brazeau, P., and Freychet, P. (1979): Endocrinology, 105:256-260.
16. Dorsa, D., deKloet, E., Mezey, E., and deWied, D. (1979): Endocrinology, 104:1663-1666.
17. Dupont, A., and Merand, Y. (1978): Life Sci., 22:1623-1630.
18. Fields, H.L., and Basbaum, A.I. (1978): Ann. Rev. Physiol., 40:217-248.
19. Garcia-Sevilla, J.A., Magnusson, T., Carlsson, A., Leban, J., and Folkers, K. (1978): Naunyn-Schmiedberg's Arch. Pharmacol., 305:213-218.
20. Gobel, S. (1975): J. Comp. Neurol., 162:397-416.
21. Iversen, L., Iversen, S., Bloom, F. Douglas, C., Brown, M., and Vale, W. Nature, 273:161-163.
22. Kerr, F. (1975): Pain, 1: 325-356.
23. Kitabgi, P., and Freychet, P. (1979): Eur. J. Pharm., 55: 35-42.
24. Kitabgi, P., and Freychet, P. (1979): Eur. J. Pharm., 56: 403-406.
25. Kitabgi, P., Haman, J., and Worcel, M. (1979): Eur. J. Pharm., 56:87-93.
26. Kitabgi, P., and Freychet, P. (1978): Eur. J. Pharm., 50: 349-357.
27. Kitabgi, P., Carraway, R., and Leeman, S. (1976): J. Biol. Chem., 251:7053-7058.
28. Kitabgi, P., Carraway, R., Von Rietschafen, J., Grauier, C., Movgat, J., Menez, A., Leeman, S., and Freychet, P. (1977): Proc. Natl. Acad. Sci. USA, 74:1846-1850.

29. Kobayasht, R.M., Brown, M., and Vale, W. (1977): Brain Res., 126:584-588.
30. Lazarus, L., Brown, M., and Pevrin, M. (1977): Neuropharmacology, 16:625-629.
31. Leeman,S., Mroz, E., and Carraway, R. (1977): In: Neurobiology of Peptides, edited by H. Gainer, pp. 99-144. Plenum Press, New York.
32. Loosen, P.T., Nemeroff, C.B., Bissette, G., Burnett, G.B., Prange, A.J. Jr., and Lipton, M.A. (1978): Neuropharmacology, 17:109-113.
33. Maeda, K., and Frohman, L. (1978): Endocrinology, 103:1903-1909.
34. Makin, T., Carraway, R., Leeman, S. and Grup, R. (1973): Proc. Soc. Study Repro., 26.
35. Malthe-Sovenssen, D., Wood, P., Cheney, D., and Costa, E. (1978): J. Neurochem., 31:685-691.
36. Mashford, M., Nilsson, G., Rokaeus, A., and Rosell, S. (1978): Acta. Physiol. Scand., 104:244-246.
37. Mashford, M., Nelsson, G., Rokaeus, A., and Rosell, S. (1978): Acta. Physiol. Scand., 104:375-376.
38. McCarthy, P., Walker, R. Yajima, H., Kitagawa, K., and Woodruff, G. (1979): Gen. Pharmacol., 10:331-333.
39. Miletic, V., and Randic, M. (1979): Brain Res., 169:600-604.
40. Nagai, K., and Frohman, L. (1976): Life Sci., 19:273-280.
41. Nasai, K., and Frohman, L. (1978): Diabetes, 27:577-582.
42. Nemeroff, C., Bissette, G., Prange, A., Loosen, P., Barlow, T., and Lipton, M. (1977): Brain Res., 128:485-496.
43. Nemeroff, C., Osbahr, A., Mauberg, P., Ervin, G., and Prange, A. (1979): Proc. Natl. Acad. Sci. USA, 76:5368-5371.
44. Novgren, R., and Leonard, C. (1973): J. Comp. Neurol., 150:217-232.
45. Orci, L., Baetens, O., Rufener, C., Brown, M., Vale, W., and Guillemin, R. (1976): Life Sci., 19:559-562.
46. Oshbahr, A.J. 3rd, Nemeroff, C.B., Manberg, P.J., and Prange, A.J. Jr. (1979): Eur. J. Pharm., 54:299-302.
47. Osumi, Y., Nagasaka, Y., Wang, Fu, L.H., and Fujiwara, M. (1978): Life Sci., 23:2275-2280.
48. Palkovitz, M., and Jacobowitz, D. (1974): J. Comp. Neurol., 157:29-42.
49. Papez, J. (1937): Arch. Neurol. Psychiat., 38:725-743.
50. Pearse, A. (1976): Nature, 262:92-94.
51. Peric-Golia, L., Gardner, C., and Peric-Golia, M. (1979): Eur. J. Pharm., 55:407-409.
52. Polak, J., Sullivan, S., Bloom, S., Buchan, A., Facer, B., Brown, M., and Pearse, A. (1977): Nature, 270: 183-184.
53. Quirion, R., Rioux, F., Regoli, D., and St.-Pierre, S. (1979): Eur. J. Pharm., 55:221-223.
54. Quirion, R., Rioux, F., and Regoli, D. (1978): Canad. J. Pharm., 56:671-673.
55. Reynolds, D.V. (1969): Science, 164:444-445.

56. Rosell, S., Burcher, E., Chang, D., and Folkers, K. (1976): Acta. Physiol. Scand., 98:484-491.
57. Sheppard, M.C., Kronheim, S., and Pimstone, B.L. (1979): J. Neurochem., 32:647-649.
58. Snyder, S.H., Uhl, G., and Miller, R. (1978): N. Engl. J. Med., 298:1259-1260.
59. Snyder, S.H., and Bennett, J.P. Jr. (1976): Ann. Rev. Physiol., 38:153-175.
60. Steiner, T.J., and Turner, L.M. (1972): J. Physiol., 222: 123P-125P.
61. Sundler, F. Hakanson, R. Hammer, R. Alumets, J. Carraway, R., Leeman, S., and Zimmerman, E. (1977): Cell Tiss. Res., 178:313-321.
62. Uhl, G. (1978): Doctoral Thesis, Johns Hopkins University.
63. Uhl, G., and Snyder, S.H. (1976): Life Sci., 19:1827-1832.
64. Uhl, G., Kuhar, M., and Snyder, S.H. (1977): Proc. Natl. Acad. Sci. USA, 74:4059-4063.
65. Uhl, G., Goodman, R., and Snyder, S.H. (1979): Brain Res., 167:77-91.
66. Uhl, G. Bennett, J., and Snyder, S.H. (1976): Abstr. Soc. Neurosci., 2:803.
67. Uhl, G., Bennett, J., and Snyder, S.H. (1977): Brain Res., 130:299-313.
68. Uhl, G., and Snyder, S.H. (1977): Eur. J. Pharm., 41: 88-91.
69. Uhl, G.R., Goodman, R.R., Kuhar, M.J., Childers, S.R., and Snyder, S.H. (1979): Brain Res., 166:75-94.
70. Uhl, G., Goodman, R., Kuhar, M., and Snyder, S.H. (1978): In: Advances in Biochemical Psycho-Pharmacology, Vol. 18, edited by E. Costa, and M. Trabucci, pp. 71-87. Raven Press, New York.
71. Uhl, G., and Snyder, S.H. (1979): Brain Res., 161:522-526.
72. Uhl, G., and Snyder, S.H. (1980): In: Role of Peptides in Neuronal Function, edited by J. Barker and T. Smith. M. Dekker, New York. (In press).
73. Ukai, M., Inoue, I., and Itatsu, T. (1977): Endocrinology, 100:1284-1286.
74. Vijayan, E., and McCann, S. (1979): Endocrinology,105: 64-88.
75. Young, W.S. 3rd, and Kuhar, M. (1979): Brain Res., 179: 255-270.
76. Young, W.D., 3rd and Kuhar, M. (1979): Eur. J. Pharm., 59: 161-163.
77. Young, W.D. 3rd, Uhl, G.R., and Kuhar, M.J. (1978): Brain Res., 150:431-435.

Neurosecretion and Brain Peptides,
edited by J. B. Martin, S. Reichlin, and K. L. Bick.
Raven Press, New York © 1981.

Introduction to Section III:
Mechanisms of Peptide Actions on Neurons

Leo P. Renaud

Division of Neurology, Montreal General Hospital, Montreal, Quebec H3A 2B4, Canada

The notion that polypeptides may mediate or modify synaptic transmission in the central and peripheral nervous systems has been entertained for more than 30 years. The techniques and approaches that were developed to determine the effects of low molecular weight endogenous neural substances, considered to be the "classical neurotransmitters", have now carried over into efforts directed towards understanding the mechanisms of action of larger molecules, i.e., peptides.

Observations on specific receptor binding of drugs and chemicals to cell membranes has promoted the development of criteria for the functional and anatomic characterization of peptide receptors in a variety of neural and non-neural tissues. The potential scope of receptor classification now extends from a description of factors that determine peptide binding at the molecular level to explanations of the pathophysiology in certain human disease states. In the area of opiate receptors alone, one can now differentiate a pharmacological and functional variety of receptors, of which only some are involved in analgesia, while others may be involved in non-analgesic functions.

Examination of possible mechanisms and consequences of peptide receptor interactions is perhaps best approached with electrophysiologic techniques. In vivo studies have demonstrated that peptides can modify neuronal excitability, that their effects are site and neuron specific and that they may behave not only as neurotransmitters but may also act through non-conventional mechanisms that are poorly understood, which for the present are aptly described as "neuromodulatory". In part due to the frustrations associated with the complexity of the intact nervous system, and an inability to obtain stable long-term intracellular recordings, neurobiologists have turned to a variety of in vitro preparations (e.g., invertebrate nervous system, myenteric plexus, frog and rat spinal cord, brain slices and tissue culture) in order to obtain a more detailed characterization of peptide neural interactions. The electrophysiologic studies reported by Kelly and Dodd on the responses of hippocampal pyramidal neurons

107

to cholecystokinin (CCK) illustrate the specificity of a CCK receptor, and the differences and similarities of such postulated neural receptors with those on endocrine cells. Differences in the functional responses of some peptides in tissue cultures from different regions are intriguing. For example, somatostatin appears to have a direct post-synaptic inhibitory effect on spinal cord neuronal cultures, whereas there is little evidence for a direct effect of somatostatin on neurons in cortical cell cultures. These differences are likely a reflection of the presence or absence of postsynaptic (and possibly extrasynaptic) receptors on neurons in tissue cultures obtained from different regions. Of great interest is the evidence for an effect of somatostatin on presynaptic terminals in both types of cultures. When these results are compared with the presumed post-synaptic excitatory effects reported for somatostatin in the hippocampal slice preparation, and the depressant or excitatory effects observed in in vivo experiments, one is convinced that much remains to be investigated regarding the neurobiologic action of this particular peptide alone. To this can be added the interesting voltage clamp analysis described by Fischbach and colleagues, which provides a new avenue for investigation by suggesting that certain peptides may effect synaptic transmission through modification of the inward calcium current associated with generation of the action potential.

It is humbling to think that we are only beginning to touch the tip of an iceberg in this dynamic and exciting area of neurobiology.

Neurosecretion and Brain Peptides,
edited by J. B. Martin, S. Reichlin, and K. L. Bick.
Raven Press, New York © 1981.

Principles of Receptor Identification

Jeffrey S. Flier

*Department of Medicine, Harvard Medical School and Beth Israel Hospital,
Boston, Massachusetts 02115*

INTRODUCTION

For a ligand to produce its biologic effect, it must first bind to its specific receptors on the target cell (23). The receptor serves two major functions. First, it must recognize the biologically active ligand by binding it. Second, the interaction of the ligand with its receptor must initiate the characteristic biochemical consequences of the specific ligand. In addition to these fundamental actions of recognition and activation, ligand-receptor interactions may also have a variety of other biologic functions (4). The hormone-receptor interaction may be an important determinant of hormone degradation (26), and may regulate receptor concentration and receptor affinity as well (5). Further, binding of the hormone to its receptor may lead to cross-linking and translocation of the hormone receptor complex to one or more compartments within the cell (13); it may also tonically regulate the activity of a variety of post-receptor events. On a more speculative level, studies of the insulin receptor suggest that the receptor may also serve as a reservoir for intact hormone which may be released into the plasma, thereby serving a function analogous to the interaction of steroid and thyroid hormones with circulating binding proteins (29).

Although both hormones and receptors were clearly conceptualized in the first decade of this century, the study of hormones advanced much more rapidly during the next sixty years. In large measure, this may be viewed as a consequence of the fact that membrane receptors are more difficult to study. Thus, unlike most hormones, receptors are not highly concentrated in a localized region of the body, are not soluble in aqueous solvents, and produce no biologic effect when introduced <u>in vivo</u> and <u>in vitro.</u> The first attempts to measure hormone receptors directly, reported thirty years ago, were unsuccessful. This is probably a consequence of the fact that labelled hormone was biologically inactive, degradation of both hormone and receptor

109

were excessive, and the demonstrated binding was not clearly shown to be relevant to any biologic consequence of the hormone.

The modern era of receptors began in the 1960s, when studies were aimed at overcoming these problems. Current methods for detecting cell surface receptors are based on the pioneer methods for studying ACTH and angiotensin binding to their specific receptors (17,18). In these studies, radioactive ligand at high specific activity was shown to retain biologic activity and attempts were made to minimize non-specific binding and degradation of both hormone and receptor. Most importantly, the binding of labelled hormone was shown to possess the precise specificity of the biologic system, with bioactive analogues demonstrating binding affinity in proportion to their biologic potency and irrelevant materials having no effect. These techniques have proven widely applicable to other peptide hormones and catecholamines, as well as to neurotransmitters, prostaglandins, lectins, toxins, lipoproteins, microbes, and other agents that have cell surface receptors. A number of additional techniques are now available to detect receptors, in addition to these widely applied methods employing radiolabelled hormone. These include the use of specific antibodies to receptors (8), EM autoradiography (1), and studies with ferritin-labelled and fluorescent-labelled ligands (9).

Demonstration of a Receptor

The demonstration that a hormone binding site is indeed a receptor ultimately requires that it be purified and shown to result in a typical physiologic response after binding the hormone in a carefully defined system. At the present time, we must settle for much more indirect demonstrations. The primary method rby which a hormone binding site is shown to be a receptor is the demonstration of a correlation between hormone binding parameters and one or more biologic effects. Two other powerful tools for the demonstration that a measured binding site is a biologically relevant receptor have derived from studies of hormone receptors in disease states. Thus, in several diseases wherein hormone effectiveness is severely diminished, hormone receptors on target cells have been shown to be specifically and markedly reduced in concentration. In addition, the demonstration that specific antireceptor antibodies or receptor antagonists of other kinds can abolish hormone responsiveness has been used to demonstrate thre biological relevance of hormone binding sites. Despite the indirect nature of these approaches, convincing demonstrations of many receptors have been achieved.

FUNCTIONAL CHARACTERISTICS OF RECEPTORS

Receptors for peptide hormones, catecholamines, neurotransmitters, and prostaglandins have certain functional characteristics in common (10). These include: a) primary localization on the plasma membrane of the cell; b) limited number (i.e., saturable) and high affinity; c) characteristic relationship between receptor binding and the biological response. In recent years, a great deal has been learned about each of these functional characteristics; in many cases, they are now known to be considerably more complex than was initally believed.

Subcellular Localization of Receptors

It is clear from many independent lines of evidence that peptide hormones first interact with receptors on the outer surface of the plasma membrane. Indirect studies favoring a plasma membrane location have included the demonstration of rapid reversibility of hormone effect using antihormone antibodies, and an inhibition of hormone action by brief treatment of intact cells with proteolytic enzymes. Direct visualization of hormone binding with EM autoradiography (1) and studies employing anti-receptor antibodies (2) also support this notion. In addition, early studies of subcellular fractionation revealed significant enrichment of hormone binding activity in plasma membrane fractions (10). Although initially ignored, it is now clear that receptors for insulin and other peptides are also present on intracellular organelles including golgi, rough and smooth endoplasmic reticulum, and probably nuclear membrane (7). In the case of insulin receptors, the intracellular receptors are very similar to the plasma membrane receptors in their specificity and in most other characteristics. Whether the intracellular receptors are merely precursors for plasma membrane receptors, or whether they have in addition a specific role in the mediation of hormone action is unknown at this time.

Reversibility of Ligand Binding

Initially, the hormone-receptor interaction appears to be fully reversible. This reversibility may be maintained for long periods of time when the binding interaction is carried out at lower temperatures. In several systems studied at 37°, some hormone molecules appear to become irreversibly bound or are only slowly reversible from the receptor site (11). Concurrently, rsome hormone can be observed at sites internal to the plasma membrane (1). Much of this non-membrane bound radioactivity is found in association with lysosomes, but a substantial fraction may be recovered intact. It is speculated that these transfers may play a role in the degradation of hormone and receptor, as well as in the initiation and termination of hormone action at the target cell. Definitive answers regarding the functional correlates of these processes are not available at this time.

Number and Affinity of Receptor Sites

Biologically relevant receptors are of limited number (i.e. are saturable), and are of high affinity. A major advance in this area, however, has been the demonstration that both the con-

centration and affinity of receptors are variables which are under complex and specific metabolic control. A commonly observed
pattern of receptor regulation is the ability of a hormone to decrease the levels of its own receptor. This phenomenon, first
demonstrated by Roth and colleagues with the insulin receptor,
was referred to as down regulation. In other systems, it is described as desensitization and tachyphylaxis. This phenomenon is
agonist dependent, and may be due to a variety of different cellular mechanisms. The process of insulin-induced receptor loss
requires energy, and involves changes in the rate of receptor
synthesis or degradation (5,15). In other systems, such as β-
adrenergic desensitization, such phenomena as irreversible binding or internalization of hormone receptor complexes may be involved (20). Whatever the mechanism, this phenomenon serves as a
feedback loop to modulate the effect of continuous presence of
hormone. In other situations, a ligand may increase the concentration of its own receptors (21), or alter the concentration of
receptors for another ligand (28). A variety of conditions may
also influence the affinity of receptor binding. The insulin receptor has been particularly well studied in this respect. Table
1 is a partial listing of biologically relevant factors known to
influence either the number or affinity of insulin receptors.

TABLE 1.   Biologically relevant factors which influence insulin
                          receptors

1.   Hormones
         Insulin, glucocorticoids
2.   Physiologic states
         Exercise
         Diet - timing, composition
         Age, cell cycle
3.   Other factors
         Cyclic AMP
         Ketones
         pH
4.   Drugs
         Sulfonylureas
5.   Disease mechanisms
         Down regulation
         Autoimmune
         Genetic

   Thus, the responsiveness of the target cell,like plasma hormone concentration, must be viewed as a potentially variable
entity. These receptor fluctuations can contribute in a major
way to physiologic control, as well as to the pathogenesis of
disease and refractoriness to pharmacologic therapy.

Specificity of Hormone Binding

Ideally, each hormone should be a unique entity capable of binding to a unique set of receptors, thereby conferring absolute specificity on the system. In many cases, a given hormone is not totally specific either chemically or in its receptor-binding characteristics. Thus, a hormone may bind strongly to its own receptors but, in addition, bind with low affinity to receptors of a closely related hormone. This may result from the fact that the total group of hormones appears to be divided into a smaller number of hormone families, most likely evolved from an extinct group of primordial hormones. Presumably, the weak affinity of one hormone for receptors of another hormone is usually not of much importance. However, when the concentration of a hormone is markedly increased, it may now act to stimulate the second pathway in addition to stimulation of its own pathway. An example of this phenomenon is the ability of insulin at high concentration to occupy receptors for somatomedin, and thereby to have growth promoting actions (19). In addition, mineralocorticoid receptors may be occupied to a significant degree by cortisol, with resulting sodium retaining effects (16). The extent to which such "specificity spillover" occurs at physiologic hormone concentrations is not known, but a potential role for such mechanisms in the generation of disease states would appear to be significant. Such phenomena may also raise important considerations for the design of pharmacologic therapy. An example would be the unwanted effect of spironolactone administered for its anti-mineralocorticoid effect to occupy and inhibit androgenic receptors (4).

Relationship Between Receptor Occupancy and Target Cell Activation

A comparison of hormone binding and hormone effect under the same conditions may yield a variety of results. In some cases (*i.e.*, AIB transport in thymocytes in response to insulin aldosterone producton in adrenal cells in response to angiotensin II) (6,25), hormone binding and response can be closely correlated over the entire range of receptor occupancy. In many other cases, the dose response for bioeffect is markedly more sensitive than the dose response for binding, and complete bio-response is evoked by occupancy of a small portion of the available receptors. When this situation occurs, it is frequently said that there are "spare receptors". In this case, one of the post-receptor events limits the response to receptor occupancy. Much evidence indicates that spare receptors are not truly spare, or useless, however, Firstly, the degree of spareness observed in a given target tissue in response to a given hormone varies depending upon the specific biolgoical effect which is measured. Thus, whereas there are many spare receptors for stimulation of glucose oxidation by insulin in adipocytes, stimulation of amino acid transport by insulin in the same cell has few, if any, spare

receptors (14). Analysis of the law of mass action as applied to
the hormone-receptor interaction suggests that the existence of
spare receptors should alter three separate response characteristics of the target cell. Thus, "spare receptors" increase the
rate of response to the hormone, the cellular sensitivity to low
concentrations of the hormone, and the duration of response to a
high concentration of hormone when compared to the response to a
low concentration of hormone. In addition, the existence of
spare receptors creates a situation in which reduction of receptor concentration results in a leftward shift in sensitivity to
hormone without reducing maximal hormonal effect.

RECEPTOR DISEASES

The study of receptors in disease states has been valuable for
at least two reasons. First, the demonstration that certain
conditions of hormone resistance are associated with severe
receptor deficiency, strongly supports the notion that the receptors which we measure are necessary intermediates in the scheme
of hormone action. Second, studies of receptors in disease have
lead to the elaboration of several new disease states in which
novel pathogenetic mechanisms are at work. As seen in Table 2,
receptor diseases can be divided into five distinct categories.

TABLE 2. Outline of receptor disease

---

1. "Down Regulation" - Receptor concentration reduced due to
   exposure to high levels of hormone itself.

2. Heterologous hormone - One hormone regulates receptor for
   another hormone.

3. Autoantibodies to receptor - may be agonists or antagonists.

4. Genetic - abnormal concentration or structure.

5. Specificity spillover - Receptor stimulated by high levels
   of a ligand which does not bind to receptor with sufficient
   affinity to stimulate receptor under normal circumstances.

In the first category, deficiency of receptors is accounted for
through the process of down regulation due to high circulating
levels of hormone. The best example is the insulin resistance of
obesity (9). The second mechanism involves modulation by one
hormone of the receptor for another hormone. An excellent example is modulation of insulin receptor affinity by both glucocorticoid excess and deficiency (12). Another example is regulation by thyroid hormone of concentration of beta-adrenergic
receptors (27). A third category involves development of autoantibodies against membrane receptors (3). Three well defined
situations in which this mechanism occurs are the antibodies to

insulin receptors in the syndrome of insulin resistance with acanthosis nigricans, antibodies to the TSH receptor in Graves' disease, and antibodies to the acetylcholine receptor in myasthenia gravis. A fourth category is genetic disorders of the receptor, such as is seen in some cases of leprechaunism in which insulin receptors are virtually absent (22). A final category of receptor disease is that which is consequent to the imperfect specificity which exists within families of closely related hormones. Perhaps the clearest example is the situation in which certain tumors produce large amounts of somatomedin-like substance which bind to and activate insulin receptors, causing severe hypoglycemia (19).

ACKNOWLEDGEMENTS

The author wishes to acknowledge Dr. Jesse Roth for his central role in the development of many of these concepts.

REFERENCES

1. Carpentier, J.L., Gorden, P., Barazzone, P., Freychet, P., Le Cam, A., and Orci, L. (1979): Proc. Natl. Acad. Sci. USA, 76: 2803-2807.
2. Flier, J.S., Kahn, C.R., Jarrett, D.B., and Roth, J. (1977): J. Clin. Invest., 60:784-794.
3. Flier, J.S., Kahn, C.R., and Roth, J. (1979): N. Engl. J. Med., 300:413-419.
4. Funder, J.W., Mercer, J.,Hood, J. (1976): Clin. Sci. Mol. Med., 51:(3):333s-334s.
5. Gavin, J.R., Roth, J., Neville, D.M., De Meyts, P.D., Buell, D.N. (1974): Proc. Natl. Acad. Sci. USA, 71:84-88.
6. Goldfine, I.D., Gardner, J.D., Neville, D.M. (1972): J. Biol. Chem., 247:6919-6926.
7. Goldfine, I.D., and Smith, G.J. (1976): Proc. Natl. Acad. Sci. USA, 73:1427-1431.
8. Jarrett, D.B., Roth, J., Kahn, C.R., Flier, J.S. (1976): Proc. Natl. Acad. Sci. USA, 73:4115-4119.
9. Jarrett, L., and Smith, R.M. (1977): J. Supramol. Struct., 6: 45-59.
10. Kahn, C.R., Neville, D.M., Roth, J. (1973): J. Biol. Chem., 248:244-250.
11. Kahn, C.R. (1976): J. Cell Biol., 70:261-286.
12. Kahn, C.R. and Baird, K. (1978): J. Biol. Chem., 253:4900-4906.
13. Kahn, C.R., Goldfine, I.D., Neville, D.M., Jr. and De Meyts, P.D. (1978): Endocrinology, 103:1054-1066.
14. Kahn, C.R., Baird, K.L., Jarrett, D.B., and Flier, J.S. (1978): Proc. Natl. Acad. Sci. USA, 75:4209-4213.
15. Kono, T., Barham, F.W. (1971): J. Biol. Chem., 246: 6210-6216.
16. Kosmakos, F.C., and Roth, J. (1976): Endocrinology, 98:69.

17. Lan, N.C., Matulich, D., Biglieri, E.G. and Funder, J.W. Cir. Res., (in press).
18. Lin, S.Y., and Goodfriend, T.L. (1970): Am. J. Physiol., 218: 1319-1328.
19. Lefkowitz, R.J., Roth, J., Pricer, W., and Pastan, I. (1970): Proc. Natl. Acad. Sci., USA, 65:745-752.
20. Megyesi, K., Kahn, C.R., Roth, J., Gorden, P. (1974): J. Clin. Endocrinol. Metab., 38:931-934.
21. Mukherjee, C., Caron, M.G., and Lefkowitz, R.J. (1975): Proc. Natl. Acad. Sci., USA, 72:1945-1949.
22. Posner, B.I., Kelly, P.A., Friesen, H.G. (1974): Proc. Natl. Acad. Sci. USA, 71:2407-2410.
23. Schilling, E.E., Rechler, M.M., Grunfeld,C. and Rosenberg, A.M. (1979): Proc. Natl. Acad. Sci. USA, 76:5877-5881.
24. Roth, J. (1973): Metabolism, 22:1059-1073.
25. Roth, J., Lesniak, M.A., Bar, R.S., Muggeo, M., Megyesi, K., Harrison, L.C., Flier, J.S., Wachslicht-Rodbard, H. and Gorden, P. (1979): Proc. Soc. Exp. Biol. Med., 162:3-12.
26. Saltmann, S., Fredlund, P., Catt, K.J. (1976): Endocrinology 98:894-903.
27. Terris, S. and Steiner, D.F. (1975): J. Biol. Chem., 250: 8389-8398.
28. Williams, L.T., Lefkowitz, R.J., Wataube, A.M., Hathaway, D.R., and Besch, H.R., Jr. (1977): J. Biol. Chem., 252:2787-2789.
29. Zeleznik, A.J., Midgley, A.R., Reichert, L.E., Jr. (1974): Endocrinology, 95:818-825.
30. Zeleznik, A. and Roth, J. (1978): J. Clin. Invest., 62:1363-1374.

Neurosecretion and Brain Peptides,
edited by J. B. Martin, S. Reichlin, and K. L. Bick.
Raven Press, New York © 1981.

Type 1 and Type 2 Opiate Receptor Distribution in Brain—What Does It Tell Us?

Candace B. Pert

*Section on Biochemistry and Pharmacology, Biological Psychiatry Branch,
National Institute of Mental Health, National Institutes of Health,
Bethesda, Maryland 20205*

EVOLUTION OF METHODS: FROM "GRIND AND BIND" TO IN VITRO AUTORADIOGRAPHY

After several months of negative trials, on September 22, 1972, I chose the fortunate combination of variables which permitted the detection of specific binding sites for opiates in nervous tissue. Scarcely three weeks later in the first crude "grind and bind" analysis of a drug or neurotransmitter receptor distribution, I dissected a fresh rat brain into six regions and homogenized each in an equivalent volume of buffer according to weight, in order to compare the quantity of radioactive naloxone bound stereospecifically to each region, i.e., readily displaceable by a low (10^{-6} M) concentration of the potent opiate analgesic levorphanol while unresponsive to the identical concentration of dextrorphan. The latter is an inert mirror-image enantiomer that has the wrong shape for attaching to opiate receptors (8), for competing for [^{3}H] naloxone binding or for eliciting any of the myriad opiate pharmacological phenomena.

The results of this experiment revealed a strikingly heterogeneous--if incomprehensible--distribution of opiate receptor binding sites in rat brain: "striatum" was almost 8 times higher than "brainstem", while cerebellar levels were undetectable. "Midbrain", "diencephalon" and "cortex" were about equivalent with intermediate values. This finding survived replications with rat after rat and was duly reported (24) in our first paper describing opiate receptor binding.

Since that first experiment concerning opiate receptor distribution in brain which was carried out nearly eight years ago, My colleagues and I have devoted a substantial portion of our research efforts to this subject, first dissecting monkey and later human brain into ever smaller and more neuroanatomically

117

defined pieces, then developing increasingly refined, autoradiographic methods for studying receptor distribution. After so
much effort, what have we really learned? What does the distribution of opiate receptors in brain mean? I am convinced that
careful study of the patterns of opiate receptor distribution can
reveal much about brain organization and function. The preparation of this chapter provides a unique opportunity for sharing
some "notions" which have been sufficiently compelling to guide
the directions of my own research.

Early attempts at deciphering the significance of the observed
heterogeneity concentrated on seeking similarities to the
markedly heterogeneous distributions of previously described
neurotransmitters and their synthetic and degradative enzymes.
When numerous statistical analyses failed to yield significant
correlations, however, we were forced to argue against "an exclusive association with cholinergic, serotonergic or noradrenergic
neurons" (15). Then, at a neurosciences research program held in
Boston in 1974 (21), John Hughes and Lars Terenius reported the
first evidence that a morphine-like peptide of yet-to-be determined structure could be extracted from brain. They reported
that the regional distribution of morphine-like material--highest
in striatum, low in brainstem, absent in cerebellum and intermediate elsewhere--was strikingly concordant with opiate receptor
distribution! Obviously, opiate receptors were the target sites
where "opiatergic" neurons directed their secretion, and a new
neurotransmitter candidate, which Hughes called "enkephalin", was
the endogenous ligand for opiate receptors.

Mapping opiate receptor distribution in fine detail took on
new importance. We were no longer studying merely a "brain component whose identification would shed light on the mechanism of
opiate action," but mapping target areas where occupancy with the
"brain's own morphine" could modulate brain function even in the
absence of exogenous opiates. With this in mind, Michael Kuhar
and I developed a method for visualizing opiate receptors in rat
brain by autoradiography (27). Radioactive diprenorphine, an
extremely potent opiate, accumulated selectively and stereospecifically in the vicinity of opiate receptor-rich areas one hour
after intraperitoneal injection of trace doses were given to
rats. Although diprenorphine bound to opiate receptors with a
very high affinity (10^{-10} M), the binding was reversible and so
could not survive traditional high resolution "wet" autoradiographic procedures which involved dipping slides with brain sections
into liquid radio-sensitive emulsion and waiting several hours
for them to dry. The solution was to adopt the "dry" autoradiographic methods devised by Roth and Stumpf (35), which involved
thaw-mounting frozen sections from brains of [^{3}H] diprenorphine-
injected rats onto pre-dried emulsion-coated slides.

The method was exceedingly laborious and "pressure artifacts"
arose with frustrating frequency. Still the first real "look" at
opiate receptor distribution revealed exciting features that
"grind and bind" was incapable of detecting. Striatal opiate
receptors were found in extraordinarily dense "patches" with low

levels between them and in a dense "streak" which lined the underside of the corpus callosum at every level (27,28). Opiate receptor values for whole homogenized spinal cord were low and unremarkable, but autoradiographic visualization revealed a dense streak in the substantia gelatinosa (27,1) where the afferent neurons of the dorsal root ganglia carry somatosensory information from the periphery to the first synapse in the central nervous system.

During the past two years Miles Herkenham, in the Neurophysiology Laboratory of the NIMH, and I have worked to perfect a technique for visualizing opiate and other brain receptors with high resolution at the light level (10). The method uses fresh-frozen, unfixed tissue and features intact cell morphology. Opiate receptors are labeled by incubations of slide-mounted sections in vitro, so that opiate receptor subtypes on adjacent sections can be labeled under different conditions or with different ligands for comparisons. Traditional "wet" autoradiography with its high resolution and reduced inclination toward artifacts is used to visualize binding sites after fixation of tritiated ligands by exposure for two hours to hot (80°C) paraformaldehyde vapors. The vapors do not form selective covalent bonds between tritiated ligands and receptors, but rather copiously cross-link all free amino groups with "molecular chickenwire," not unlike a finishing coat of varnish that seals in radioactive ligands and prevents their diffusion. Some results obtained by this new method will be described later, after first considering the issue of "multiple opiate receptors."

OPIATE RECEPTOR SUBTYPES: μ and δ

Electrically-induced contractions of both the guinea pig ileum and mouse vas deferens in vitro can be suppressed in a dose-dependent, naloxone reversible fashion by opiates. However, with the discovery of the structure of enkephalin (12) and the synthesis of literally hundreds of analogs, it became apparent that these two smooth muscle preparations display quite different sensitivities to the various opiate ligands. Kosterlitz termed the ileal receptor in which morphine is more potent than leu-enkephalin "μ" and termed the vas deferens receptor in which leu-enkephalin is more potent than morphine "δ" (after deferens) (18). Martin (20), on the basis of experiments with unusual mixed agonist-antagonist opiates in spinal dogs, had, in fact, ten years previously devised a tripartite opiate receptor sub-classification scheme which featured a "μ" receptor displaying a ligand selectively so similar to the ileum that Kosterlitz's adoption of the μ nomenclature was, in fact, based on these experiments. The two other opiate receptor subtypes in Martin's scheme, unfortunately, do not mesh with Kosterlitz's classification, a finding that may be confounded by the high affinity that benzomorphans have for another receptor, the phencyclidine or "angel-dust" receptor (40).

Since the mouse vas deferens and guinea pig ileum have become
the de rigeur bioassay "screens" of the pharmaceutical companies,
they have been by far the most extensively studied in terms of
their opiate ligand selectivities. However, a number of other
peripheral opiate receptors present in the rat vas deferens, the
rabbit central ear artery and the cat nictitating membrane, to
name just three, have been sufficiently well-characterized that
it is obvious that their ligand selectivity is neither "μ" nor
"δ" (see, for example ref. 14). It seems intemperate to invoke a
new opiate receptor subtype for every opiate-responsive tissue
with a distinct ligand selectivity pattern, especially since
opiate ligand selectivity patterns in a given tissue can change
from species to species.

The second distinguishing difference between the μ ileal
opiate receptor and the δ deferens receptor (18) is the concentration of naloxone required to antagonize opiate effects in
each: the δ deferens receptor is about 10-fold less sensitive to
naloxone's potency as an antagonist than the μ ileal receptor.
Opiate receptors unfortunately fail to fall neatly into two
groups by this criterion as well. Opiate agonist effects at
their receptors are reversed by naloxone over a very broad spectrum of antagonist potencies which differs dramatically depending
upon which opiate receptors are being considered (20). For example, opiate receptors in neuroblastoma glioma cells require 10
μM naloxone to antagonize stereospecific opiate agonist effects on
cyclic AMP production (33). At the other end of the spectrum,
mice which have been chronically implanted with morphine pellets
become progressively more exquisitely sensitive to naloxone's
ability to precipitate withdrawal signs, an anti-morphine effect
elicited by less than 0.1 nM concentration of naloxone in brain
(37). With such an enormous (100,000 fold!) range of sensitivity
to opiate antagonists already well documented, perhaps a stereo-
selective opiate receptor displaying "complete" naloxone resistance should not be ruled out.

TYPE 1 AND TYPE 2 OPIATE RECEPTORS

Spurred by the hope of finding a more universally applicable
subclassification scheme which would reflect the basic biochemical components composing opiate receptors, Duncan Taylor, Agu
Pert and I compared the opiate receptor binding properties of
very small "punches" of rat brain derived from periaqueductal
gray and the gigantocellular region of the pontine reticular
formation [brain regions known to mediate antinociception in the
rat (23)] to other punches from assorted regions. Since
analgesic opiate receptors from these brain regions had been very
well characterized behaviorally and had been shown to display a
ligand selectivity pattern and marked naloxone sensitivity not
unlike   opiate receptors in the ileum, we sought incubation
variables which could discriminate analgesic, "μ-oid" receptors
from other, putative receptor subtypes in brain. Under the
conditions we developed, opiate receptors fell clearly into two

subgroups, GTP-sensitive and GTP-insensitive (30). Moreover, the ligand selectivity pattern of these two subtypes determined in binding experiments was clearly different: morphine was twice as potent in displacing [³H] diprenorphine from GTP-sensitive opiate receptors in the midbrain, while leu-enkephalin was four times as potent on GTP-insensitive receptors in the frontal cortex (30).

The effect of GTP on a number of membrane-bound receptors has been studied extensively in numerous laboratories in highly refined biochemical experiments (32). Receptors (i.e., recognition units) have been shown to float laterally through membranes (34), "coupling up" to their "effectors" [e.g., adenylate cyclase (6)] via "GTP regulatory subunits," floating protein "couplers" that can link a recognition unit to an effector in a stimulatory or inhibitory mode. Thus, our subclassification scheme based upon differential GTP sensitivity might reflect the relative ability of opiate receptors to "couple up" to other membrane components.

Is it possible that apparent "multiple opiate receptors" are the result of a variable deformation of the recognition unit depending on the closeness of its coupling with other membrane components? Studies with other membrane-bound receptors raise the possibility that, at the molecular level, there may be only one opiate receptor after all. The membrane-bound muscarinic receptor exhibits differing ratios of "subtypes" with different ligand selectivity patterns in each brain region, but the purified muscarinic recognition unit appears homogeneous (2). The ligand selectivity pattern of the dopamine receptor changes depends upon whether it is coupled to adenylate cyclase (Type 1) or uncoupled (Type 2) (13). The nicotinic cholinergic "receptor" consists of circular clusters of chemically homogeneous subunits which display differential ligand selectivity patterns and sodium conductance efficiencies depending on how closely they are packed together (7).

DISTRIBUTION OF TYPE 1 AND TYPE 2 OPIATE RECEPTORS IN RAT BRAIN VISUALIZED BY IN VITRO AUTORADIOGRAPHY

Even though future biochemical experiments may well reveal a single opiate receptor recognition unit, we have been able to visualize two clearly different patterns of opiate receptor distribution in adjacent sections of rat brain depending upon the ligands and conditions used (3). Opiate alkaloid-like [³H]morphine and [³H]naloxone with their higher affinity for Type 1 receptors are prone to produce the typical Type 1 pattern of striatal islands and clusters shown in Fig. 1. On the other hand, we have been able to use [³H] diprenorphine, [³H] opiate peptides, and even [³H] morphine to produce the typical Type 2 striatal pattern (31) of rather diffuse, even labeling of the striatum. Critical variables for visualizing Type 1 or Type 2 receptors are the presence or absence of GTP (2 μM) and NaCl (50 mM) in the incubation medium as well as the incubation temperature employed (5). The inclusion of sodium ion and GTP in the

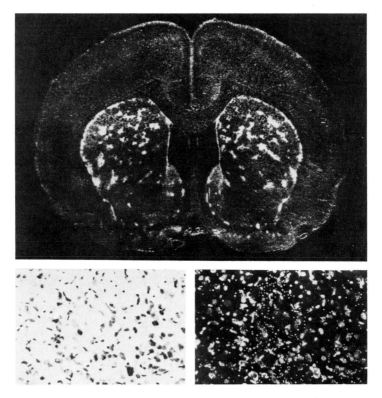

FIG. 1.   (Top) Low-power (x 5.4) darkfield photomicrograph of an
autoradiogram (6 weeks exposure) of a [³H]naloxone-incubated rat
brain section at the level of the caudate and accumbens nuclei.
Opiate receptor-rich areas appear as white patches and streaks.
More diffuse labeling appears as lighter shades of gray. (Bottom)
High-power (x 80) brightfield (left) and darkfield (right)
photomicrographs of a striatal section exposed to the emulsion
for 6 days.   In both high-power views, normal morphology of
neurons and glia is seen in relation to fiber fascicles that
appear as clear areas surrounded by the neurons.   In darkfield,
the silver grains appear as small white dots, situated over cells
and neuropil, composing a discrete cluster in the center of the
field.   No grains overlie the fiber fascicles (From ref. (9),
with permission).

incubation medium selectively reduces the binding of opiate
agonists to Type 1, GTP-sensitive receptors, and thus results in
a Type 2 pattern whether [³H] morphine (10 nM) or [³H] D-Ala-D-
Leuenkephalinamide (10 nM) is employed.   In the absence of GTP
and sodium ion, however, the same concentrations of [³H]morphine
and [³H] D-Ala-D-Leu-enkephalinamide give rise to the clustered
Type 1 striatal pattern shown in Fig. 1.

Not only in the striatum, but at other levels of rat brain, Type 1 opiate receptor distribtution patterns have sharp, discrete borders, while Type 2 receptors are rather diffuse. For example, the sharp streak in the substantia gelatinosa of the spinal cord is composed of Type 1 opiate receptors, while Type 2 conditions label the entire dorsal horn with a diffuse gradient which is densest at the most superficial laminae (31). At the midbrain and frontal cortex levels of brain, shown in Fig. 2, Type 1 and 2 receptors show many similarities and one striking difference. The similarities of Type 1 and Type 2 binding include the overall comparable density (provided by the areas under the curves), the relatively greater density in limbic (medial) as opposed to sensorimotor (lateral) cortex, and the peaks of high density in the deepest cortical layers. The remarkable difference is the existence of the dense band of Type 1 receptors-- absent under Type 2 conditions--localized to the outer edge of layer 1 of frontal cortex. This exclusive accumulation of Type 1 receptors in layer I can be followed caudally along the medial wall into cingulate cortex and is visible in Fig. 1. Regional variations in receptor subtype densities are also apparent in the midbrain (Fig. 2). The independence of Type 1 and Type 2 binding is revealed in the ratios of counts taken from identical loci in the adjacent sections. These ratios range from unity in regions containing mutually low or nonexistent receptors [e.g., the pontine nuclei (PN) and the trigeminal nerve (V)] to a ratio orf 5.3 in the periaqueductal gray (PAG), where there are moderate levels of Type 1 binding and virtually no Type 2 binding. Not only the PAG, but several other brain loci implicated in analgesic mechanisms show a characteristic high Type 1:Type 2 opiate receptor ratio (31). Note that for most of the midbrain, Type 1 binding is two to three times as great as Type 2 binding, whereas in frontal cortex the levels of binding are equivalent for both types. Comparing frontal cortex to midbrain for the same receptor subtypes, Type 1 receptors appear equally dense (or slightly more dense in midbrain regions), whereas Type 2 receptors are more dense in frontal cortex than in all but discretely small areas of midbrain [such as the interpeduncular nucleus (IPN)].

FROM TYPE 1 OPIATE RECEPTORS TO OPIATERGIC NEUROCIRCUITRY

In contrast to Type 2 opiate receptors in rat brain with their selective but rather diffuse distribution, Type 1 opiate receptors form a number of very dense clusters with highly discrete edges. These clusters are striking and reproducible from one rat to another. To the extent that postsynaptic receptors are selectively labeled by Type 1 conditions, these receptor patterns may bear striking resemblances to patterns of terminal endings of neural pathways marked by radiolabeled amino acids injected into nuclei of living rats and transported intra-axonally to their endings. Concordances noticed between receptor distributions and tract terminal distributions, visualized by the appropriate autoradiographic methods to demonstrate them, allow us to surmise

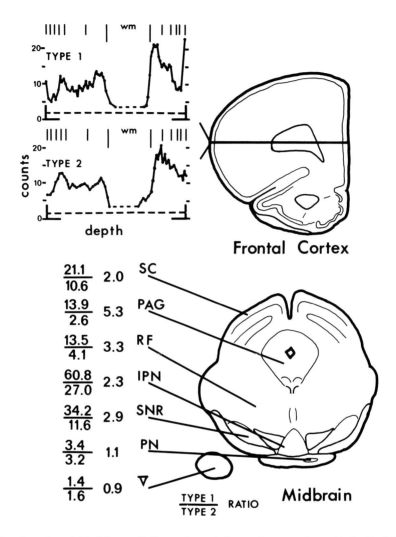

FIG. 2. Quantitation of Type 1 and Type 2 receptor distribution determined by counts of autoradiographic grain density in two regions of rat brain: frontal cortex and midbrain. Adjacent sections from the same brain were incubated in either [³H] naloxone (50,000 cpm/ml) under Type 1 conditions as described (9), or [³H]D-Ala²-D-Leu-enkephalins (50,000 cpm/ml) under Type 2 conditions (25°C, 30 min. in the presence of NaCl (50 mM) and GTP (2 μM)). After washing and drying, the sections were fixed, defatted in xylene, autoradiographed, developed and stained together. Counts of grain density were made at 500 x magnification under darkfield illumination using an eyepiece reticule. For each determination, grains were totaled within grid squares measuring 28 μm on a side. The numbers represent the average of ten

FIG. 2. Caption continued.

such determinations. After one-month exposure, background counts of grains in the emulsion overlying glass, pia, choroid plexus or the trigeminal nerve (V) for both ligands were remarkably similar, ranging from 1.2 over glass to 2.1 over tissue. The heavy dashed lines in the graphs and the measures over V represent the background determinations, equal for both ligands. At the frontal cortical level counts were made within a horizontal 280 μm-wide strip taken through the limbic prefrontal cortex on the medial wall, the subcortical white matter (wm) and the anterior sensorimotor cortex in the lateral wall. The counts are plotted as a function of depth on the left side. The cortical layers are marked by vertical stripes. Other abbreviations, not mentioned in the text include: superior colliculus (SC), reticular formation (RF) and substantia nigra pars reticulata (SNR).

---

that the labeled tract represents the presynaptic link to these receptors and, therefore, is an "opiatergic" tract. The site of amino-acid injection reveals the source neurons of the opiatergic pathway.

We have, in fact, seen many striking concordances, surmised several opiatergic pathways, and have noticed further that they are often linked sequentially: neural connections appear to interconnect opiate-rich areas, forming a potential five-link opiatergic pathway through olfactory and limbic structures in the rat (Fig. 3). Experiments in which putative opiatergic cell bodies are destroyed by electrolytic lesions and putative opiatergic axonal projections are severed by a knife cut in order to produce the expected depletions of endogenous enkephalin-like material measured by a radioreceptor assay (22) in the putative terminal fields of opiatergic projections are in progress. While already extensive, the opiatergic circuitry which we have proposed thus far (10) is still incomplete; the remainder should yield to a vigorous pursuit by this strategy.

A number of other findings are consistent with the notion that Type 1 opiate receptors mark sites which are postsynaptic relative to opiatergic neurons which synthesize and secrete enkephalins. Only Type 1 opiate receptors show the marked sodium sensitivity which discriminates between agonists and antagonists (26). Only visualization of Type 1 opiate receptors can be disrupted by including GTP and sodium chloride in the incubation medium with tritiated opiate agonists. Only Type 1 opiate receptors are highly dependent on appropriate incubation temperatures [warm for agonists, cold for antagonists (5)] for their visualization. All this is consistent with the notion of a postsynaptic receptor capable of coupling closely to form a sodium ionophore regulating sodium conductance (39) and requiring warm temperatures for membrane fluidity sufficient to permit the floating of agonist-occupied recognition units and GTP-regulatory units to their effectors.

If Type 1 opiate receptors are postsynaptic, does any evidence suggest how Type 2 opiate receptors are distributed at the cellular level? First, preliminary experiments from our laboratory indicate that unilateral kainic acid lesions of rat striatum reduce the content of Type 2 opiate receptors while selectively sparing Type 1 (to be published). These experiments suggest that Type 2 receptors are on cell bodies and/or presynaptic elements of intrinsic striatal neurons. Second, the opiate receptors in mouse vas deferens must be presynaptic since this simple tissue is thought to contain only noradrenergic axons, in contrast to the complex neuronal plexus contained within the excised strips of guinea pig ileum. As has been pointed out, the ligand selectivity of Type 2 receptors determined in binding studies resembles that determined by pharmacological studies in the vas deferens (18). Still, the notion that Type 2 receptors are presynaptic must remain a matter for speculation until visualization of the two receptor subtypes has been accomplished in neuronal cell cultures where axons, cell bodies and points of interneuronal contact are clearly visible.

SPECULATIONS ON INTERCONVERSION OF OPIATE RECEPTOR SUBTYPES

Are the couplers, as Rodbell suggests (32), limiting components so that various recognition units (for opiates, norepinephrine, dopamine, etc.) float upon their shared membrane in a constant struggle for the chance to regulate the effectors? Carpenter (36,38) has, in fact, produced neurophysiological evidence that various neurotransmitter receptors regulate the same ionophore. The interactions between GABA receptors and benzodiazepine receptors are greatly augmented by the presence of anion channel-specific ions in the incubation medium, suggesting that these recognition units share and regulate a common chloride ionophore (4). Can this be the way, i.e., through shared couplers and effectors, that opiates and dopamine produce their interactive effects on motor behavior (29)? Over time, through use and disuse, do the recognition units which are "plugged into" the couplers and effectors shift around, creating now "dopamine receptor supersensitivity", then "tolerance to opiates"? In such a dynamic vision, it is easy to imagine synapses with the whole spectrum of Type 1/Type 2 opiate receptor ratios, each giving rise to a unique ligand selectivity pattern and a unique sensitivity to naloxone. Only such a dynamic model can accommodate the plethora of well documented and varied pharmacological effects produced by exogenous and endogenous opiates.

OPIATE RECEPTORS IN MONKEY CORTEX, AN ORDERLY PROGRESSION

A recent study of over 40 regions dissected from monkey has revealed an amazingly orderly progressive increase in Type 1 receptor density related to sensory brain function (17). Electrophysiological, neuroanatomical and behavioral approaches have revealed that visual, auditory and probably somatosensory

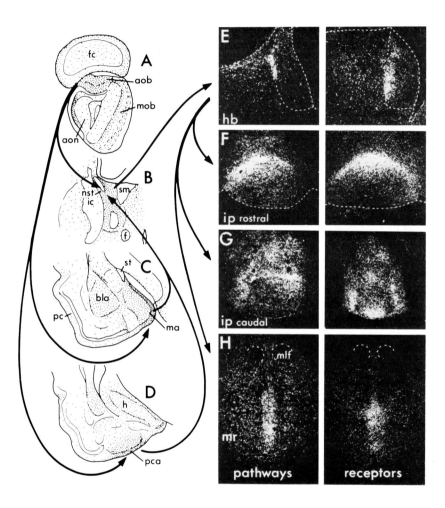

FIG. 3. A-D at left are projection drawings of receptor-labeled
sections at the levels of (A) the frontal pole showing olfactory
structures, (B) the posterior bed nucleus of the stria terminalis
and (C) anterior and (D) posterior levels of the amygdala. Stip-
pling represents opiate receptor distributions in these sections.
Arrows indicate neural pathways, derived from other studies, con-
necting the accessory olfactory bulb (oab) with the bed nucleus

FIG. 3. Caption Continued.

of the stria terminalis (nst) and the medial (ma) and the posterior cortical (pca) nuclei of the amygdala. These amygdaloid
nuclei project to the nst which, in turn, projects to the border
zone between the medial and lateral habenular nuclei (M.
Herkenham, unpublished data). E-F are paired photomicrographs of
two kinds of autoradiography. In pairs shown on the left transport of [^{3}H] amino acids has marked neural pathways from the nst
to the habenula (hb), shown in (E), and from this border zone to
the interpeduncular nucleus (ip) and the median raphe (mr) as
shown in F-H [also illustrated) by Figs. 7 and 16 of Herkenham
and Nauta (40)]. The identical corresponding levels shown in the
right-hand pairs illustate Type 1 opiate receptor distributions
marked by [^{3}H] naloxone. The striking concordances between the
terminal distribution of the nst-hb, hb-ip and hb-mr paths and
the receptor distributions suggest that these may be "opiatergic"
tracts. Other abbreviations in this figure are: aon, anterior
olfactory nuclei; bla, basolateral nucleus of amygdala; f. fornix; h. hippocampus; ic, internal capsule; mob, main olfactory
bulb; mlf, medial longitudinal fasciculus; pc, piriform cortex;
sm, stria medullaris, st, stria terminalis, (From ref. 9, with
permission).

----

information is processed sequentially along a series of cortical
"way-stations." While Type 2 opiate receptors show similar
densities throughout the cortex, Type 1 receptors increase
exponentially along the cortex as the functional hierarchy is
ascended and progressively more complex sensory information is
processed.

We have taken note of the striking association of Type 1 opiate
receptors with primary afferent input zones of sensory neurons
carrying information from all modalities (i.e., sight, sound,
smell, taste and touch) (10). Generally, the distribution of
opiate receptors in mammalian brain is consistent with the notion
that these modulatory components filter all incoming sensory
information along some biophysical dimension that ultimately
becomes perceived as "pleasure-pain." Meanwhile, it shapes which
sensory information is permitted to gain access to our attention,
selecting inputs on the basis of criteria stored in the limbic
circuitry. Thus, the very highest levels of opiate receptors in
brain are in the amygdala and other closely associated parts of
the limbic system where emotional behavior is encoded (19), not
only in monkeys (15,16), but in humans as well (11). It seems
that not merely our decisions, but even our very perceptions are
at the mercy of emotional circuitry selected purely for survival
value.

TABLE 1. Characteristics of Type 1 and Type 2 opiate receptors

Type 1	Type 2
GTP (2 µM) sensitive	GTP (2 µM) resistant
µ-oid ligand selectivity	δ-oid ligand selectivity
Morphine more potent than leu-enkephalin	Leu-enkephalin more potent than morphine
Sodium discriminates between agonist/antagonist	No sodium effect
Warm (25-37°C) incubation temperature required to visualize [³H]agonist binding	Temperature insensitive
Sensitive to opiate antagonists	Insensitive to opiate antagonists
Postsynaptic to opiatergic neurons	Extrasynaptic, presynaptic?
Capacity for coupling to effectors through GTP-sensitive "N" (19) subunit	Lacks "N" subunit coupler
Barely detectable in invertebrates	Predominate in invertebrates
Heterogeneous, discrete distribution	Diffuse distribution
Increases in cortex with increasing functional complexity	No progression, marks ancient limbic areas
Flexible modulation capacity?	Inflexible, hard-wired?

Progressively increasing ratios of Type 1 to Type 2 opiate receptors characterizes not only regions of monkey cortex with increasing functional complexity, but also characterizes the vertebrate as opposed to the invertebrate CNS. Stereospecific [³H] morphine and [³H] naloxone binding (Type 1 receptors) can be demonstrated in numerous vertebrate brains surveyed, but not in invertebrates (25). Recently, Linda Hall and I (unpublished) have shown that membranes derived from the heads of the fruitfly, Drosophila melanogaster, contain Type 2 opiate receptors. Thus, there may also be a phylogenetic progression toward increasing Type 1:Type 2 opiate receptor ratios. Moreover, the Type 2 opiate receptor, since it lacks the GTP-sensitive subunit normally required for membrane modulation via adenylate cyclase regulation, is a reasonable candidate for association with "primitive", "hard-wired", "old" brain areas. The possibility that the evolution of nervous systems of increasing complexity has featured the gradual addition of opiatergic neurocircuitry bestowing enhanced modulatory capacity is a fascinating one.

ACKNOWLEDGEMENTS

I wish to express my gratitude to my colleagues, Agu Pert and
Miles Herkenham, not only for reviewing this manuscript, but also
for providing me with a never-ending behavioral and neuroanatomical education. I also thank Dawn McBrien for expert secretarial
assistance.

REFERENCES

1. Atweh, S.F., and Kuhar, M.J. (1977): Brain Res., 124:53–67.
2. Birdsall, N., and Hulme, E. (1976): J. Neurochem., 27:7–16.
3. Bowen, W., Herkenham, M., and Pert, C.B. In preparation.
4. Costa, T., Rodbard, D., and Pert, C.B. (1979) In: Physi-
    cal-Chemical Aspects of Cell Surface Events in Biological
    Regulation, edited by C. DeLisi and R. Blumenthal, pp. 37–
    52. Elsevier/North Holland, Amsterdam.
5. Creese, I., Pasternak, G.W., Pert, C.B., and Snyder, S.H.
    (1975): Life Sci., 16:1837–1842.
6. Cuatrecasas, P. (1974): Annu. Rev. Biochem. 43:169–214.
7. Giraudet, J., and Changeux, J.P. (1980): In: Trends in
    Pharmacological Science, pp. 198. Amsterdam, Elsevier/North
    Holland.
8. Goldstein, A., Aronow, L., and Kalman, M. (1974): Principles
    of Drug Action: The Basis of Pharmacology, pp. 33. Harper
    and Row, New York.
9. Herkenham, M., and Nauta, W.J.H. (1979): J. Comp. Neurol.,
    187:19–48.
10. Herkenham, M., and Pert, C.B. (1980): Proc. Natl. Acad. Sci.
    USA, in press.
11. Hiller, J.M., Pearson, J., and Simon, E.J. (1973): Commun.
    Chem. Path. Pharmacol., 6:1052–1061.
12. Hughes, J., Smith, T.W., Kosterlitz, H.W., Fothergill, L.A.,
    Morgan, B.A., and Morris, H.R. (1975): Nature, 258:577–599.
13. Kebabian, J., and Calne, D. (1979): Nature, 277:93–96.
14. Knoll, J. (1976): Eur. J. Pharmacol., 39:403–407.
15. Kuhar, M.J., Pert, C.B., and Snyder, S.H. (1973): Nature,
    245:447–450.
16. LaMotte, C., Snowman, A., Pert, C.B., and Snyder, S.H.
    (1978): Brain Res., 155:374–379.
17. Lewis, M.E., Mishkin, M., Bragin, E., Brown, R.M., Pert,
    C.B., and Pert A.: Science, submitted.
18. Lord, J.A.H., Waterfield, A.A., Hughes, J. and Kosterlitz,
    H.W. (1977): Nature, 267:495–499.
19. MacLean, P.D. (1972): Ann. NY Acad. Sci., 193:137–149.
20. Martin, W.R. (1967): Pharmacol. Rev., 19:463–522.
21. Matthysse, S., and Snyder, S. (1975): Neurosci. Res. Prog.
    Bull., 13.
22. Naber, D., Pickar, D., Dionne, R.A., Bowie, D.L., Ewels, B.,
    Moody, T.W., Soble, M.G., and Pert, C.B. (1980): Substance
    and Alcohol Action/Misuse, 1:113.

23. Pert, A. (1978): In: Bases of Addiction, edited by J. Fishman, pp. 299-332. Abakon, Berlin.
24. Pert, C.B., and Snyder, S.H. (1973): Science, 179:1011-1014.
25. Pert, C.B., Aposhian, D., and Snyder, S.H. (1974): Brain Res., 75:356-361.
26. Pert, C.B., and Snyder, S.H. (1974): Mol. Pharmacol., 10:868-879.
27. Pert, C.B., Kuhar, M.J., and Snyder, S.H. (1975): Life Sci., 16:1849-1853.
28. Pert, C.B., Kuhar, M.J., and Snyder, S.H. (1976): Proc. Natl. Acad. Sci. USA, 73:3729-3733.
29. Pert, A. (1978): In: Characteristics and Function of Opioids, edited by J. Van Ree, and L. Terenius, pp. 389-401. Elsevier/North Holland, Amsterdam.
30. Pert, C.B. and Taylor, D. (1980) In: Endogenous and Exogenous Opiate Agonists and Antagonists, edited by E.L. Way, pp. 87-90. Pergamon Press, New York.
31. Pert, C.B., Taylor, D.P., Pert, A., Herkenham, M.A., and Kent, J.L. (1980): In: Advances in Biochemical Psychopharmacology, Neural Peptides and Neuronal Communication, edited by E. Costa, and M. Trabucchi, pp. 581-589. Raven Press, New York.
32. Rodbell, M. (1980): Nature, 284:17-22.
33. Sharma, S.K., Klee, W.A., and Nirenberg, M. (1977): Proc. Natl. Acad. Sci. USA, 74:3365-3369.
34. Singer, S.J., and Nicolson, G.L. (1972): Science, 175:720-731.
35. Stumpf, W.E., and Roth, L.G. (1966): J. Histochem. Cytochem. 14:274-287.
36. Swann, J.W., and Carpenter, D.O. (1975): Nature, 258:751-754.
37. Way, E.L., Loh, H.H., and Shen, F.H. (1969): J. Pharmacol. Exp. Ther., 167:1-8.
38. Yardwsky, P.J., and Carpenter, D.O. (1978): J. Neurophysiol., 41:531-541.
39. Zieglgansberger, W., and Bayerl, H. (1976): Brain Res., 115:111-128.
40. Zukin, S.R., and Zukin, R.S. In: PCP: Historical and Current Perspectives, edited by E. Domino. NPP Books, Ann Arbor, Michigan. (In press.)

Neurosecretion and Brain Peptides,
edited by J. B. Martin, S. Reichlin, and K. L. Bick.
Raven Press, New York © 1981.

Cholecystokinin and Gastrin as Transmitters in the Mammalian Central Nervous System

J. S. Kelly and *Jane Dodd

*Department of Pharmacology, St. George's Hospital Medical School, London SW17 0RE, England; and *Department of Neurobiology, Harvard Medical School, Boston, Massachusetts 02115*

In 1975, Vanderhaeghen, Signeau and Gepts (26) demonstrated the presence of a peptide in nervous tissue which was smaller than gastrin-17 but cross-reacted with an antibody raised against gastrin. Later, Dockray (4), using six different sequence-specific gastrin antisera, suggested that the immunoreactivity present in the brain may represent the C-terminal fragment of the gastrointestinal hormone cholecystokinin (CCK), whose C-terminal pentapeptide sequence is identical to that of gastrin. Similarly, Rehfeld (21), using a number of CCK sequence specific antibodies, was able to show that the peptide present in the brain did indeed appear to be the C-terminal octapeptide fragment of CCK.

Cholecystokinin immunoreactivity appears to account for almost all that previously reported as being due to the presence of gastrin in the central nervous system (22). High levels of CCK immunoreactivity have been found in the neocortex, hippocampus, amygdala, hypothalamus (12,15), and spinal cord (15). In the spinal cord, the location of the immunoreactivity in the substantia gelatinosa is reminiscent of the distribution of Substance P in the fibers thought to be involved in nociception. In the rat hippocampus the CCK immunoreactivity appears to be concentrated in the nerve terminals which innervate the cell body layer and proximal dendrites of the pyramidal cells (15).

In the peripheral nervous system, CCK immunoreactivity is found in mesenteric ganglia, in both Meissner's and Auerbach's plexuses, in the walls of the colon and in the nerves to the pancreatic islet cells.

Gel chromatography of nervous tissue homogenates (15) confirmed the presence of the parent triacontatripeptide CCK-33 and at least 3 C-terminal fragments dodeca (CCK-12), octa-(CCK-8) and tetra-

133

(CCK-4) peptides. The two smallest peptide fragments, CCK-8 and CCK-4, account for 77% of the total immunoreactivity present in nervous tissue. Subcellular fractionation has shown CCK-8 and CCK-4 present in the rat cerebral cortex or hypothalamus to be concentrated in the vesicular fraction (23). In slices of these tissues depolarizing stimuli evoke a calcium-dependent release of the fragments.

On the isolated perfused porcine pancreas, Rehfeld and colleagues (24) have shown the tetrapeptide to be highly potent and a specific releaser of insulin, glucagon, somatostatin and pancreatic polypeptide. They suggest that the tetrapeptide may be the neurotransmitter of the nerves which innervate the pancreatic islet cells.

On the frog spinal cord, Phillis and Kirkpatrick (20) have shown that the application of CCK to the isolated hemisected toad spinal cord causes a tetrodotoxin sensitive depolarization of both roots. On this preparation the CCK affects primary afferent terminals and motoneurones in the same way as Substance P. In the gut (19) caerulein, a CCK analogue, causes a rapid depolarizatin of pancreatic acinar cells which in many ways resembles that evoked by ACh and bombesin. Although only the ACh mediated response is blocked by atropine, all three responses can be mimicked by the intracellular injection of calcium (13). The responses to all three secretagogues have a similar reversal level and latency. Because of the structural similarity between their C-terminal peptide sequences, CCK, gastrin and caerulein (1,17) are believed to share a common receptor on the acinar cell membrane.

The aim of the present study was to determine the actions of CCK-8, CCK-8 less the sulphated group of the tyrosine (8) residue (CCK-NS), CCK-4, gastrin G-13 and G-14 and caerulein on pyramidal cells of the CA1 region of the hippocampus, using intracellular recording electrodes in a hippocampal slice preparation. On the same cell, the characteristics of these responses were compared with the response to glutamate, released from adjacent barrels of the same multibarrelled micropipete (3,14).

In each experiment the tip of the multibarrelled microiontophoretic pipette, containing ACh, glutamate and NaCl, was placed just below the surface of a hippocampal slice in the region of the pyramidal cell body layer, and a second, intracellular micropipette was used to penetrate the soma of a nearby neuron. The intracellular electrode was then used to record changes in membrane potential. Excitability was tested by passing depolarizing ramps of current through the electrode of just sufficient magnitude to evoke one or two action potentials during the control period. Membrane resistance was monitored using rectangular and ramp-shaped hyperpolarizing current pulses of similar amplitude and duration (Fig. 1).

Subsequently, the data was digitalized on a PDP-12 computer and the peak responses to glutamate and the peptides were determined by plotting against time the membrane potential, membrane resistance, the number of spikes evoked by the depolarizing ramp and the latency and threshold of the first spike evoked by the ramp.

The proximity of the iontophoretic pipette to the cell under study was tested by showing the cell to be depolarized and excited by relatively small, brief applications of L-glutamate. In the majority of cells a significant degree of excitation and depolarization could be attained with glutamate-ejecting currents in the region of 50nA applied for 20s.

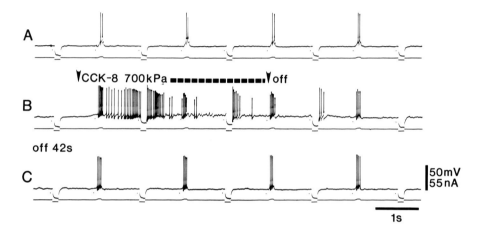

FIG. 1. The excitatory action of CCK-8 on a CA1 pyramidal cell of the isolated rat hippocampal slice preparation. In this and subsequent figures, the upper trace shows the voltage response of the neuronal membrane to the intracellular injection of alternating pulses of current shown on the lower trace. Rectangular hyperpolarizing pulses were used to test the membrane resistance and depolarizing ramps to monitor the excitability of the cell. In B the first downward facing arrow indicates the onset of an application of CCK-8 from the second intracellular microelectrode whose tip was located as close as possible to the intracellular electrode. The broken line and second downward-facing arrow show the duration of the peptide application. Clearly CCK-8 rapidly depolarized and increased excitability of the cell under study and full recovery in this instance was delayed. In record B, the clear-cut decrease in amplitude of the voltage response of the second injection of hyperpolarizing current shows the change in membrane potential and excitability to be accompanied by an increase in membrane conductance.

ACTIONS OF CCK-8 AND CCK-4

Effects on Excitability

As shown in Fig. 1, the pressure ejection of CCK-8 into the immediate vicinity of the pyramidal cells of the CA1 region of the rat hippocampus caused an abrupt, short latency depolarization accompanied by a marked increase in excitability. Not only was there an increase in the number of spikes evoked by a depolarizing ramp, but spikes were seen to occur between the ramps and immediately after the rectangular hyperpolarizing pulses. The depolarization was also accompanied by an increase in the "noisiness" of the baseline and a decrease in membrane resistance. The magnitude of the decrease in membrane resistance was determined by measuring the amplitude of the hyperpolarizing response evoked by injecting a constant amount of hyperpolarizing current through the electrode. In record B of Fig. 1 the second hyperpolarizing response is clearly reduced in magnitude.

In much the same way, Fig. 2 shows a similar application of CCK-4 to excite another CA1 cell of the rat hippocampus. Again, the depolarization is associated with a "noisy" baseline. In this case, however, this cell repolarization and the excitatory action disappeared after 11 sec., long before the application of CCK-4 was terminated. This desensitization was not an uncommon occurrence. Following a desensitization of this magnitude, a further response to CCK-4 could only be assured by delaying the next application of peptide for approximately 15 minutes.

In similar experiments on 21 slices of the rat hippocampus, CCK-8 and CCK-4 were found to be potent excitants on 55 CA1 pyramidal cells when injected on average with a pressure of 647kPa, which was shown by radioimmunoassay to release fmolar quantities of CCK-8 per second. In no case was inhibition observed. Although on 40 cells, CCK-8 was excitatory on 97 occasions, on 7 cells, which had been shown to be excited by glutamate, applied iontophoretically from the same multi-barrelled pipette, CCK-8 had no effect. Similarly, CCK-4 was excitatory on 7 cells on 26 occasions, yet had no effect on 1 cell, excited by glutamate. On 12 cells which had already been shown to be excited by CCK-8 or CCK-4, a further 19 doses had no effect. The sensitivity of a number of these cells to CCK returned when subsequent doses were tested less frequently, at intervals of about 15 minutes.

Below is a more detailed account of the excitation evoked by CCK-8 and CCK-4, respectively, on 9 cells with a mean resting potential of -71mV and 4 cells with a mean resting membrane potential of -16mV.

Effects on Membrane Potential

Both CCK-8 and CCK-4 evoked depolarizations of similar magnitude to those evoked by glutamate on the same cells. Both the onset and the rate at which the CCK-8 evoked depolarization developed were as fast as those evoked by glutamate. However, on average, the actions of CCK-4 appeared to develop sooner and faster.

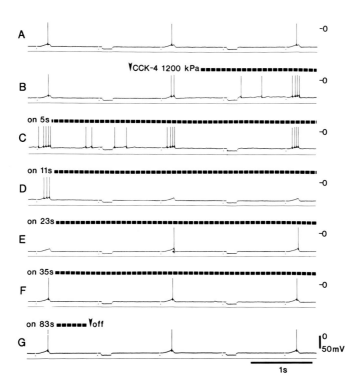

FIG. 2. The excitatory action of CCK-4 on another CAl pyramidal neurone. The depolarization evoked by the peptide is again associated with an increase in excitability and an increase in baseline "noise". After about 11 seconds (D) the cell repolarized and became less excitable, even although the application of peptide was continued.

Effects on Membrane Resistance

On 9 occasions CCK-8 caused a clear decrease in membrane resistance in 6 cells from a control level of 65 ± 6M Ω to 49 ± 6 MΩ . On four occasions on 3 cells in which the mean resting membrane resistance was 41 ± 5MΩ, CCK-4 evoked a mean decrease of 9.5 ± 1MΩ. An increase in resistance was never observed. In contrast, glutamate on the same cells evoked either no change, or a small increase in membrane resistance.

Reversal Potential of the Cholecystokinin Effects

Using the single pulse method of Ginsborg (9), the mean reversal potential of the action of CCK-8 on 8 cells determined on 9 occasions was 21 $\pm$ 1.0mV, depolarizing with respect to the resting membrane potential, and was not significantly different from that of CCK-4 estimated on 3 cells to be 38 $\pm$ 13mV.

Apparent Desensitization

As mentioned earlier, the termination of an application of either CCK-8 or CCK-4 (Fig. 2) caused the membrane potential and the excitability of the cell to return rapidly to the resting level. However, in a number of neurons the membrane potential, the excitability of the cell and the membrane resistance returned to resting levels prematurely and the remainder of the dose was ineffective (Fig. 1). In addition, on 9 occasions on 12 cells a subsequent application of CCK-4 or CCK-8 failed to evoke a response. Sometimes this apparent desensitization could be overcome by delaying the next application of the peptide to the cell for at least 15 minutes. During the apparent desensitization the membrane potential and resistance remained constant and the response evoked by glutamate appeared to be unaltered.

EFFECTS OF CCK-RELATED PEPTIDES

Non-Sulphated CCK-8 (CCK-NS)

An interesting feature of naturally occurring CCK and caerulein is the presence of the sulphate group on tyrosine (8) whereas both sulphated and unsulphated forms of gastrin have been isolated. CCK-8 is 5-8 times as active as the parent molecule as a stimulant of the guinea-pig gall bladder, and the non-sulphated forms of CCK-8 and caerulein appear to be 300 times less active (cf. review by Thompson, ref. 25).

On 32 out of 35 occasions on 10 cells CCK-NS was totally inactive, although on the same cell glutamate applied from the same multibarrelled micropipette was excitatory. On 3 occasions on 3 cells CCK-NS was excitatory.

Fragments of Gastrin G-13 and G-14)

G-13 and G-14, the C-terminal sequences of the larger gastrointestinal hormone, gastrin-34, were synthesized, as the leucine analogues, by Dr. Brian Williams at the Molecular Biology Laboratory in Cambridge. Both contained the C-terminal pentapeptide sequence of CCK. However, the tyrosine residue (7) was non-sulphated and lay one amino acid sequence nearer to the C-terminal end of the peptide.

As shown in Fig. 3, the excitation evoked by G-13 was also accompanied by a depolarization and a decrease in membrane resistance. In record B the response to the ejection of a constant rectangular-hyperpolarizing pulse of current is clearly reduced in amplitude. An unusual feature of the depolarization and excitation evoked by both gastrins was the way in which the excitation continued for several seconds after the application of peptide was terminated.

In contrast, the termination of the excitation evoked by glutamate, CCK-4 and CCK-8 coincided with the end of glutamate or peptide application or, as mentioned earlier, even sooner.

In all, 9 cells were strongly excited by either G-13 or G-14. One cell was unaffected by 5 separate applications of G-13, although it was readily excited by glutamate, applied by microiontophoresis from an adjacent barrel of the same multibarrelled micropipette. On 2 cells, 3 doses of G-13 and G-14 were ineffective, although earlier and later doses from the same micropipette were excitatory. On 3 cells, the excitations evoked by both G-13 and G-14 were indistinguishable when applied to the same cell from the same multibarrelled micropipette.

Effects on Membrane Potential

Both gastrins evoked a substantial depolarization with a mean value of $13 \pm 6mV$. On 9 occasions, on 4 cells, the mean latency of the depolarization evoked by either gastrin was not significantly different from that evoked by glutamate on the same cells. However, on 9 other occasions, on 5 different cells, the onset was very much slower.

Attenuation of the response to gastrin before the termination of the application was rarely observed and recovery was usually delayed when compared to that following glutamate on the same cell or CCK-8 and CCK-4 on other cells. On several occasions the cells appeared to recover quickly, but then went on to display intermittent bursts of depolarization and increased excitability for up to 5 min after initial recovery. This phenomenon was never observed following excitations evoked by glutamate on the same cells, or CCK-4 or CCK-8 on other cells.

Caerulein

Caerulein is a decapeptide found in the skins of some South American and Australian amphibians, and it was first isolated from the skin of Hyla caerula. Both natural and synthetic caerulein exhibit the entire range of CCK activity on vascular and extravascular smooth muscles, gastrin secretion, amylase secretion from the pancreas and gall bladder contractility (7). The structure of caerulein is very similar to that of CCK-8. It contains the pentapeptide sequence common to both gastrin and CCK and the sulphate group on the tyrosine residue (8) which appears to be essential for its CCK-like activity (25).

Caerulein, applied by pressure ejection was excitatory on 2 cells, on 4 out of 4 occasions and the effects appeared to be indistinguishable for those evoked by CCK-8.

Bombesin

Bombesin, a tetradecapeptide originally isolated from extracts of the skin of the frog, <u>Bombina bombina,</u> has been shown to depolarize pancreatic acinar cells in a manner very similar to caerulein (19), the two yielding almost identical reversal potentials. However, bombesin does not bear any obvious structural similarity to caerulein and it has been postulated that, although they share a final common pathway in the acinar cells, the actions of bombesin and caerulein are mediated by different receptors (2,19). Bombesin was applied to 2 neurons and on 4 out of 6 occasions it was excitatory. On one cell, although excitation was evoked by the first application, bombesin was ineffective on two subsequent occasions. Glutamate remained excitatory throughout the experiment on this cell. Later, the response to bombesin recovered. Although the increase in excitability evoked by bombesin was accompanied by a clear depolarization, there was no detectable change in membrane resistance.

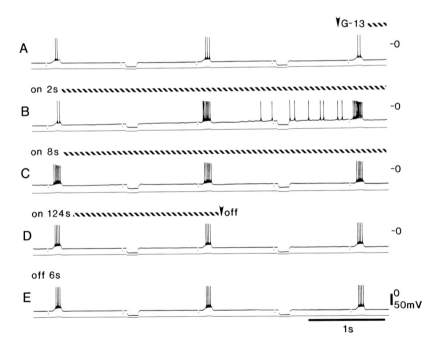

FIG. 3. The excitatory action of G-13 on another CA1 neuron. Again, the increase in excitability is associated with a depolarization and a decrease in membrane resistance.

The time-course of the response to bombesin differed from those to CCK-4 and CCK-8, described earlier, on other cells and to the application of glutamate on the same cell. Only after the pressure ejected application of bombesin had continued for over one minute did the number of spikes evoked by the depolarizing ramp of current begin to increase. This increase in firing continued to increase for a further three minutes and recovery was not complete 30 seconds after the application was terminated. Although the slowly developing excitation evoked by bombesin was quite unlike that evoked by glutamate on the same cell, it was not unlike that evoked by acetylcholine (14). In the pancreas, the depolarization evoked by bombesin is said to differ from that evoked by the analogues of CCK and to depend upon the generation of an intracellular messenger, such as cyclic GMP.

Substance P

Substance P was applied by iontophoresis and pressure ejection to a total of 14 neurons and on 45 separate occasions was without effect. On 17 occasions, on 7 cells, iontophoresis of Substance P was without effect. Pressure application of Substance P had no effect on 28 occasions on 7 cells which were shown repeatedly to be excited by glutamate iontophoretically applied from an adjacent barrel of the same multibarrelled pipette.

CONTROLS

Omission of Glutamate from Pipettes

CCK-8 excited 6 cells when applied by pressure ejection from single and multibarrelled pipettes which did not contain glutamate or any other putative excitatory substance. The excitation evoked by CCK-8 was indistinguishable from that evoked by CCK-8 released from pipettes with a glutamate-containing barrel.

Pressure Ejection of Peptide-Free Vehicle Solutions

The effect of pressure application of peptide-free vehicle solution from an adjacent barrel of the peptide containing a multibarrelled pipette was examined on 50 occasions, on 12 cells which had been shown to be excited by one or another of the peptides, or not affected by CCK-NS or Substance P.

On no occasion did this procedure have any effect on the electrical characteristics of the cell under study.

In the gastrointestinal tract all the various actions of gastrin are possessed by the C-terminal peptide, TRP-MET-ASP-PHE-HN$_2$ and therefore shared with CCK and caerulein. Of the four amino acids present in the gastrin C-terminal peptide, only ASP appears to be essential for activity and even in this situation substitution is possible (Fig. 4). In the same system, optimal CCK-like activity requires a longer peptide and the presence of the TYR-O-SO$_4$ residue ensures optimal activity. However, the three N-terminal amino acids of CCK-8 appear to be essential for contraction of the gall bladder (25) and the 5-8 tetrapeptide is inactive. In the brain, therefore, the excitatory actions of the CCK-4, CCK-8, G-13 and G-14 on pyramidal cells could be due to the interaction of the peptide with a specific receptor.

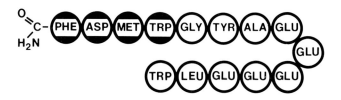

FIG. 4. Diagram to show the amino acid sequence of CCK-8 (upper) and G-14 (lower) and the presence of the tetrapeptide common terminal.

Since the same cells are also excited by bombesin, the receptor may also be said to resemble that of the pancreatic acinar cells, which are activated by both caerulein and bombesin. Like the pancreatic acinar cells, the pyramidal cells of the hippocampus also respond to acetylcholine (14). The failure of pyramidal cells to respond to Substance P is also interesting since this region of the brain appears to be devoid of Substance P. In other regions of the brain, such as the amygdala, a number of authors have correlated the high levels of Substance P with the ability of iontophoretically-applied Substance P to excite cells (10,11,16). Thus, despite a certain sequence of homology, the hippocampal receptor would appear to distinguish between Substance P and bombesin. In the salivary galnds, Gallagher and Petersen (8) were able to show that Substance P, but not caerulein or bombesin, causes a marked increase in membrane conductance. Since hippocampal cells also appear to be excited by at least two more unrelated peptide sequences, vasoactive intestinal polypeptide (VIP) (6) and somatostatin (5), the failure of the cells to respond to Substance P is of added significance.

Since in a number of preliminary experiments we have been able to show that the excitation evoked by CCK-4 was unaffected by a medium containing sufficient atropine to block the excitation of the same cell by acetylcholine, we can exclude the possibility that the peptides act indirectly to cause the release of acetylcholine from adjacent nerve terminals.

ACKNOWLEDGEMENTS

We are grateful to George Marshall for technical support, Jens Rehfeld for radioimmunoassays, Dr. J.S. Morley of ICI, Dr. M. Ondetti of the Squibb Institute, and Dr. de Castiglione of Farmitalia for supplies of peptides. Jane Dodd was in receipt of an MRC Scholarship. The experimental work was carried out at the MRC Neurochemical Pharmacology Unit, Hills Road, Cambridge.

REFERENCES

1. Anastasi, A., Erspamer, v., and Endean, R. (1968): Archs. Biochem. Biophys., 125:57-68.
2. Deschodt-Lanckman, M., Robberecht, P., de Nect, P., Lammens, M., and Christophe, J. (1976): J. Clin. Invest., 58:891-898.
3. Dingledine, R., Dodd, J., and Kelly, J.S. (1980):L J. Neurosci. Methods., (in press).
4. Dockray, G.J. (1976): Nature, 264:568-570.
5. Dodd, J., and Kelly, J.S. (1978): Nature, 273:674-675.
6. Dodd, J., Kelly, J.S. and Said, S.I. (1979): Brit. J. Pharmacol., 66:125p.
7. Erspamer, V., Bertaccini, G., de Caro, G., Endean, R. and Impicciatore, M. (1967): Experientia, 23:702-703.
8. Gallagher, P.V., and Petersen, O.H. (1980): Nature, 283:393-395.
9. Ginsborg, B.L. (1967): Pharmacol. Rev., 19:289-316.
10. Guyenet, P.G. and Aghajanian, G.K. (1977): Brain Res., 136: 178-184.
11. Henry, J.L. (1976): Brain Res., 114:439-451.
12. Innis, B.R., Correa, F.M.A., Uli, G.R., Schneider, B. and Snyder, S.H. (1979): Proc. Natl. Acad. Sci. USA, 76:521-525.
13. Iwatsuki, N. and Petersen, O.H. (1979): Nature, 268:147-149.
14. Kelly, J.S., Dodd, J. and Dingledine, R. (1979): Prog. in Brain Res., 49:253-266.
15. Larsson, L.I., Rehfeld, J.F. (1979): Brain Res., 165:201-218.
16. Le Gal La Salle, G., Ben-Ari, Y. (1977): Brain Res., 135: 174-179.
17. Morley, J.S., Tracey, H.J. and Gregory, R.A. (1965): Nature, 207:1356-1359.

18. Peterson, O.H. and Iwatski, N. (1978): <u>Ann. N.Y. Acad. Sci.</u>, 307:599-618.
19. Petersen, O.H. and Philpott, H.G. (1979): <u>J. Physiol.</u>, 290: 305-315.
20. Phillis, J.W. and Kirkpatrick, J.R. (1969): <u>Can. J. Physiol. and Pharmacol.</u>, 57:887-899.
21. Behfeld, J.F. (1978a): <u>J. Biol. Chem.</u>, 253:4016-4021.
22. Rehfeld, J.F. (1978b): <u>J. Biol. Chem.</u> 253: 4022-4030.
23. Rehfeld, J.F., Coltermann, N., Larsson, L.I., Emson, P.M. and Lee, C.M. (1979): <u>Federation Proc.</u> 38:2325-2329.
24. Rehfeld, J.F., Larsson, L.I., Goltermann, N.R., Schwartz, T.W., Holst, J.J., Jensen, S.L., Morley, J.S. (1980): <u>Nature</u>, 284:33-38.
25. Thompson, J.C. (1973): In <u>"Pharmacology of Gastrointestinal Motility and Secretion:</u>, edited by P. Holton, <u>Internat. Encyclopedia Pharmacol. and Ther.</u> Section 39A, pp. 261-286. Pergamon Press, Oxford.
26. Vanderhaeghen, J.J., Signeau, J.C. and Gepts, W. (1975): <u>Nature</u> 257:604-605.

Neurosecretion and Brain Peptides,
edited by J. B. Martin, S. Reichlin, and K. L. Bick.
Raven Press, New York © 1981.

Somatostatin and Cortical Neurons in Cell Culture

Marc A. Dichter and John R. Delfs

*Department of Neurology, Harvard Medical School, Children's Hospital Medical Center
and Beth Israel Hospital, Boston, Massachusetts 02115*

The discovery that several small peptides, which are known to be either hormones or hypothalamic releasing factors for hormones, are found in many parts of the central nervous system (CNS) presumably having little to do with hormonal regulation, has stimulated a search for more general neurobiological roles for these substances. A number of behavioral studies indicate that neuropeptides may play a role in both normal and pathophysiological cortical function (2,3,6,16,22,23), and additional pharmacological and neurophysiological studies have shown that some peptides exhibit transmitter-like properties in the CNS (1,18,19,25,33). Several peptides have been found in neuron cell bodies within the mammalian cortex, while others have so far been noted only in axons that are presumed to arise from neurons in subcortical structures and to terminate in cortex.

Our laboratory has been studying the physiology and pharmacology of mammalian cortical neurons grown in cell culture. This chapter will review the results of our studies on the synthesis and cellular localization of several neuropeptides found in these cultures and will describe our efforts to understand the role of one particular neuropeptide, somatostatin (SOM), in modulating the activity of cortical neurons.

The cell cultures are derived from the brains of 15 to 16 day rat embryos. The cortical mantle is dissected and other parts including striatum, diencephalon, cerebellum, and brainstem are discarded. The tissue is minced, dissociated by trypsin and trituration, filtered for cell clumps, and plated at 150-250,000 cells per collagen-polylysine coated 35 mm petri dish (7,27). By three to four weeks in culture, the neurons are remarkably similar in their morphologic and electrophysiologic properties to comparable rat cortical neurons in situ (7). They have spontaneous and evokable action potentials and have formed extensive excitatory and inhibitory synaptic connections with one another. Fig. 1 illustrates several neurons from rat cortical cell cultures.

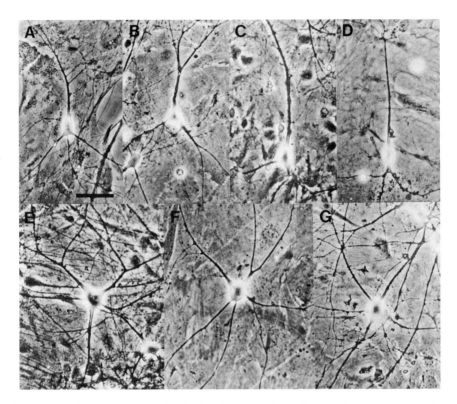

FIG. 1. Phase contrast photomicrographs of cortical neurons in dissociated cell cultures, 18 to 40 days *in vitro*. Calibration bar in A is 50 μm.

Cortical cell cultures allow study of neuropeptides in several different ways. The synthesis of peptides can be studied in an environment where the influences of subcortical and extra-CNS tissues have been removed. Immunohistochemical staining can identify both the cells of origin and the intracellular distribution of a particular neuropeptide. Other parameters such as receptor binding, uptake, storage, and release can also be studied in a system where cells remain intact and viable. Finally, a given neuron can be studied with intracellular microelectrode recording during application of peptides directly onto that neuron. The study of neuropeptides began by assaying the cultures for several neuropeptides which had been found in cortex.

PRESENCE OF NEUROPEPTIDES IN CORTICAL NEURONS

Various prior studies have revealed the presence of many neuropeptides in extrahypothalamic brain, and a number of these are found in cortex. Table 1 indicates the peptides found in extra-hypothalamic brain and in neocortex by radioimmunoassay (RIA) or by immunohistochemistry and notes those which have been localized by immunohistochemistry to neuronal cell bodies. Table 1 also indicates those peptides found in the cortical cell cultures to date. Among peptides for which the cultures have been assayed, SOM is present in highest amounts. Substance P is present in smaller amounts (10); LHRH and vasopressin are not detectable. TRH has not been detected with RIA, but preliminary immunofluorescent studies suggest the presence of occasional neurons containing this peptide.

TABLE 1.   Peptides in neocortex in vivo compared with cortical cell cultures[a]

Peptide	Extra-hypothalamic brain (any area)	Neocortex	Cortical cultures
Somatostatin	+(+)[b]	+(+)	+(+)[c]
Substance P (13)	+(+)	+	+(+)
TRH (17)	+	+	?(+)
LHRH	+	0	0
Vasopressin	+	0	0
ACTH (31)	+	0	
VIP	+(+)	+(+)	
CCK (11,20,24)	+(+)	+(+)	
Gastrin	+(+)	+(+)	
Glucagon (4)	+	+	
Bombesin (30)	+	+	
Neurotensin (29)	+	+	
Enkephalin	+	0	
Angiotensin II	+	0	
Oxytocin	+	0	

[a] Data from Hokfelt et al. (14) unless otherwise noted.
[b] + or 0 represents presence or absence respectively by RIA or immunohistochemical determination, (+) represents demonstration in neuronal cell bodies.
[c] In cortical culture column, +, 0, or (indeterminate) represents results from RIA only.

The large amounts of SOM in these cultures prompted us to pursue the role of this peptide in greatest detail. Most of the data presented in this chapter focuses on SOM as a prototype neuropeptide.

The first set of experiments involved the production of SOM by the cultures. Cells and media were extracted separately at various time points after plating, and in collaboratin with Robbins and Reichlin (5), samples were assayed for SOM using a well-characterized RIA (21). No SOM was detected in initial cell suspensions, but by one week in culture both media and cell fractions contained significant amounts (Fig. 2). Concentrations in both fractions peaked at approximately three weeks in vitro and thereafter decreased, probably due to declining neuronal survival. Chromatography showed the SOM from both fractions to elute identically with a cyclic SOM standard. Interestingly, 8% of the SOM from the cell extracts, but none from the media extracts, migrated in a separate peak corresponding to a larger form of SOM. This larger form, a probable prohormone or precursor, has also been noted in assays of brain tissue obtained from intact animals (28).

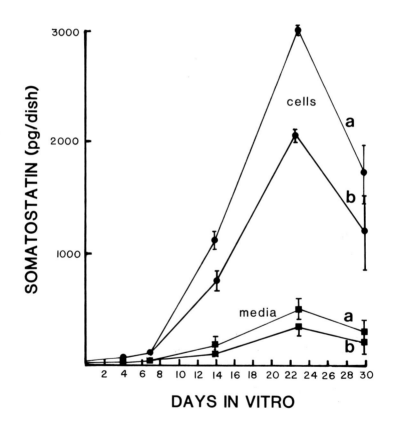

FIG. 2. Immunoreactive-SOM in cells (●) and media (■) per culture (mean of three 35 mm dishes ± s.e.) as a function of days in vitro. Data from two separate cultures, a and b.

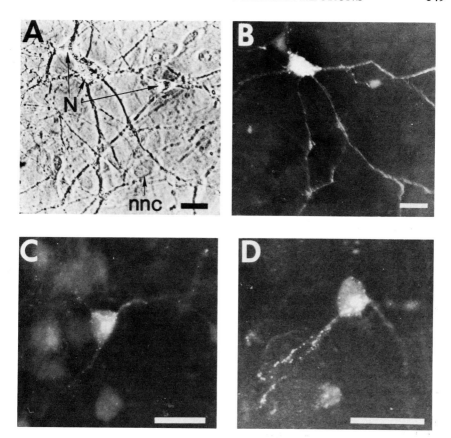

FIG. 3. Phase contrast and indirect immunofluorescent microscopy of neurons in the dissociated cell cultures. A. Dispersed cerebral cell culture at 32 days seen under phase-contrast microscopy. Note flat non-neuronal cells (nnc) as well as neuron cell bodies (N) and neuronal processes. B, C, and D. Fluorescence photomicrographs showing cultures stained with anti-SOM using indirect immunofluorescent technique. B. Same field as A, showing positive staining of the perikaryon and its neuronal processes. Note the selective staining of this neuron as distinguished from the several unstained neuron cell bodies and many unstained neuronal processes, prominently seen in A by phase-contrast. C. Pyramidal-shaped neuron positively stained by anti-SOM in presence of more dimly-seen negatively-staining neurons of roughly the same size in culture at 18 days. Staining in perikaryon of neurons often appeared continuous with staining in proximal part of cell processes. D. Neuron positively stained by anti-SOM with smaller collections of staining trailing out a neuronal process in 18-day culture. Scale bars, 25 μm. (From Delfs et al. (5), used with permission of Nature.)

Immunofluorescent studies carried out in collaboration with Connolly (5) demonstrated SOM to be localized to neurons, with approximately 3-5% of all neurons staining positively in each culture (Fig. 3). No non-neuronal cells stained. Specificity was demonstrated by absence of staining when the anti-SOM antibody was reacted with SOM prior to use, while reaction with ACTH and LHRH had no effect. The neurons which stained for SOM were of varying shapes and sizes and were not morphologically distinct from many unstained neurons. Positive staining for SOM was seen in the soma and in varicosities along the processes, some of which appeared to contact other cells.

PHYSIOLOGICAL EFFECTS OF SOM ON CORTICAL NEURONS

Several groups have reported that SOM affects neuronal firing rates in mammalian cerebral cortex in vitro. Renaud et al. (26) noted a decrease in firing in 23 of 28 cortical neurons (82%) in anesthetized rats in response to SOM iontophoresis. On the other hand, Ioffe et al. (15) reported increased firing in 35 of 60 neurons (58%) in sensorimotor cortex of unanesthetized rabbits. They also noted that SOM appeared to potentiate the effects of glutamate on firing rates. In the only published study utilizing intracellular recording, Dodd and Kelly (12) recorded from pyramidal neurons in rat hippocampal slice preparations and administered SOM either by iontophoresis, pressure injection, or local application of 3 mM SOM. They observed a gradual depolarization of about 20 mV and an increase in the number of action potentials evoked by a ramp of depolarizing current.

One of the main advantages of the cortical cell culture system is that the properties of the neurons and the affects of neurotransmitters or other agents can be determined using intracellular recording from identified neurons, and the concentration of the agent can be controlled. Agents to be tested are dissolved in recording media at known concentrations and placed into a micropipette (tip diameter ca. 5-15 μm). The tip of this micropipette can be positioned under visual control to deliver the agent locally. By the application of air pressure to the shaft of the micropipette, the media containing the test agent is gently perfused over the neuron being studied. When the perfusion stops, the test agent rapidly diffuses into the large volume of bathing media and the environment around the cell returns to baseline.

Previous work with this system has shown that the neurons have electrophysiological properties which are very similar to those of comparable neurons in situ, and they form extensive excitatory and inhibitory synaptic connections (7). More recently, in a series of electrophysiological, biochemical and autoradiographic experiments, it has been shown that gamma-aminobutyric acid (GABA) acts as the inhibitory transmitter between the cortical neurons (9,27,32). Studies with acetylcholine indicate that this agent does not mimic the endogenous excitatory neurotransmitter

(8), and experiments are currently underway to determine the effects of the putative excitatory transmitters, glutamate and aspartate.

Effects of SOM on Neuronal Membrane Properties

In order to determine the effects of SOM, the peptide was applied by miniperfusion to individual neurons which were impaled with microelectrodes. The SOM was dissolved in recording media at its final concentration (e.g., 10 μM) and gently perfused over the neuron. During the SOM application, neuron membrane potential (MP), membrane resistance (the MP change in response to a constant hyperpolarizing current pulse), action potential size, shape and threshold and incoming spontaneous synaptic potentials were monitored.

BEFORE   SOMATOSTATIN PERFUSION   AFTER

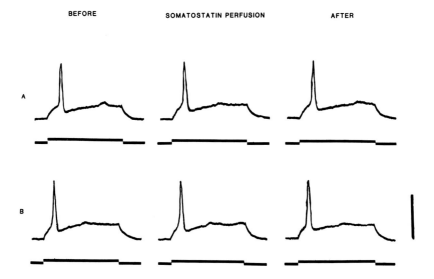

FIG. 4. Oscilloscope records of action potentials evoked by a 50 msec depolarizing current pulse. No change is seen in shape, size, rate of rise, or threshold during or after application of SOM. Vertical calibration 50 mV.

In over 100 neurons (with MP larger than -40 mV and action potentials larger than 50 mV) from 20 different cultures, SOM in concentrations from 1 nM to 100 μM produced no signficant change in membrane potential of the neurons. The cells were neither depolarized nor hyperpolarized. In addition, no change in membrane resistance was seen. This is illustrated in Fig. 5A and 7A-D. Action potential size, shape and threshold also appeared unchanged (Fig. 4). When applied at 1 mM, some of the neurons irreversibly depolarized and appeared to degenerate morphologically. We consider this a possible non-physiologic response to

SOM and mention it because it illustrates the importance for control of the concentration of agents during electrophysiological studies.

Effects of SOM on Neuronal Activity

In many neurons (but less than 50% of the overall group) SOM produced a dramatic enhancement of the frequency of incoming spontaneous postsynaptic potentials (Figs. 5, 6A). In a responsive neuron there could be an enhancement of excitatory or inhibitory postsynaptic potentials (EPSPs or IPSPs) or both. The EPSPs sometimes triggered action potentials (Fig. 5A-C). The effects often had a latency of several seconds and outlasted the SOM application by many seconds. Thus, SOM seemed to have an effect on the neuron or axon terminal which was <u>presynaptic</u> to the cell being studied.

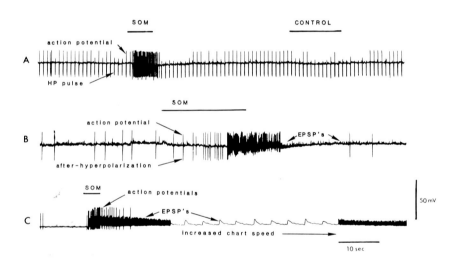

FIG. 5. SOM increases excitatory synaptic activity without affecting MP or input resistance. In this and all subsequent physiology figures traces represent penwriter records of intracellular recordings. A. SOM (1 μ m) induces increased EPSPs and action potential (AP) firing with no change in MP or resistance. (Downward deflections in trace represent MP changes induced by 50 msec constant hyperpolarizing current pulses and are proportional to input resistance. Attenuated anode break APs are also seen.) Control perfusion induces very slight hyperpolarization with loss of anode break AP. B. Same cell as A. Another perfusion with SOM (10 μ M) produces enhanced frequency of EPSPs (seen better with increased chart speed).

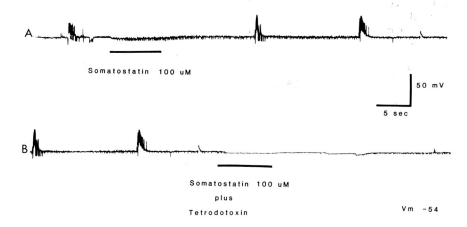

FIG. 6. A. SOM application results in an increase in the frequency of incoming IPSPS. B. This response is not seen when SOM is applied with terodotoxin.

In order to determine the mechanism for this increase in frequency of discrete PSPs, SOM was applied with tetrodotoxin (TTX), an agent known to block Na^+-dependent action potentials (such as are found in the cortical neurons). The experiment was performed by positioning two miniperfusion pipettes near a cell to be studied. One contained SOM, the other, the same concentration of SOM plus TTX (10^{-7} g/ml). The TTX completely blocked the enhanced PSPs as illustrated in Fig. 6B. Thus, this SOM effect of increasing the frequency of PSPs appears to require active Na^+ channels in the presynaptic neuron or axon terminal.

Effects of SOM on Glutamate and Aspartate Responses

During the course of our studies of the mechanisms of action of the putative excitatory amino acids, glutamate and aspartate, we noted that application of these agents to the neurons often had two kinds of effects. Glutamate can depolarize the cell being studied and it also can cause an increase in incoming PSPs. The presynaptic effect of increasing PSPs is often seen at concentrations of glutamate which are just below threshold for directly depolarizing the postsynaptic cells. (The exact threshold concentration of glutamate varies from 10 µ M to as high as 100 µM from neuron to neuron and appears to increase with age in culture). The presynaptic effect can also be seen when glutamate miniperfusion, even with suprathreshold concentrations, is directed slightly away from the neuron being studied and onto a nearby group of neurons. Since all the neurons in the cortical cultures are sensitive to glutamate, it is not necessary to postulate two different receptors, one on the soma producing the direct depolarization and one on the presynaptic axon terminals. The two different responses may simply represent slightly different sensitivities of different cells in the area of the glutamate application. It should be emphasized that the glutamate response is different from the SOM response in that membrane depolarization is not seen with SOM.

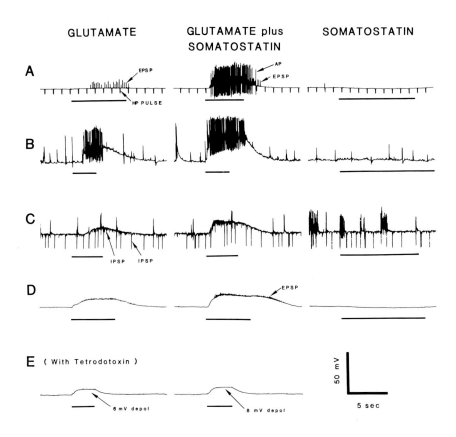

FIG. 7. SOM potentiates the effects of glutamate on 5 cortical neurons in culture. In each cell glutamate (25 µM) produces an increase in synaptic potentials (A-D) and/or a direct depolarization (B-E). In E the presynaptic effect is blocked by tetrodotoxin at 10⁻⁷ g/ml. Cell A illustrates no change in input resistance to hyperpolarizing (HP) current pulses when glutamate produces only a presynaptic effect. In each neuron, application of glutamate (25 µM) plus SOM (50 µM) produces an enhanced presynaptic effect and an enhanced direct depolarization as compared with application of glutamate alone. SOM (50 µm) alone has no obvious effect in these particular neurons.

When SOM is applied together with glutamate the response to glutamate is enhanced, even when the SOM has no effect on its own. Fig. 7 shows five different responses to 25 μM glutamate (first column) including pure presynaptic (Fig. 7A), combined pre- and postsynaptic (Fig. 7B-D), and postsynaptic alone in the presence of 10^{-7}g/ml TTX (Fig. 7E). The second column shows that SOM increased both the presynaptic response (increased PSPs) and the postsynaptic response (membrane depolarization), the latter even in the presence of TTX. The third column indicates that SOM had no detectable direct effects on these particular cells when applied alone at the same concentrations. Similar potentiating effects of SOM on aspartate responses have been seen.

The mechanism by which SOM potentiates the excitatory effects of glutamate and aspartate is not clear. Since glutamate and aspartate have been proposed as neurotransmitters mediating the EPSPs seen in these neurons, it was thought that SOM might potentiate these EPSPs. This remains a possibility and more detailed analysis needs to be performed.

GABA is the inhibitory neurotransmitter utilized by the cortical neurons (9). GABA produces a hyperpolarization of the membrane potential by increasing Cl- conductance. SOM at 50 M had no effect on GABA responses in these neurons, indicating that its effect on the excitatory amino acids is somewhat specific and does not represent an enhancement of all neurotransmitter function.

The results of our physiological studies to date indicate that SOM does not depolarize neocortical neurons in culture and, thus, does not mimic the excitatory neurotransmitter. This action is different from that which has been reported in hippocampal neurons in slice preparations (12). In the cortical cell cultures the application of SOM alone increases the frequency of incoming PSPs in some individual neurons but has no direct effects on the neural membrane. SOM does, however, dramatically enhance the actions of the excitatory neurotransmitter candidates glutamate and aspartate. In these latter cases it enhances the presynaptic effect of these excitatory agents and also potentiates their direct membrane effects. Further studies will be needed to determine if this potentiation of the effects of excitatory amino acids can explain the effect of SOM on endogenous neuronal activity.

SUMMARY

Rat cortical neurons in culture have morphological, physiological and biochemical properties similar to their counterparts in situ. Neuropeptides are synthesized by these cultures and can be localized to individual neurons. One such peptide, somatostatin, is produced in relatively large amounts, exists in a discrete neuronal population, and is released in the culture medium. Somatostatin has no detectable effect on membrane potential, resistance, or action potential of cells to which it is applied but does not appear to increase incoming spontaneous postsynaptic potentials in some neurons. Somatostatin potentiates the direct membrane effects of gluatamte and aspartate. Somatostatin may play a role as a neuromodulator in mammalian cerebral cortex by potentiating the effects of excitatory neurotransmitter agents.

ACKNOWLEDEGMENTS

This work has been supported by The Esther A. and Joseph Klingenstein Fund, NS15362, NS00130, NS06320 and the Children's Hospital Mental Retardation Core Grant HD06276. We thank Mr. Bernard Biales and Ms. Sara Vasquez for expert technical assistance and Ms. Diane Kilday for help with preparation of the manuscript.

REFERENCES

1. Berelowitz, M., Matthews, J.,Pimstone, B.L., Kronheim, S., and Sacks, H. (1978): Metabolism, 27:1171-1173.
2. Bloom, F., Segal, D., Ling, N., and Guillemin, R. (1976): Science, 194:630-632.
3. Brown, M. and Vale, W. (1975): Endocrinology, 96:1333-1336.
4. Conlon, J.M., Samson, W.K., Dobbs, R.E., Orci, L. and Unger, R.H., (1979): Diabetes, 28:700-702.
5. Delfs, J., Robbins, R., Connolly, J.L., Dichter, M., and Reichlin, S. (1980): Nature, 283:676-677.
6. de Wied, D. (1977): Life Sci., 20:195-204.
7. Dichter, M.A. (1978): Brain Res., 149:279-293.
8. Dichter, M. (1979): Neurosci. Abst., 5:1990.
9. Dichter, M.A. (1980): Brain Res., (in press).
10. Dichter, M., Delfs, J., and Leeman, S.: Unpublished observations.
11. Dockray, G.J. (1976): Nature, 264:568-570.
12. Dodd, J., and Kelly, J.S. (1978): Nature, 273:674-675.
13. Gilbert, R.F.T., and Emson, P.C. (1979): Brain Res., 171: 166-170.
14. Hokfelt, T., Elde, R., Johansson, O., Ljungdahl, A., Schultzberg, K., Fuxe, K., Goldstein, M., Nilsson, G., Pernow, B., Terenius, L., Ganten, D., Jeffcoatte, S.L., Rehfeld, J. and Said,S. (1978): In: Psychopharmacology: A Generation of Progress, edited by M.A. Lipton, A. DiMascio, and K.F. Killam, pp. 39-66. Raven Press, New York.
15. Ioffe, S.,Havlick, V., Friesen, H., and Chernick, V. (1978): Brain Res., 153:414-418.
16. Kastin, A.J., Olson, R.D., Schally, A.V., and Coy, D.H. (1979): Life Sci., 25:401-414.
17. Kreider, M.S., Winokur, A.,and Utiger, R.D. (1979): Brain Res., 171:161-165.

18. Lee, S.L., Havlicek, V., Panerai, A.E., and Friesen, H.G. (1979): Experientia, 35:351-352.
19. Martin, J.B., Renaud, L.P., and Brazeau, P. (1975): Lancet, 2:393-395.
20. Muller, J.E., Straus, E., and Yalow, R.S. (1977): Proc. Natl. Acad. Sci. USA, 74:3035-3037.
21. Patel, Y.C., and Reichlin, S. (1978): Endocrinology, 102: 523-530.
22. Plotnikoff, N.P., Kastin, A.J., and Schally, A.V. (1974): Pharmacol. Biochem. Behav., 2:693-696.
23. Prange, A.J., Nemeroff, C.B., and Lipton, M.A. (1978): In: Psychopharmacology: A Generation of Progress, edited by M.A. Lipton, A. DiMascio, and K.F. Killam, pp. 441-458. Raven Press, New York.
24. Rehfeld, J.F., Goltermann, N., Larsson, L.I., Emson, P.M., and Lee, C.M. (1979): Fed. Proc., 38:2325-2329.
25. Renaud, L.P. (1978): In: Psychopharmacology: A Generation of Progress, edited by M.A. Lipton, A. DiMascio, and K.F. Killam, pp. 423-430. Raven Press, New York.
26. Renaud, L.P., Martin, J.B., and Brazeau, P. (1975): Nature, 255:233-235.
27. Snodgrass, S.R., White, W.F., Biales, B., and Dichter, M. (1980): Brain Res., (in press).
28. Spiess, J., and Vale, W. (1978): Metabolism, 27:1175-1178.
29. Uhl, G., Kuhar, M.J., and Snyder, S.H. (1977): Proc. Natl. Acad. Sci. USA, 74:4059-4063.
30. Walsh, J.H., Wong, H.C., and Dockray, G.J. (1979): Fed. Proc., 38:2315-2319.
31. Watson, S.J., Richard, C.W. III, and Barchas, J.D. (1978): Science, 200:1180-1182.
32. White, W.F., Snodgrass, S.R., and Dichter, M. (1980): Brain Res., (in press).
33. Zetler, G. (1976): Biochem. Pharmacol., 25:1817-1818.

Neurosecretion and Brain Peptides,
edited by J. B. Martin, S. Reichlin, and K. L. Bick.
Raven Press, New York © 1981.

Substance P and Somatostatin Actions on Spinal Cord Neurons in Primary Dissociated Cell Culture

Robert L. Macdonald and *Linda M. Nowak

*Department of Neurology and *Neurosciences Program, University of Michigan, Ann Arbor, Michigan 48109*

Substance P (SP) and somatostatin (somatotropin-release inhibiting factor: SRIF) are peptides with widespread regional distribution in the central nervous system (CNS) (9,10,33,55,63,70,72), peripheral nervous system (PNS) (34-38) and gut (1,32,56,62,70,71) which have been localized to neuronal cell bodies, fibers, synaptic terminals and synaptic vesicles (6,15,20,22,39,75). Calcium-dependent release of the peptides from various CNS regions using electrical stimulation or potassium-induced polarization has been demonstrated (5,7,41,68,87). Thus, the available evidence suggests that SP and SRIF may be neurotransmitters. SP and SRIF have been extensively investigated in the spinal cord where both peptides are neurotransmitter candidates for small-diameter primary afferent fibers (34,36,54,73). The peptides have been found in separate populations of small dorsal root ganglion (DRG) neurons, and nerve fibers and synaptic terminals containing SP-like and SRIF-like immunoreactivity have been localized to the dorsal horn of the spinal cord (2,13,34,36,38,55). SP and SRIF have been released from dorsal root afferent terminals in vivo (43) and in vitro (59,60,68) using electrical stimulation or potassium depolarization, and the release was calcium-dependent. Dorsal root and dorsal horn concentrations of SP were greater than ventral root and ventral horn concentrations respectively (2,29,54,73,85), and transection or ligation of dorsal roots reduced the SP content of spinal cord (13,37,38,44,85). A role for SP and possibly SRIF in transmitting noxious stimuli is particuarly likely (30,38). When applied iontophoretically, SP and SRIF stimulated dorsal horn neurons receiving nociceptive input but not those receiving mechanoreceptor input (30,76,77). Stimulation of small-diameter C and A fibers, but not low threshold rapidly conducting afferents, released both peptides (43). Capsaicin administration also released SP from dorsal root afferents (87), resulting in depletion of SP (44,92) and thermal analgesia (92). Thus, there is considerable evidence suggesting that SRIF and SP are released from dorsal root afferent synaptic terminals, and that they may be involved in mediating afferent information carried by small-diameter primary afferents.

The mechanism of action of these peptides on spinal cord neurons remains uncertain. It is difficult, therefore, to directly assign them a specific synpatic function. We have investigated the actions of SP and SRIF on spinal cord neurons in primary dissociated cell (PDC) culture and found that the peptides had direct excitatory postsynaptic actions (SP) and altered transmitter release from presynaptic terminals (SRIF,SP).

<u>Substance P</u>

Substance P is an undecapeptide discovered in 1931 by von Euler and Gaddum (23) when they found that extracts of equine brain and intestine contracted an isolated intestine preparation. Research into the pharmacological actions and CNS distribution of SP was hampered by unavailability of the purified peptide until Leeman and her colleagues (53) isolated and purified a "sialogogic" peptide which was identical to SP (52). Subsequent amino acid sequencing (12) and synthesis (88) made SP available for study. The physiological and pharmacological actions of SP have been studied in both CNS and PNS, and SP has been shown to be excitatory (16, 17,21,25,27,28,30,31,45-58,50,51,61,64-67,74,76,82,90,93). When applied to spinal cord neurons by iontophoresis or bath profusion, SP depolarized membrane potential and increased firing rate with a long latency (0.5-15 sec) and duration (3 sec-20 min) (45-48,61, 66,67,82,90,93). During SP responses, membrane conductance decreased (48,82), increased (61,90) or did not change (82,83). Based on the apparent SP reversal potential of -86 to -88 mV and an increase in membrane resistance in cat spinal motor neurons (48), it was suggested that SP reduced either resting potassium or chloride conductance. In guinea pig myenteric neurons, SP decreased potassium conductance (45,56). In contrast, in frog spinal cord SP depolarized motor neurons with either no conductance change or a slight conductance increase (12) leading to the suggestion that SP increased sodium conductance. In mouse spinal cord neurons in PDC culture, a different type of SP response was described (90) which had a rapid onset and a short duration (<100 msec), was rapidly desensitized and was associated with an increase in membrane conductance. The response had a reversal potential more positive than 0 mV suggesting that sodium and/or calcium conductance was increased. Thus while SP had been shown to be excitatory, uncertainty exists concerning the ionic mechanisms underlying this action.

Presynaptic actions of SP have also been reported. At the Mauthner fiber-giant synapse, low SP concentrations reduced postsynaptic potential (PSP) amplitude without reducing quantal content (presumably a postsynaptic effect), but at higher concentrations, miniature PSP frequency and PSP quantal content were reduced (presynaptic actions) (84). Calcium-dependent presynaptic actions of SP were reported at the frog neuromuscular junction (83) where SP initially depressed endplate potentials (EPP) by reducing quantal content. After this initial depression, SP increased EPP amplitude and quantal content to five times the control values. In addition to having pre- and postsynaptic actions, SP antagonized nicotinic cholinergic responses recorded from cat Renshaw cells (4,17,49,81) and glutamate responses recorded from cat dorsal horn neurons (48), mouse spinal cord neurons in PDC culture (90) and rat substantia nigra neurons (16). Thus, SP may modulate specific excitatory neurotransmitter responses.

We have studied the actions of SP and a structurally similar peptide, eledoisin-related peptide (ERP), on spinal cord neurons in PDC culture and found that both peptides produce excitatory responses by decreasing resting potassium and sodium conductances.

Somatostatin

Somatostatin is a tetradecapeptide originally isolated from hypothalamus where it was demonstrated to inhibit growth hormone release (8,11). Studies of the pharmacological and physiological actions of SRIF on CNS and PNS neurons have been conflicting. SRIF inhibited spontaneous or evoked activity in cerebellar (80), cortical (80), hypothalamic (80), spinal cord dorsal horn (77) and myenteric (91) neurons, but was excitatory on hippocampal pyramidal (19) and sensorimotor cortical (40) neurons. SRIF could have altered neuronal activity via pre- or postsynaptic mechanisms. SRIF had had excitatory presynpatic action on the dorsal root-ventral root potential of the isolated frog spinal cord (69) and increased the frequency of postsynaptic potentials recorded in cortical neurons in PDC culture (18). In contrast, SRIF decreased chick DRG calcium action potential duration (6) which would have reduced transmitter release (if occurring in DRG synaptic terminals). In the periphery, SRIF acted presynaptically to antagonize adrenergic transmission in the rat vas deferens (14,58) and cholinergic transmission in the guinea pig ileum (14,24,26). Postsynaptic actions of SRIF have also been reported. SRIF inhibited guinea pig myenteric neurons (91), hyperpolarized frog motor neurons and stimulated cholinergic nerves in guinea pig ileum (24). However, SRIF depolarized cat hippocampal pyramidal neurons (19). Thus while both pre- and postsynpatic actions of SRIF have been described, variations in preparations and techniques used make interpretation of CNS SRIF actions difficult.

In addition to the above synpatic actions, SRIF selectively depressed glutamate, but not GABA, responses evoked on dorsal and ventral roots of the frog spinal cord (69) and augmented glutamate responses in rabbit sensorimotor cortex (40). Thus SRIF may modulate excitatory glutamate responses.

We have studied the action of SRIF on murine spinal cord neurons grown in PDC culture and have found that SRIF evoked both inhibitory and excitatory PSPs in a dose-dependent fashion, suggesting a presynpatic action of SRIF.

METHODS

Neuronal cell cultures were prepared from spinal cords and attached dorsal root ganglia removed from 12.5 to 14-day old fetal mice as described previously (79). Intracellular recordings were made from spinal cord neurons using high resistance (25-50 M Ω) glass micropipettes on the modified stage of an inverted phase microscope which was heated to maintain a culture temperature of 35^{0}-37^{0}. SRIF, SP and ERP were applied to the surface of neurons (which had been impaled by a recording micropipette) using blunt (5-10 μ) micropipettes [miniperfusion (MP) pipettes] filled with peptides (200 nM-25 μ M). The open end of the MP pipettes was connected to a pressure regulator via a voltage-activated three-way valve. By using an electronic timer, the peptides could be applied for known durations. Peptides were purchased from several sources (Beckman, Bachem, Sigma and Cal Biochem), dissolved in weakly acidic buffers, some containing physiological saline (0.9%) and bovine serum albumin (0.1%) (BSA), and then frozen. Dithiothreitol (Cleland's reagent, 6 μ M) was added to the SP in some experiments to reduce oxidation. Glassware was precoated with a BSA buffer to reduce peptide binding to the glass. For each experiment, peptide stock solutions were diluted with the appropriate bathing medium, and pH was adjusted to between 7.2 and 7.4.

RESULTS AND DISCUSSION

Substance P

Application of either SP (Fig. 1A) or ERP (Fig. 1B) to spinal cord neurons depolarized membrane potential and increased firing rate. With relatively high peptide concentrations, a bursting pattern occurred frequently (Fig. 1A,B). The responses were of long latency (0.7-1.5 sec) and long duration (30-120 sec) and were similar for both ERP and SP. The peptides were usually applied for one second but similar responses were produced with pulses as short as 0.1 sec. Membrane resistance was assessed by applying frequent hyperpolarizing constant current pulses, and when either SP or ERP (Fig. 1C) was applied, membrane resistance either increased or did not change. For all large (>2 mV) responses, membrane resistance increased. The peptides produced dose-dependent depolarizations (Fig. 2A,B) which were unaltered or increased by substitution of 3M KCl-containing for 4M KAc-containing recording micropipettes (Fig. 2C). Recording with KCl-containing micropipettes increased the intracellular chloride concentration, and decreased the chloride equilibrium potential to about -15 mV. Under these circumstances, a decrease in chloride conductance would have produced hyperpolarizing responses. Since KCl-containing recording electrodes did not invert the ERP responses, they were probably due to a reduction in chloride conductance. To identify

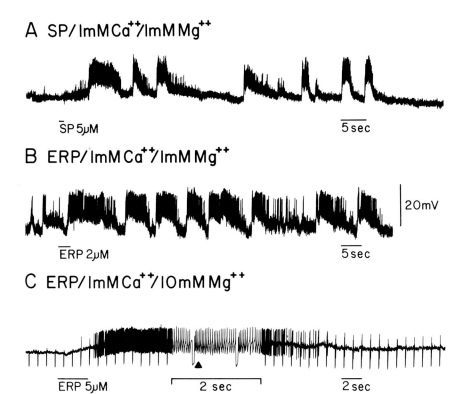

FIG. 1. Substance P (SP) and eledoisin-related peptide (ERP) evoked excitatory responses from cultured mouse spinal cord neurons. A) SP produced a slow depolarization (0.75 sec latency) and increased activity including paroxysmal bursts. Resting membrane potential (RMP) was -48 mV. B) ERP produced a similar increase of activity (RMP -56 mV). C) ERP slowly and reversibly depolarized membrane potential and decreased membrane conductance (RMP -50 mV). Magnesium concentration was increased to 10 mM to reduce the spontaneous potentials. Membrane conductance was measured by passing 0.7 nA, 3 msec constant current pulses at 800 msec intervals through the recording electrode to produce brief membrane hyperpolarizations. Larger hyperpolarizations (filled triangles) during the ERP-produced depolarization indicated an increase of cell membrane resistance. The expanded portion of the record shows a high rate of action potentials during the depolarization that decreased as the cell repolarized. Action potentials were truncated by the limited frequency response of the pen recorder. [Intracellular recordings were made with 4M KAc filled micropipettes in phosphate buffered saline containing 1 mM Mg^{++} (except C)].

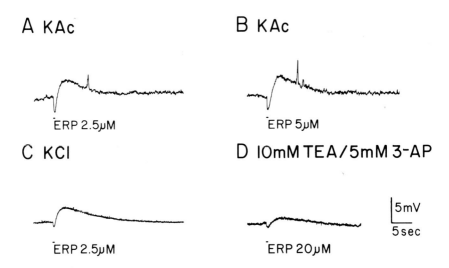

FIG. 2. ERP responses were dependent upon potassium conductance.
A,B) ERP responses increased as ERP concentration was increased
from 2.5 µ M (A) to 5 µM (B). PSPs were also evoked during the
falling phase of the response, and PSP amplitude and frequency
were dose-dependent when observed (RMP -67 mV). C) Changing from
4M KAc-containing to 3M KAc-containing recording micropipettes had
little effect on the response to ERP (compare A and C) (RMP -64
mV). D) The ERP response was attenuated by addition of 10 mM
tetraethylammonium bromide (TEA) and 5 mM 3-aminopyridine (3-AP)
(RMP -59 mV). Signals were filtered (15 Hz $\pm$ 10%, -3db) so that
high gain recordings could be made. The early hyperpolarizations
seen in these records were pressure artifacts. Recordings were
made in phosphate buffered saline with 10 mM Mg^{++}/1 mM Ca^{++} (A,B
and C) and in tris-HCl buffered saline with 0.8 mM Mg^{++}/5 mM Ca^{++}
(D).

whether potassium conductance changes were involved, the potassium
channel blockers tetraethylammonium (TEA) and 3-aminopyridine
(3-AP) were added to the recording medium and the ERP response was
reduced (Fig. 2D). Thus it was probable that SP and ERP reduced
resting potassium conductance. To confirm this, we determined reversal potentials for the ERP responses (E_{erp}) by penetrating
single neurons with two recording micropipettes, one for recording
and the other for passing current to vary membrane potential. The
ERP response had an extrapolated E_{erp} of -76 mV in 5 mM potassium
(Fig. 3), and increasing the extracellular potassium to 20 and

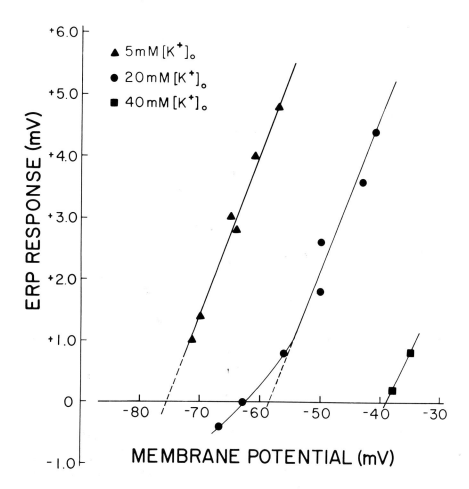

FIG. 3. The reversal potential of ERP responses was dependent upon extracellular potassium concentration ($[K^+]_o$). ERP was applied to neurons impaled with two intracellular recording micropipettes, one to apply current and the other to record membrane voltage. ERP was applied to each cell at different membrane potentials and extrapolated (to correct for anomalous rectification) reversal potentials (E_{erp}) were obtained (RMP -70 mV in 5 mM $[K^+]_o$; -58 mV in 20 mM $[K^+]_o$; -38 mV in 40 mM $[K^+]_o$). Recordings were made in tris-HCl buffered saline containing 5 mM Ca^{++} to facilitate two electrode penetrations.

40 mM shifted the extrapolated E_{erp}s to the lower values of -58 and -39 mV. When the extrapolated E_{erp}s were plotted against the

log of the extracellular potassium concentration, a linear plot was not obtained. At the higher potassium concentrations, the E_{erp}s formed a line with a slope of -60 mV/decade change in potassium concentration which is the slope predicted from the Nernst equation for potassium. At lower potassium concentrations, however, the E_{erp}s were more depolarized than predicted by the Nernst equation suggesting that ERP and SP decreased sodium conductance as well as potassium conductance. The low E_{erp} values at low potassium concentrations also suggest that potassium conductance was diminished relatively more than sodium conductance leading to an increase in the sodium to potassium permeability ratio. Thus, SP and ERP produced excitatory responses on spinal cord neurons and depolarized membrane potential by blocking resting potassium and sodium conductances. In addition, SP and ERP evoked release of transmitter from presynaptic terminals (Fig. 2A,B) which was blocked by TTX (not illustrated). Whether this was a direct action of ERP on the terminals or a consequence of diffusion of ERP to and stimulation of adjacent neurons presynaptic to the neuron under study could not be determined from our data.

Thus we have demonstrated directly that SP and ERP had dose-dependent pre- and postsynaptic actions on spinal cord neurons in PDC culture. The postsynpatic excitatory action was due to decreases in resting potassium and sodium conductances and was consistent with findings of decreased membrane conductance in cat anterior horn cells (48) and guinea pig myenteric neurons (45,46). In other studies of cat dorsal horn (82) and motor (93) neurons, no consistent change in conductance was reported. This may have resulted from either a difference in the experimental preparations or technique. Both of the latter studies used single recording electrodes with iontophoretic application of SP, and resistance was measured using the single electrode bridge technique. In our experiments, two independent intracellular electrodes were employed and SP was applied by MP. Thus a larger portion of the neuronal surface area was bathed in SP, and our resistance measurements were independent of bridge technique. Small increases in membrane resistance sufficient to produce a depolarization might have been inapparent in the studies reporting no change in resistance. In frog spinal cord neurons, slight increases in conductance were noted following SP perfusion (61). Again, this may reflect a difference between preparations, but an alternate explanation is that delayed rectification (voltage-dependent increase in potassium conductance) may have masked a concomitant SP-induced decrease in potassium conductance. Finally, the rapidly desensitizing SP response associated with an increase in membrane conductance described in mouse spinal cord neurons in PDC culture (90) was not observed in these experiments. Recent findings by the same investigators that similar responses were produced by application of solutions with low pH suggest that these responses might have been a consequence of reduced pH and not SP (57).

The prolonged time course and hyperpolarized reversal potential of SP reponses make it unlikely that SP mediates the rapid transmitter responses with depolarized reversal potential elicited in anterior horn cells by Ia afferent stimulation. However, stimulation of small diameter fibers mediating nociceptive information characteristically produced prolonged responses (30,31,76,82) with after-discharges consistent with these SP responses. Capsaicin treatment has been shown to release SP from primary afferents (92), to deplete spinal cord SP (42), and to produce analgesia (87) consistent with a role for SP in mediating nociceptive input via small diameter primary effects.

Reduction of resting conductance by SP might have a number of consequences for neuronal function. For example, a decrease in conductance in the postsynaptic membrane beneath a single synaptic terminal would have little effect on overall membrane conductance. Therefore, to have a major impact, the SP afferent fiber must either: 1) ramify and provide many terminals, 2) terminate on small cells, 3) release relatively large amounts of SP for diffusion to other membrane sites or 4) activate an intracellular "second messenger" which would decrease potassium and sodium conductance throughout the neuron. Alternatively, SP could have a significant action if it decreased dendritic conductance and thus increased space constant allowing previously ineffective synaptic potentials to reach the soma by electrotonic spread. Such an action might have little impact on overall neuronal conductance but would have a significant effect on synaptic inputs to the involved dendrite. Finally, the excitatory action of SP would depend very much on neuronal resting membrane potential. If the cell was hyperpolarized to about the potassium equilibrium potential, then little alteration in membrane potential would occur following release of SP. If, however, the cell was depolarized, the response to SP release would increase. Thus, modification of resting membrane potential by other synaptic input, for example, would have a marked action on the effectiveness of synaptically released SP.

Somatostatin

When applied to SC neurons at concentrations up to 20 μM, SRIF usually did not alter membrane potential or input resistance but occasionally hyperpolarized membrane potential. On other cells, SRIF application evoked volleys of large PSPs (Fig. 4). This effect of SRIF was rapid in on- and offset, reversible and dose-dependent (Fig. 4A) and was not produced by application of control medium. PSPs were not evoked by SRIF when sodium-dependent action potentials were blocked by addition of TTX (Fig. 4B) and SRIF evoked both EPSPs and IPSPs (Fig. 5). In addition to variable sensitivity to SRIF between neurons, there was variable release on single spinal cord neurons. Regions of high sensitivity were found where SRIF evoked brisk release of neurotransmitter while applications of SRIF to other regions of the neuron did not

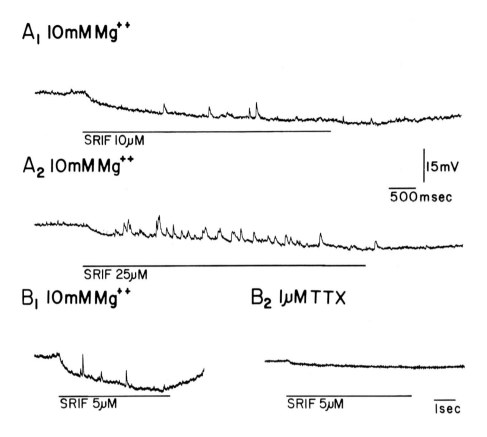

FIG. 4. Somatostatin (SRIF) evoked transmitter release. A) SRIF evoked dose-dependent release of neurotransmitter from synaptic terminals demonstrated by an increase of PSP frequency and amplitude in responses to 10 µM SRIF (A_1) and 25 µM SRIF (A_2). B) SRIF-evoked transmitter (B_1) release was not observed when 1 µM TTX was added to the recording medium (B_2). The same MP pipette was used to apply peptide to successive neurons in B_1 and B_2.

evoke PSPs. These findings suggested that SRIF at relatively high concentrations (1-20 µM) applied by MP to the surface of spinal cord neurons evoked release of neurotransmitter from synaptic terminals contacting the SC neuronal surface. Specificity of action of SRIF was suggested by the lack of uniform sensitivity between neurons and on individual neurons. Thus SRIF may act on specific "SRIF sensitive" presynaptic terminals to modify neurotransmitter release. A similar presynaptic action of SRIF has also recently been described using cortical neurons in PDC culture (18).

The mechanism of action of SRIF was not clearly defined in these experiments but it is most likely that the peptide modified

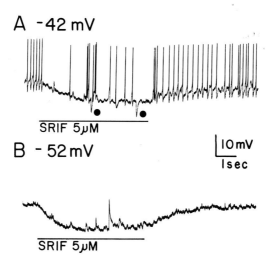

FIG. 5. SRIF evoked both inhibitory and excitatory postsynaptic potentials. A) SRIF application produced by EPSPs and IPSPs (filled circles) at a membrane potential of -42 mV. The decreased spike activity observed and hyperpolarization were due to a pressure artifact. B) Following hyperpolarization of the cell membrane potential to -52 mV, only depolarizing synaptic potentials were evoked (EPSPs and inverted IPSPs).

calcium entry into presynaptic terminals thereby affecting release of neurotransmitter. SRIF in low concentrations decreased neurotransmitter release from neurons in guinea pig ileum (14,24,26) and rat vas deferens (58) but at high concentrations caused contractions in distal colon (24). Since these actions were blocked by TTX and postsynaptic sensitivity to neurotransmitter was unaltered, SRIF likely had a presynaptic site of action. Since the actions of SRIF were sensitive to extracellular calcium concentration (24) and SRIF altered calcium uptake and release (96) from hippocampal synaptosomes, it is also likely that SRIF regulates the release of neurotransmitter (or hormone) by modifying calcium entry into presynaptic terminals.

Elucidation of the physiological actions of synpatically released SRIF, however, will require the study of synaptic transmission between SRIF-containing DRG and spinal cord neurons.

CONCLUSION

In this paper, we have demonstrated that two peptides, SP and SRIF, which have been localized to and are released from small DRG neurons, have specific but different actions on mouse spinal cord neurons in PDC culture. SP has been shown to be excitatory by decreasing potassium conductance and to a lesser extent sodium conductance; SP also appeared to evoke release of neurotransmitter. SRIF has been shown to modify neurotransmitter release by a presynaptic action. The extensive experimental control that is possible over neurons in vitro permits precise measurements of specific membrane actions to be made and thus allows hypotheses of cellular mechanisms of peptide actions to be advanced. Predictions made based on these hypotheses should then be tested using intact preparations. A reasonable goal for such studies is a detailed understanding of the cellular and ionic mechanisms underlying actions in the CNS, PNS and gut.

REFERENCES

1. Arimura, A., Sato, H., Dupont, A. Nishi, N., and Schally, A.V. (1975): Science, 189:1007-1009.
2. Barber, R.P. Vaughn, J.E., Slemmon, J.R., Salvaterra, P.M., Roberts, E., and Leeman, S.E. (1979): J. Comp. Neurol., 184: 331-351.
3. Barker, J.L., and Ransom, B.R. (1978): J. Physiol. (London), 280:331-354.
4. Belcher, G., and Ryall, R.W. (1977): J. Physiol. (London), 272:105-119.
5. Bennett, G.W., Edwardson, J.A., Marcano de Cotte, M., Berelowitz, M., Pimstone, B.L., and Kronheim, S. (1979): J. Neurochem., 32:1127-1130.
6. Berelowitz, M., Hudson, A. Pimstone, B., Kronheim, S., and Bennett, G.W. (1978): J. Neurochem., 31:751-753.
7. Berelowitz, M., Kronheim, S., Pimstone, B., and Sheppard, M. (1978): J. Neurochem., 31:1537-1539.
8. Brazeau, P., Vale, W., Burgus, R., Ling, N., Butcher, M., Rivier, J., and Guillemin, R. (1973): Science, 179:77-99.
9. Brownstein, M., Arimura, A., Sato, H., Schally, A.V. and Kizer, J.S. (1975): Endocrinology, 96:1456-1461.
10. Brownstein, M.J., Morz, E.A., Kizer, J.S., Palkovits, M., and Leeman, S.E. (1976): Brain Res., 116:299-305.
11. Burgus, R., Ling, N., Butcher, M., and Guillemin, R. (1973): Proc. Natl. Acad. Sci. USA, 70:684-688.
12. Chang, M.M., Leeman, S.E., and Niall, H.D. (1971): Nature New Biol., 232:86-87.
13. Chan-Palay, V., and Palay, S.L. (1977): Proc. Natl. Acad. Sci. USA, 74:3597-3601.

14. Cohen, M.L., Rosina, E., Wiley, K.S., and Slater, I.H. (1978): Life Sci., 23:1659-1664.
15. Cuello, A.C. Jessell, T., Kanazawa, I., and Iversen, L.L. (1977): J. Neurochem., 29:747-751.
16. Davies, J., and Dray, A. (1976): Brain Res., 107:623-627.
17. Davies, J., and Dray, A. (1977): Nature (London) 268:351352.
18. Delfs, J., Dichter, M., Robbins, R., Connolly, J., and Reichlin, S. (1979): Neurosci. Abstr., 5:526.
19. Dodd, J., and Kelly, J.S. (1978): Nature (London), 273:674-675.
20. Duffy, M.J., Mulhall, D., and Powell, D. (1975): J. Neurochem., 25:305-207.
21. Dun, N.J., and Karczmar, A.G. (1979): Neuropharmacology, 18:215-218.
22. Epelbaum, J., Brazeau, P., Tsang, D., Brawer, J., and Martin, J.B. (1977): Brain Res., 126:309-323.
23. Euler, V.S. von, and Gaddum, J.H. (1931): J. Physiol. (London), 72:74-87.
24. Furness, J.B., and Costa, M. (1979): Eur. J. Pharmacol., 56:69-74.
25. Grafe, P., Mayer, C.J., and Wood, J.D. (1979): Nature (London), 279:720-721.
26. Guillemin, R. (1976): Endocrinology, 99:1653-1654.
27. Guyenet, P.B., and Aghajanian, G.K. (1977): Brain Res.,136:178-184.
28. Guyenet, P.G., and Aghajanian, G.K. (1979): Eur. J. Pharmacol., 53:319-328.
29. Hellauer, H. (1953): Naumyn-Schmiedeberg's Arch. Exp. Path. Pharmak., 219:234-241.
30. Henry, J.L. (1976): Brain Res., 114:439-451.
31. Henry, J.L., Krnjevic, K., and Morris, M.E. (1975): Can. J. Physiol. Pharmacol., 53:423-432.
32. Hokfelt, T., Efendic, S., Hellerstrom, C., Johansson, O., Luft, R., and Arimura, A. (1975): Acta Endocrinol. Cophn., 80 Suppl. 200: 5-41.
33. Hokfelt, T., Efendic, S., Johansson, O., Luft, R. and Arimura, A. (1974): Brain Res., 80:165-169.
34. Hokfelt, T., Elde, R., Johansson, O., Luft, R., Nilsson, G., and Arimura, A. (1976): Neuroscience, 1:131-136.
35. Hokfelt, T., Elfvin, L.G., Elde, R., Schultzberg, M., Goldstein, M., and Luft R. (1977): Proc. Natl. Acad. Sci. USA, 74: 3587-3591.
36. Hokfelt, T., Elde, R., Johansson, O., Luft, R., and Arimura, A. (1975): Neurosci. Lett., 1:231-235.
37. Hokfelt, T., Kellerth, J.O., Nilsson, G., and Pernow, B. (1975): Science, 190:889-890.
38. Hokfelt, T., Kellerth, J.O., Nilsson, G., and Pernow, B. (1975): Brain Res., 100:235-252.
39. Inoue, A., and Kataoka, K. (1962): Nature (London), 193: 585.

40. Ioffe, S., Havlicek, V., Friesen, H., and Chernick V. (1978): Brain Res., 153:414-418.
41. Iversen, L.L., Iversen, S.D., Bloom, F., Douglas, C., Brown, M., and Vale, W. (1978): Nature (London), 273: 161-163.
42. Jessell, T.M., Iversen, L.L., and Cuello, A.C. (1978): Brain Res., 152:183-188.
43. Jessell, T.M., Mudge, A.W., Leeman, S.E., and Yaksh, T.L. (1979): Neurosci. Abstr., 5:611.
44. Jessell, T.M., Tsunoo, A., Kanazawa, I., and Otsuka, M. Brain Res., 168:247-259.
45. Katayama, Y., and North, R.A. (1978):   Nature (London), 274:387-388.
46. Katayama, Y., North, R.A., and Williams, J.T. (1979):   Proc. R. Soc. Lond., 206:191-208.
47. Konishi, S., and Otsuka, M. (1974):   Brain Res., 65:397-410.
48. Krnjevic, K. (1977): In: Substance P, edited by U.S. von Euler, and B. Pernow, pp. 217-230. Raven Press, New York.
49. Krnjevic, K., and Lekic, D. (1977): Can. J. Physiol. Pharmacol., 55:958-961.
50. Krnjevic, K., and Morris, M.E. (1974): Can. J. Physiol. Pharmacol., 52:736-744.
51. LaSalle, G., and Ben-Ari, Y. (1977): Brain Res., 135: 174-179.
52. Leeman, S.E., and Hammerschlag, R. (1967): Endocrinology, 81:803-810.
53. Leeman, S.E., and Mroz, E.A. (1975): Life Sci., 15: 2033-2044.
54. Lembeck, F. (1953): Arch. Exp. Path. Pharmak., 219:197-213.
55. Ljungdahl, A., Hokfelt, T., and Nilsson, G. (1978): Neuroscience, 3:861-943.
56. Luft, R., Efendic, S., Hokfelt, T., Johansson, O., and Arimura, A. (1974): Med. Biol., 52:428-430.
57. MacDonald, J.F., Barker, J.L., Gruol, D.L., Huang, L.M., and Smith, T.G. (1979): Neurosci. Abstr., 5:592.
58. Magnan, J., Regoli, D., Quirion, R., Lemaire, S., St. Pierre, S., and Rioux, F. (1979): Eur. J. Pharmacol., 55: 347-354.
59. Mudge, A.W., Fischbach, G.D., and Leeman, S.E. (1977): Neurosci. Abstr., 3: 410.
60. Mudge, A.W., Leeman, S.E., and Fischbach, G.D. (1977): Proc. Natl. Acad. Sci. USA, 76:526-530.
61. Nicoll, R.A. (1978): J. Pharmacol. Exp. Ther., 207:817-824.
62. Nilsson, G. (1975): Histochemistry, 43:97-99.
63. Nilsson, G., Hokfelt, T., and Pernow, B. (1974): Med. Biol., 52:424-427.
64. Ogata, N. (1979): Nature (London), 277:480-481.
65. Ogata, N. (1979): Brain Res., 176:395-400.
66. Otsuka, M., and Konishi, S. (1976): Cold Spring Harbor Symp. Quant. Biol., 40:135-143.
67. Otsuka, M., and Konishi, S. (1977): In: Substance P, edited by U.S. von Euler, and B. Pernow, pp. 207-214. Raven Press, New York.

68. Otsuka, M., and Konishi, S. (1976): Nature (London), 264: 83-84.
69. Padjen, A.L. (1977): Neurosci. Abstr., 3:411.
70. Parsons, J.A., Erlandsen, S.L., Hegre, O.D., McEvoy, R.C., and Elde, R.P. (1976): J. Histochem. Cytochem., 24: 872-882.
71. Pearse, A.G.E., and Polak, J.M. (1975): Histochemistry, 41:373-375.
72. Pelletier, G., Leclerc, R., Dube, D., Labrie, F., Puviani, R., Arimura, A., and Schally, A.V. (1975): Am. J. Anat., 142:397-401.
73. Pernow, B. (1953): Acta Physiol. Scand., 29, Suppl. 105: 1-90.
74. Phillis, J.W., and Limacher, J.J. (1974): Brain Res., 69: 158-163.
75. Pickel, V.M., Reis, D.J. and Leeman, S.E. (1977): Brain Res., 122:534-540.
76. Randic, M., and Miletic, V. (1977): Brain Res., 128:164-169.
77. Randic M., and Miletic, V. (1978): Brain Res., 152: 196-202.
78. Ransom, B.R., Bullock, P.N., and Nelson, P.G. (1977): J. Neurophysiol., 40:1163-1177.
79. Ransom, B.R., Neale, E., Henkart, M., Bullock, P.N., and Nelson, P.G. (1977): J. Neurophysiol., 40:1132-1150.
80. Renaud, L.P., Martin, J.B., and Brazeau, P. (1975): Nature (London), 255:233-235.
81. Ryall, R.W., and Belcher, G. (1977): Brain Res., 137: 376-380.
82. Sastry, B.R. (1979): Life Sci., 24:2169-2178.
83. Steinacker, A. (1977): Nature (London)., 267:268-270.
84. Steinacker, A., and Highstein, S.M. (1976): Brain Res., 114:128-133.
85. Takahashi, T., and Otsuka, M. (1975): Brain Res., 87:1-11.
86. Tan, A.T., Tsang, D., Renaud, L.P., and Martin, J.B. (1977): Brain Res., 123:193-196.
87. Theriault, E., Otsuka, M., and Jessel, T.M. (1979): Brain Res., 170:209-213.
88. Tregear, G.W., Niall, H.D., Potts, J.T., Leeman, S.E., and Chang, M.M. (1971): Nature New Biol., 232:87-88.
89. Vale, W., Brazeau, P., Rivier, C., Brown, M., Boss, B., Rivier, J., Burgus, R., Ling, N., and Guillemin, R. (1975): Recent. Prog. Horm. Res., 31:365-397.
90. Vincent, J.D., and Barker, J.L. (1979): Science, 205: 1409-1412.
91. Williams, J.T., and North, R.A. (1978): Brain Res., 155: 165-168.
92. Yaksh, T.L., Farb, D.H., Leeman, S.E., and Jessell, T.M. (1979): Science, 206:481-483.
93. Zieglgansberger, W., and Tulloch, I.F. (1979): Brain Res., 166:273-282.

Neurosecretion and Brain Peptides,
edited by J. B. Martin, S. Reichlin, and K. L. Bick.
Raven Press, New York © 1981.

Peptide and Amine Transmitter Effect on Embryonic Chick Sensory Neurons *In Vitro*

Gerald D. Fischbach, Kathleen Dunlap, Anne Mudge, and *Susan Leeman

*Departments of Pharmacology and *Physiology, Harvard Medical School, Boston, Massachusetts 02115*

Sensory neurons can be dissociated from 10 day old embryonic chick dorsal root ganglia (DRG) and grown in cell culture in the absence of peripheral or central targets or ganglionic non-neuronal cells. Such "pure" neuronal cultures were first described by Okun (18) who eliminated fibroblasts and glia by allowing these relatively adherent cells to attach to glass beads prior to plating the neurons. We have achieved the same result by plating cells dissociated from isolated lumbar and thoracic sensory ganglia in medium containing 5 x 10^6M cytosine arabinoside. This drug kills dividing cells and few, if any, fibroblasts or glia survive a 48-72 hour exposure. Naked sensory neurons form an extensive network of processes and remain healthy on a collagen substrate for several weeks when nurtured with Eagles Minimum Essential Medium supplemented with horse serum (10% v/v), embryo extract (5% v/v), antibiotics and NGF (3) (Fig. 1). One obvious advantage of this simplified culture system in studies of transmitter synthesis, storage and release is that there can be little doubt concerning the type of neuron responsible, the role of target tissue, or the role of glia.

A sensitive radioimmunoassay that can detect as little as 1.5 fmol of synthetic Substance P (14) was used to assay extracts of DRG cultures. At the time of plating, cells dissociated from 5 ganglia contain about 20 fmol of Substance P-like immunoreactivity (SPLI). Two weeks later culture dishes seeded with the same number of cells contain between 1 and 3 pmol of SPLI. Thus, embryonic neurons synthesize and rapidly accumulate SPLI in sparse cell culture. It is significant that the amount of SPLI present after 2 weeks in vitro is comparable to the amounts of SPLI found at hatching in a segment of spinal cord innervated by 5 DRGs.

Antibodies in the serum (R6P) used in the radioimmunoassay, recognize the carboxyterminal end of Substance P but they exhibit about 50% cross-reactivity with the carboxyterminal octapeptide

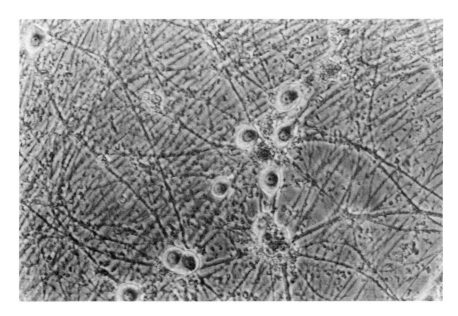

FIG. 1. Phase contrast micrograph of dissociated sensory neurons after 14 days in culture. These cultures were treated with cytosine arabinoside and few, if any, non-neuronal cells are visible.

(residues 4-11) of Substance P. The material in DRG cultures recognized by R6P migrated with synthetic Substance P during Sephadex G-25 gel filtration and eluted from sulfopropyl-Sephadex ion exchange resin at the same ionic strength (15). Moreover, SPLI in DRG cells displaced radiolabeled Substance P from the antibodies in parallel to the displacement caused by synthetic Substance P, whereas the octapeptide did not. Therefore, the SPLI in cultured sensory neurons is probably the undecapeptide.

Many but not all of the DRG cell bodies were darkly stained by the Sternberger peroxidase-antiperoxidase technique following incubation with a Substance P antiserum (RD2) (Fig. 2). In different cultures between 40-60% of the nerve cell bodies were judged to be labeled above background. This percentage of labeled cells could not be increased by prior incubation in colchicine. Although it is not certain that a lightly stained, "negative" cell contains no Substance P at all, it seems likely that the population of neurons present in 2-week old cultures is a heterogeneous one. To date, we have only used ganglia dissected from 9 or 10 day old embryos. Cell division within the neuronal precursor population has nearly ceased by this time and at least two groups of neurons distinguished by size and position (large, medio-dorsal cells and small, ventrolateral cells) are present in the ganglia (1,3). The round cell bodies all look alike in vitro, albeit with some variation in cell diameter, and they all generate the same broad action potentials (see below) but their biosynthetic preferences may be quite different.

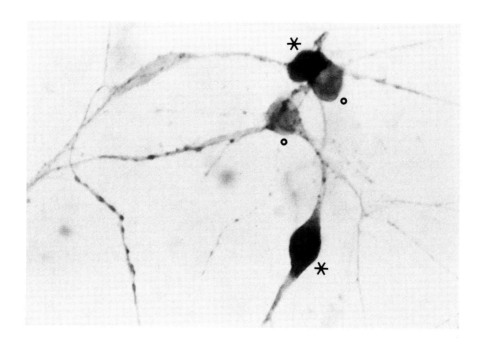

FIG. 2. Substance P immunohistochemistry: a bright field micrograph. Two neurons are stained for SPLI (*) and two neurons are
not (O). Cultures were fixed with paraformaldehyde and incubated
with a Substance P antibody (RD2) at 1:1000 dilution. The cultures were then washed and exposed successively to goat anti-rab-
bit IgG (1:100) and rabbit antiperoxidase conjugated to peroxidase (1:200). Bound antibody was visualized by reaction with
diaminobenzidine and H_2O_2. Control cultures were exposed to RD2
(1:1000) which had previously been adsorbed with excess Substance
P (1 µg/ml). All neurons in control plates appeared light brown.
They resembled the cells considered negative (O) in this figure.

It is significant in this regard that somatostatin has been
detected in 2-week old neuron-alone cultures by radioimmunoassay.
In addition, at least some of the cells can synthesize acetylcholine (ACh) and norepinephrine (NE) from radiolabeled precursors
(A. Mudge and P.H. Patterson, unpublished data). Substance P and
somatostatin are both present in adult sensory ganglia but the
two peptides are apparently not present in the same neurons (6).
Substance P may be restricted to the smallest neurons. Indeed,

this is one of the observations taken as evidence that Substance P is involved in signaling noxious stimuli. We are currently trying to determine if the same histochemical picture applies _in vitro_ and also to define conditions that regulate the synthesis and accumulation of the various transmitter candidates. The synthesis of somatostatin in DRG cell culture is greatly increased when the neurons are cocultured with non-neuronal cells (15). These developmental questions are important and answerable but it is the release of Substance P and the regulation of that release by neurotransmitters that is the issue in this report.

When the neurons were depolarized in high K^+ (60-120 mM) medium about 5% of the total SPLI was released into the medium during a 5 minute collection period (Fig. 3). Little, if any, SPLI was released during the same interval at normal membrane potential (K = 6 mM). The amount of SPLI released varied directly with extracellular Ca^{++}. No release could be evoked in 0 mM Ca^{++} or when 5 mM Co^{++} was added to the bath. Additional SPLI could be released when high K^+ pulses were repeated but the amount released in 5 min was decreased (Fig. 3). We have no sure explanation for this fall off. It may be that some aspect of the release mechanism, perhaps the influx or the mobilization of Ca^{++}, inactivates with time. A similar phenomenon has been observed in studies of amino acid release from intact nerves and from synaptosomes so it is not unique to the release of peptides.

We have recently found that much larger amounts (up to 15%) of SPLI can be released during 5 minute periods of electrical stimulation. In these experiments two large (2-5 mm) electrodes were fixed in the dish and the cells were activated by brief electrical pulses delivered at frequencies between 1/sec-10/sec. Intracellular recordings from neurons located in different regions of the 35 mm culture dish indicate that all of the cells are activated by each stimulus even at frequencies as high as 20/sec. The fact that more SPLI can be released by intermittent low frequency electrical pulses than by sustained depolarization serves tro emphasize the unphysiological nature of the commonly used high K^+ stimulation.

Evidence that Substance P is involved in some way in nociception includes the immunohistochemical demonstration of Substance P in small sensory neuron cell bodies and in fine unmyelinated afferents that terminate in the dorsal horn of the spinal cord (5,6) and the ability of Substance P to excite dorsal horn interneurons that also respond to noxious stimuli (4). Enkephalin is present in some dorsal horn interneurons (5) and opiate receptors are present on afferent fibers in the dorsal horn (11). This scenario suggests that one mechanism whereby opiates interrupt incoming impulses conducted along the pain pathway is by preventing the release of Substance P in the dorsal horn.

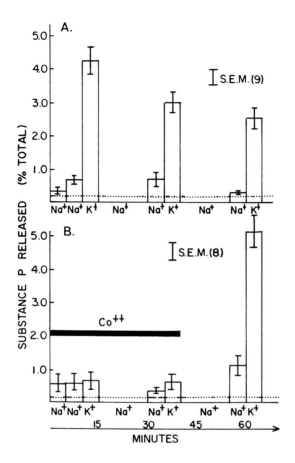

FIG. 3.   Potassium-evoked release of Substance P.   In A, 12
day-old cultures were incubated for successive 5-min periods in
Hepes-buffered salt solutions (pH 7.4) containing the usual
cation concentration indicated as Na^+ (Na^+, 130mM; K^+ 6mM; Ca^{2+},
18mM; Mg^{2+} 0.8mM) or high K^+ indicated as K^+ (Na^+, 17mM; K^+, 119
mM; Ca^{2+}, 18mM; Mg^{2+}, 0.8mM). The amount of peptide present in
the solution at the end of each 5-min incubation is expressed as
percent total of the peptide present in the cells.   The dashed
line gives the limit of detection of the assay. In B, 5mM Co^{2+}
was added to all solutions for the first 40 min of incubation.
The Ca^{2+} concentration was lowered to 0.2mM and Mg^{2+} was omitted
when Co^{2+} was present.

It has been known for some time that opiates can decrease the release of ACh from ganglionic neurons or preganglionic fibers in the gut (19). Konishi et al. (10) recently showed that enkephalin can decrease the mean quantum content of fast, cholinergic epsps recorded in inferior mesenteric ganglion neurons following stimulation of preganglionic nerves. McDonald and Nelson (12) found that the quantum content of epsps recorded in cultured mouse spinal cord neurons following stimulation of co-cultured DRG neurons was decreased by etorphine. The transmitter in this case is unknown. A comparable quantal analysis of putative Substance P synapses has not yet been performed. This is hardly surprising considering that no fast (msec time scale) peptide-mediated, synaptic responses have been described and it is not known if peptides are released in a quantal or graded manner. However, Jessell and Iversen (8) and we (16) have reported that

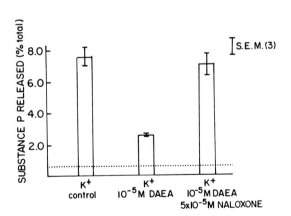

FIG. 4. Inhibition of Substance P release by D Ala$_2$-enkephalin amide (DAEA). Release of Substance P from sibling cultures during depolarization for 5 min. All three groups were depolarized by 30mM potassium; the second group was depolarized in the presence of 10^{-5}M DAEA; the third group was depolarized in the presence of 10^{-5}M DAEA plus 5 x 10^{-5}M naloxone.

enkephalin at 10^{-6}M to 10^{-5}M can inhibit the K^{+}-evoked release of Substance P from sensory neurons. Jessell and Iversen studied slices through the trigeminal nucleus.

Fig. 4 shows the effect of an enkephalin analog on SPLI release from cultured chick DRG neurons. Release was inhibited by 60% at maximal doses of enkephalin and the inhibition was completely blocked by naloxone. Enkephalin did not inhibit release in every experiment. The failures are not easily explained by the "unphysiological" nature of the K^{+} depolarization because we have encountered the same variability in more recent experiments in which release was evoked by electrical pulses. It is possible that opiate receptors are regulated in an as yet unpredictable way as the sensory neurons mature <u>in vitro</u>. All cultures from a

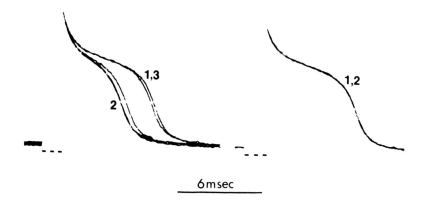

6 msec

FIG. 5. The effect of enkephalin on DRG soma action potentials. On the left the number 1 denotes the control spike; 2 indicates the spike after pressure ejection of 0.1μM enkephalin; 3 shows return to control spike duration about one minute after the ejection pulse. The superimposed traces on the right (recorded in the same cell from which the traces on the left were obtained) show that 1 μM naloxone ejected along with 0.1μM enkephalin blocked the peptide effect on spike duration.

given plating shared the same fate, i.e., either they were all inhibited by enkephalin or none were affected, and this might be taken as an argument that small changes in the conditions of dissociation or growth in vitro are crucial. We plan to investigate this question using a more defined medium and radiolabeled opiate receptor ligands.

In addition to enkephalin, NE, serotonin (5-HT) and gamma amino butyric acid (GABA) have also been implicated as presynaptic "modulators" of the pain pathway at the spinal cord level. Each of these transmitters (and also somatostatin) inhibited the K^+ evoked release of Substance P in vitro. Their effects were additive: inhibition was complete when DRG neurons were exposed to a cocktail of all four agents.

Enkephalin, somatostatin, NE, GABA and 5-HT all decrease the duration of the action potential recorded in DRG neuron cell bodies. The effect of enkephalin applied at 10^{-6}M by pressure ejection from a nearby, blunt tip pipette is shown in Fig. 5. No affect on the spike was evident when naloxone was ejected along with the enkephalin.

The broad action potential characteristic of embryonic chick sensory neurons in vitro is a mixed Na^+ -Ca^{++} spike (2). That is, inward current activated by depolarization is carried by both Na^+ and Ca^{++} ions. Na^+ and Ca^{++} voltage-sensitive channels can be distinguished by physiological and pharmacological criteria. The relatively large Na^+ current (I_{Na}) is responsible for the rapid upstroke of the action potential; it turns on and off rapidly and it is blocked by tetrodotoxin (TTX). The smaller Ca^{++} current (I_{Ca}) waxes and wanes relatively slowly. It is, in part, responsible for the plateau phase of the spike. Repolarization, of course, is ordinarily triggered by a delayed, voltage-dependent increase in outward K^+ current (I_{K^+}). Thus, a decrease in spike duration implies a decrease in Ca^{++} influx. The importance of this observation on the soma spike for the studies of SPLI release outlined above is that the same or similar voltage-dependent Ca^{++} channels are probably present along the axon at sites of transmitter release. If the several modulators of I_{Ca} in the soma exert the same action at synapses, then the decrease in Ca^{++} influx might account for their ability to inhibit Substance P release. Unfortunately, it is impossible, at present, to penetrate DRG nerve terminals (or axon varicosities) reliably with microelectrodes. Therefore, while recognizing that the relevance of the soma Ca^{++} spike to transmitter release remains an assumption, we have investigated the cell body spike in some detail.

A decrease in spike duration and I_{Ca}, might be due to a direct effect of the drugs on inward I_{Ca} or it might be due to a drug induced increase (or early activation) of an outward I_K. In the latter case I_{Ca} would be reduced secondarily, because the membrane repolarized at an earlier time. These alternatives were explored by recording membrane currents in voltage-clamped DRG cell bodies (Dunlap and Fischbach, unpublished). Most voltage-clamp experiments have been performed with NE, but preliminary experiments indicate that enkephalin, GABA, and 5HT act in a similar, if not identical, way.

Relatively large neurons (> 35 μm across) were penetrated with two microelectrodes, and a high-gain feedback amplifier was used to drive current through one microelectrode while recording the membrane potential with the other. In adequately clamped cells it is possible to shift the membrane potential to any desired level and then hold the potential constant (clamped) by feeding back current that is exactly equal but opposite to the membrane current activated at the new potential. The effect of membrane capacitance is minimized so the feedback current provides an accurate estimate of the magnitude and time course of the net ionic current. A typical record obtained in control medium is shown in Fig. 6. A command pulse that depolarized the membrane from -50 mV to +10 mV activated a transient inward current that passed smoothly into a large and maintained outward current. As noted above, the inward current is carried mostly by Na^+ but Ca^{++} ions also contribute. The outward K^+ current is made up of at least two components. One, I_K (V) is activated by depolarization. The exact voltage dependence on this channel in chick DRG neurons has not been investigated in detail. The other, I_K (Ca) is independent of membrane potential but is activated by the preceding entry of Ca^{++}. Such Ca^{++} activated, K^+ channels have been detected in neurons of several species.

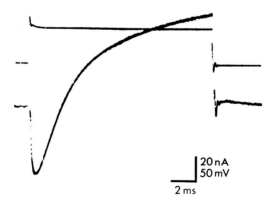

20 nA
50 mV
2 ms

FIG. 6. A typical record of membrane current obtained under voltage clamp conditions. The upper trace shows a command voltage step from -50mV to about +10mV. The net current is inward at first, but it passes smoothly into a large outward current.

When I_{Ca} was eliminated with Co^{++}, NE did not affect the remaining inward current or the delayed outward current. Thus, NE probably does not affect I_{Na} or $I_K(V)$. A direct effect of the drugs on I_{Ca} was demonstrated by examining membrane currents immediately after the termination of a depolarizing pulse. These "tail" currents reflect the fact that voltage-dependent channels that open during the preceding depolarization take time to close. This ability to shift Vm fast enough to serve the relaxation of open channels is one feature of the voltage-clamp technique that makes it such a powerful tool. When a cell bathed in TTX is depolarized to +10mV, K^+ with Ca^{++} channels opened and if the membrane is clamped back to the usual holding potential of -50mV, both K^+ and Ca^{++} contribute to the tail current. However, if the membrane is clamped back to a more hyperpolarized level equal to E_K, then the driving force on K^+ ions ($Vm-E_K$) is zero and all K^+ current, whether voltage-dependent or $Ca^{++}-K$ activated, is eliminated. NE invariably decreased TTX-resistant tail currents recorded at E_K (Fig. 7). This strongly implies a direct effect on I_{Ca}.

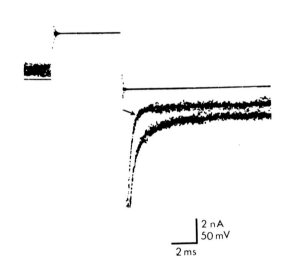

FIG. 7. TTX resistant-tail currents recorded at E_K are decreased by NE. TTX was added to the bath at 10^{-7}g/ml. The command voltage pulse from -50mV to +10mV and then back to about -75mV (E_K) is shown in the thin trace. Two current traces are superimposed. The arrow points to the tail current (at E_K) following pressure ejection of 10^{-7}M NE.

The ionic current carried by Ca^{++} can be described as

$$(1) \quad I_{Ca} = g_{Ca} \, (Vm - E_{Ca})$$

where g_{Ca} is the time and voltage dependent Ca^{++} conductance, Vm is the membrane potential, and E_{Ca} is the Ca^{++} equilibrium potential. Thus, a decrease in I_{Ca} (measured at +10mV or at E_K) might be due to a decrease in g_{Ca} at all potentials, to a shift in the voltage dependence of g_{Ca}, or to a decrease in E_{Ca}. To investigate the voltage dependence of g_{Ca}, it was necessary to eliminate I_K at all membrane potentials--not only at E_K. This was accomplished with tetraethyl ammonium (TEA). At low doses, TEA blocks $I_K(V)$ selectively, but in <u>Helix</u> and <u>Aplysia</u> neurons, high doses of TEA blocked I_K (Ca) as well (13). We obtained the same result with chick DRG neurons.

When 125mM TEA was applied by pressure ejection and the membrane potential was stepped from -50mV to +10mV, the net current remained inward for several hundred msec. At early times there was no sag in the trace suggestive of an opposed outward current.

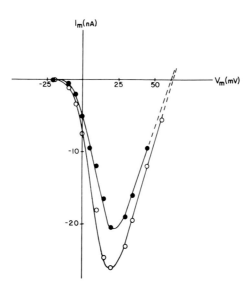

FIG. 8. NE decreases I_{Ca} without shifting the voltage dependence of the Ca^{++} channels or the apparent Ca^{++} reversal potential. I_m represents the peak inward current recorded after stepping the membrane potential from -50mV to +10mV. 0 = current following application of 125mM TEA. ● = current following application of 125mM TEA plus 5×10^{-6}M NE. TTX was present in the bath at 10^{-7} g/ml.

NE decreased I_{Ca} at all potentials above the threshold for activation of Ca^{++} channels. As shown in Fig. 9, there was no obvious shift in the voltage-current relation. Fig. 8 also shows that NE apparently did not shift the I_{Ca} reversal potential (E_{Ca}). This result must be interpreted with caution, since the null potential is probably not a precise estimate of E_{Ca}. Na^+ and other cations may flow, albeit with some difficulty, through TTX and TEA-resistance channels at positive membrane potentials so the null potential may reflect a composite equilibrium. Nevertheless, a large change in E_{Ca} is required to explain the observed decrease in I_{Ca} and this would have been detected.

Other experiments showed that NE did not affect the rate of rise or the rate of fall of I_{Ca}. The rise and fall of I_{Ca} followed simple exponential curves so the opening and closing of Ca^{++} channels can be represented as a two-state first order process

$$\text{closed} \quad \xrightarrow{\quad \alpha \quad} \quad \text{open}$$

$$(1-y) \quad \xleftarrow{\quad \beta \quad} \quad (y)$$

where y is the fraction of channels in the open state: α and β are voltage dependent rate constants. In this kinetic scheme, the time constant of the activation process is $\tau = 1/(\alpha + \beta)$ and in the steady state: $y_{ss} = \alpha/(\alpha + \beta)$. Our findings that NE does not affect the shape of the steady state voltage-current curve or the time constant of I_{Ca} activation, implies that this drug does not alter α or β.

A more explicit form of equation 1 is

$$(2) \quad I_{Ca} = ng_o \, y \, (Vm - E_K)$$

where n is the number of Ca^{++} channels and g_o is the single Ca^{++} channel conductance. Thus, our results to date indicate that NE decreases n or g_o.

Enkephalin, somatostatin, GABA, and 5HT can all decrease the inward current recorded in the presence of 10^{-7} g/ml TTX and 125 mM TEA, but it is premature to suggest that each of these agents act in exactly the same way as NE. One experiment with the potent opiate agonist, etorphine, illustrates the complexities that might be encountered. A brief pulse (1-2 sec) of 10^{-6} M etorphine decreased action potential duration and inward current. During a prolonged pulse (30 sec), spike duration decreased at first but then returned toward control values, and, when the long pulse was terminated, the action potential gradually became longer than control. Voltage clamp records showed that the late prolongation was due to a decrease in outward K^+ current (Fig. 9). This rebound "facilitation" merits further study. It is not yet clear if the initial spike shortening caused by etorphine is due to a decrease in I_{Ca} or an increase I_K.

SUMMARY

It has been known for some time that NE and ACh can affect voltage-sensitive channels in the heart but it has only recently been appreciated that neurotransmitters (and certain peptides) can modulate voltage-sensitive channels in neurons. In addition to the effect on the action potential of embryonic chick sensory neurons described here, NE decreases the duration of spikes in rat superior cervical ganglion neurons [7]. Serotonin prolongs action potentials recorded in <u>Aplysia</u> sensory neurons [9] and an as yet unidentified transmitter decreases an inward Ca^{++} current in the same cells [20].

In the heart, one important consequence of the effect of NE and ACh on the action potential is a change in the strength and/or duration of contraction. In neurons, attention has been focused on the possibility that modulation of voltage-sensitive channels might result in a change in transmitter release. In <u>Aplysia</u>, the 5-HT induced prolongation of sensory neuron soma spikes is associated with a dramatic augmentation of transmitter release at sensory nerve-motorneurone synapses [9] and the decrease in soma inward Ca^{++} current is associated with presynaptic inhibition [20]. Enkephalin, NE, GABA, and 5-HT can inhibit the evoked release of Substance P from cultured embryonic chick sensory neurons. These same drugs decrease I_{Ca}, apparently by decreasing the number or the conductance of voltage-sensitive Ca^{++} channels. The two phenomena may be related. It is significant in this regard that the same variability (between platings) in the ability of enkephalin to reduce Substance P release was also observed in the effect of enkephalin on action potential duration. Cells that released normal amounts of Substance P in the presence of enkephalin also exhibit spikes of normal duration in the presence of the peptide.

REFERENCES

1. Carr, V.M., and Simpson, S.B. (1978): <u>J. Comp. Neurol.</u>, 182: 727-740.
2. Dichter, M.A., and Fischbach, G.D. (1977): <u>J. Physiol. (Lond.)</u>, 267:281-298.
3. Hamburger, V., and Levi-Montalcini, R. (1949): <u>J. Exp. Zool.</u>, 111:457-501.
4. Henry, J.L. (1976): <u>Brain Res.</u>, 114:438-451.
5. Hokfelt, T., Ljungdahl, A., Terenius, L., Elde, R., and Nilsson, G. (1977): <u>Proc. Natl. Acad. Sci. USA</u>, 74:3081-3085.
6. Hokfelt, T., Elde, R., Johannson, O., Luft, R., Nilsson, G., and Arimura, A. (1976): <u>Neuroscience</u>, 1:131-136.
7. Horn, J., and McAffee, D. (1980): <u>J. Physiol.</u>, 301:191-204.
8. Jessell, T.M., and Iversen, L.L. (1977): <u>Nature (Lond.)</u>, 268: 549-551.
9. Klein, M. and Kandel, E. (1978): <u>Proc. Natl. Acad. Sci. USA</u>, 75:3512-3516.

10. Konishi, S., Tsundo, A., and Orsuka, M. (1979): Nature 282: 515-516.
11. Lamotte, C., Pert, C.B., and Snyder, S.S. (1976): Brain Res., 112:407-412.
12. MacDonald, R.L., and Nelson, P.G. (1978): Science, 199:1449-1450.
13. Meech, R.W., and Standen, N.B. (1975): J. Physiol., 249:211-239.
14. Mroz, E.A. and Leeman, S.E. (1978): In: Methods of Hormone Radioimmunoassay, edited by B.M. Jaffe and H.R.Behrnen, pp. 121-137. Academic Press, New York.
15. Mudge, A.M. (1979): Soc. for Neurosci. Abstr., 1803.
16. Mudge, A.W., Leeman, S.E., and Fischbach, G.D. (1979): Proc. Natl. Acad. Sci. USA, 76:523-530.
17. Noble, D. (1975): The Initiation of the Heartbeat. Clarendon Press, Oxford, England.
18. Okun, L.M. (1972): J. Neurobiol., 3:111-151.
19. Paton, W.D.M. (1975): Brit. J. Pharmacol., 12:119-127.
20. Shapiro, E., Cartellucci, V., and Kandel, E. (1980): Proc. Natl. Acad. Sci. USA, 77:1185-1189.

Neurosecretion and Brain Peptides,
edited by J. B. Martin, S. Reichlin, and K. L. Bick.
Raven Press, New York © 1981.

The Role of Substance P in Sensory Transmission and Pain Perception

Thomas M. Jessell

*Department of Pharmacology, St. George's Hospital Medical School,
London SW17 0RE, England*

Substance P is one of at least three neurally active peptides known to be present within dorsal root ganglion neurones (23). The earliest evidence implicating Substance P in sensory transmission was based on the observations of Lembeck (33) that Substance P-like biological activity was concentrated within dorsal, but not ventral, roots. Identification of the motoneurone depolarizing peptide characterized by Otsuka and Konishi (1976) as Substance P provided the first electrophysiological evidence that Substance P might play a role in sensory transmission.

By adopting Werman's (55) guidelines for the identification of central nervous system transmitters it is possible to derive a number of criteria that a sensory transmitter of noxious stimuli would be expected to fulfill: 1. The transmitter should be synthesized and stored within primary sensory neurones that are activated by noxious peripheral stimuli. 2. Noxious stimuli should evoke the release of the transmitter from primary sensory terminals within the dorsal horn of the spinal cord. 3. Application of the transmitter should excite nociceptive dorsal horn neurones (for definition of nociceptive neurones see 43). 4. Preventing the actions of the transmitter by pre- or post-synaptic mechanisms might be expected to alter pain threshold. In this article I shall use these criteria as a framework for discussing recent studies on the role of Substance P in sensory transmission.

SYNTHESIS AND STORAGE OF SUBSTANCE P

Anatomical studies have provided the most compelling evidence that Substance P may be a sensory transmitter. About 20% of all spinal ganglion neurones in the rat and cat exhibit Substance P immunofluorescence (19), almost exclusively within small (15-30 μm diameter) B-type cells. Since small cells represent 60-70% of the total population of dorsal root ganglion neurones, then Substance P can be present in only a minority of small sensory neurones.

The central process of small sensory ganglion cells represents the original axon of the neurone and enters the dorsal root either unmyelinated or thinly myelinated. Developmentally, the peripheral process represents the sensory neurone dendrite which grows into the receptive field and adopts the morphological characteristics and conduction features of an axon. Substance P is found in both the central and peripheral processes of the sensory neurone (19,9,30). Substance P fibers enter the dorsal horn along the tract of Lissauer and terminate in a dense immunoreactive plexus within the most superficial laminae (I and II) of the dorsal horn, with occasional fibers penetrating into deeper dorsal horn laminae (19,10). On purely anatomical grounds it is difficult to determine if Substance P is restricted to specific populations of small sensory neurones. Light and Perl (34) have mapped the terminal fields of high threshold mechanoreceptive afferents conducting in the A range and shown them to be located predominantly within lamina I, although some collaterals penetrate as deep as lamina V. Rethelyi (48) and Ralston and Ralston (45) have suggested that unmyelinated afferents arborize and terminate within the substantia gelatinosa (lamina II). Since Substance P is clearly found within lamina I and II of the dorsal horn, it is possible that the peptide is located in both A δ and C nociceptive afferents. However, such interpretation is complicated by the presence of intrinsic Substance P-containing dorsal horn neurones and a significant Substance P projection to the dorsal horn originating from the raphe nuclei (21,6). Immunohistochemical localization of Substance P in the dorsal root might enable a correlation to be made between the diameter and degree of myelination of sensory axons and the presence of the peptide. The localization of Substance P in physiologically characterized dorsal root ganglion neurones would also help to clarify the relationship between Substance P and specific sensory modalities.

There have been no studies on the ultrastructural characteristics of the peripheral terminals of primary neurones that contain Substance P. Within central terminals there is also some doubt about the precise subcellular location of Substance P. Immunoperoxidase reaction product has been described in association with large (100-120 nm) dense core vesicles (42,5,2) within smaller, clear vesicles (5) and even distributed within the cytoplasm (5).

Little information is known about the synthesis of Substance P in sensory neurones. However, Harmar et al. (15) have recently described the incorporation of ^{35}S-methionine into immunoreactive Substance P in isolated dorsal root ganglia.

RELEASE OF SUBSTANCE P FROM PRIMARY SENSORY NEURONES

Soon after the suggestion by Lembeck (33) that Substance P might represent the dorsal root transmitter, Angelucci (1) reported that material with biological activity similar to Substance P was released from exposed frog spinal cord. These observations were extended by Otsuka and Konishi (41) who showed that Substance P was released from the isolated newborn rat spinal cord after potassium depolarization or electrical stimulation of the dorsal roots. The release of Substance P *in vitro* has been detected from the terminal regions of primary sensory neurones in the spinal trigeminal nucleus (28) and spinal cord (Gamse *et al*; 1979) and from embryonic dorsal root ganglion neurones grown in dissociated cell culture (37). Recent experiments by Yaksh *et al*. (58) have demonstrated that Substance P-like immunoreactivity can be released from cat spinal cord *in vivo* following stimulation of A δ and C fiber afferents. Stimulation of sensory afferents at intensities sufficient to recruit only $\alpha\beta$ fibers did not evoke the release of Substance P. Substance P can also be released from the peripheral terminals of primary sensory neurones located in the tooth pulp after antidromic activation of sensory fibers (39).

POST-SYNAPTIC ACTIONS OF SUBSTANCE P RELEASED FROM
PRIMARY SENSORY NEURONES

The physiological actions of Substance P on postsynaptic neurones in the dorsal horn are difficult to examine in detail. Very little progress can be made until the identity of dorsal horn neurones that receive a direct Substance P afferent input is established. At present it appears that the iontophoretic application of Substance P results in a prolonged excitation of some dorsal horn neurones that are also activated by noxious peripheral stimuli (17,46,59). Most of these dorsal horn neurones have soma located within deeper laminae. There have been no studies on the responses to Substance P of nociceptive neurones located in lamina I of the dorsal horn (7).

In view of the duration of excitation of dorsal horn neurones evoked by Substance P, it may be misleading to compare these effects with those of "classical" transmitters. Substance P may exert more prolonged changes in the excitability of dorsal horn neurones receiving sensory input. Many substantia gelatinosa neurones, for example, exhibit extremely long periods of excitation following brief stimuli applied to the peripheral receptive field (18,54,4). It is also possible that Substance P may be involved in the temporal summation of dorsal horn neuronal responses (36,43) that occurs after stimulation of C-fiber afferents. Some precedent for this type of peptide action may already exist. An LHRH-like peptide is thought to be responsible for the late, slow depolarization of post-ganglionic sympathetic neurones that occurs after pre-ganglionic nerve stimulation (25).

PRE-AND POST-SYNAPTIC REGULATION OF SUBSTANCE P
IN SENSORY SYSTEMS

Pharmacological approaches to the study of Substance P at primary afferent synapses are extremely limited. No Substance P antagonists have been reported (with the possible exception of baclofen) (49). However, two methods of modifying the actions of Substance P by pre-synaptic mechanisms have recently been described.

There is a striking overlap of enkephalin and Substance P nerve terminal fields in the superficial laminae of the dorsal horn (20). Moreover, the spinal analgesic actions of the opiates and opioid peptides (56) may be mediated, at least in part, by the activation of opiate receptors located on primary afferent terminals (32,30). Release of Substance P from sensory terminals in the rat trigeminal nucleus maintained in vitro can be inhibited by opiates and opioid peptides (28) at concentrations similar to those needed to depress synaptic transmission in the isolated spinal cord. Furthermore, the release of Substance P from chick dorsal root ganglion neurones grown in cell culture is inhibited by opioid peptides (37). In these experiments it is clear that opiate receptors mediating inhibition of Substance P release are located directly on the sensory neurones. Opiates have also been reported to inhibit the release of an unidentified transmitter in co-cultures of mouse dorsal root ganglion and spinal cord neurones (35). An inhibition of Substane P release via pre-synaptic opiate receptors is also consistent with a presynaptic action of opiates on transmitter release in symapthetic ganglia (31,3), and in a variety of other central and peripheral tissues (24).

The precise relationship between enkephalin and Substance P-containing neuronal elements in the dorsal horn is still unclear. Primary afferent terminals in laminae I and II are post-synaptic in axo-axonal contacts (45). Furthermore, enkephalin-containing terminals within the dorsal horn may not originate solely from dorsal horn interneurones. Glazer and Basbaum (14) have reported the presence of enkephalin immunoreactivity in lamina I cells possibly within spino-thalamic projection neurones. In addition, Hokfelt et al. (22) have reported the presence of a descending enkephalin projection to the spinal cord.

Substance P-containing primary sensory neurones are sensitive to the homovanyllic acid derivative, capsaicin. This compound may represent the first example of a neurotoxin that is specific to certain classes of sensory afferents. Peripheral application of capsaicin produces a severe burning pain (26) that probably reflects the activation of C-fibers in sensory nerves (50). Administration of capsaicin directly to the spinal cord in vitro produces a calcium-dependent release of Substance P (53,11) (see Table 1).

Capsaicin also depolarizes motoneurones in the newborn rat spinal cord preparation at concentrations similar to those used to evoke Substance P release (Fig. 1).

The depolarization of motoneurones produced by capsaicin is abolished by removing calcium from the superfusing medium, whereas Substance P-induced depolarization is unaffected by the removal of calcium (Fig. 1). These results suggest that the depolarization of spinal motoneurones by capsaicin may be mediated by the release of Substance P from nearby nerve terminals (53).

TABLE 1. Capsaicin-evoked release of Substance P from newborn rat spinal cord

Incubation solution	Substance P release/30 min (fmol)
artificial CSF	56.5 + 29.6
artificial CSF + 1 µm capsaicin	287.3 + 33.9*
artificial CSF (0.1 mMCa^{2+}/ 2.0 mM Mg^{2+}) + 1 µm capsaicin	60.8 + 24.2

Each value is the mean + SEM of 3 separate experiments. *p<0.01 when compared with the value in the absence of capsaicin. Reprinted with permission from "ref. 53".

In chronic studies, the subcutaneous administration of capsaicin results in the degeneration of small diameter sensory neurones (27) that contain Substance P (13). Capsaicin also depletes Substance P from the dorsal horn in adult rats (29) as illustrated in Fig. 2.

Administration of capsaicin directly into the subarachnoid space depletes Substance P from primary sensory neurones and produces a prolonged elevation in thermal and chemical pain threshold with no discernible change in response to noxious pinch (57).

Newborn rats treated with capsaicin exhibit an elevation in threshold to noxious thermal stimuli when tested some months later (11,39). Opiate receptor binding in the spinal cords of the same animals was also significantly reduced. Hayes and Tyers (16) however, have reported that subcutaneous capsaicin administration in adult rats leads to a reduction in response to certain types of noxious mechanical stimuli. Clearly, the effect of capsaicin on different sensory modalities requires further study.

Thermal analgesia in rats elicited by a single intrathecal injection of capsaicin (30 µg) lasts for many months (57) suggesting that some degenerative changes may occur following intrathecal capsaicin in adult animals.

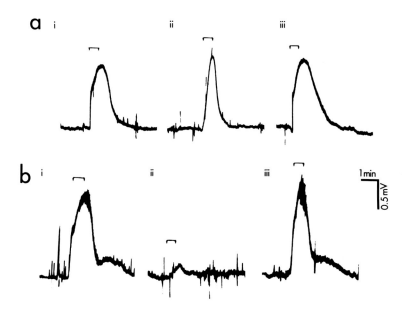

FIG. 1.   The effect of Substance P and capsaicin on potentials recorded from the ventral root of the isolated rat spinal cord. Depolarization of ventral roots induced by (a): Substance P (5 x 10^{-7} M) and (b) capsaicin (5 x 10^{-7} M).   (i) response in artificial CSF (ii) response in artificial CSF containing 0.1 mM calcium and 2.0 mM magnesium (iii) response after return to normal CSF. Reprinted with permission from (53).

Most of the Substance P within the dorsal horn is located in the terminals of primary afferent neurones.   However, there is a significant contribution from some raphe-spinal neurones, in which Substance P seems to co-exist with serotonin (6,21).

The effects of capsaicin in the spinal cord are probably restricted to sensory neurones since capsaicin treated animals show no loss of serotonin from the spinal cord, although Substance P levels are reduced by 50% (57).   Intrathecal application of the serotonergic neurotoxin 5,6-dihydroxytryptamine produces a 40% depletion of Substance P from the spinal cord.   However, this depletion was not associated with thermal analgesia: in fact, animals exhibited a significant decrease in nociceptive threshold, implying that the sensory and raphe-spinal Substance P systems exert opposite effects on responses to thermal pain (57).

Capsaicin is apparently without effect on other transmitters in the spinal cord (29,38), although it is not clear whether other peptides in primary afferents are affected.

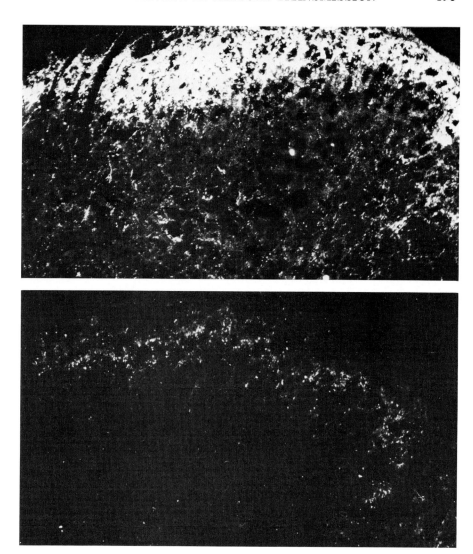

FIG. 2. Depletion of Sustance P from the dorsal horn of adult rat spinal cord following subcutaneous capsaicin treatment. Upper panel shows immunofluorescence of micrograph of the dorsal horn of the spinal cord of a vehicle-treated rat. Lower panel shows immunoreactivity in spinal cord of capsaicin-treated rat processed on the same slide (29) (x 170).

The analgesia resulting from intrathecal capsaicin may, therefore, be the consequence of a series of events in which capsaicin rapidly liberates the majority of releasable stores of Substance P from primary afferent terminals and subsequently induces a prolonged and possibly permanent depletion of Substance P from primary sensory neurones associated with the transmission of certain types of noxious stimuli. It follows then, that the central terminals of small diameter primary sensory neurones may possess receptors for capsaicin similar to those that are thought to exist on the peripheral terminals of the same neurones (51).

Much of the work discussed in this article has provided indirect evidence that Substance P acts as a transmitter of nociceptive sensory afferents. Despite this, none of the criteria originally outlined have been fully satisfied. Somatostatin and a gastrin-like peptide, possibly cholecystokinin, are also contained within small diameter sensory ganglion cells (23). These peptides are equally likely to be involved in sensory processing. Since some dorsal horn neurones are inhibited by the iontophoretic application of somatostatin (47) it is possible that the transmitter of at least one population of sensory afferents may be inhibitory. Further studies on the physiological actions of the peptides on identified dorsal horn neurones are essential before any real understanding of peptide function in sensory transmission can be claimed.

ACKNOWLEDGEMENTS

This work was supported by grants from The Royal Society and The Wellcome Trust and by A Harkness Fellowship.

REFERENCES

1. Angelucci, L. (1956) Brit. J. Pharmacol., 11:161-170.
2. Barber, R.P., Vaughn, J.E., Slemmon, J.R., Salvaterra, P., Roberts, E. and Leeman, S.E. (1979): J. Comp. Neurol. 184:331-351.
3. Bornstein, J.C. and Fields, H.L. (1979): Neurosci. Letters, 15:77-82.
4. Cevero, F., Iggo, A. and Molony, V. (1979): Quat. J. Exp. Physiol., 64:297-314.
5. Chan-Palay, V. and Palay, S.L. (1977): Proc. Natl. Acad. Sci. USA, 74:4050-4054.
6. Chan-Palay, V., Palay, S.L. and Jonsson, G. (1978): Proc. Natl. Acad. Sci. USA, 75:1582-1586.
7. Christensen, B.N. and Perl, E.R. (1970): J. Neurophysiol., 33:293-307.
8. Cuello, A.C., Jessell, T.M., Kanazawa, I. and Iversen, L.L. (1977): J. Neurochem., 29:747-751.
9. Cuello, A.C., del Fiacco, M. and Paxinos, G. (1978): Brain Res. 152:499-500.

10. Cuello, A.C. and Kanazawa, I. (1978): J. Comp. Neurol., 178: 129-156.
11. Gamse, R., Molnar, A. and Lembeck, F. (1979a): Life Sci. 25:629-636.
12. Gamse, R., Holtzer, P. and Lembeck, F. (1979b): Naunyn. Schmeideberg. Archiv. Pharmacol., (in press).
13. Gamse, R., Holtzer, P. and Lembeck, F. (1980): Brit. J. Pharmacol., 68:207-213.
14. Glazer, E.J. and Basbaum, A.I. (1979): Soc. Neuroscience Abst., 5:723.
15. Hamar, A., Schofield, J.G. and Keen, P. (1980): Nature, (in press).
16. Hayes, A. and Tyers, M. (1980): J. Physiol., (in press).
17. Henry, J.L. (1976): Brain Res., 114:439-451.
18. Hentall, I. (1977): Exp. Neurol., 57:792-806.
19. Hokfelt, T., Johansson, O., Kellerth, J.O., Ljungdahl, A., Nilsson, G., Nygards, A. and Pernow, B. (1977a): In: Substance P, edited by U. von Euler and B. Pernow, pp. 117-145. Raven Press, New York.
20. Hokfelt, T., Ljungdahl, A., Terenius, L., Elde, R. and Nilsson, G. (1977b): Proc. Natl. Acad. Sci. USA, 74:3081-3085.
21. Hokfelt, T., Ljungdahl, A., Steinbusch, H., verhofstad, A. Nilsson, G., Brodin, E., Pernow, B. and Goldstein, M. (1978): Neuroscience, 3:517-538.
22. Hokfelt, T., Terenius, L., Kuypers, H.G. and Dann, O. (1979): Neurosci. Lett., 14:55-60.
23. Hokfelt, T. (1980): Neurosci. Res. Prog. Bull., (in press).
24. Iversen, L.L. and Jessell, T.M. (1979): In: Central Regulation of the Endocrine System, edited by K. Fuxe, T. Hokfelt, and R. Luft, pp. 189-207. Plenum Press, New York.
25. Jan, Y.N., Jan, L.Y. and Kuffler, S.W. (1979): Proc. Natl. Acad. Sci. USA, 76:1501-1505.
26. Jancso, N. (1966): Proc. 3rd. Int. Pharmac. Meeting, pp. 33-55. Pergamon Press, Inc., New York.
27. Jancso, G., Kiraly, E. and Jancso-Gabor, A. (1977): Nature, 270: 741-743.
28. Jessell, T.M. and Inversen, L.L. (1977): Nature, 268:549-551.
29. Jessell, T.M., Iversen, L.L. and Cuello, A.C. (1978): Brain Res., 152:183-188.
30. Jessell, T.M., Tsunoo, A., Kanazawa, I. and Otsuka, M. (1979): Brain Res., 168:247-259.
31. Konishi, S., Tsunoo, A. and Otsuka, M. (1979): Nature, 282:515-516.
32. LaMotte, C., Pert, C.B. and Snyder, S.H. (1976): Brain Res., 112:407-412.
33. Lembeck, F. (1953): Naunyn. Schmeideberg. Archiv. Pharmacol., 219:197-213.
34. Light, A.R. and Perl, E.R. (1979): J. Comp. Neurol., 186:117-132.

35. Macdonald, R.L. and Nelson, P.G. (1978): Science, 199:1449-1450.
36. Mendell, L.M. and Wall, P.D. (1956): Nature, 106:97-99.
37. Mudge, A.W., Leeman, S.E. and Fischbach, G.D. (1979): Proc. Natl. Acad. Sci. USA, 76:526-530.
38. Nagy, J.I., Vincent, S.R., Staines, W.A., Fibiger, H.C., Reisine, T.D. and Yamamura, H.I. (1980): Brain Res., (in press).
39. Olgart, L., Gazelius, B., Brodin, E. and Nilsson, G. (1977): Acta. Physiol. Scand., 101:510-512.
40. Otsuka, M. and Konishi, S. (1976a): Cold. Spring Harbour, Symp. Quant. Biol., 40:135-143.
41. Otsuka, M. and Konishi, S. (1976h): Nature, 264:83-84.
42. Pickel, V., Reis, D.J. and Leeman, S.E. (1976): Brain Res., 122:534-540.
43. Price, D.D., Hayes, R.L., Ruda, M.A. and Dubner, R.J. (1978): J. Neurophysiol., 41:933-947.
44. Price, D.D. (1978): Neurosci. Res. Prog. Bull., 16:82-83.
45. Ralston, H.J. and Ralston, D.D. (1979): J. Comp. Neurol., 184:643-684.
46. Randic, M. and Miletic,V. (1977): Brain Res., 128:164-169.
47. Randic, M. and Miletic, V. (1978): Brain Res., 152:196-202.
48. Rethelyi, M. (1977): J. Comp. Neurol., 172:511-521.
49. Saito, K., Konishi, S. and Otsuka, M. (1975): Brain Res., 97:77-180.
50. Szolcsanyi, J. (1977): J. Physiol. Paris, 73:251-259.
51. Szolcsanyi, J. and Jancso-Gabor, A. (1975): Artzneimittel Forschung, 25:1877-1881.
52. Suzue, T. and Jessell, T.M. (1980): Neurosci. Lett., 16:161-166.
53. Theriault, E., Otsuka, M. and Jessell, T.M. (1979): Brain Res., 170:209-213.
54. Wall, P.D., Merrill, E.G. and Yaksh, T.L. (1979): Brain Res., 160:245-260.
55. Werman, R. (1966): Comp. Biochem. Physiol. 18:745-766.
56. Yaksh, T.L. and Rudy, T.A. (1978): Pain, 4:299-359.
57. Yaksh, T.L., Farb, D., Leeman, S.E. and Jessell, T.M. (1979): Science, 206:481-483.
58. Yaksh, T.L., Gamse, R., Mudge, A.W., Leeman, S.E. and Jessell, T.M. (1980): In preparation.
59. Zieglgansberger, W. and Tulloch, I. (1979): Brain Res., 166:273-282.

Neurosecretion and Brain Peptides,
edited by J. B. Martin, S. Reichlin, and K. L. Bick.
Raven Press, New York © 1981.

An Endorphin-Mediated Analgesia System: Experimental and Clinical Observations

Howard L. Fields

*Departments of Neurology and Physiology, University of California, San Francisco,
San Francisco, California 94153*

Over the past few years remarkable progress has been made in
our understanding of pain modulation within the central nervous
system. Several comprehensive reviews of this subject have
recently been published (14,37,69,108), so the present review
will emphasize newer developments. For the sake of brevity in
this review, "endorphin" will be used as a generic term referring
to endogenous opioids including β-endorphin and leucine- and
methionine-enkephalin. When referring to animal studies "analgesia" will be used loosely to encompass suppression of reflexes
or behavior usually elicited by noxious stimulation.

A PAIN-SUPPRESSING SYSTEM

Although the brain may have more than one pain-modulating
mechanism (46), we know most about an analgesia system with links
in midbrain, medulla and spinal cord. Early studies had shown
that the midbrain periaqueductal gray (PAG) is a site which when
electrically stimulated, can produce suppression of responses to
noxious stimuli. This "stimulation-produced analgesia" (SPA),
from PAG or from the rostrally contiguous diencephalic periventricular gray (PVG) (84,), has subsequently been observed in a
variety of species (68,69), including human patients with painful
conditions (2,52,74,85,86).
Stimulation at sites effective for SPA can produce highly
selective effects. Animals manifesting analgesia from PAG
stimulation are not sedated, they move about freely and respond
to innocuous environmental stimuli, yet they appear to have profound analgesia. Furthermore, patients who obtain analgesia from
electrodes implanted in or near the PVG may report mild thermal
sensations but have no other consistent sensory or motor phenomena during stimulation (74,86). The observation that analgesia
can be produced selectively by electrical stimulation of certain
discrete sites is of fundamental importance in establishing pain

199

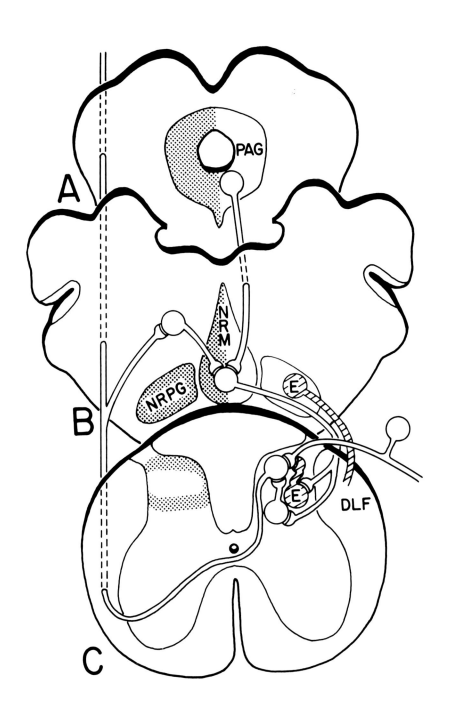

FIG. 1. The endorphin mediated analgesia system.

This system has three important levels of organization. Endogenous opioid peptides (stippled regions) have been localized immunocytochemically at all three levels.

A. The periaqueductal gray (PAG) has been shown to contain enkephalin perikarya and both enkephalin and β-endorphin terminals.

B. In the ventromedial medulla of the rat, enkephalin peri-karya and terminals are present in both nucleus raphe magnus (NRM) and nucleus reticularis paragigantocelularis (NRPG). Some of these enkephalinergic neurons have been shown to project to the spinal cord.

C. At the level of the spinal cord enkephalinergic inter-neurons have been demonstrated in the superficial layers of the dorsal horn. There is some evidence that these interneurons may induce presynaptic inhibition.

The diagram indicates that in this system, the analgesic actions of the PAG may be exerted upon the spinal cord through a link in the ventromedial medulla.

(Modified from Basebaum and Fields, ref. 14.)

---

modulation as a separate physiological function of the central nervous system.

In animals, analgesia is inferred by the absence of behaviors usually elicited by noxious stimulation. Some of these behaviors, such as tail-flick and paw withdrawal are spinally mediated. The suppression of these spinal reflexes by PAG stimulation must re-quire a descending pathway in the spinal cord. In fact, Basbaum and colleagues (12,16) have shown that spinal dorsolateral funi-culus (DLF) lesions are very effective in blocking SPA from PAG. Furthermore, DLF lesions block PAG effects on more complex "pain" behaviors such as orienting, vocalization and escape. This implies that descending pathways in the DLF are involved in suppressing pain sensation as well as spinal reflexes induced by noxious stimuli.

Several descriptions of supraspinal structures contributing descending axons in the DLF have recently appeared (15,67,97, 100). These studies are based on retrograde transport of horse-radish peroxidase (HRP) by DLF axons to their cell somata in the brainstem. The picture that has emerged is quite consistent; surprisingly few PAG neurons project via the DLF to the spinal cord. A very substantial number of descending DLF axons arise from two major groups of neurons; one in the rostral ventromedial medulla (the serotonergic nucleus raphe magnus (NRM) and the adjacent paragigantocellular and magnocellular reticular nuclei) and the second in the dorsolateral tegmentum of the pons.

Although other nuclei may be of importance in linking PAG to
cord via the DLF, interest has focused on the region of the ventromedial medulla.   There is evidence that 5-hydroxytryptamine
(5HT) is involved in analgesic mechanisms (73) including SPA from
PAG (3).   The midline NRM is thus of particular interest because
it is the major 5HT-containing nucleus which projects to the cord
via the DLF.   Autoradiographic techniques have demonstrated that
neurons in this region project via the DLF and terminate in spinal cord regions which contain cells that respond maximally to
noxious stimuli (12,13).   Furthermore, electrical stimulation of
NRM produces potent "analgesia" (77,82) and inhibits spinal
neurons with nociceptive inputs (11,32,38,71,103).   That NRM is a
link between PAG and the cord is supported by anatomical evidence
for a large, direct PAG to NRM projection (1,41,90) and by physiological data that PAG stimulation excites raphe-spinal neurons
(17,36,80).   Furthermore, acute lesions of the ventromedial medulla (NRM plus adjacent reticular formation) block the analgesia
elicited by glutamate injections into PAG (17).

In summary, the outline has emerged for an intrinsic neural
network for controlling pain transmission.   This system can be
activated in the midbrain PAG, and has an excitatory connection
to NRM.   NRM, in turn, produces inhibition of spinal pain transmission neurons.   Evidence that endorphin-containing neurons play
an integral role in the operation of this system will now be
considered.

DISTRIBUTION OF ENDOGENOUS OPIOIDS (ENDORPHINS) IN RELATION TO THE INTRINSIC ANALGESIA SYSTEM

Periaqueductal-Periventricular Gray

The most convincing evidence for neuroanatomical localization
of endorphins comes from immunohistochemical studies.   Although
there may be crossreactivity with related peptides, this technique does provide a high degree of accuracy for anatomical
localization. Both β-endorphin-like (23,101,102) and enkephalin-
like (34,42,48,49,51,91) immunoreactivity are present in the PAG.
Enkephalin-like immunoreactivity (ELI) is found in both cells and
fibers in PAG.   β-endorphin-like immunoreactive cells have not
been found in PAG but are restricted to the medial basal hypothalamic region which includes the arcuate nucleus (23,101).
Glutamate injection into PAG (which should activate cells in
preference to fibers of passage) produces an NRM-mediated, naloxone-sensitive suppression of spinal nociceptive reflexes (17).
Since enkephalin but not β-endorphin cell bodies are present in
PAG, this observation supports the concept that enkephalinergic
neurons in the PAG can activate a descending pain modulatory
system.

Other evidence supporting a central role for the PAG-PVG
region in analgesic mechanisms is:
1. This region contains a relatively high concentration of
   opiate receptors (9,47,48).

2. Opiate microinjection in this area produces analgesia that can be blocked by naloxone (107,108) or by lesions of the DLF. Met-enkephalin produces weak but definite potent analgesia in mouse, rat (66) and cat (72) when injected intraventricularly.
3. Intraventricular injections of morphine or met-enkephalin produce a marked excitation of PAG neurons associated with analgesia (98).
4. Lesions of PAG block analgesia from systemic opiates (31).
5. Naloxone reverses SPA from PAG in rats (4,58,81) and from PVG in man (2,52).
6. SPA from PVG sites in man is associated with an elevation of CSF β-endorphin (5,53) and ELI (6) in third ventricular cerebrospinal fluid.
7. Analgesic doses of opiate alkaloids given systemically induce a rise in endorphins in the CSF (20).

Thus endorphins are present in PAG and can be shown to be released upon electrical stimulation. When applied locally they produce analgesia mediated by descending pathways. Specific antagonists block endorphin action at PAG. Thus the PAG-PVG region is a critical activation site for responses to noxious stimuli. At least part of this pain-modulating function requires a link in the ventromedial medulla.

Ventromedial Medulla

Although it is well established that this region makes important contributions to the descending pathways controlling pain transmission, the role of particular cell groups and the transmitters involved is still to be worked out. Terminals with ELI have been demonstrated in this region in rat (48,50). Neuronal perikarya with ELI have also been demonstrated in NRM and adjacent reticular formation (including the nucleus paragigantocellularis (NPGC)) (49,50). In the cat, ELI positive fibers are not present in NRM or NPGC, but NPGC does have ELI positive cells (42). These enkephalinergic neurons in NPGC may, in fact, contribute to descending medullospinal pathways (49).

Despite the lack of positive evidence of opiate receptor in the ventromedial medulla (9), microinjection of opiates into either NRM (29,30) or the adjacent NPGC (10,95,96) has been reported to produce analgesia. It is not clear whether these opiate microinjections simply produce interesting pharmacological actions or actually mimic the action of an endorphin transmitter which "normally" acts at these sites. It is of interest that the ventromedial medulla is one of the few sites in the brainstem at which opiate iontophoresis has been demonstrated to produce a naloxone-reversible excitatory effect upon neurons (75).

Spinal Cord

The descending projections to spinal cord from ventromedial medulla via the DLF terminate most densely in those layers of the

dorsal horn which contain neruons maximally sensitive to noxious
stimuli (Rexed's layers, I, II, V). These same layers contain
numerous enkephalin-positive terminals (34,42,49,91) and perikarya (42,49). Layers I and II are also densely invested with
opiate receptors (8,78), a significant proportion of which is
associated with the terminals of primary afferent fibers (57,60).
The relevance of this spinal cord dorsal horn enkephalin system
to analgesia is suggested by several findings.

1. Direct intrathecal application of opiates produces anal-
    gesia in a variety of animals (105,106,197) and in man
    (99).

2. Iontophoresis of opiates into the superficial layers of
    dorsal horn produces a selective inhibition of nociceptive
    input to presumed projection neurons (33).

3. In spinal animals, systemic opiates inhibit spinal inter-
    neurons (61) and nociceptive reflexes (54).

Although some of the enkephalinergic dorsal horn neurons have
the location and shape of projection cells rather than local interneurons (42), the simplest hypothesis is that they are inhibitory interneurons activated by descending medullospinal pathways
(e.g., raphe-spinal). Supporting evidence for this hypothesis is
that naloxone blocks the analgesia produced by electrical stimulation of NRM (76) and partially blocks NRM-induced inhibition of
dorsal horn lamina V cells (87).

Certain segmental inputs have also been reported capable of
producing naloxone-reversible inhibition of nociceptive reflexes
(25,94,104). Since this effect can be seen in spinal animals
(104) the possibility of an intraspinal enkephalinergic network
for analgesia must be considered. The enkephalin-positive dorsal
horn cells are obvious candidates for this segmental inhibition.

Table 1 summarizes the relationship that the intrinsic analgesic system has with endorphins, opiate receptor and analgesia-
producing sites. Despite the information gaps concerning certain
links in the descending pathway, the evidence reviewed above jus-
tifies the concept of an endorphin-mediated analgesia system
(EMAS).

OPERATING CHARACTERISTICS OF EMAS

One of the most important things to stress about EMAS is that
it is a modulatory system. It is an identifiable neural network
that has been designed not to transmit but to control afferent
transmission. Thus, when activated by itself, it will produce
neither sensation nor motor behavior. Clearly, if no noxious
stimulation is being applied, i.e., if there is no afferent input to be modulated, there should be no behavioral or subjective
consequence of EMAS activation. It is interesting in this regard
that pain patients with effectively placed periventricular electrodes report little except a fading of pain during stimulation
(74,86).

TABLE 1. Endorphins and analgesia: anatomical relationships

Anatomical Site	β-endorphin fibers[a]	enkephalin cells	enkephalin fibers	SPA	Opiate receptor	Opiate microinjection Analgesia
Periventricular diencephlon	+	+	+	+	+	+
Periaqueductal gray	+	+	++	+	+	+
N. Raphe Magnus	o	+	+r	+	o	+
N. Paragiganto-cellularis	o	+	+r	+	o	+
Dorsal horn	o	+	++	+	++	+

+ present
++ large amounts
o not demonstrated
? unknown
+r present in rat, not demonstrated in cat

[a] β-endorphin perikarya have only been found in the anterobasal hypothalamus

SPA stimulation produced analgesia

Thus operation of a modulatory system is initially difficult to study because of the indirect and inconsistent relationship between its activity and the observable consequences of its activity. On the other hand, because of our present detailed knowledge of its anatomy and its susceptibility to fairly specific pharmacologic manipulation, EMAS offers us a unique opportunity to study a modulatory system.

Two complementary approaches can be used to determine factors that activate EMAS. One is to record from the component neurons of EMAS. A second method is to apply a noxious stimulus, determine the environmental variables that reduce the response and establish whether the reduction is blocked by disruption of EMAS.

Neurons in the ventromedial medulla including those demonstrated to project to the spinal cord are most consistently excited by stimuli in the noxious range (7,36,80). This indicates that EMAS can be activated by the very input it modulates. This observation leads to the prediction that selective disruption of EMAS will cause intensification of ongoing pain.

The most commonly used method to disrupt EMAS has been to give a narcotic antagonist. The most specific antagonist presently available is naloxone. The rationale for most of the naloxone studies is that if endogenous opiates are producing analgesia, giving naloxone should cause hyperalgesia.

Jacob et al.(56) first demonstrated that naloxone by itself enhances responsiveness to intense thermal stimuli. The effect was only seen with stimuli of gradually increasing intensity. Later studies demonstrated that other opiate antagonists produce "hyperalgesia" and that this action represents a stereospecific antagonism at the opiate receptor (55,83,93). The original rfinding of Jacob et al. has been supported by other workers using different noxious stimuli (22,27,59).

Initial observations in humans failed to demonstrate naloxone hyperalgesia (35,45). However, using a double-blind crossover design in patients with postoperative dental pain, Levine et al. (65) demonstrated that naloxone produces a significant increase rin reported pain severity compared to placebo. In that study, patients were premedicated with diazepam and the dental extraction procedure was carried out under nitrous oxide (N_2O) and xylocaine block. Pain ratings were made no sooner than 90 minutes following discontinuation of the N_2O, and only when the xylocaine block began to wear off. Differences between placebo and naloxone groups were greater at 4 than at 3 hours. Although naloxone can partially reverse N_2O analgesia (21), naloxone hyperalgesia has recently been reported in dental postoperative patients who received neither N_2O nor diazepam (44). The effect in this clinical situation appears to require a relatively higher dose of naloxone (8 mg i.v. per patient) (64) than is usual for treating narcotic overdose (0.4-0.8 mg).

Although the naloxone hyperalgesia described above is seen when data from individuals is pooled, more striking changes are revealed when subjects are categorized (24). Thus, Levine et al.

(63) have presented evidence that naloxone produces significant rhyperalgesia in placebo responders, but has no effect in placebo non-responders. Furthermore, there is some evidence that a positive placebo response (which may represent activation of EMAS) is more likely to occur in patients reporting more severe pain levels just prior to receiving a placebo (62).

Other manipulations which have been reported to produce naloxone-reversible analgesia include acupuncture in man (70) and animals (79), transcutaneous electrical stimulation in man (26, 94) and animals (104) and hypnosis (40). It is of interest that in human volunteers with experimental ischemic pain, naloxone (2 mg) only seems to reverse hypnotic analgesia in subjects reporting high levels of "stress" (40). At the present time stress is a very hazy concept and the most one can say is that complex sensory, motivational and environmental cues contribute to activating EMAS.

In addition to the manipulation discussed above there is evidence that pain sensitivity undergoes a regular "spontaneous" diurnal variation that reflects changes in EMAS. Thus patients with chronic pain from a variety of causes report a reproducible rise in pain severity between 0800 and 2200 hours (43). When pain threshold is measured in normal volunteers it is found to be significantly lower in the afternoon than in the morning (88). Circadian fluctuations in latencies elicited by noxious heat have been well documented in mice that have been entrained by appropriate light-dark cycles (28,39). That this circadian variation is due to EMAS activity is indicated by the observation that it is markedly suppressed by naloxone (39).

The observations of naloxone-induced hyperalgesia described above support the concept that endorphins modulate pain under "physiological" conditions. However, endorphins and opiate receptors are present in regions not known to be part of EMAS. Systemically administered naloxone could work at any endorphin synapse. What is the evidence that the naloxone hyperalgesia described above results from an action upon the endorphin links of EMAS? The only available evidence is indirect:

1. There is an anatomical homology between the effective meso-diencephalic sites for SPA in man and animal;
2. There is opiate receptor in PAG-PVG, and SPA from these sites is naloxone-reversible in man and animals;
3. The behavioral effects produced in man (i.e., selectivity for analgesia) are similar to those produced in animals by stimulation at PAG-PVG sites;
4. Intrathecal opiates can act directly upon the spinal cord in man, a finding consistent with the spinal cord opiate receptor for analgesia that has been demonstrated in animals.

A major gap in present knowledge is the lack of direct evidence that descending pathways are involved in the analgesia produced by stimulation in PVG, by placebo administration or by acupuncture in human subjects. Further relevant information may come from clinicopathologic correlations in patients with various lesions of EMAS. One would predict that certain small spinal cord lesions might impair EMAS and lead to hyperalgesia. Lesions involving critical brainstem regions for EMAS would tend to produce other devastating neurological impairments.

In summary, recent clinical and experimental observations have established the importance of an endorphin-mediated pain modulating system with links in the midbrain, medulla and spinal cord. The activity of this system is enhanced by noxious stimulation and by other less well defined environmental cues. In addition, it seems to undergo a circadian fluctuation. Further studies of the operation of this system will improve our understanding of the variability of pain responses and provide a model for the study of modulatory networks in the brain.

ACKNOWLEDGEMENTS

I thank Dr. A.I. Basbaum for useful comments on this manuscript, and Mr. Robert McClary and Ms. M. Nugent for editorial assistance.

Supported by P.H.S. research grant DA 01949.

REFERENCES

1. Abols, I.E., and Basbaum, A.I. (1979): Anat. Rec., 193:467.
2. Adams, J.E. (1976): Pain, 2:161-166.
3. Akil, H., and Liebeskind, J.C. (1975): Brain Res., 94:279-296.
4. Akil, H., Mayer, D.J., and Liebeskind, J.C. (1976): Science, 191:961-962.
5. Akil, H., Richardson, D.E., Barchas, J.D., and Li, C.H. (1978): Proc. Natl. Acad. Sci. USA, 75:5170-5172.
6. Akil, H., Richardson, D.E., Hughes, J., and Barchas, J.D. (1978): Science, 201:463-465.
7. Anderson, S.D., Basbaum, A.I., and Fields, H.L. (1977): Brain Res., 123:363-368.
8. Atweh, S.F., and Kuhar, M.J. (1977a): Brain Res., 124:53-67.
9. Atweh, S.F., and Kuhar, M.J. (1977): Brain Res., 129:1-12.
10. Azami,J., Roberts, M.H.T., and Wright, D.M. (1979): J. Physiol., 293:63P-64P.
11. Basbaum, A.I., Clanton, C.H., and Fields, H.L. (1976): Proc. Natl. Acad. Sci. USA, 73:4685-4688.
12. Basbaum, A.I., Clanton, C.H., and Fields, H.L.(1978): J. Comp. Neurol., 178:209-224.
13. Basbaum, A.I., and Fields, H.L. (1978): Ann. Neurol., 4:451-462.
14. Basbaum, A.I., and Fields, H.L. (1979): J. Comp. Neurol., 187:513-532.
15. Basbaum, A.I., Marley, N.J.E., O'Keefe, J., and Clanton, C.H. (1977): Pain, 3:43-56.

16. Beall, J.E., Martin, R.F., Applebaum, A.E., and Willis, W.D. (1976): Brain Res., 114:328-333.
17. Behbehani, M.M., and Fields, H.L. (1979): Brain Res., 170:85-93.
18. Beitz, A.J. (1979): Neurosci. Abstr., 5:605.
19. Belluzzi,J.D., Grant, N., Garsky, V., Sarantakis, D., Wise, C.D., and Stein, L. (1976): Nature, 260:625-626.
20. Bergmann, F., Altstetter, R., and Weissman, B.A. (1978): Life Sci., 23:2601-2608.
21. Berkowitz, B.A., Finck, A.D., and Ngai, S.H. (1977): J. Pharmacol. Exp. Ther., 203:539-547.
22. Berntson, G.G., and Walker, J.M. (1977): Brain Res. Bull., 2:157-159.
23. Bloom, F., Battenberg, E., Rossier, J., Ling, N., and Guillemin, R. (1978): Proc. Natl. Acad. Sci. USA, 75:1591-1595.
24. Buchsbaum, M.S., Davis, G.C., and Bunney, W.E., Jr. (1977): Nature, 270:620-622.
25. Buckett, W.R. (1979): Eur. J. Pharmacol., 58:169-178.
26. Chapman, C.R., and Benedetti, C. (1977): Life Sci., 21:16451648.
27. Chesher, G.B., and Chan, B. (1977): Life Sci., 21:1569-1574.
28. Crockett, R.S., Bornschein, R.L., and Smith, R.P. (1977): Physiol. Behav., 18:193-196.
29. Dickenson, A.H., Fardin, V., LeBars, D., and Besson,J. (1979): Neurosci. Lett., 15:265-270.
30. Dickenson, A.H., Oliveras, J.L., and Besson, J.M. (1979): Brain Res., 170:95-111.
31. Dostrovsky, J.O., and Deakin, J.F.W. (1977): Neurosci. Lett., 4:99-103.
32. Duggan, A.W., and Griersmith, B. (1979): Pain, 6:149-161.
33. Duggan, A.W., Hall, J.G., and Headley, P.M. (1976): Nature, 265:456-458.
34. Elde, R., Hokfelt, T., Johansson, O., and Terenius, L. (1976): Neuroscience, 1:349-351.
35. El-Sobky, A., Dostrovsky, J.O., and Wall, P.D. (1976): Nature, 263:783-784.
36. Fields, H.L., and Anderson, S.D. (1978): Pain, 5:33-349.
37. Fields,H.L., and Basbaum, A.I. (1978): Ann. Rev. Physiol., 40:217-248.
38. Fields, H.L., Basbaum, A.I., Clanton, C.H., and Anderson, S.D. (1977): Brain Res., 126:441-453.
39. Frederickson, R.C.A., Burgis, V. and Edwards, J.D. (1977): Science, 198:756-758.
40. Fried, M., and Singer, G. (1979): Psychopharmacology, 63:211-215.
41. Gallager, D.W., and Pert, A. (1978): Brain Res., 144:257-275.
42. Glazer, E.J., and Basbaum, A.I. (1979): Anat. Rec., 193:549.

43. Glynn, C.J., and Lloyd, J.W. (1976): Proc. R. Soc. Med. 69:369-372.
44. Gracely, R.H., Deeter, W.R., Wolskee, P.J., Wear, B.L., Sayer, J.S., Heft, M.W., Sweet, J., Butler, D., and Dubner, R. (1979): Neurosci. Abstr., 5:609.
45. Grevert, P., and Goldstein, A. (1977): Proc. Natl. Acad. Sci. USA, 74:1291-1294.
46. Hayes, R.L., Price, D.D., Bennett, G.L., Wilcox, G.L., and Mayer, D.J. (1978): Brain Res., 155:91-101.
47. Hiller, J.M., Pearson, J., and Simon, E.J. (1973): Res. Commun. Chem. Pathol. Pharmacol., 6:1052-1062.
48. Hokfelt, T., Elde, R., Johansson, O., Terenius, L., and Stein, I. (1977): Neurosci. Lett., 5:25-31.
49. Hokfelt,T., Ljungdahl, A., Terenius, L., Elde, R., and Nilsson, G. (1977): Proc. Natl. Acad. Sci. USA, 74:3081-3085.
50. Hokfelt, T., Terenius, L., Kuypers, H.G.J.M., and Dann, O. (1979): Neurosci. Lett., 14:55-60.
51. Hong, J.S., Yang, H.Y.T., Fratta, W., and Costa, A. (1977): Brain Res., 134:383-386.
52. Hosobuchi,Y., Adams, J.E., and Linchitz, R. (1977): Science, 197:183-187.
53. Hosobuchi, Y., Rossier, J., Bloom, F.E., and Guillemin, R. (1979): Science, 203:279-281.
54. Irwin, S., Houde, R.W., Bennett, D.R., Hendershot, L.C., and Seevers, M.H. (1950): J. Pharmacol. Exp. Ther., 101:132-143.
55. Jacob, J.J.C., and Ramabadran, K. (1978): Br. J. Pharmacol., 64:91-98.
56. Jacob, J.J.C., Tremblay, E.C., and Colombel, M. (1974): Psychopharmacologia, 37:217-223.
57. Jessell, T., Tsunoo, A., Kanazawa, I., and Otsuka, M. (1979): Brain Res., 168:247-259.
58. Kelly, D.D., Such, S.K., Brutus, M., Glusman, M., and Bodnar, R.J. (1978): Neurosci. Abstr., 4:460.
59. Kokka, N., and Fairhust, A.S. (1977): Life Sci., 21:975-980.
60. Lamotte, C., Pert, C.B., and Snyder, S.H. (1976): Brain Res., 112:407-412.
61. LeBars, D., Guilbaud, G., Jurna, I., and Besson, J.M. Brain Res., 115:518-524.
62. Levine, J.D., Gordon, N.C., Bornstein, J.C. and Fields, H.L. (1979): Proc. Natl. Acad. Sci. USA, 76:3528-3531.
63. Levine, J.D., Gordon, N.C., and Fields, H.L. (1978): Lancet, II:654-657.
64. Levine, J.D., Gordon, N.C., and Fields, H.L. (1979): Nature, 278:740-741.
65. Levine, J.D., Gordon, N.C., Jones, R.T., and Fields, H.L. (1978): Nature, 272:826-827.
66. Loh, H.H., Tseng, L.F., Wei, E., and Li, C.H. (1976): Proc. Acad. Natl. Acad. Sci. USA, 73:2895-2898.

67. McCreery, D.B., Bloedel, J.R., and Hames, E.F. (1979): J. Neurophysiol., 42:166-182.
68. Martin, R.F., Jordan, L.M., and Willis, W.D. (1978): J. Comp. Neurol., 182:77-88,.
69. Mayer, D.J., and Liebeskind, J.C. (1974): Brain Res., 68: 73-93.
70. Mayer, D.J., and Price, D.D. (1976): Pain, 2:379-404.
71. Mayer, D.J., Price, D.D., and Rafii, A. (1977): Brain Res., 121:368-373.
72. Meglio, M., Hosobuchi, Y., Loh, H.H., Adams, J.E., and Li, C.H. (1977): Proc. Natl. Acad. Sci. USA, 74:774-776.
73. Messing, R.B., and Lytle, L.D. (1977): Pain, 4:1-21.
74. Meyerson, B.A., Boethius, J., and Carlsson, A.M. (1979): In: Advances in Pain Research and Therapy, Vol. 3, edited by J.J. Bonica, J.C. Liebeskind, and D.G. Albe-Fessard, pp. 525-533. Raven Press, New York.
75. Mohrland, J.S., and Gebhart, G.F. (1979): Neurosci. Abstr., 5:63.
76. Oliveras, J.L., Hosobuchi, Y., Redjemi, F., Guilbaud, G., and Besson, J.M. (1977): Brain Res., 120:221-229.
77. Oliveras, J.L., Redjemi, F., Guilbaud, G., and Besson, J.M. (1977): Pain, 1:139-145.
78. Pert, C.B., Kuhar, M.J., and Synder, S.H. (1976): Proc. Natl. Acad. Sci. USA, 73:3729-3733.
79. Pomeranz, B. and Chiu, D. (1976): Life Sci., 19:1757-1762.
80. Pomeroy, S.L., and Behbehani, M.M. (1979): Brain Res., 176: r143-147.
81. Prieto, G.J., Giesler, G.J., Jr., and Cannon, J.T. (1979): Neurosci. Abstr., 5:614.
82. Proudfit, H.K., and Anderson, E.G. (1975): Brain Res., 98: 612-618.
83. Ramabadran, K., and Jacob, J.J.C. (1979): Life Sci., 24: 1959-1970.
84. Rhodes, D.L. (1979): Pain, 7:51-63.
85. Richardson, D.E., and Akil, H. (1977): J. Neurosurg. 47: 178-183.
86. Richardson, D.E. and Akil, H. (1977a): J. Neurosurg. 47: 184-191.
87. Rivot, J.P., Chaouch, A., and Besson, J.M. (1979): Brain Res., 176:355-364.
88. Rogers, E.J., and Vilkin, B. (1978): J. Clin. Psych., 39: 431-438.
89. Rossier, J., Guillemin, R., and Bloom, F. (1978): Eur. J. Pharmacol., 48:465-466.
90. Ruda, M. (1975): Ph.D. Dissertation, University of Penn.
91. Sar, M., Stumpf, W.E., Miller, R.J., Chang,, K.J., and Cuatrecasas, P. (1978): J. Comp. Neurol., 182:17- 37.
92. Satoh, M., Akaike, A., and Takagi, H. (1979): Brain Res., 169:406-410.
93. Satoh, M., Kawajiri, S., Yamamoto, M., Makino, H., and Takagi, H. (1979): Life Sci., 24:685-690.

94. Sjolund, B.H., and Eriksson, M.B.E. (1979): Brain Res., 173: 295-301.
95. Takagi, H., Satoh, M., Akaike, A., Shibata, T., and Kuraishi,Y. (1977): Eur. J. Pharmacol., 45:91-92.
96. Takagi, H., Satoh, M., Akaike, A., Shibata, T., Yajima, H. and Ogawa, H. (1978): Eur. J. Pharmacol., 149:113-116.
97. Tohyama, M., Sakei, K., Salvert, D., Touret, M., and Jouvet, M. (1979): Brain Res., 173:383-403.
98. Urca, G., Frenk, H., Liebeskind,J.C., and Taylor, A. (1977): Science, 197:83-86.
99. Wang, J.K., L. Nauss, and Thomas, J., (1979): Anesthesiology, 50:149-151.
100. Watkins, L.R., Griffin, G., Leichnetz, G.R. and Mayer, D.J. (1980): Brain Res., 181:1-15.
101. Watson, S.J., Akil, H.,Richard, C.W., and Barchas, J.D. Nature, 275:226-228.
102. Watson, S.J., Akil, H., Sullivan, S., and Barchas, J.D. (1977): Life Sci., 21:733-738.
103. Willis, W.D., Haber, L.H., and Martin, R.F. (1977): J. Neurophysiol., 40:968-981.
104. Woolf, C.J., Barrett, G.D., Mitchell, D., and Myers, R.A. (1977): Eur. J. Pharmacol., 45:311-314.
105. Yaksh, T.L. (1978): Brain Res., 153:204-211.
106. Yaksh, T.L., and Rudy, T.A. (1976): Science, 192: 1357-1358.
107. Yaksh, T.L., and Rudy, T.A. (1978): Pain, 4:299-359.
108. Yaksh, T.L., Yeung, J.C., and Rudy, T.A. (1976): Brain Res., 114:83-103.

Neurosecretion and Brain Peptides,
edited by J. B. Martin, S. Reichlin, and K. L. Bick.
Raven Press, New York © 1981.

Possible Role of Opioid Peptides in Pain Inhibition and Seizures

J. W. Lewis, S. Caldecott-Hazard, J. T. Cannon, and J. C. Liebeskind

*Department of Psychology and Brain Research Institute, University of California,
Los Angeles, Los Angeles, California 90024*

Important developments in opiate pharmacology (57,64,69) and in our work (50) and that of Reynolds (55) on the phenomenon of stimulation-produced analgesia (SPA) led us (50) almost a decade ago to propose the existence of an endogenous pain-inhibitory system. This system was seen to originate in the medial brainstem, to be activated by direct electrical stimulation and by the administration of opiate drugs, and to operate by the reinforcement of descending controls on spinal nociceptive mechanisms. A great deal of evidence is now available supporting this hypothesis (cf. the following recent reviews 14,42,48; also, see Fields, this volume). None of the evidence is more exciting nor seemingly direct than the discovery of opiate binding sites and opioid peptides throughout much of this brainstem and spinal cord circuitry.

OPIOID PEPTIDES AND ANALGESIA

Naloxone Antagonism of Stimulation-Produced Analesia

An early finding of considerable interest was the demonstration by Akil et al. (5,6) that naloxone at least partially blocks electrically induced analgesia from the rat periaqueductal gray matter. This observation supported our earlier contention (50) that SPA and opiate analgesia share common sites and mechanisms of action. Since then, some workers have confirmed this finding (1,34,52); others have not (53,70). Recent work in our laboratory appears to identify some sources of the variance responsible for these discrepant results.

We now find (cf. 16,54) clear site specificity within the periaqueductal gray matter and subjacent tegmentum of the rat for naloxone-sensitive and naloxone-insensitive analgesic effects. In this work we have carefully mapped the vertical dimension of midline electrode placements from dorsal periaqueductal gray, through the dorsal raphe nucleus in ventral periaqueductal gray, to the

median raphe nucleus lying ventral to the periaqueductal substance. Threshold currents for securing complete analgesia in the
tail-flick test were established with a modified ascending method
of limits. The brain was stimulated with intermittently applied
rectangular pulses for 15 seconds, behavioral tests being conducted during the last 5 seconds. Control and post-naloxone tests
were carried out within the same session. For electrode sites
lying above the dorsal raphe nucleus (dorsal placements), as well
as for those lying within or between dorsal and median raphe
nuclei (ventral placements), potent analgesic effects were rout-
inely observed at current thresholds that did not systematically
vary from site to site. Nonetheless, the SPA threshold was raised
by naloxone (but not by saline) for virtually every animal having
a ventral electrode placement and was completely unchanged by this
drug (and by saline) for virtually every animal in the dorsal
placement group.

Furthermore, although within a broad range the dose of naloxone
seems not to be an important variable, the time between drug
administration and SPA testing can be crucial (16). Tests were
conducted at naloxone doses ranging from 0.01 to 10.0 mg/kg.
Marginal effects were seen at 0.01 mg/kg; clear but still incomplete effects were evident at the 0.10 mg/kg dose with but little
increase in drug efficacy at the 2.0 and 10.0 mg/kg dose levels.
In these experiments, drug effects were sought at 10, 18, and 32
minutes post-naloxone. Blocking was almost never seen before the
18 minute test, and there was no tendency for higher naloxone
doses to induce more rapid effects.

Thus, for naloxone to interfere with SPA, it is evident that
ventral but not dorsal electrode placements need be employed.
Taken together with the fact that naloxone blockade of SPA is
rarely complete, these findings suggest, in accordance with a
growing body of recent evidence (cf. 29,41,49), that analgesic
mechanisms independent of opioid peptides also surely exist. In
addition, to demonstrate naloxone blockade of SPA rather substantial time must be allowed for the drug to be active (somewhere
between 10 and 18 minutes under our test conditions). This
observation is in striking parallel with recent reports that
low-dose naloxone blockade of analgesia induced by periaqueductal
microinjection of glutamate is not evident until 20 minutes or
more have elapsed (9,67). Cannon et al. (16) have suggested that
such findings may bear upon the question of which opioid peptide
is involved in these forms of analgesia. First, β-endorphin-
containing fibers are found in much heavier concentration in ventral than in dorsal periaqueductal gray (12,68), the ventral area
being the one from which our naloxone-sensitive SPA derives.
Moreover, substantial evidence has accumulated suggesting the
occurrence of multiple opiate receptor types (cf. 37), and β-
endorphin and naloxone are known to have higher affinities for the
so-called "μ" opiate receptor than do the enkephalins (45). It has
already been tentatively suggested that μ receptors in the periaqueductal gray play a significant role in opiate analgesia (24).
Certainly our present finding (16) that systemic injection of as

little as 5 μg (0.01 mg/kg) of naloxone can begin to antagonize SPA argues strongly for a high-affinity receptor effect of this drug. Finally, Akil et al. (2,3) have recently reported that β-endorphin dissociates from the opiate receptor at a much slower rate than do the enkephalins. This finding may account for the unusually long latencies required for naloxone antagonism of SPA and glutamate analgesia in contrast to its rapid antagonism of opiate analgesia. In light of the above considerations, we have suggested (16) that naloxone's antagonism of SPA and glutamate analgesia results from the drug's interaction with μ opiate receptor sites already occupied by β-endorphin.

Stress Analgesia and Opioid Peptides

A good deal of thought has been given to the question "If the central nervous system really possesses a natural pain-inhibitory capacity, when is it normally used?". Certainly, as already suggested (44), the pain-inhibitory system must not be easily accessed since noxious stimuli are generally perceived as such and the warning signals they provide are clearly adaptive. On the other hand, it would be adaptive to inhibit pain under those emergency conditions when feeling pain could be disruptive to effective defense or escape behavior.

In this context, we have watched with interest a rapidly growing literature on the analgesic effects of certain painful and/or stressful conditions. A critical question has been whether or not opioid peptides mediate such analgesia. The results have, however, been controversial (4,29). Recent work in our laboratory (41) again seems to identify a possible source of the variance in these conflicting earlier reports.

We find (41) that, depending only on its temporal parameters, inescapable footshock can cause analgesia that either is or is not sensitive to naloxone blockade. In the rat, brief, continuous footshock (3 ma sine waves delivered for 3 minutes) and prolonged, intermittent footshock (3 ma sine waves; 1 second pulses delivered every 5 seconds for 30 minutes) provide comparable profound analgesia in the tail-flick test lasting several minutes. Naloxone (10 mg/kg) was injected twice, once just after baseline testing and again, 30 minutes between baseline and post-stress tests (prolonged stress) or during the last 3 minutes of this interval (brief stess). Under these conditions, naloxone completely blocked analgesia caused by prolonged stress without affecting that caused by brief stress. Thus again, it seems apparent that opioid and non-opioid mechanisms of analgesia exist. In more recent work (43), doses of naloxone as low as 0.1 but not 0.01 mg/kg have proven as effective as 10 mg/kg against the prolonged stress.

We have also found (41) that the synthetic glucocorticoid, dexamethasone, when administered prior to stress according to a schedule in which it is known to block the stress-induced rise in plasma β-endorphin/β-lipotropin (22), blocks analgesia from prolonged but not brief footshock, just as naloxone does. This fact, taken together with the finding that hypophysectomy blocks stress analgesia (7,13), suggests that pituitary ß-endorphin is significantly involved in this phenomenon. However, as has been pointed out by others (56), plasma concentrations of ß-endorphin after stress are far below those required to produce analgesia with systemic ß-endorphin injections. In view of recent evidence suggesting the existence of blood flow to the brain in the hypophyseal portal system (11), we have suggested that opioids of pituitary origin reach brain areas mediating stress analgesia via this route, perhaps by transport through the ventricular system (41). Preliminary evidence from our laboratory is consistent with this hypothesis. We find that ventral periaqueductal gray lesions, in the area from which naloxone-sensitive SPA derives, block analgesia from prolonged footshock.

OPIOID PEPTIDES AND SEIZURES

Recent data suggest that opiate receptors and opioid peptides play a role in certain seizure phenomena. That opioids might serve normally in mediating or modulating functions other than pain inhibition seemed likely (27) from the known multiplicity of opiate drug effects and from the occurrence of opiate receptors and endogenous opioids in brain regions not known to be implicated in nociceptive or anti-nociceptive mechanisms.

Seizures Induced by Opioid Peptides

In 1977 we first reported (66) that 200 µg methionine-enkephalin injected into the lateral ventricle of awake rats elicted profound epileptiform activity in the cortical EEG. These effects were more reliably seen than was analgesia from this ventricular injection site. The EEG manifestations were accompanied by intermittent "wet-dog" shakes and myoclonic twitches but almost never by full motor convulsions. In subsequent work (26), we found that 100 µg of either leucine- (leu-) or methionine- (met-) enkephalin caused EEG seizures of very similar electrographic character; in fact, morphine at this same dose injected into the lateral ventricle also caused seizures very much resembling those caused by the two enkephalins, but longer lasting. Importantly, systemic injections of naloxone blocked all seizure manifestations at a dose of 10 µg but not 2 mg/kg (26,66). Seizures were still provoked by the enkephalins at doses as low as 25 µg, and even at 10 µg in some animals. Non-pathological, rhythmic spindling was seen at 1.0 µg and at 10 µg in animals not manifesting seizures. Suggesting that leu-enkephalin has greater epileptic potency than met-enkephalin, it was found that the EEG effects of leu-enkephalin lasted significantly longer at all doses above 1.0 µg. Analgesia was virtually never obtained at the 100 µg dose of the enkephalins and was never seen below that dose level. By contrast, morphine at an intraventricular dose of 30 µg caused profound analgesia but no epileptiform activity whatsoever. Thus, the enkephalins cause seizures from the lateral ventricle at roughly 1/10 or 1/20 of their analgesic dose from this injection site. Morphine, on the other hand, causes seizures only at doses 10-20 times those required for analgesia.

Pursuing these effects further, we next showed (24) that 100 μg of met-enkephalin injected directly into ventral but not dorsal periaqueductal gray matter caused analgesia. Injections at none of these midbrain placements, however, altered the EEG. On the other hand, the same enkephalin injections directed at the dorsomedial thalamus caused seizures in some animals identical to those seen with intraventricualr administration. Other placements were ineffective. No thalamic injections of enkephalin, however, caused any evidence of analgesia. Thus, an anatomical dissociation between enkepahlin's analgesic and epileptic effects seems clear. We suggested that the analgesic and epileptic effects of the enkephalins are not only mediated by opiate binding sites in different brain areas but by opiate binding sites possessing different pharmacological properties. This hypothesis is consistent with recent evidence on multiple opiate receptor types (37,45; see Pert, this volume). It is supported by the different rank orders of potency of the enkephalins and morphine in producing analgesia and seizures from the same ventricular injection site, by the apparently greater dose of naloxone needed to block seizures than needed to block analgesia, and by the greater epileptic potency of leucine- than methionine-enkephalin (26). We suggested that seizures are mediated by enkephalin's interaction with δ receptors in the forebrain and analgesia by enkephalin's interaction with μ receptors in the periaqueductal midbrain region.

Several other laboratories have made similar observations. β-endorphin, D-ala^2-methionine-enkephalinamide, and the enkephalins all appear to provoke similar, non-convulsive EEG seizures that can be blocked by moderate to high doses of naloxone (19,21,-38,59,62). In fact, another brain peptide, arginine vasopressin, at a dose of 1.0 μg in the lateral ventricle of the rat has recently been reported to cause immobility and staring after initial injection and convulsions after a second injection two days later (36). Tolerance and cross-tolerance have been found for the seizure-producing effects of enkephalin and morphine (19,63). Prior administration of various drugs effective in the treatment of human petit mal epilepsy (e.g., ethosuximide, sodium valproate) have recently been shown to abort completely the EEG effects, the immobility, and the myoclonic jerks caused by 100 μg of leu-enkephalin injected into the rat lateral ventricle (59).

Findings such as these lend support to our initial hypothesis that when normal regulatory processes are disrupted, endogenous enkephalin may play a significant role in the etiology or elaboration of epileptic phenomena (66). Much stronger support would derive from the demonstration that naloxone blocks seizure manifestations in some models not relying on opiate or opioid drug administration. Until very recently, such evidence has been lacking. We and others have been completely unsuccessful in blocking with naloxone seizures caused by metrazol, electroconvulsive shock, or amygdaloid stimulation (23,32,46). Similarly, the development of amygdaloid kindling is not retarded by naloxone (18,28) nor does this drug prevent audiogenic seizures in mice (58) or spontaneously occurring seizures in gerbils (8,25). Furthermore, naloxone (4 mg, i.v.) had no apparent effect on inter-ictal spiking in 7 adults with major motor epilepsies of varying etiology (Engel, Frenk & Liebeskind, unpublished observations). Of great interest, however, is the recent demonstration by Snead and Bearden (60) that seizures of a petit mal type, provoked in rats by systemic administratin of the GABA metabolite, γ-hydroxybutyrate, or a pharmacological precursor, γ-butyrolactone, are blocked by either naloxone or naltrexone at a moderate high dose (> 2 mg/kg). This finding, together with their observation (59) that drugs effective against human petit mal epilepsy are particularly potent in blocking enkephalin-induced seizures in the rat, prompted Snead and Bearden to suggest the enkephalins are involved in the pathogenesis of this disease (60).

Opioid Peptides Released by Seizures

In our first work investigating the effects of naloxone on seizures caused by amygdaloid stimulation in the rat, we found (23) the drug had no effect at 1.0 or 10 mg/kg on either seizure threshold or the intensity or duration of the seizure's electrographic manifestations. However, naloxone did significantly shorten the period of behavioral depression normally following ictus at the 1.0 mg/kg dose, and it eliminated this period altogether at the higher dose level. Morphine, on the other hand, greatly augmented the duration of post-ictal depression at a dose of 10 mg/kg. The extreme depression of activity and the abnormal postures seen post-ictally in morphine treated rats bear considerable resemblance to the cataleptic state caused by much higher doses of morphine given alone. Independently, Holaday and coworkers reported very similar findings with naloxone studied in relation to electroconvulsive shock (32). Again, naloxone (10 mg/kg) did not affect seizure duration but significantly shortened the duration of the post-ictal cataleptic state. That naloxone blocks post-ictal catalepsy suggests that opioid peptides released during ictus play a role in this post-ictal behavioral depression (23,32). The fact that morphine treated rats show such pronounced catalepsy also suggests that opioid peptides are released during ictus and act synergistically with a moderate dose of morphine to protentiate greatly its depressive action (23). Direct evidence for opioid release by seizures has recently become available. In rats, Hong and co-workers (33) found that repeated, but not single electroconvulsive shocks caused up to a 100% increase of met-enkephalin content in certain brain areas (hypothalamus, n. accumbens, septum, amygdala). In other regions (e.g., hippocampus, lower brain stem), no changes were seen; and in the hypothalamus, where met-enkephalin levels were doubled, β-endorphin levels were unchanged. Also, but using a very different paradigm, Bergland et al. (10) find that a single electroconvulsive shock causes a 200-

fold rise in sagittal sinus β-endorphin/β-lipotropin concentrations in the intact but not hypophysectomized sheep. Enkephalin concentrations were not measured. Because sagittal sinus endorphin levels were typically higher than carotid artery levels after the convulsion, and an intact pituitary was required for the endorphin increase, it was concluded that pituitary endorphin released by the seizure flows directly to the brain and may be responsible for post-ictal behavioral changes.

We speculated (23) that opioid-induced post-ictal depression may have the effect of forestallng the onset or reducing the severity of subsequent seizures. Engel and Katzman (20) had earlier reported that stria terminalis lesions facilitate amygdaloid kindling in the rat. We suggested (23) that such lesions, by interrupting the amygdalofugal enkephalin-containing fibers found in the rat (65), block the opioid-induced process of post-ictal depression and thus enable kindling to proceed at a more rapid pace. Supporting this hypothesis, Hardy et al. (28) have recenty found that 5 mg/kg naloxone facilitates amygdaloid kindling in rats (cf. also 18). These authors also report that naloxone can reduce and morphine prolong post-ictal behavioral depression. Thus, for amygdaloid kindled seizures, which (unlike γ-hydroxybutyrate) may be an animal model for motor convulsions, it seems clear that opioid peptides do not serve an excitatory role in seizure initiating or regenerative processes. On the contrary, the available evidence suggests that opioid peptides are released by such seizures and exert a behavioral depressive and anticonvulsant action (cf. also 61).

Seizures and Opiate Receptors

Starting from the premise that opiate and opioid seizures are mediated by opiate receptors that are pharmacologically different from those mediating analgesia, we sought to determine if the "κ" and δ opiate receptor agonists, the benzomorphans (cf. 47), had clear seizure-producing properties. We found (15) that intraventricular injections in the rat of ketocyclazocine, cyclazocine, and WIN-35, 197-2 failed to cause seizures throughout a dose range in which EEG spindles, analgesia, and death could all be obtained. Similar results were reported by Henriksen and Bloom (30). We conclude that κ and δ receptors are not responsible for opiate seizures. As Kosterlitz has pointed out (37), analysis of receptor affinities of opioid peptides in his smooth muscle assays does not yield results fitting well with Martin's receptor classification. Thus, the δ receptor of Kosterlitz (cf. 45), for which enkephalins have particlularly high affinity, appears not to be equated with either κ or δ receptors. We suggest once again, therefore, that δ receptors mediate opioid seizures.

Finally, in an attempt to meet another criterion for opiate receptor involvement, that of stereospecificity, we sought to compare the epileptic potencies of the stereoisomers, levorphanol and dextrorphan tartrate. To our surprise, neither agonist injected into the rat lateral ventricle caused seizures throughout a range of doses (50-300 µg) that, like the benzomorphans, caused EEG synchrony, analgesia, and sometimes death (15). Recent work by LaBella et al. (38,39) provided an important clue that has now permitted us to demonstrate the stereospecific effect we sought. These investigators showed an impressive array of opiate-like effects, including seizures, with intraventricular administration of androsterone sulfate (39); and in subsequent work it became apparent that it was the sulfate moiety that was active (38). They postulated that sulfate, acting as a calcium chelator, was importantly involved in the opiate-like effects they observed (38). In fact, high doses of sodium sulfate (> 200 µg) or low doses of the powerful calcium complexing agent, EGTA, produced the same effects when given alone, including seizures resembling those provoked by opioid peptides and morphine (38). The important role of calcium in opiate mechanisms has been recognized for several years (cf. the recent review by Chapman and Way [17]).

With such findings in mind, we showed (15) that if either sodium or potassium sulfate (26 or 52 µg) were added to the levorphanol solution to make the sulfate concentration the same or twice that of an equimolar solution of morphine sulfate, intraventricular administration of 100 µg levorphanol was then capable of eliciting seizures indistinguishable from those caused by the enkephalins or morphine at this dose level. Adding sulfate in this fashion to the same concentration of dextrorphan never causes seizures. Naloxone (10 mg/kg, i.p.) blocked levorphanol sulfate seizures completely. An intriguing additional finding was that the same 26 and 52 µg doses of potassium sulfate that unmasked levorphanol seizures reduced the analgesic effect of levorphanol in a dose dependent manner. By contrast, potassium sulfate had no effect on the considerably weaker analgesic action of dextrorphan seen at this 200 µg dose, just as it failed to alter dextrorphan's inability to cause seizures.

That levorphanol lacks some of the excitatory actions of morphine has been pointed out in several recent studies (35,40). We have shown in this regard that morphine hydrochloride is at least as effective as morphine sulfate in provoking seizures at a dose of 100 µg (unpublished findings); and, of course, opioid peptides by themselves are potent epileptogens. Thus, unlike levorphanol, morphine and opioid peptides have epileptic effects independent of sulfate ions. On the other hand, Mudge et al. (51) recently reported that enkephalin decreases calcium influx in dorsal root ganglion neurons in culture; and Chapman and Way (17) discuss a still controversial literature on the calcium depleting action of opioid peptides and morphine in the brain. Thus, it seems at least plausible that the epileptic properties of morphine and opioid peptides, like levorphanol sulfate, derive from their ability to reduce calcium influx and hence alter membrane excitability and/or transmitter release in neurons influencing seizure production. It may be that the difference between levorphanol and these other agonists lies in its relative inability to affect nerve membrane calcium flux at sub-lethal doses. The intriguing observation that sulfate ions diminish levorphanol analgesia while unmasking levorphanol seizures is suggestive of yet another pharmacological difference between opiate receptors mediating seizures and analgesia.

ACKNOWLEDGEMENTS

The authors express their appreciation to Ms. Sheila Roberts for her careful attention to preparing this manuscript. JWL and SCH were supported by USPHS Training grant MH 15345-02. Our reseach is funded by NIH grant NS 07628.

REFERENCES

1. Adams, J.E. (1976): Pain, 2:161-166.
2. Akil, H., Hewlett, W., Barchas, J.D., and Li, C.H. (1980): Life Sci., (in press).
3. Akil, H., Hewlett, W.A., Barchas, J.D., and Li, C.H. (1980): Eur. J. Pharm., (in press).
4. Akil, H., Madden, J., Patrick, R.L., and Barchas, J.D. (1976): In: Opiates and Endogenous Opioid Peptides, edited by H.W. Kosterlitz, pp. 63-70. Elsevier, Amsterdam.
5. Akil, H., Mayer, D.J., and Liebeskind, J.C. (1972): C.R. Acad. Sci. (Paris), 274:3603-3605.
6. Akil, H., Mayer, D.J., and Liebeskind, J.C. (1976): Science, 191:961-962.
7. Amir, S. and Amit, Z. (1979): Life Sci., 24:439-448.
8. Bajorek, J.G., Chesarek, W., Felmer, M., and Lomax, P. (1978): Proc. West. Pharmacol. Soc., 21:365-370.
9. Behbehani, M.M. and Fields, H.L. (1979): Brain Res., 170:85-93.
10. Bergland, R.M., Blume, H.W., Hamilton, A., Monica, P., Paterson, R., and Wright, R.D. (1980): Science, (in press).
11. Bergland, R.M. and Page, R.B. (1979): Science, 204:18-24.
12. Bloom, F.E., Battenberg, E., Rossier, J., Ling, N., and Guillemin, R. (1978): Proc. Natl. Acad. Sci. (USA), 75:1591-1595.
13. Bodnar, R.J., Glusman, M., Brutus, M., Spiaggia, A., and Kelly, D.D. (1979): Physiol. and Behav., 23:53-62.
14. Cannon, J.T. and Liebeskind, J.C. (1979): In: Mechanisms of Pain and Analgesic Compounds, edited by R.F. Beers, Jr. and E.G. Bassett, pp. 171-184. Raven Press, New York.
15. Cannon, J.T., Nahin, R.L., Ryan, S.M., Moskowitz, A.S., and Liebeskind, J.C. (1979): Neurosci. Abstr., 5:190.
16. Cannon, J.T., Prieto, G.J., Lee, A., and Liebeskind, J.C. (1980): Fed. Proc., 39:603.
17. Chapman, D.B. and Way, E.L. (1980): Ann. Rev. Pharmacol. Toxicol., 20:553-579.
18. Corcoran, M.E. and Wada, J.A. (1979): Life Sci., 24:791-796.

19. Elazar, Z., Motles, E., Ely, Y., and Simantov, R. (1979): Life Sci., 24:241-248.
20. Engel, J., Jr. and Katzman, R. (1977): Brain Res., 122:137-142.
21. Firemark, H.M. and Weitzman, R.E. (1979): Neuroscience, 4:1895-1902.
22. French, E.D., Bloom, F.E., Rivier, C., Guillemin, R., and Rossier, J. (1978): Neurosci. Abstr., 4:408.
23. Frenk, H., Engel, J., Jr., Ackermann, R.F., Shavit, Y., and Liebeskind, J.C. (1979): Brain Res., 167:435-440.
24. Frenk, H., McCarty, B.C., and Liebeskind, J.C. (1978): Science, 200:335-337.
25. Frenk, H., Paul, L., Diaz, J., and Bailey, B. (1978): Soc. Neurosci. Abstr., 4:142.
26. Frenk, H., Urca, G., and Liebeskind, J.C. (1978): Brain Res., 147:327-337.
27. Goldstein, A. (1976): Science, 193:1081-1086.
28. Hardy, C., Panksepp, J., Rossi, J., III, and Zolovick, A.J. (1980): Brain Res., (in press).
29. Hayes, R.L., Bennett, G.J., Newlon, P.G. and Mayer, D.J. (1978): Brain Res., 155:69-90.
30. Henriksen, S.J. and Bloom, F.E. (1980): Frontiers in Hormone Research, Vol. 5., (in press).
31. Henriksen, S.J., Bloom, F.E., McCoy, F., Ling, N., and Guillemin, R. (1978): Proc. Natl. Acad. Sci. (USA), 75:5221-5225.
32. Holaday, J.W., Belenky, G.L., Loh, H.H., and Meyerhoff, J.L. (1978): Soc. Neurosci. Abstrs., 4:409.
33. Hong, J.S., Gillin, J.C., Yang, H.-Y.T., and Costa, E., (1979): Brain Res., 177:273-278.
34. Hosobuchi, Y., Adams, J.E., and Linchitz, R. (1977): Science, 197:183-186.
35. Jacquet, Y.F. and Lajtha, A. (1974): Science, 185:1055-1057.
36. Kasting, N.W., Veale, W.L., and Cooper, K.E. (1980): Can. J. Physiol. Pharmacol., (in press).
37. Kosterlitz, H.W. (1979): In: Mechanisms of Pain and Analgesic Compounds, edited by R.F. Beers, Jr. and E.G. Bassett, pp. 207-214. Raven Press, New York.
38. LaBella, F.S., Havlicek, V., and Pinsky, C. (1979): Brain Res., 160:295-305.
39. LaBella, F., Havlicek, V., Pinsky, C., and Leybin, L. (1978): Can. J. Physiol. Pharmacol., 56:940-944.
40. LaBella, F.S., Pinsky, C., and Havlicek, V. (1979): Brain Res., 174:263-271.
41. Lewis, J.W., Cannon, J.T., and Liebeskind, J.C. (1980): Science, (in press).
42. Lewis, J.W., Cannon, J.T., Ryan, S.M., and Liebeskind, J.C. (1980): In: Role of Peptides in Neuronal Function, edited by J.L. Barker, (in press). Marcel Dekker, Inc., New York.
43. Lewis, J.W., Cannon, J.T., Stapleton, J.M., and Liebeskind, J.C. (1980): Proc. West. Pharmacol. Soc., (in press).

44. Liebeskind, J.C., Giesler, G.J., Jr., and Urca, G. (1976): In: Sensory Functions of the Skin in Primates, edited by Y. Zotterman, pp. 561-573. Pergamon Press, Oxford.
45. Lord, J.A.H., Waterfield, A.A., Hughes, J., and Kosterlitz, H.W. (1977): Nature (London), 267:495-499.
46. Mannino, R.A. and Wolf, H.H. (1974): Life Sci., 15:2069-2096.
47. Martin, W.R., Eades, C.G., Thompson, J.A., Huppler, R.E., and Gilbert, P.E. (1976): J. Pharmacol. Exp. Ther., 197:517-532.
48. Mayer, D.J. and Price, D.D. (1976): Pain, 2:379-404.
49. Mayer, D.J., Price, D.D., Rafii, A., and Barber, J. (1976): In: Advances in Pain Research and Therapy, Vol. 1, edited by J.J. Bonica and D. Albe-Fessard, pp. 751-754. Raven Press, New York.
50. Mayer, D.J., Wolfle, T.L., Akil, H., Carder, B., and Liebskind, J.C. (1971): Science, 174:1351-1354.
51. Mudge, A.W., Leeman, S.E., and Fischbach, G.D. (1979): In: Endorphins in Mental Health Research, edited by E. Usdin, W.E. Bunney, Jr., and N.S. Kline, pp. 344-351. Oxford University Press, New York.
52. Oliveras, J.L., Hosobuchi, Y., Redjemi, F., Guilbaud, G., and Besson, J.M. (1977): Brain Res., 120:221-229.
53. Pert, A. and Walter, M. (1976): Life Sci., 19:1023-1032.
54. Prieto, G.J., Giesler, G.J., Jr., and Cannon, J.T. (1979): Neurosci. Abstr., 5:614.
55. Reynolds, D.V. (1969): Science, 164:444-445.
56. Rossier, J., French, E.D., Rivier, C., Ling, N., Guillemin, R., and Bloom, F.E. (1977): Nature (London), 270: 618-620.
57. Satoh, M. and Takagi, H. (1971): Europ. J. Pharmacol., 14:60-65.
58. Schreiber, R.A. (1979): Psychopharmacology, 66:205-206.
59. Snead, O.C., III and Bearden, L.J. (1980): Neurology, (in press).
60. Snead, O.C., III and Bearden, L.J. (1980): Neurology, (in press).
61. Tortella, F.C., Cowan, A., and Adler, M.W. (1979): Neurosci. Abstr., 5:542.
62. Tortella, F.C., Moreton, J.E., and Khazan, N. (1978): J. Pharmacol. Exp. Ther., 206:636-643.
63. Tortella, F.C., Moreton, J.E., and Khazan, N. (1979): J. Pharmacol. Exp. Ther., 210:174-179.
64. Tsou, K. and Jang, C.S. (1964): Scientia Sinica, 13: 1099-1109.
65. Uhl, G.R., Kuhar, M.J., and Snyder, S.H. (1978): Brain Res., 149:223-228.
66. Urca, G., Frenk, H., Liebeskind, J.C., and Taylor, A.N. (1977): Science, 197:83-86.
67. Urca, G., Nahin, R.L., and Liebeskind, J.C. (1980): Brain Res., (in press).
68. Watson, S.J., Akil, H., Richard, C.W., and Barchas, J.D. (1978): Nature (London), 275:226-228.

69. Way, E.L. and Shen, F. (1971): In: <u>Narcotic Drugs:</u> <u>Biochemical Pharmacology</u>, edited by D.H. Clouet, pp. 229-253. Plenum Press, New York.
70. Yaksh, T.L., Yeung, J.C., and Rudy, T.A. (1976): <u>Life Sci.</u>, 18:1193-1198.

Neurosecretion and Brain Peptides,
edited by J. B. Martin, S. Reichlin, and K. L. Bick.
Raven Press, New York © 1981.

Introduction to Section IV:
Principles of Neuronal Growth and Differentiation

Dennis M. D. Landis

*Department of Neurology, Massachusetts General Hospital and Harvard Medical School,
Boston, Massachusetts 02115*

There is widespread interest in the possibility that a variety of peptides may be involved in information processing, either directly as synaptic neurotransmitters or as modulators in synaptic function.   In this session, however, the speakers drew attention to a different set of cell interactions: those occurring during development.   The precise nature of the signalling process in cellular interactions during development is entirely unknown, and participation by peptides in such interactions is an intriguing possibility.   It is more certain that differentiation during development contributes to defining the types of precursor molecule processing that will characterize a given peptide-containing neuronal population.   Moreover, the sensitivity of techniques for detection of peptides may be exploited in future studies to define precisely steps in the differentiation of peptide-containing neurons.

Dr. Black noted several processes regulating differentiation of neural crest cells which give rise to peripheral sympathetic, parasympathetic, and sensory systems.   For example, differentiating adrenergic neurons in the superior cervical ganglion (SCG) suffer a prompt and persistent loss of tyrosine hydroxylase activity if deprived of spinal cord cholinergic input by mechanical denervation or by long-acting nicotinic blockade.   Extirpation of target tissue results in exaggerated cell death in developing SCG, and this effect can be blocked by addition of exogenous nerve growth factor (NGF).   The nature of the target tissue may also contribute to the regulation of neurotransmitter content in parasympathetic ganglia.   Neural crest cells migrating into the gut during embryogenesis transiently express immunoreactive tyrosine hydroxylase and dopamine beta hydroxylase, characteristics of adrenergic cells, but in the adult are cholinergic. This in vivo transition from adrenergic to cholinergic function is surprisingly gradual.   Even in the absence of demonstrable endogenous catecholamine content or immunoreactive tyrosine

225

hydroxylase and dopamine ß-hydroxylase, presumably cholinergic cells show for weeks a residual high infinity uptake of norepine-phrine.

NGF is the best characterized trophic substance influencing sympathetic system embryogenesis. Dr. Thoenen noted that it is produced in target tissues, taken up specifically, is transported neurons. NGF is required for survival and growth of sympathetic neurons in vitro. It appears to enhance the survival of sensory neurons grown from dorsal root ganglia (DRG). Dr. Black had found that accumulation of Substance P by DRG cells in culture was increased by NGF, but not blocked by antisera to NGF. Recent experiments in Dr. Thoenen's laboratory have found that medium conditioned by glial or heart cells and tissue extracts from brain and peripheral nervous system also influence the survival of sensory cells in vitro and the effect is not blocked by anti-sera to NGF. Cell survival promoted by glial-conditioned medium is progressively greater during later stages of embryogenesis, but it is not yet clear whether a specific subset of DRG neurons is being affected. Dr. Thoenen predicted that judicious balanc-ing of glial-conditioned medium and NGF would yield nearly com-plete survival of DRG cells in culture.

Potter and his colleagues have found that neurons derived from SCG and culture in the presence of NGF will synthesize, store, release, and take up catecholamines, but if similar cells are co-cultured with non-neuronal cells or in medium conditioned by non-neuronal cells, these adrenergic characteristics are prog-ressively replaced by cholinergic functions. There is an inter-val during the transition from adrenergic to cholingeric function when cells manifest stimulus-evoked release of both neurotrans-mitters, and have morphological characteristics intermediate between those of cholinergic and adrenergic cells. Sequential electrophysiological study of single neurons has clearly docu-mented a progression from adrenergic to dual function and in other cells from dual function to cholinergic. In this system, NGF supports the survival of the neurons regardless of the neuro-transmitter phenotype. An intensive effort, led by P. Patterson, is underway to characterize the factor directly influencing neurotransmitter choice.

The extra cellular matrix (ECM), or basal lamina, represents yet another potent regulator of cell survival, division, growth, and differentiation in vitro. Dr. D. Gospodarowicz emphasized that the presence or absence of a basal lamina clearly alters cellular response to exogenous serum or plasma factors, such as fibroblast growth factor (FGF). In turn, such factors may inter-act with cellular production of basal lamina. Dr. R.P. Bunge has found that myelination of axons by Schwann cells in vitro pro-ceeds only when a collagen substrate has been provided. By devising methods which reduce or eliminate fibroblasts in cul-ture, it has been found that Schwann cells in the presence of neurons will generate basal lamina and small extracellular fibrils. However, in the same culture system grown in a serum-free, defined medium, Schwann cells did not produce morphologically evident basal lamina and do not myelinate axons. Dr. Bunge argues that there may be a signal (provided by serum) which leads to production of collagen-like materials by Schwann cells, which in turn may modify interaction of Schwann cells and axons. The possibility that the serum-free medium acts more directly on the axons remains to be explored.

Neurosecretion and Brain Peptides,
edited by J. B. Martin, S. Reichlin, and K. L. Bick.
Raven Press, New York © 1981.

Observations on the Role of Schwann Cell Secretion in Schwann Cell–Axon Interactions

R. Bunge, F. Moya, and M. Bunge

Department of Anatomy and Neurobiology, Washington University School of Medicine, St. Louis, Missouri 63110

An important component of the normal development of peripheral nerves is the apposition of Schwann cells to axonal processes to form either a myelinated or unmyelinated investment. Recent tissue culture observations indicate that the cell of Schwann both secretes active substances and depends upon local secretions for completion of this ensheathment process. A deficiency of secretion by normal Schwann cells under abnormal tissue culture conditions leads to a distinctive failure of Schwann cell function. This defect is reminiscent of that seen in portions of the peripheral nerve roots of the dystrophic mouse. In this paper, evidence that Schwann cells produce extracellular matrix components is described in relation to complementary observations that these cells also require contact with certain types of extracellular matrix components for normal differentiation to occur.

SCHWANN CELL-AXONAL INTERACTION DURING DEVELOPMENT

Schwann cells appear among developing neurons of neural crest origin during the embryonic period. As axons grow out from the neuronal somata of sensory and autonomic ganglia, Schwann cells accompany them to provide a Schwann cell population for the extensive neuritic networks of the peripheral nervous system.

There is now evidence that Schwann cell development during this period is directly influenced by the axon. At least three aspects of this interaction are known; 1) the axon provides a mitogenic signal resulting in Schwann cell proliferation, 2) the axon signals certain Schwann cells to form myelin sheaths and 3) the axon induces the Schwann cell to produce connective tissue components of the extracellular matrix.

The Mitogenic and the Myelinogenic Signal

In the earliest stages of Schwann cell development (when the population of Schwann cells is not yet large enough to provide

ensheathment for all of the axons growing into the periphery)
intense Schwann cell proliferation occurs until the axons are
"saturated" with Schwann cell ensheathment. It is possible to
demonstrate in tissue culture preparations undergoing rapid axonal
and Schwann cell growth that Schwann cell proliferation drops
rapidly to reach very low levels only two days after axonal withdrawal. Schwann cell proliferation may be rapidly resumed if
ingrowing axons again contact the Schwann cell population (30).
It is also possible to demonstrate that the application of particulate fractions of axonal membranes to mitotically "quiescent"
isolated Schwann cells in culture causes a dramatic increase in
radioactive thymidine labelling. If axons have been treated with
trypsin prior to harvest for the preparation of a particulate
fraction, the ability of the axolemmal fraction to stimulate
Schwann cell proliferation is lost. These and related observations have been reviewed and documented in a recent series of
papers (21-24). We conclude that the axon provides a signal
(believed to be located on its surface) which stimulates Schwann
cell proliferation upon contact.

It should be added that several types of experiments indicate
that a signal travels from certain axons to Schwann cells to in-
struct that myelin sheaths be formed. Both in vitro and in vivo
observations also indicate that the specificity for myelin formation resides within the axon and not within the accompanying
Schwann cell, for it can be clearly demonstrated that Schwann
cells related to unmyelinated nerve fibers will, upon contacting
larger nerve fibers competent to engender myelin formation, form
myelin sheaths around these axons (for recent review see ref. 1).

The Demonstration of Secretory Activity
by Mammalian Schwann Cells

The availability of tissue culture preparations in which
Schwann cells divide and differentiate in relation to normal nerve
cells has allowed identification of extracellular matrix components generated in the absence of the normal fibroblast component
of peripheral nerve. This tissue culture approach allows direct
observations on the sources of endoneurial and perineurial constituents of peripheral nerve, particularly the basal lamina which
in all cases surrounds the axon–Schwann cell unit of peripheral
nerve, and the prominent and copious collagen fibrils which constitute the main visible structural elements within the endoneurial compartment of the peripheral nerve fascicle. There has, in
the past, been no general agreement regarding the source of the
collagenous components of endoneurium, i.e., whether these are
derived from the fibroblast population of the endoneurium or from
the more frequently occurring and ubiquitous cells of Schwann
(26).

Our observations derive from the study of tissue cultures
prepared from dorsal root ganglia of fetal or newborn rodents,
established and maintained in long-term culture. Cultures con-

taining only nerve cells may be obtained if, during the early days
in culture, the preparation is subjected to a series of pulses of

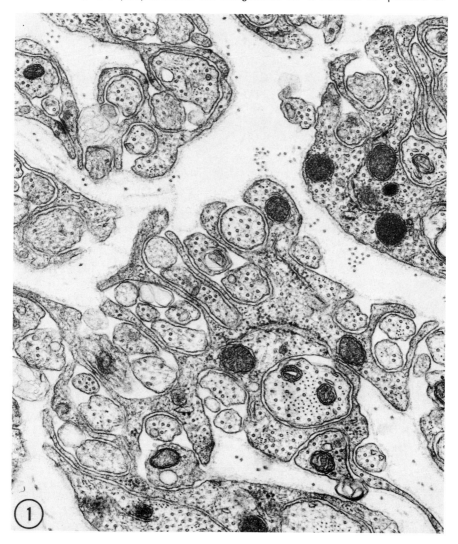

FIG. 1. Electron micrograph of outgrowth region in a control
neuron–Schwann cell culture without fibroblasts after 4 weeks in
vitro. Most of the nerve fibers are ensheathed by Schwann cell
cytoplasm. The exterior of the Schwann cell is covered by basal
lamina, and external to these laminae are small diameter collagen
fibrils. Tissue for the electron micrographs shown was fixed in
buffered glutaraldehyde followed by OsO_4, embedded in Epon-
Araldite and stained with uranyl acetate and lead citrate. X
36,000.

antimitotic agents to suppress the proliferation of both Schwann cells and fibroblasts (29). In the presence of appropriate trophic factors, these sensory neurons survive in long-term culture, but they lack, both along the course of their nerve cell processes and in relation to the cell soma, the connective tissue components and Schwann cells present in the peripheral nerve in vivo (7). The lack of basal lamina and adjacent extracellular fibrillar material allows nerve cell bodies and fibers to come into direct apposition to one another, as do the cellular elements of the central nervous system. It is possible, by adjusting the initial antimitotic treatment, to obtain cultures that contain nerve cells and also retain a small population of Schwann cells (29). With time a substantial Schwann cell population is generated but, in most instances, fibroblasts are not, and a culture is obtained containing only neurons and Schwann cells. If these cultures are allowed to mature for several weeks, the Schwann cells proliferate and ensheathe; the smaller nerve fibers are surrounded by cytoplasmic extensions of Schwann cells, and larger nerve fibers competent to become myelinated are provided with sequential segments myelin (7). As this ensheathment occurs, there is deposited around each axon-Schwann unit a characteristic basal lamina and a population of distinctive thin extracellular fibrils that average 18 nm in diameter and exhibit a repeating banding pattern (Fig. 1). There are also formed thin, ruthenium red-stained strands, which are seen to interconnect the basal laminae of adjacent axon-Schwann cell units. The 18 nm fibrils are observed to be resistent to digestion by trypsin but are removed by purified preparations of collagenase. After the initial few weeks in culture, no substantial additional fibrillar material is formed.

Biochemical analysis of the collagenous components formed by nerve cell-Schwann cell cultures in the absence of fibroblasts has now been reported (7). Collagen synthesis is indicated by the incorporation of labelled proline into peptide-bound hydroxyproline. Studies of the protein formed (determined by pepsin digestion-ammonium sulfate precipitation-polyacrylamide gel electrophoresis techniques) indicate that these cultures synthesize hydroxyproline-containing polypeptides that migrate to positions characteristic for the $\alpha 1$ (I), $\alpha 2$, $\alpha 1$ (III), and $\alpha\beta$ chains of type I, type III and A-B collagens, respectively. Type 1 collagen is one of the primary components of the extracellular matrix in a variety of tissues. Type III, considered to be found in reticulin, is present in vivo as a delicate sheath intimately applied to each axon-Schwann cell unit external to the basal lamina; A-B collagens have been found in basement membranes (for discussion see ref. 7). We emphasize that the analytical techniques employed detect only labelled peptide-bound hydroxyproline which may be only a fraction of the secretory products of the Schwann cell. It seems reasonable to assume that the Schwann cell secretes additional materials which may be instrumental in the trophic support of the neurons, as will be discussed below.

If fibroblasts are either retained or added to these neuron-Schwann cell cultures, a substantial augmentation of organization is observed (Fig. 2). After several weeks of co-culture, there is

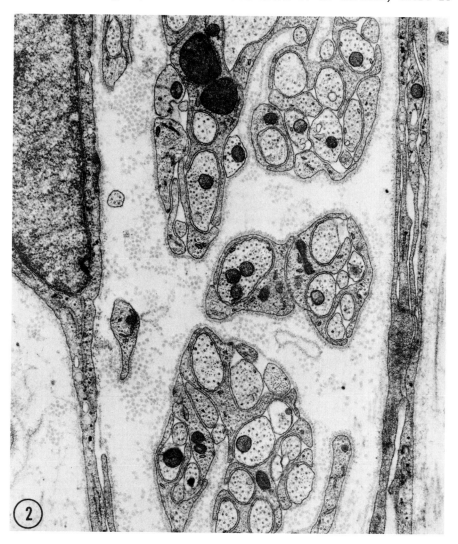

FIG. 2. Typical fascicle region in the outgrowth of a control neuron–Schwann cell preparation containing fibroblasts after 16 weeks in culture. Nerve fibers, Schwann cells and basal laminae are illustrated. The collagen fibrils in the extracellular space are larger in diameter than those found in Fig. 1. Another difference is the presence of perineurial cells; their flattened processes form partitions around the Schwann cell-neurite partners. X 22,000.

formed around axon–Schwann cell units a perineurial ensheathment (i.e., a layer of flattened epithelial-like cells circumferentially disposed around the axon–Schwann cell units). Accompanying this perineurial ensheathment is a large number of substantially thicker collagen fibrils (averaging 35 nm in diameter) that more closely resemble the copious collagenous fibrillar component occurring normally in the endoneurial compartment of in situ peripheral nerve (5).

Electron microscopic analysis of neuron–Schwann cell cultures, demonstrates that the basal lamina of the Schwann cell–axon unit is reformed (after removal by trypsin treatment) only if axons are retained in the culture system (6). Cultures have been prepared in which the basal lamina is removed from the axon–Schwann cell unit with trypsin and, at the same time, neurons have been excised from the cultures; cultures bereft of neurons retain a substantial population of Schwann cells, but basal lamina is not regenerated. In contrast, control cultures in which neurons are retained reform basal lamina around each axon–Schwann cell cultures of short-term duration, the more recently formed basal lamina vanishes from the Schwann cell surface; if neurons are re-introduced into these Schwann cell preparations and axons establish contact with the Schwann cells, a basal lamina reappears around the axon–Schwann cell complex (28). Thus, by two different culture strategies, it has been shown that the presence of nerve cells is required for the generation of a basal lamina on the Schwann cell exterior.

It should be noted, however, that the failure to deposit a morphologically visible basal lamina on the Schwann cell surface may not indicate the complete failure of Schwann cell secretory activity.

Whereas basal lamina construction ceases when axon contact is lost, it is possible to demonstrate (by studying the amount of labelled hydroxyproline appearing in the culture following the addition of labelled proline) that Schwann cells in the absence of nerve cells continue to produce some hydroxyproline-containing proteins (7). Considering the series of events which must occur for the normal secretion and disposition of collagenous components, this observation may indicate that, although Schwann cell–axon contact may not be required for the secretion of collagen components per se, it may be required for the simultaneous secretion of necessary enzymes for collagen polymerization. Subsequent to its secretion as procollagen, the procollagen molecule is acted upon by procollagen peptidases which remove terminal portions of the polypeptide chain; the subsequent polymerization of the tropocollagen molecules into ordered strands requires the action of lysyl oxidases. Taken together, the experiments above point to the Schwann cell as the source of a portion of the components of the endoneurium of peripheral nerve, i.e., particularly of at least a portion of the basal lamina and a portion of the relatively thin collagen fibers of the endoneurium. Furthermore, these studies show that the secretion of these materials is influenced by contact with an axon.

SCHWANN CELL INTERACTION WITH EXTRACELLULAR MATRIX

There is now evidence that the complex set of interactions between axon and Schwann cell outlined above are not in themselves sufficient to allow full Schwann cell differentiation. Certain recent observations have been interpreted as indicating that a "third element" (in addition to axon and Schwann cell) is involved in the genesis of normal axonal ensheathment by Schwann cells (9-10). This "third element" is considered to be a component of extracellular matrix which provides a critical inductive signal at an early stage of Schwann cell differentiation. This suggestion derives from tissue culture preparations in which Schwann cells are related to normal mature axons but are not in contact with the substratum provided in the culture dish. Under these conditions Schwann cells fail to relate normally to axons in order to en-sheathe or myelinate axonal elements. These observations were made on axons (in the outgrowth from sensory ganglion explants) which are suspended in tissue culture medium for part of their course between the bulbous explant and the underlying collagen substratum. These suspended fascicles often contain a number of Schwann cells but these do not display the normal disposition in relation to the suspended nerve fibers. They perch precariously on the surface of the aggregated axons establishing only tenuous contact with the more peripherally located nerve fibers. If the Schwann cells in the suspended nerve fascicles are forced into contact with the collagen substratum, the abnormal Schwann cell ensheathment is corrected. During this correction, the Schwann cells proliferate, ensheathe axons rapidly, and begin to form myelin around the largest axons within several days. Thus, con-tact with a surface in addition to that normally provided by the axon appears to be needed for the development of normal axon-Schwann cell relationships. This may appear to be a puzzling observation in light of the experiments discussed above, indi-cating that Schwann cells are capable of producing collagenous components in the presence of axonal contact. It would seem reasonable to suggest that, in fact, what is needed in terms of a "third element" by the developing Schwann cell is not those collagenous components which are produced by axon–Schwann cell cultures, but an alternate type of collagen or collagen-associated molecule which is normally provided by other cell types in the environment of the developing nerve. There are in embryonic development many instances of cell types which themselves produce extracellular matrix components but which require contact with (presumably other) extracellular matrix components for full expression of their functional capacities. An example is provided by the cells of the corneal epithelium; the collagen and glyco-saminoglycan production of these cells is greatly enhanced by contact with extracellular matrix components containing collagen and glycosaminoglycans (13).

CONSEQUENCES OF FAILURE OF SCHWANN CELL SECRETION

In the various experiments described above, cultures were grown in the presence of a standard medium which contains vitamins, amino acids, minerals, and cofactors, as well as two undefined components - serum and chick embryo extract. There are now available a number of newly designed defined media, some of which are known to support neuronal development in culture (3). We have recently studied the development of cultures prepared in a manner similar to the cultures described above, but maintained for substantial periods of time in the medium N2 (3). This medium contains a rather standard repertoire of vitamins, amino acids and inorganic salts plus the additives putrescine, insulin, NGF, progesterone, selenium and transferrin. Confirming results first reported by others (25), we have observed that neurite extension from sensory ganglion cells occurs in this serum-free medium and that, in addition, a substantial amount of Schwann cell proliferation is observed in relation to these growing neurites. We have also observed (as has been previously reported) that fibroblast proliferation in these preparations is suppressed without the use of specifc antimitotic agents (25). Thus, over a course of 2-3 weeks a fetal dorsal root ganglion explant in culture generates a substantial neuritic outgrowth which becomes populated by a large complement of Schwann cells. In control cultures (which contain serum and embryo extract components added to this medium), myelination begins at this time (Fig. 3).

In cultures that are maintained in N2 medium Schwann cell ensheathment appears to falter subsequent to the proliferation phase (Fig. 4). Large numbers of Schwann cells are observed in relation to neurites; at times Schwann cell numbers appear to be in excess of those seen in the control cultures. Even after several additional weeks, however, no myelination is observed in cultures maintained in N2 medium. When these cultures are examined in the electron microscope, the lack of myelination is confirmed and, in addition, it is apparent that Schwann cell development has been arrested at a much earlier stage of differentiation; the Schwann cells typically do not surround the axons, and basal lamina and the distinctive thin extracellular fibris seen in normal Schwann cell-axon cultures are not formed (Fig. 5). This arrested configuration of the Schwann cell is expressed in conditions in which there has been contact with both the axon and the collagen substratum. The absence of basal lamina and extracellular fibrillar material suggest that the neuron-related Schwann cell fails to secrete certain of the products which normally appear at this stage of differentiation. That secretion is halted is further suggested by the appearance of the Schwann cell cytoplasm in these cultures. The cytoplasm contains frequent substantially dilated cisterns of rough endoplasm reticulum con-

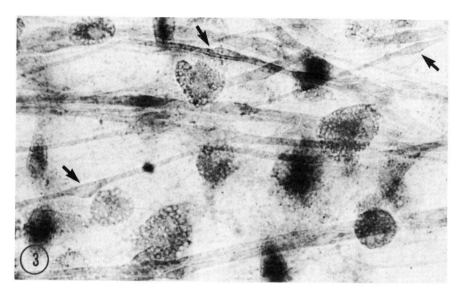

FIG. 3. Several nerve fascicles from a culture grown in non-defined medium. One myelin segment and nuclei of Schwann cells related to unmyelinated nerve fibers are indicated (arrows). Phagocytes and fibroblasts are present. Sudan Black; X 500.

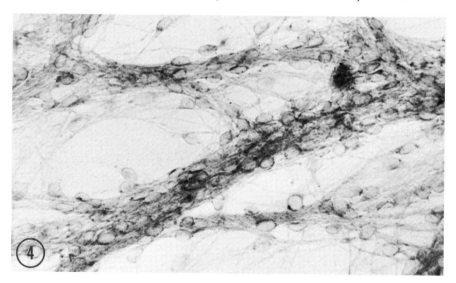

FIG. 4. Light micrograph of several nerve fascicles in the outgrowth of a culture grown in N2 medium. The nuclei of Schwann cells are more numerous and are not aligned along the neurites as in Fig. 3. No myelin is formed. Sudan Black; X 500.

taining fine fibrillar material which would presumably be destined under normal circumstances to be secreted from the cell (Fig. 6). This cytoplasmic "congestion" of Schwann cell products can also be

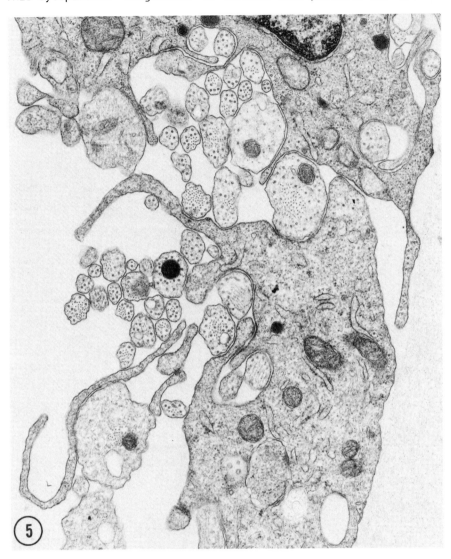

FIG. 5. Electron micrograph of a fascicle region in the outgrowth of a neuron–Schwann cell culture, maintained in N2 medium for 4 weeks <u>in vitro</u>. Neurites are not typically ensheathed, and basal lamina and extracellular collagen fibrils are lacking. The collagen substratum on which the cells are grown is shown on the right. X 24,000.

observed at the light microscope level; after Sudan Black staining of fixed cultures it is possible to detect substantial and varied granularity within the Schwann cell cytoplasm which is not normally seen in mature fully nurtured cultures. These very recent studies are being continued to determine the reversibility of this failure of Schwann cell differentiation. Preliminary experiments indicate that the subsequent addition of serum and embryo extract may lead to a rapid correction of the Schwann cell abnormality with substantial ensheathment of individual axons and the production of basal lamina material (18).

One interpretation of these results is that the defined medium (devoid of serum) allows expression of the mitogenic effect of the axon, but fails to provide humoral factors necessary to allow the Schwann cell in contact with axon and substrate to further its differentiative course. Alternately, these observations may be interpreted as indicating that Schwann cell function fails because the cell is laboring under a nutritional deficit. However, the observation that the Schwann cell is in fact producing a secretory product but failing to deliver it into the extracellular space would favor the interpretation that the axon is signalling the Schwann cell to produce, but that the act of secretion fails because of the lack of a necessary humoral agent. This stark failure of Schwann cell differentiation is remarkable in light of known Schwann cell pathology. In most pathological conditions the basic aspects of Schwann cell function, in terms of elementary axon ensheathment, are satisfactorily carried out and Schwann cell

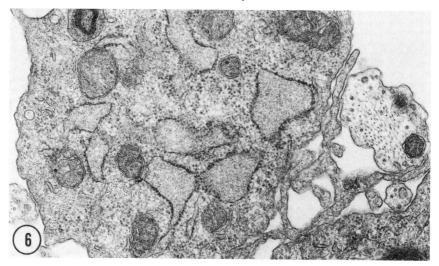

FIG. 6. Schwann cell in an outgrowth region of a culture grown in N2 medium for 4 weeks. The distinctive morphological change portrayed here is the prominence of dilated cisterns of rough endoplasmic reticulum. X 32,000.

function fails at some later developmental stage, as in failure of normal myelination. But there is one important exception. That exception is found in the nerve roots of the dystrophic mouse.

The dystrophic mouse (dy) was for some time thought to contain primarily muscle lesions, but recently it has been observed that there are abnormalities in the peripheral nervous system as well (4,14). These abnormalities are most fully expressed in the dorsal and ventral root regions of both the cranial and spinal nerves, particularly the spinal nerves of the cervical and lumbar enlargements. In these regions, there are large numbers of axons which would normally obtain ensheathment from Schwann cells but which are closely packed together without any intervening Schwann cell processes or connective tissue components. This remarkable lesion leads to complex physiological abnormalities within the peripheral nerve, which may themselves explain the muscle lesions (20). There are, in relation to these bundles of naked axons, a small number of adjacent Schwann cells which show some proclivity to contact axons but no ability to separate them and to ensheathe them individually. These have been thought to be a "stunted" form of Schwann cell. The similarity of the abnormal Schwann cell development in the nutritionally deprived cultures described above and the "Schwann cells" related to unensheathed axons in the dystrophic mouse is striking.

The relation between these two observations is not yet clear, but the similarity of observations suggests that there may be an abnormality in the formation or deposition of extracellular matrix materials in the root region of the peripheral nerves of the dystrophic mouse. We have elsewhere elaborated on this concept and provided additional evidence to point to the extracellular matrix as a possible source of abnormality within the tissues of the dystrophic mouse (19).

OTHER AXON–SCHWANN CELL INTERACTIONS

We have outlined above a rather complex repertoire of axon-Schwann cell interactions, including very recent observations which suggest that production of and interaction with extracellular matrix components are important aspects of Schwann cell biology. Our discussion has not concentrated on the question of the provision of trophic materials by Schwann cells to axons. The transfer of polypeptides from Schwann cells to axons in invertebrates is now well established (12,15). The cellular species in peripheral tissues that satisfy the well known need of neurons for trophic support are beginning to be identified. Tissue culture experiments indicate that "non-neuronal" cells from peripheral nerve can provide trophic support for peripheral neurons (27). The Schwann cell sheath must be considered along with the target tissues of neurons (11,16,17) as a source of trophic material for neurons. There is also evidence that central glia provide trophic support for neurons (2). The availability of pure Schwann cell populations should allow assessment of the specific role of Schwann cells in providing trophic support; our preliminary observations indicate neuronal types, but much more work needs to be done.

It seems very likely that we have examined, in the experiments summarized above, only a portion of the secretory products of mammalian Schwann cells. If the unidentified products include trophic materials it will be of interest to determine if secretion of these products is controlled (as is the production of extra-cellular matrix materials) by axon contact. If this should prove to be the case then an impressively regulated glandular function can be assigned to the cell of Schwann. It may secrete trophic materials needed by the contacting nerve cell for trophic support, but only on contact with an axon which engenders the requisite secretory activity.

ACKNOWLEDGEMENTS

Work in the authors' laboratory is supported by Grant NS 09923 from the National Institutes of Health and Grant RG 1118 from the National Multiple Sclerosis Society. F. Moya is a fellow of the National Multiple Sclerosis Society.

REFERENCES

1. Aguayo, A.J., Bray, G.M., Perkins, S., and Duncan, I.D. (1979): Soc. Neurosci. Symp., 4:361-383.
2. Barde, Y.A., Lindsay, R.M., Monard, D., and Thoenen, H. (1978): Nature, 274:818.
3. Bottenstein, J.E., and Sato, G.H. (1979): Proc. Natl. Acad. Sci. USA, 76:514-517.
4. Bradley, W.G., and Jenkinson, M. (1973): J. Neurol. Sci., 18:227-247.
5. Bunge, M., Jeffrey, J., and Wood, P. (1977): J. Cell Biol., 75:(2 pt. 2) 161a.
6. Bunge, M.B., Williams, A.K., and Wood, P.M. (1979): J. Cell Biol., 83:130a.
7. Bunge, M.B., Williams, A.K., Wood, P.M., Uitto, J., and Jeffrey, J.J. (1980): J. Cell Biol. 84:184-202.
8. Bunge, R.P. (1980): In: Nerve Repair and Regeneration: Its Clinical and Experimental Basis, edited by D. Jewett and H. McCarroll, pp. 58-67. C.V. Mosby, St. Louis, MO.
9. Bunge, R.P., and Bunge, M.B. (1978): J. Cell. Biol., 78:943-950.
10. Bunge, R.P., Bunge, M.B., and Cochran, M. (1978): Neurology, 28:59-67.
11. Ebendal, T., Belew, M., Jacobson, C., and Porath, J. (1979): Neurosci. Lett., 14:91-95.
12. Gainer, H., Tasaki, I. and Lasek, R.J. (1977): J. Cell Biol., 74:524-530.

13. Hay, E.D. (1977): In: <u>Cell and Tissue Interactions</u>, edited by W. Lash, and M.M. Burger, pp. 115-137. Raven Press, New York.
14. Jaros, E., and Bradley, W.G. (1979): <u>Neuropathol. and Appl. Neurobiol.</u>, 5:33-47.
15. Lasek, R.J., Gainer, H., and Barker, J.L. (1977): <u>J. Cell Biol.</u>, 74:501-523.
16. Lindsay, R.M. (1979): <u>Nature</u>, 282:80-84.
17. Lindsay, R.M. and Tarbit, J. (1979): <u>Neurosci. Lett.</u>, 12: 195-200.
18. Moya, F., Bunge, R.P., and Bunge, M. (1980): <u>Tiss. Cult. Assoc.</u>, (in press).
19. Okada, E., Bunge, R.P., and Bunge, M.B. (1980): <u>Brain Res.</u>, (in press).
20. Rasminksy, M. (1978): <u>Ann. Neurol.</u>, 3:351-357.
21. Salzer, J., Glaser, L., and Bunge, R.P. (1979): <u>Soc. Neurosci. Abstr.</u>, 5:433.
22. Salzer, J.L., and Bunge, R.P. (1980): <u>J. Cell Biol.</u>, 84:739-752.
23. Salzer, J.L., Williams, A.K., Glaser, L., and Bunge, R.P. (1980): <u>J. Cell Biol.</u>, 84:753-766.
24. Salzer, J.L., Bunge, R.P., and Glaser, L. (1980): <u>J. Cell Biol.</u>, 84:767-778.
25. Skaper, S., Manthorpe, M., Adler, A., and Varon, S. (1980): <u>J. Neurocytol.</u>, (in press).
26. Thomas, P.K., and Olsson, Y. (1975): In: <u>Peripheral Neuropathy</u>, edited by P.J. Dyck, R.K. Thomas, and E.H. Lambert, pp. 168-189. W.B. Saunders, Philadephia, P.A.
27. Varon, S., Raiborn, C.W. Jr., and Burnham, P.A. (1974): <u>Neurobiology</u>, 4:231-252.
28. Williams, A.K., Wood, P.M., and Bunge, M.B. (1976): <u>J. Cell Biol.</u>, 70:138a.
29. Wood, P.M. (1976): <u>Brain Res.</u>, 115:361-375.
30. Wood, P.M., and Bunge, R.P. (1975): <u>Nature</u>, 256:662-664.

Neurosecretion and Brain Peptides,
edited by J. B. Martin, S. Reichlin, and K. L. Bick.
Raven Press, New York © 1981.

The Extracellular Matrix and the Control of Cell Proliferation

Denis Gospodarowicz

*Cancer Research Institute, University of California Medical Center,
San Francisco, California 94143*

Current evidence suggests that the physical substrate upon which epithelia rest, both in vivo and in vitro, can modulate their response to growth factors. For example, the basal layer of the epidermis is normally columnar and mitotically active. Cellular orientation toward the epidermal-dermal interface is lost when the epidermis is separated from the dermis, and the cells lose their ability to proliferate. Maintenance or recovery of the active state occurs upon recombination with dermis or in presence of epidermal growth factor (EGF). This agent is effective only if the epidermis touches a suitable substrate, however (7). In this paper, we review recent evidence which tends to demonstrate that the extracellular matrix (ECM) or basal lamina upon which cells migrate, proliferate, and differentiate in vivo plays an important role in deciding whether cells will divide or not in response to various stimuli provided by their microenvironment, and in particular by growth factors.

FIBROBLAST GROWTH FACTOR AND THE EXTRACELLULAR MATRIX

In early studies in which the mitogenic effect of EGF on corneal epithelial cells maintained in either organ or tissue culture was compared to that of fibroblast growth factor (FGF), it was observed that, when corneas were maintained in organ culture, the basal epithelial cell layer was very sensitive to the mitogenic effect of EGF but not to serum or to FGF. When the same cell type was maintained in tissue culture, the very opposite sensitivities were observed. The cells now proliferated in response to serum or FGF but no longer responded to EGF, although they were still capable of binding, internalizing, and degrading it (14). The most obvious difference between corneal epithelial cells maintained in tissue culture and those maintained in organ culture was in the substrate upon which the cells rested. In tissue culture, the epithelial cells proliferated on plastic as a

243

collection of individual cells which had lost their orientation
to one another and had a flattened appearance, while in organ
culture, as in vivo, the basal cell layer of the corneal epithelium rested on the Bowman's membrane. This led us to conclude
that the use of corneal epithelial tissue cultures as a bioassay
for the mitogenic activity of factors purified from plasma or
serum will selectively isolate fators that are active in tissue
culture but are probably inactive either in organ culture or in
vivo. For example, although the corneal epithelium is responsive
to EGF, any attempt to purify EGF from tissue by monitoring the
extract for an effect on corneal epithelial cells would fail.
The use of organ cultures, however, would result in the purification of EGF. This demonstrates that a single cell type, maintained in vivo under two different culture conditions, could give
rise to the isolation of two completely different mitogenic
agents (14). Of these two, the one isolated by use of the organ
culture assay would be most likely to have activity in vivo.

In an effort to restore the sensitivity of corneal epithelial
cells maintained under tissue culture conditions to EGF, they
were plated on collagen-coated dishes. Although the addition of
FGF to the culture did not result in any marked increase in the
rate of cell proliferation, the cultures did respond to EGF with
both a marked increase in cell number and an increased rate of
keratinization which reflected the increase in cell number (15).
We also observed that a noticeable change in the shape of the
cells occurred when they were maintained on collagen as opposed
to plastic. On plastic, the cells remained flattened, while on
collagen they formed a monolayer of packed, tall columnar cells.
Multiple layers began to form during the second week, the second
and third cell layers having the appearance of winged cells.
During the third week, keratinization occurred in the upper cell
layers. Basal cells maintained on collagen could therefore mimic
the complete sequence of events present in the corneal epithelium
(differentiation into winged and plated cells), while they did
not progress beyond the stage of a monolayer when maintained on
plastic.

Since corneal epithelial cells proliferated in response to EGF
when maintained on collagen, it is probable that the primary factor determining the proliferation of basal cells to the mitogen
is whether or not the cells are in contact with the extracellular
matrix (ECM) which can be produced in vivo either by the mesenchymal element or by the corneal epithelium itself, when the
latter is maintained under the proper conditions. One of the
means by which the ECM could influence the rate of proliferation
of the epithelium could be by giving the cells their proper
orientation (27). The substrate upon which the corneal epithelial cells rested could therefore dictate the cellular morphology,
and cellular morphology could in turn determine the sensitivity
of cells to mitogens (15).

Among the factors which could control both the synthesis and
polarity of secretion of the ECM is FGF. This is best seen with

vascular endothelial cells repeatedly passaged at very low cell density, when their requirement for FGF to avoid precocious senescence became evident (58). Cultures grown and maintained in the presence of FGF (when cells expressed their normal phenotype and proliferated actively) synthesize primarily interstitial collagen type III and basement membrane collagen (types IV and V) at a ratio of 3:1. Collagen type V is composed of A, B, and possibly C chains (57). The ECM is localized exclusively beneath the cell layer and is closely associated with the basal cell surface. The polarity of ECM secretion is therefore the same as in vivo. The apical cell surface, which is free of ECM, is nonthrombogenic (59). Cultures grown and maintained in the absence of FGF, (conditions under which the cells proliferate poorly and no longer exhibit their normal phenotypes), synthesize predominantly collagen types I and III. The ratio of collagen types I:III:basement membrane (IV & V) is now 2:5:1, type V basement membrane collagen being composed exclusively of the A chain (56). Concomitant with these changes is a loss of polarity of secretion and distribution of the ECM. When analyzed by immunofluorescence using either fibronectin- or collagen-specific antisera, ECM is observed associated with both apical and basal cell surfaces. The appearance of an ECM on the apical cell surface correlates with a loss of its nonthrombogenic properties. This is reflected by the cells ability to bind platelets. At the same time that the phenotypic expression of the cells is altered, as reflected by the changes in cell morphology and collagen production, a noticeable decrease in their growth-rate is observed, and after a few passages the cells stop proliferating (58).

THE EXTRACELLULAR MATRIX AND CELL PROLIFERATION

Vascular endothelial cells maintained in the absence of FGF exhibit, in addition to a much slower growth rate, morphological as well as structural alterations which mostly involve changes in the composition and distribution of the ECM (56,58). This raises the possibility that the ECM produced by these cells could have an effect on their ability to proliferate and to express their phenotype once confluent.

The importance of the ECM for normal growth and development in vivo has long been recognized. Dodson (9) and Wessel (63) demonstrated that the basal cell layers of the epidermis have to be in direct contact with the ECM upon which they rest in vivo in order to retain their normal orientation and to remain mitotically active in organ culture. This substrate can be produced either by the mesenchyme which is closely associated with most epithelia or by the epithelia themselves, after they interact with the ECM produced by other tissues. Such is the case with the isolated corneal epithelium which can recreate its own stroma if cultured in vitro on isolated lens capsule but not if cultured on a noncollagenous stroma (30). Recent transfilter experiments have shown that direct contact by epithelial cells with a collagen substrate is required if they are to produce their own ECM and

that the extent of cell surface area in contact with the substrate is directly proportional to the stimulation of stroma
production (30,29,28). This newly produced ECM could in turn be
held responsible for the control of proliferation of the basal
epithelial cell layer, possibly by affecting the cell shape (30).
Investigation of the role of extracellular materials at the
epithelial-mesodermal interface has shown that glycosaminoglycans
present as major molecular species at the junction of interacting
tissues could be implicated in epithelial morphogenesis (5).
Likewise, evidence that the substrate upon which cells rest when
maintained in tissue culture is important for their proliferation
is now plentiful. The pioneering work of Ehrman and Gey (10) has
shown that various tissues demonstrate enhanced growth and differentiation when cultured on collagen gels. Recent studies have
also shown that collagen is important in promoting cell attachment (13,33,51,65,36), migration (12) and cell proliferation
(33,31,67). Of particular interest is the study of Liotta and
his colleagues (31) on the growth of fibroblasts in culture which
indicates that, even when grown on plastic, the cells deposit a
collagen substrate which is required for proliferation. This was
shown using the proline analog cis-hydroxyproline, which prevents
collagen secretion when incorporated into collagen (57).   Cultures maintained on plastic and exposed to cis-hydroxyproline did
not produce collagen and did not proliferate, while cultures
exposed to cis-hydroxyproline and provided with an artificial
collagen substrate did proliferate.   This observation linked
collagen and ECM production to cell proliferation.   It is not
known, however, if the main effect of collagen is to promote cell
attachment, thereby indirectly allowing the cells to proliferate,
or if it has a direct effect on both cell attachment and proliferation. In either case, the possibility exists that cells which
do not adapt and grow readily in tissue culture could be limited
in their collagen production and that providing them with an ECM
produced by other cell types could be one way to circumvent this
limitation.
     Among the other components of the ECM which have been studied
in regard to cell attachment and proliferation in vitro is fibronectin. Like collagen, it has been shown to promote cell attachment, migration, and proliferation (66,1,38).   Whether fibronectin directly mediates these effects or acts by stimulating the
production of ECM from cells which are exposed to it has not been
analyzed.
     The effect of the ECM on the proliferation of cells maintained
in cultures has not been studied. This is mostly because with the
exception of lens capsule, it is difficult in vivo to isolate
such material from neighboring tissues. In vitro the reconstitu-
tion of an ECM from its separate elements (collagens, proteogly-
cans, glycosaminoglycans, and glycoproteins) may be difficult, if
not impossible. Not only must the correct ratio of components
constituting the ECM be respected, but they must also be linked
in such a way that the resulting tri-dimensional structure will
be like that of the extracellular scaffolding in vivo.   The

problem in reconstituting an ECM <u>in</u> <u>vitro</u> is made even more difficult by the fact that collagen types IV and V, which <u>in</u> <u>vivo</u> are the main components of basement membrane collagens, can only be extracted from tissue following proteolysis. This could result in structural alterations and prevent their proper poly- merization <u>in</u> <u>vitro</u>. One must also consider that our knowledge of the components of the ECM is limited and that components such as laminin (55) have only recently been isolated. The number of components which remain to be isolated can only be guessed.

Corneal endothelial cells maintained in tissue culture retain their ability, in contrast to most cell types, to synthesize and secrete an ECM found underneath, but not on top of the cells (17). As shown in Fig. 1A, corneal endothelial cells upon reaching confluence form a monolayer of small, highly flattened, tightly packed (1100 cells per mm^2), and nonoverlapping cells. Secretion of an ECM takes place only underneath the cell layer (18,19), and the underlying matrix is revealed after exposing the cell layer to 0.5% Triton X-100 and subsequent washing with PBS to remove remaining nuclei and cytoskeletons (20). The matrix then appears as a uniform layer of amorphous material coating the entire area of the dish (Fig. 2B & C). The chemical composition of the ECM produced by corneal endothelial cells <u>in</u> <u>vitro</u> is currently being analyzed by our group. Based on immunofluores- cence studies, collagen types III and IV appear as the major collagen components of the matrix, forming an evenly distributed fibrillar meshwork. Fibronectin, as well as laminin, are also present in large quantities. This ECM, whose appearance has been shown to correlate with the acquisition by cultured corneal endo- thelial cells of their normal "<u>in vivo</u>" morphology, cell surface polarity, and function (18), could substitute for the ECM pro- duced by other cell types. The ability of corneal endothelial cells in tissue culture to produce an extensive ECM could provide us with a tailor-made ECM with which to test the proliferation and response of other cell types to growth factors. We have compared the rates of proliferation of vascular endothelial cells maintained on plastic versus an ECM.

GROWTH OF CULTURED BOVINE VASCULAR ENDOTHELIAL CELLS MAINTAINED ON PLASTIC VERSUS ECM AND EXPOSED OR NOT TO FGF

When the growth of bovine vascular endothelial cells main- tained on plastic versus ECM was compared, cells maintained on an ECM, whether exposed or not to FGF, divided extremely rapidly. Addition of FGF to cultures maintained on an ECM did not decrease their mean doubling time, which was already at a minimum (18 to 20 hr), nor did it result in a higher final cell density, already at a maximum (900-1000 cells/mm^2). The rate of proliferation of

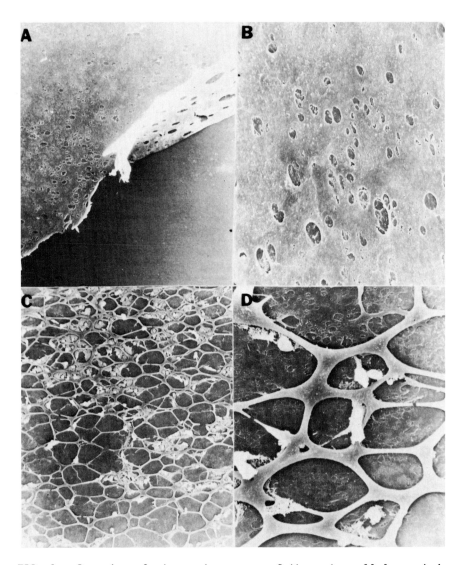

FIG. 1. Scanning electron microscopy of the extracellular matrix
produced by corneal endothelial cell cultures. (A) Area of light
deposit of extracellular matrix. The matrix has been scratched
with a needle to release it from the tissue culture dish that it
coats uniformly (X 1000). (B) (Same as A, but the area has not
been scratched (X 2000). (C) Area of heavy deposit of extracellular matrix. The extracellular matrix uniformly coats the tissue
culture dish and polygonal ridges of amorphous material can be
seen superimposed on it (X 300). (D) Same as C, but at a higher
magnification.

bovine vascular endothelial cells maintained on plastic in the presence of FGF or maintained in its absence was strictly a function of the serum concentration of the cultures. While cells maintained on an ECM and exposed to serum concentration as low as 1% proliferated actively even in the absence of FGF, cells maintained on plastic, when exposed to a serum concentration as high as 10%, proliferated poorly. In contrast, if FGF was added to such cultures, then active proliferation resumed (22).

It can therefore be concluded that bovine vascular endothelial cells maintained in low density culture on plastic proliferate poorly. FGF is needed in order for the cultures to become confluent within a few days. In contrast, when the cultures are maintained on ECM, they proliferate actively and no longer require FGF in order to become confluent. In both cases, the rate of proliferation was a direct function of the serum concentration. It is likely that the effect of the ECM is more a permissive than a direct mitogenic one, since cells still required serum in order to proliferate.

To test the possibility that collagen or fibronectin alone could be the component of the ECM responsible for the increased rate of proliferation, we have compared the growth of bovine vascular endothelial cells plated on dishes coated with purified collagen types I, II, III, and IV or with fibronectin. In no case did the cultures significantly increase their rate of growth when maintained on these different substrates (22). In all cases, an aberrant morphological appearance was observed, the cultures being composed of large cells of which a high proportion were binucleated. The possibility that the component in the ECM produced by corneal endothelial cells which could have a permissive effect on their proliferation is either collagen or fibronectin alone is therefore excluded.

Proliferation of cells in culture is not only a function of the medium, serum, or growth factor(s) to which cells are exposed; it is also a function of cell density. While at high cell density, cells can rapidly condition their medium, thereby compensating for the nutrient deficiency of the medium, when plated at clonal density they can no longer do so. Factors or nutrients required for cell survival and proliferation may be more readily apparent when cells are maintained at low rather than at high density. In particular, requirements for a proper substrate could become apparent. While cells maintained at high density could readily make a basement membrane, thereby facilitating further proliferation, at clonal density, even if every cell were to produce a basement membrane, it would be extremely difficult for them to cover the whole dish in a reasonable period of time. We have analyzed the proliferation of vascular endothelial cells plated at clonal density on plastic versus plates coated with an ECM. When cells were plated at low cell density on plastic, the plating efficiency was extremely poor or cells died rapidly, since no clones were visible after 10 days. If FGF was present in the medium, 25% of the cells gave rise to individual clones.

In contrast, when cells were plated at low density on an ECM, not only was a 90% plating efficiency observed at all cell densities (from 0.012 cells/mm^2 to 1.2 cells/mm^2) but, in addition, <u>all</u> cells gave rise to clones even in the absence of FGF. At clonal density the substrate upon which cells rest is crucial to insure both their survival and proliferation in response to serum factors (22). Since cultures maintained on ECM no longer require FGF in order to proliferate actively, we have also investigated its effect on the lifetime of vascular endothelial cells. Cultures which had been maintained on plastic in the presence of FGF for 50 generations and which had been shown to senesce rapidly as soon as FGF was no longer added to the media (58) were maintained in the absence of FGF on dishes coated with an ECM. When maintained on such a substrate the cells proliferated actively and could be passaged weekly at a split ratio of 1 to 95. It was not until 50 generations later that a detectable decrease in their average doubling time was observed. Cells maintained on ECM have a much longer lifespan in culture than do cells maintained on plastic alone when passaged at very low cell density.

The present results raise the possibilty that, although the final effect of FGF is that of a mitogen (16), its action could be indirect. It could either replace the cellular requirement for a substrate such as the ECM and thereby make the cells fully responsive to growth factors present in serum and plasma even when the cells are maintained on plastic, or alternatively, it could induce control of the synthesis and secretion of the ECM produced by the cells. Such control could in turn make the cells sensitive to factors present in serum or plasma. That the latter alternative could occur finds support in our previous observation that FGF can control the production by vascular endothelial cells of extracellular and cell surface components such as fibronectin and various types of collagen (58,26). Since sparse cultures of endothelial cells proliferate poorly when maintained on plastic but not when maintained on an ECM, it may be that at very low cell density, cultures maintained on plastic are unable to reproduce enough extracellular material to support further growth. The mitogenic effect of FGF on these cells could be the result of an increased synthesis of the ECM by vascular endothelial cells. These possibilities are currently being tested. These results therefore emphasize how drastically one can modify the proliferative response of a given cell type to serum factors depending on the substrate upon which the cells are maintained. It is possible that the lack of response of different cell types maintained under tissue culture conditions to agents responsible <u>in vivo</u> for their proliferation and differentiation could be attributed to the artificial substrate, whether plastic or glass, upon which the cells rest and which limit their ability to produce an ECM.

PLASMA VERSUS SERUM. IS THERE A DIFFERENCE IN THEIR
ABILITIES TO PROMOTE CELL GROWTH?

Culture of most cells in vitro requires the presence of serum
(6). Consequently, investigators have spent much effort in a
search to identify the various factors in serum that stimulate
cell growth in vitro. An important step in the search for serum
growth factors has been the finding that one of the most potent
mitogenic factors present in serum is derived from platelets
(2,3). While plasma was unable to support the growth of aortic
smooth muscle cells or that of BALB/c 3T3 cells (48), serum made
from the same pool of blood stimulated their proliferation.
Addition of a platelet extract to cell-free plasma-derived serum
restored the growth-promoting activity (48,31,49).   One could
therefore conclude that one of the principal mitogens responsible
for the induction of DNA synthesis present in whole blood serum
is derived from platelets. The difference in the proliferative
ability of cells exposed to plasma versus serum results from the
absence of the platelet factor in the former. However, all studies have thus far used cells maintained on plastic rather than on
an ECM.   This could have prevented their response to factors
present in plasma, thereby creating the difference in mitogenic
activity between plasma and serum. To explore the possibility
that the serum factors to which cells maintained on ECM become
sensitive are also present in plasma, we have compared the mitogenic activity of plasma versus serum, using as target cells
vascular smooth muscle cells maintained on either plastic or an
ECM (21).

Vascular smooth muscle cells maintained on plastic and
exposed to plasma (10%) proliferate poorly while when exposed to
serum (10%), the cells proliferate actively (21). In contrast,
cells maintained on an ECM and exposed to plasma proliferate
actively. Plasma was even more mitogenic for cells maintained on
an ECM than was serum for cells maintained on plastic. Growth
rate and the final cell density of cultures maintained on an ECM
and exposed to either plasma or serum were the same. The final
cell density of cultures maintained on an ECM and exposed to
either plasma or serum was a direct function of the serum or
plasma concentration. It is therefore likely that the proliferation of cells maintained on an ECM is controlled by factor(s)
present in plasma and that the ECM has only a permissive role
(21).

Maintaining vascular smooth muscle cells on an ECM and plasma
is clearly a closer approximation to physiological conditions
than exposing them to plastic and serum. Since they proliferate
at a maximal rate when maintained in this manner, it is likely
that they are responding to mitogen(s) already present in plasma
rather than to FGF or to mitogen(s) generated during the coagulation process.

The permissive effect of the ECM produced by corneal
endothelial cells on cell proliferation has also been observed

in the case of fibroblasts, granulosa and adrenal cortex cells
(20), corneal endothelial cells, and corneal and lens epithelial
cells (23,24). Although when maintained on plastic, cells required either serum and/or growth factor in order to proliferate
and did not respond to plasma factor(s), when maintained on an
ECM they no longer required growth factor and addition of plasma
alone was sufficient to make them proliferate at an optimal rate.

GROWTH FACTORS IN VIVO: DO THEY EXIST?

Our earlier hypothesis that use of tissue culture techniques
could lead to the identification of factors which are mitogenic
solely for cells maintained on plastic has therefore been fulfilled. Does this mean, however, that growth factors do not
exist in vivo? This conclusion is for the moment unwarranted,
since cells maintained on an ECM still require plasma in order to
proliferate. It is therefore likely that the use of dishes coated
with an ECM could lead to the purification from plasma of growth
factors whose existence has been previously undetected because of
the repressive conditions imposed on the cells by the substrate
(polystyrene) upon which they were maintained. In the case of
vascular endothelial cells, we have now been able to isolate from
plasma a factor which will support their proliferation and differentiation in medium free of serum, provided that the cells are
maintained on an ECM. When cells are maintained on plastic, this
factor is no longer active (54). It is likely that for other
cell types plasma factor(s) will be identified which can support
their growth when maintained on an ECM.
It is therefore possible that two classes of growth factors
exist. One consists of factors which will allow cells maintained
on plastic to proliferate. Whether such factors have a direct
mitogenic effect or whether they permit cells maintained at low
density to make an ECM and thereby restore their response to
plasma factor(s) can only be a matter of conjecture. The other
consists of factors present in plasma that will act directly on
cells which are maintained on an ECM or which can produce one.
Athough the first generation growth factors will be active only
in tissue culture, the second generation could be active both in
tissue culture, provided that cells no longer rest on plastic, as
well as in vivo.
Cells in culture are known to be capable of producing an ECM,
but this capacity is a strict function of cell density. While at
high density, cells could make an ECM readily, probably because
of their ability to condition the medium, when maintained at low
cell density they no longer have this ability. It is evident that
under these conditions cells maintained in tissue culture could
exhibit a differential response to mitogens that depends on the
density at which they are maintained on plastic. One would predict that at low cell density the first generation growth factor
will primarily be active, while at high cell density the second

generation growth factor will be active. This is in fact what is observed. FGF, a first generation growth factor, has been reported to be active on cells maintained at either clonal or low cell density. With cultures maintained at high cell density or passaged at a high split ratio, it is only weakly active, since cells proliferate readily (58). In contrast, factors such as EGF which represent second generation growth factors in the sense that they have been shown to be active in vivo or in organ culture, work best on cells maintained at very high cell density. In fact, in the case of the analysis of the effect of EGF on established cell lines such as BALB/c 3T3, most studies have been done on subconfluent or confluent cultures. In sparse culture (less than 10 cells/mm^2) EGF has little or no effect.

GROWTH FACTORS IN VIVO: WHAT ARE THEY?

The factors which control cell proliferation during fetal and postnatal development are still largely unknown. It is likely that they vary greatly, depending on the developmental phase of the individual. In early embryonic development increase in cell number occurs primarily through cleavage of the egg. Although active DNA replication takes place, no G_1 phase is necessary since there is no total increase in cell mass. It is likely that at this stage there is no control of cell proliferation other than that provided by the decreasing mass of each blastomer. During gastrulation and the subsequent phase of embryonic development, it is likely that cell proliferation is governed by short-range interactions between the various tissues which became associated within an organ. This could be mediated by the basal lamina or the ECM upon which the cell layers are resting. This is best seen in the case of lobular structures whose cell proliferation has to be localized in very precise regions if correct morphogenesis, as reflected by lobule formation, is to be observed. As demonstrated by Banerjee et al. (4) in the salivary gland, cell proliferation has to be restricted to specific areas localized in the distal end of the lobules. It is in those areas that the fastest turnover of the glycosaminoglycans (GAG) present in the basal lamina can be observed. In intact glands it correlates with the sites where the greatest changes in cell shape and cell proliferation can be observed. Although the exact relationship between high turnover of the GAG present in the basal lamina and cell proliferation has not yet been established, it has been suggested that degradation of the GAG by the underlying mesenchyme will reduce cell anchorage to the disappearing basal lamina. This will lead to changes in cell shape and the cell population will invaginate because of the cell junctions (5). The beginning cleft becomes deeper due to the enlargement by cell division of the intervening lobules and, as it matures, becomes stabilized at its base by extracellular bundles of fibrous collagen. Within the cleft, laminar GAG degradation decreases due to reduced proximity of the lamina to the mesenchyme, allowing the laminar GAG to reaccumulate into an ordered array. The lobular morphology of the epithelium is maintained by adhesive junctions between the cells and the anchorage of the cells to the basal lamina. With continued mitosis, GAG turnover, and new cleft formation, a tree-like epithelium gradually emerges (5). It is therefore clear that the substrate upon which cells rest is already important in deciding which cell population will divide and which will not. It is only during the late phase of embryonic development that growth factors such as nerve growth factor or EGF, which are long-range effectors, could govern the survival, proliferation, and differentiation of various cell populations, so that either remodeling of a given tissue or organ or physiological integration of the different functons performed by various organs is made possible. Such control by growth factors can only occur when specific receptor binding sites are made within the cells and expressed on their cell surface.

In the neonate and the adult, a second type of control for cellular proliferation is superimposed on the control already exerted by the ECM. It corresponds to the development of endocrine organs and involves humoral factors (hormones) in long-range interactions with various tissues as well as in the activation of structures already formed in the embryo. These structures, although committed to give rise to definite organs, remain dormant until late in the neonatal development. This is the case for the reproductive organs and secondary sex organs. This second type of control is best seen when bursts of mitotic activity have to take place in a precise and highly coordinated time sequence in organs located at a great distance from one another. Such is the case of the cell proliferation which occurs during the ovarian cycles in organs as different as the ovaries, mammary glands, and uterus and which is controlled by hormones as different in structure as steroids (estrogen) or polypeptides (FSH or prolactin).

However, during the whole of embryogenesis, as well as in neonatal and adult life, an intact ECM or basal lamina scaffold will be required for the maintenance of orderly tissue structure (67). By its presence it defines the spatial relationships among similar and dissimilar types of cells and between these cells and the space occupied by connective and supportive tissues. Replenishment of cells which have died during normal functioning or have become damaged occurs with new cells in an orderly way along the framework of the basal lamina scaffold (62). This process appears to be aided by the polarity of the basal lamina and by an apparent specificity for cell types, and it enables multicellular organisms to reconstitute histologic structures of most tissues and organs to what they were prior to loss of cells. If the basal lamina is destroyed, the healing in most tissues results in formation of a scar and loss of function (62).

EVENTS WHICH MEDIATE CELL PROLIFERATION AS A
FUNCTION OF THE SUBSTRATE UPON WHICH CELL REST

The substrate upon which cells rest affects not only the ways in which cells respond to a mitogen. It could also affect the mechanisms which lead cells to replicate. There have been numerous and conflicting reports on the involvement of cyclic nucleotides in the control of cell growth (52,37,35). This could be a result of the various growth conditions used and of the substrate upon which cells were maintained. In vivo, cellular proliferation triggered by trophic hormones such as ACTH results in an increased cAMP content of the adrenal cortex cells (52,37,35,47). Yet in vitro exposure of adrenal cortex cells which are anchorage-dependent to agents such as ACTH or direct exposure to cAMP analog leads to an inhibition of cell growth (46,34). Likewise, in fibroblasts an increase in cAMP has usually been thought to bring an arrest of cell proliferation (39). In contrast, in the case of cells such as lymphocytes, which have no need for an ECM and whose growth is not anchorage-dependent, cAMP as well as hormones capable of activating adenylate cyclase are mitogenic (64). Although one could make a case that a cAMP increase in cells which are anchorage-dependent is growth inhibitory while the opposite is true for cells which are no longer anchorage-dependent, this does not seem to be the case. In at least four different cases of anchorage-dependent cells, viz. melanoma cells (40), Schwann cells (44), epidermal cells (25), and pancreatic cells (11), a number of agents which are known to increase intracellular cAMP level have also been shown to be mitogenic. The case of epidermal cells is particularly interesting, since it has been shown that addition of cAMP to briefly cultured mouse epidermis inhibits cell division. In isolated epidermis, the addition of cAMP (61) or β-adrenergic stimulation, which results in an increase in intracellular cAMP, led to a marked reduction in a number of divisions (43). In both cases, the cell layers were separated from their underlying dermis. Yet, when epidermal cells are maintained in tissue culture on a feeder layer of 3T3 cells, cholera toxin, which increases the cAMP content of the cells, can markedly increase the overall rate of cell proliferation in the colonies (25). Similar effects can also be observed on plastic dishes, but only in areas of high cell density (25). It is in those areas that cells will most likely produce their ECM.

The involvement of cAMP in control of cell proliferation is even more evident in the case of the pancreatic epithelia (50). While at 11 days of gestation the rat pancreas is composed of a few hundred cells, 9 to 10 days later it is composed of 7 million cells. Transfilter experiments have shown that, as in other systems, pancreatic epithelial anlagen separated from their mesenchymal components fail to develop, but co-culture of the recombined tissues leads to normal development. An extract of embryos rich in mesenchymal tissues can replace the mesenchymal requirement for pancreas development. This mesenchymal factor (MF) is neither species nor tissue specific (50). In the absence of mesenchyme or MF, pancreatic epithelial primordia do not proliferate. In the presence of MF, there is active cell proliferation and later acinar cell differentiation as in vivo. MF apparently acts on the epithelial cell surface and could be one of the ECM components, as indicated by its ability to bind Ca^{++} and by its insolubility. Purified MF can be bound to Sepharose. Pancreatic epithelial buds adhere to the MF-Sepharose beads at their basal surface (which normally interacts with the mesenchymal cells (50). The result is that the pancreatic primordia is everted and the apical surface, that normally faces the lumen, is now in contact with the culture medium. Rapid DNA synthesis occurs in cells directly attached to the MF-Sepharose beads and later acinar cells containing zymogen granules are observed (50). These experiments suggest that the mesenchymal-epithelial interaction is mediated by a molecular species that acts at the level of the cell membrane. If one partially denatures MF with periodate, addition of cAMP analog fully restores its mitogenic activity, thereby suggesting that MF activity is cAMP-dependent and that the periodate-sensitive moiety could be the adenylate cyclase stimulating site (11). In the case of pancreatic epithelial cells, therefore, cAMP is a positive signal for cell proliferation.

THE COMMITMENT AND THE PROGRESSION FACTORS

Soluble factors which control the production of the ECM have not previously been reported. Until now, production of ECM was thought to be an automatic process which was mostly a function of the substrate upon which cells rest and of cell density. Although it is quite possible that in vivo the production of an ECM is not under any control other than that provided by the substrate upon which cells migrate during the early embryological phase, our results suggest that, at least in vitro, factors such as FGF could influence the formation of an ECM scaffolding. We do not know the extent of this influence. Our observations that FGF could control the ECM production, thereby making cells sensitive to plasma factor(s), is consistent with the findings of others that there could be two sets of growth factors in vitro, one of which, the commitment factor(s), could be involved in the formation of the ECM.

Earlier studies done on the control of cell cycle by growth factors have developed the concept that it could be controlled by two independent sets of factors present in serum, each of which controls a different phase of the cell cycle (60,41). One set is composed of heat-stable ($100^{o}C$) factor(s) that are released into serum during the clotting process and induces BALB/c 3T3 cells to become capable of synthesizing DNA. A second set of components, found in defibrinogenated platelet-poor plasma allows competent cells to progress through G_0/G_1 and to synthesize DNA (60,42,41). Stiles and his colleagues have further developed this observeration by looking at the dual control of cell growth by somatomedins, PDGF, and FGF (53). Quiescent BALB/c 3T3 cells exposed briefly to PDGF or FGF become "competent" to replicate their DNA but do not "progress" into S phase unless exposed to somatomedin C, which is required for progression (53). Since neither FGF nor PDGF is required to commit vascular smooth muscle cells to proliferation when they are maintained on an ECM and exposed to plasma, it is likely that competence factors would no longer be required if cells are maintained on an ECM.

THE IMPLICATIONS OF GROWING CELLS ON AN ECM INSTEAD OF ON PLASTIC

If one starts with a primary culture, it is likely that cells are selected which in the subsequent passages retain their ability to produce an ECM. Alteration in their phenotypic expression could be the direct result of an alteration in the type of ECM produced. This is best seen in the case of vascular endothelial cells, which express their normal phenotype when producing an ECM composed of collagen type III versus types IV and V at a ratio of 3 to 1. In contrast, when these cells no longer express their normal phenotype, the type of collagen produced is altered. Collagen type I begins to be synthesized, while the B and C chains of collagen type V are no longer produced. This results in aberrant ECM production parallel with a loss of phenotypic expression (56). It is therefore possible that the widely acknowledged instability of the phenotypic expression of cultured cells could be due to their inability, when maintained on plastic, to continue to produce a normal ECM. If this should be the case, providing the cells with an artificial substrate closely resembling that produced in vivo should stabilize their phenotypic expression.

In the field of aging, it is generally recognized that senescent cells stop making an ECM (8). Whether this is the result or the cause of senescence has not been investigated. It is to be suspected, however, that alterations in production or a loss of ability to produce a normal ECM could be directly linked to cell senescence, since this will result in a loss of proliferative ability. In fact, it has been our experience that when cells previously maintained on an ECM, conditions under which they proliferate actively, are cultured on plastic, they become senescent within a few passages. This is correlative with their inability to produce an ECM, suggesting that the substrate upon which cells are maintained is an important factor in determining the culture lifetime.

In the field of tumor cell biology, the growing recognition that the substrate upon which cells are maintained could modify their phenotypic expression is also important. One of the main characteristics of tumor cells grown in tissue culture is their loss of anchorage-dependence, as reflected by their ability to

grow either in soft agar or in suspension when maintained on
plastic. Yet in vivo, tumor cells from solid tumors, although
they can adhere loosely to one another, can adhere tenaciously to
the substrate which is provided by the host tissue or which they
themselves produce. This is best reflected in the phenomenon of
metastasis, where tumor cells which are carried away by the
bloodstream can stick to basement membrane, infiltrate through
it, and form secondary tumors in organs located far from the
original tumor. If tumor cells were anchorage-independent, one
would expect them to proliferate freely in the bloodstream, but
this rarely happens. It is therefore likely that, if one
provides tumor cells in culture with an adequate substrate, they
could totally shift their pattern of growth. This is in fact
what we have observed with hepatocarcinoma cells, Ewing's tumor
cells, or melanoma tumor cells. It should also be pointed out
that maintaining active proliferation of epithelial cell cultures
of either normal or tumoral origin is a challenge. For example,
in the case of carcinoma cells, less than 5% of the original
cells put in culture give rise to cell lines. Using an ECM as a
natural substrate could change the figures and a higher percentage of either normal or tumor epithelial cells could be established in culture. It has also been shown by others that the ECM
can control the morphogenetic and phenotypic expression of the
tissue associated with it (45). Use of ECM as a natural substrate
for culturing epithelial cells could therefore provide an opportunity to study not only their proliferation and the physiological factors controlling it but their differentiation as well.
Although in the present study the ECM produced by corneal endothelial cells was used, other extracellular matrices produced by
other cell types could also be used and would allow one to study
the effects of various extracellular matrices on cell migration,
proliferation, and differentiation.

Since cells maintained on an ECM now respond to plasma growth
factors instead of to serum factors, one may wonder what these
factors are. They could possibly be physiological agents such as
trophic hormones which modulate cell proliferation in vivo but
are inactive in vitro. Alternatively, they could be factors
which have gone undetected because of the lack of sensitivity to
them of cultured cells maintained on plastic. If this were so,
one would now have an ideal substrate for restoring the normal
growth response of many tissues to these naturally occurring
factors. One might also suspect that conclusions concerning the
mechanisms and controls of cell proliferation and cell migration
of normal cells maintained on plastic may somehow differ from
those which can be reached when cells are maintained on ECM.

REFERENCES

1. Ali, I.U., Mautner, V., Lanza, R. and Hynes, R.O. (1977):
    Cell, 11:115-126.
2. Balk, S.D., (1971): Proc. Natl. Acad.Sci. USA, 68:1689-1692.

3. Balk, S.D., Whitfield, J.F., Youdale, T. and Braun, A.C. (1973): Proc. Natl. Acad.Sci. USA, 70:675-679.
4. Banerjee, S.D., Cohn, R.H., and Bernfield, M.R. (1977): J. Cell Biol., 73:445-463.
5. Bernfield, M.R. (1978): In: Birth Defects, edited by J.W. Littlefield and J. de Grouchy, pp. 111-125. Amsterdam, Oxford.
6. Carrel, A. (1912): J. Exp. Med., 15:516-528.
7. Cohen, S. (1965): Devel. Biol., 12:394-407.
8. Cristofalo, F.J.(1972): Adv. Gerontol. Res., 4:45-79.
9. Dodson, J.S. (1963): Exp. Cell Res., 31:233-235.
10. Ehrmann, R.L. and Gey, G.O. (1956): J. Nat. Cancer Inst. 162:1375-1402.
11. Filosa, S., Pictet, R. and Rutter, W.J. (1975): Nature, 257:702-704.
12. Fisher, M. and Solursh, M. (1979): Exptl. Cell Res. 123:1-14.
13. Gey, G.O., Svotelis, M., Foard, M. and Band, F.B. (1974): Exptl. Cell Res., 84:63-71.
14. Gospodarowicz, D., Mescher, A.L., Brown, K., and Birdwell, C.R. (1977): Exp. Eye Res., 25:631-642.
15. Gospodarowicz, D., Greenburg, G. and Birdwell, C.R. (1978): Cancer Res., 38:4155-4171.
16. Gospodarowicz, D., Mescher, A.L., and Birdwell, C.R. (1978): National Cancer Institute Monographs, 48:109-130.
17. Gospodarowicz, D., Greenburg, G., Vlodavsky, I., Alvarado, J. and Johnson, L.K. (1979): Exptl. Eye Res., 29:485-509.
18. Gospodarowicz, D., Vlodavsky, I., Greenburg, G., Alvarado, J., Johnson, L.K. and Moran, J. (1979): Rec. Prog. in Hormone Res., 35:375-448.
19. Gospodarowicz, D., Vlodavsky, I., Greenburg, G. and Johnson, L.K. (1979): In: Cold Spring Harbor Conferences on Cell Proliferation, Hormones and Cell Culture, Vol. 6, edited by R. Ross and G. Sato, pp. 561-592. Cold Spring Harbor Press, New York.
20. Gospodarowicz, D., Delgado, D. and Vlodavsky, I. (1980): Proc. Natl. Acad. Sci. USA, (in press).
21. Gospodarowicz, D. and Ill, C.R. (1980): Proc. Natl. Acad. Sci. USA, 77:2726-2730.
22. Gospodarowicz, D. and Ill, C.R. (1980): J. Clin. Inv., (in press),.
23. Gospodarowicz, D., Savion, N. and Vlodavsky, I. (1980): Vision Res., (in press).
24. Gospodarowicz, D. and Ill, C.R. (1980): Exptl. Eye Res., (in press).
25. Green, H. (1979): Cell, 15:801-811.
26. Greenburg, G., Vlodavsky, I., and Gospodarowicz, D. (1979): J. Cell Physiol., (in press).
27. Grobstein, C. (1967): Natl. Cancer Inst. Monograph, 26:279-299.
28. Hay, E.D. and Meier, S. (1976): Develop. Biol., 52:141-157.

29. Hay, E.D. (1977): In: Cell and Tissue Interactions, edited by J.W. Lash and M.M. Burger, pp.115-138. Raven Press, New York.
30. Hay, E.D. (1978): In: Birth Defects, edited by J.W. Littlefield and J. de Grouchy, pp. 126-140, Excerpta Medica, Amsterdam.
31. Kohler, N. and Lipton, A. (1974): Exp. Cell Res., 87:297-301.
32. Liotta, L.A., Vembu, D., Kleinman, H., Martin, G.R. and Boone, C. (1978): Nature, 272:622-624.
33. Liu, S.C., and Karasek, M. (1978): J. Invest. Dermatol., 71:157-162.
34. Masui, H. and Garren, L.D. (1971): Proc. Natl. Acad. Sci. USA, 68:3206-3210.
35. Moens, W., Vokaer, A. and Kram, R. (1975): Proc. Natl. Acad. Sci. USA, 72:1063-1067.
36. Murray, J.C., Stingl, G., Kleinman, H.K., Martin, G.R., and Katz, S.I. (1979): J. Cell Biol., 80:197-202.
37. Oey, J., Vogel, A. and Pollack, R. (1974): Proc. Natl. Acad. Sci. USA, 71:694-698.
38. Orly, J. and Sato, G. (1979): Cell, 17:295-305.
39. Otten, J., Johnson, G.S. and Pastan, I. (1972): J. Biol. Chem., 247:7082-7087.
40. Pawelek, J., Halaban, R. and Christie, G.(1975): Nature, 258:539-540.
41. Pledger, W.J., Stiles, C.D., Antoniades, H.N. and Scher, C.D.(1977): Proc. Natl. Acad. Sci. USA, 74:4481-4485.
42. Pledger, W.J., Stiles, C.D., Antoniades, H.N. and Scher, C.D. (1978): Proc. Natl. Acad. Sci. USA 75:2839-2843.
43. Powell, J.A., Duell, E.A. and Voorhees, J.J. (1971): Arch. Dermatol., 104:359-365.
44. Raff, M.C., Hornby-Smith, A. and Brockes, J.P. (1978): Nature, 273:672-673.
45. Rafferty, K.A. (1975): Adv. Cancer Res., 21:249-272.
46. Ramachadran, J. and Suyama, A.T. (1975): Proc. Natl. Acad. Sci. USA, 72:113-117.
47. Robinson, G.A., Butcher, R.W. and Sutherland, E.W. (1971): In: Cyclic AMP, pp. 17-46 and 339-398. Academic Press, New York.
48. Ross, R., Glomset, J., Kariya, B., and Harker, L. (1974): Proc. Natl. Acad. Sci. USA, 71:1207-1210.
49. Ross, R. and Vogel, A. (1978): Cell, 14:203-210.
50. Rutter, W.J., Pictet, R.L., Harding, J.D., Chirgwin, J.M., MacDonald, R.J. and Przybyla, A.E. (1978): In: Molecular Controls of Cell Proliferation and Cytodifferentiation, edited by J. Papaconstantinou and W.J. Ritter, pp. 205-227. Academic Press, New York.
51. Schor, S.L. and Court, J. (1972): J. Cell Sci., 38:267-281.
52. Sheppard, J.R.(1972): Nature New Biol., 236:14-16.
53. Stiles, C.D., Capone, G.T., Scher, C.D., Antoniades, H.N., Van Wyk, J.J. and Pledger, W.J. (1979): Proc. Natl. Acad. Sci. USA, 76:1279-1283.

54. Tauber, J.P. and Gospodarowicz, D. (1980): Proc. Natl. Acad. Sci. USA, (in press).
55. Timpl, R., Rohde, H., Gehran-Robey, P., Rennard, S.I., Foidart, J.M. and Martin, G.R. (1979): J. Biol. Chem., 254:-9933-9937.
56. Tseng, S., Savion, N., Stern, R. and Gospodarowicz, D. (1980): Proc. Natl. Acad. Sci. USA, (in press).
57. Uitto, J. and Prockop, D.Y. (1974): Biochim. Biophys. Acta, 336:234-251.
58. Vlodavsky, I., Johnson, L.K., Greenburg, G., and Gospodarowicz D. (1979): J. Cell Biol., 83:468-486.
59. Vlodavsky, I., Liu, G.M. and Gospodarowicz, D. (1980): Cell, (in press).
60. Vogel, A., Raines, E., Kariya, B., Rivest, M.J. and Ross, R. (1978): Proc. Natl. Acad. Sci. USA, 75:2810-2814.
61. Voorhees, J.J., Duell, E.A.and Kelsey, W.H. (1972): Arch. Dermatol., 105:384-386.
62. Vracko, R. (1974): Am. J. Path., 77:314-334.
63. Wessels, N.K. (1964): Proc. Natl. Acad. Sci. USA, 52:252-259.
64. Whitfield, J.F., MacManus, J.P., and Gillan, D.J. (1970): J. Cell Physiol., 76:65-76.
65. Wicha, M.S., Lio;tta, L.A., Garbisa, S., and Kidwell, W.R. (1979): Exptl. Cell Res., 124:181-190.
66. Yamada, K.M., Yamada, S.S., and Pastan, I. (1976): Proc. Natl. Acad. Sci. USA, 73:1217-1221.
67. Yang, J., Richards, J., Bowman, P., Guzman, R., Enami, J., McCormick, K., Hamamoto, S., Pitelka, D. and Nandi,S. (1979): Proc. Natl. Acad. Sci. USA, 76:3401-3405.

Neurosecretion and Brain Peptides,
edited by J. B. Martin, S. Reichlin, and K. L. Bick.
Raven Press, New York © 1981.

The Role of Nerve Growth Factor (NGF) and Related Factors for the Survival of Peripheral Neurons

H. Thoenen, Y.-A. Barde, and D. Edgar

Department of Neurochemistry, Max-Planck Institute for Psychiatry,
D-8033 Martinsried, West Germany

Regulation of the survival of neurons during ontogenesis is one of the most important mechanisms determining the final organization of the nervous system (16,39). Generally, neurons are produced in excess early in development and the proportion surviving is regulated by the target tissues innervated (cf. 28). Ablation and transplantation experiments (17,27,63), together with the phenomenon of retrograde transsynaptic degeneration (cf. 16) provided the initial evidence for the importance of target cells for the survival of innervating neurons.

The classical experiments of Shorey (63), Detwiler (17), Hamburger (27,28) and others (cf. 28) demonstrated that after ablation and/or transplantation of chick and Amblystoma limb buds at early embryonic stages, there was a correlation between the number of surviving spinal sensory and motor neurons and the volume of the peripheral tissues innervated by those neurons. The conclusion drawn from these experiments (that target cells represent an important factor in the regulation of the survival of the innervating neurons) has been confirmed by a great variety of experiments (cf. 28,39). However, such experiments do not allow conclusions as to the mechanism underlying these retrograde effects. Experiments designed to further establish and expand the concept that the volume of peripheral tissue determines the extent of the survival of the innervating neurons (10) provided a key to the molecular understanding of at least one mechanism by which effector cells can influence innervating neurons. The transplantation of mouse sarcoma tissue into the body wall of young chick embryos led not only to an increased survival of sensory neurons innervating the tumor but also of those out of contact with the tumor (10,45,46). Moreover, a careful inspection of the embryos led to the surprising observation that the sympathetic ganglia (including those far away from the tumor) were even more affected than the sensory ganglia (45,46). These seminal observations by Levi-Montalcini and Hamburger (45,46) led to the detection of nerve growth factor (NGF), identified as the agent mediating the effects on the nervous system of the transplanted tumor (cf. 44,53). The biologically active NGF isolated from mouse salivary gland consists of two noncovalently linked, identical peptide chains, each chain containing 118 amino acid residues.

Below we summarize the evidence that NGF acts as a survival factor for sensory and sympathetic neurons in vivo and in vitro. More recent investigations demonstrate that NGF is not the only macromolecule able to support their survival and that the relative effects of the different factors change during ontogenesis.

EVIDENCE THAT THE ACTION OF NGF AS A SURVIVAL FACTOR
DEPENDS ON THE STAGE OF DEVELOPMENT

In Vivo Experiments

Evidence for the physiological importance of NGF for the survival of neurons during ontogenesis is based mainly on the effect of antibodies to NGF on the developing sympathetic nervous system (15,43,44,53). In this system where well-defined markers exist, such as specific enzymes synthesizing catecholamines, NGF-antibodies lead to a complete destruction or an impaired function depending on the developmental stage (24,43). There is convincing evidence that the destructive action of NGF antibodies in early developmental stages does not result from a complement-mediated cytotoxic effect but on the neutralization of endogenous NGF or immunologically crossreacting molecules (4,23).

NGF has been shown to be taken up with high selectivity by sympathetic nerve terminals and to be transported retrogradely (18,37,66) in membrane-confined compartments (61,71) to the perikaryon. There specific biochemical changes are triggered, in particular, the regulation of the synthesis of the enzymes characteristic of adrenergic neurons (59,69). The general growth promoting effect of NGF is dependent on the stage of development, being more prominent at early stages when NGF is responsible for the survival of sympathetic cells (43). This age-dependent difference is reflected by a 50-fold increase of ornithine decarboxylase activity in sympathetic ganglia of newborn rats (50,56) whereas in adult animals the increase of this enzyme (which accompanies processes of general growth and regeneration) is only 2 to 3-fold (56,69). In contrast, the selective induction by NGF of tyrosine hydroxylase (TH) and dopamine β-hydroxylase (DBH), enzymes characteristic for adrenergic neurons and adrenal chromaffin cells, remains virtually constant from birth to adulthood (56,69). At our present state of knowledge we do not know whether the survival effect is pleiotypic, depending on general actions of NGF such as an enhanced transport of glucose and amino acids (64,72,73), or whether it depends on a more specific single action. A direct membrane effect of NGF on the sodium dependent transport (65) of glucose, protein and RNA-precursors seems to be improbable since it has been demonstrated that the retrogradely transported moiety of NGF or any hypothetical second messenger is sufficient to maintain neuronal survival (12,33). However, the small smooth membrane-limited compartment which confines retrogradely transported NGF (61,69,70) would be barely sufficient to supply the perikaryon with the necessary glucose, nucleotides and amino acids. Therefore, if such an effect as that of NGF on sodium flux (65) should be responsible for survival, one must assume an indirect mechanism via second messenger(s), acting on the plasma membrane of the perikaryon.

The suggestion that the physiologically important sources of NGF for peripheral sympathetic neurons are the nonneuronal effector tissue stems from the observation that the interruption of the connection between the sympathetic cell body and the nerve terminals has the same effect as the neutralization of endogenous NGF by NGF-antibodies (24,35,36,40,42,52). This interruption between the periphery and the cell bodies can be achieved in different ways i.e., by destruction of adrenergic nerve terminals by 6-hydroxydopamine (43), by blockade of retrograde axonal transport by colchicine or vinblastin (40,52,60) and by surgical interruption (axotomy) (35,36). All these procedures result in a degeneration of the cell body in early postnatal stages (24,35, 40,52) and an impaired synthesis of TH and DBH in adult animals (24). The functional consequences of the interruption between the periphery and the perikaryon of the sympathetic neurons can be avoided by the administration of exogenous NGF (36,42).

The dependence of adrenal medullary cells on NGF for survival is confined to the prenatal period (2), although they remain responsive to NGF throughout life (57). However, this responsiveness is restricted to the selective induction of TH and DBH (57). There is no evidence for a general hypertrophic action of NGF in the postnatal phase (3).

Although Substance P has been localized in spinal sensory neurons and changes in the level of this peptide have been demonstrated both after axotomy and administration of NGF (41; see Black, this volume), it is not yet clear whether this undecapeptide is an exclusive marker for neurons which depend for survival on NGF alone. It may also be contained in neurons which depend on other factors or depend on a combination of macromolecular factors.

Unfortunately, it has not been possible to determine reliably the level of NGF in sympathetically innervated effector organs (31,32,68), not to speak of a correlation between the NGF content of the organs and the density of sympathetic innervation. Thus, the evidence remains indirect. Failure to reliably determine NGF levels in tissues and serum is due to the insufficient sensitivity of the available immunological and biological assays coupled with the very low quantities of NGF present in tissue and serum (32,68). A possible explanation for the very low levels of NGF present in sympathetic effector tissues may be that NGF is present largely in the form of an inactive precursor molecule which is cleaved to its active form in restricted quantities. These may then be removed specifically and efficiently by nerve terminals and transported retrogradely to the perikaryon. Observations made in the male mouse salivary gland (62) support the assumption that precursor molecules exist which are not recognized by antibodies raised against the active NGF molecule. Electron microscopic immunological investigations on the salivary gland have demonstrated that only the secretory granules close to the lumen of the secretory ducts and the released material react with the antibodies raised against active NGF (62). No positive reaction could be detected in secretory granules in the basal parts of the secretory cell where the synthesis and compartmentalization of macromolecules occurs (62).

In this context it is noteworthy that in several sympathetically innervated organs a small but consistent quanitity of NGF could be determined provided these organs were kept in culture for at least 24 hours (29,30). In contrast, the same tissues assayed for NGF immediately after removal from the animal did not contain NGF levels above the limits of the sensitivity of the available biological and immunological assays.

It remains to be established whether the presence of NGF in tissues in culture is an artefact (resulting from an enhanced synthesis or activation under culture conditions) or whether it is a manifestation of the removal of neuronal regulatory mechanisms which in situ inhibit either the synthesis of the precursor molecule or the processing of the NGF precursor, or remove the small amount of active NGF by uptake by the nerve endings.

NGF as a Survival Factor In Vitro

Depending on their developmental stage, sympathetic and sensory neurons require for their survival the addition of NGF to the culture medium (5,6,13,14,19,25,26,44). The survival effect of NGF is restricted to sensory and sympathetic neurons: parasympathetic neurons are not supported by NGF (cf. 34,72,73). Recently, however, other factors have been described which also support the survival of parasympathetic neurons (1,7,34,51).

Many experiments have been performed on ganglionic explants which allow only an indirect estimation of neuronal survival indicated by the extent of fiber outgrowth (19,51). Moreover, these experiments do not indicate to what extent nonneuronal cells contribute to neuronal survival. Thus, in order to obtain reliable quantitative information on the modification of the neuronal survival relative to their developmental stage it is necessary to use neurons dissociated from ganglia and from which the nonneuronal cells have been removed (5,6,13,14,25,26). The latter is of particular importance since a large number of tissues including glial cells in culture can produce NGF or NGF-like macromolecules (11,20,30,47,54,55,72,73,74) which support survival of neurons.

EFFECT OF NGF ON THE SURVIVAL OF CHICK SPINAL SENSORY NEURONS OF DIFFERENT EMBRYONIC AGES

Before embryonic day 15 it is possible to distinguish morphologically two different populations of dorsal root ganglion neurons, namely large ventro-lateral neurons and smaller medio-dorsal neurons (39,44). Only the latter have been shown to respond to NGF in vivo and in explants of dorsal root ganglia in vitro (44).

Under conditions where neurons are plated at low density on an adequate substrate such as polyornithine which also suppresses the proliferation of any residual nonneuronal cells (see Barde, et al. (5,6) for details), it is easy to quantify the number of neurons which survive after 2 days in culture. The surviving neurons are characterized by a clearly delineated, strongly refractile cell body and the production of neurites. As shown in Fig. 1, in the absence of nonneuronal cells virtually no neurons survive if the culture medium (containing serum) is not supplied with NGF or other factors (Figs. 1,2). The addition of a saturating concentration of NGF (5 ng/ml) enhanced the survival of neurons depending on the developmental stage. The maximal survival of 40% of the neurons was reached between day 10 and 12 (Fig. 2). Thereafter, survival gradually decreased and reached low values by day 16 when essentially no neurons survived. These results are concordant with the marked decrease in NGF receptors with increasing embryonic age and the failure of NGF to evoke neurite outgrowth from explants of dorsal root ganglia of embryos in an advanced developmental stage (44,53). Starting the cultures with embryos of day 12, the number of surviving neurons was the same between 48 h and 1 week in culture. This result is surprising because after 7 days in culture these neurons might be expected to behave as those from embryos of 19 days of age. In that case it would be expected from Fig. 2 that no neurons would survive with NGF. The fact that neurons from 12 day old embryos do survive in culture with NGF for at least a week indicates that the maturation of neurons maintained in vitro with NGF does not parallel the maturation of sensory neurons in vivo.

EFFECT OF GLIAL CONDITIONED MEDIUM AND BRAIN EXTRACT

If the cultures of dissociated sensory neurons were supplied with concentrated media conditioned by C-6 glioma cells the number of surviving neurons was very small at early embryological stages, amounting to about 10% at day 8 (Figs.1,2). However, in contrast to NGF, there was a continuous increase in the proportion of surviving neurons estimated to be about 80% at day 16 (Fig. 2). The real number is most probably even higher, since the percentage of surviving neurons is based on the total number of plated cells including a residual, variable number of nonneuronal cells which represent about 5 to 20% of the total cell number.

As with NGF, the concentration of the survival factor in glial conditioned medium used was saturating. No greater survival was achieved by further concentration of the conditioned medium. Since the heat- and protease-sensitive, non-dialyzable factor(s)

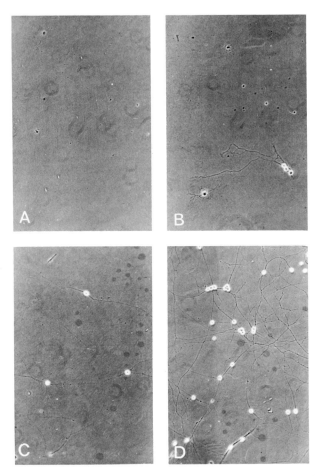

FIG. 1.   Effect of NGF and glial conditioned medium (GCM) on the survival of chick dorsal root neurons.   Neurons dissociated from chick embryos (day 8) were incubated in F-14 medium supplemented with 10% (V/V) horse serum.   Pictures taken 48 h after plating; of the cells.   A = Control. B = NGF (5 ng/ml). C = glial conditioned medium + affinity chromatography purified NGF-antibodies (67) (500 ng/ml).   The antibodies were added to exclude any possible effect of small quantities of NGF produced by glial cells. D = GCM + NGF (5 ng/ml).

was produced by glial tumor cells (6), it was of interest to see if similar activities could be detected in tissues rich in glia. Indeed, the effect of brain extract on the survival of sensory

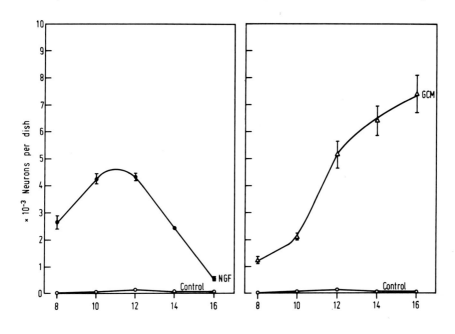

Embryonic age [days]

FIG. 2. Effect of NGF and glial conditioned medium on the survival of chick sensory neurons of different embryonic ages. Neurons from dorsal root ganglia of chick embryos of different embryonic ages were grown with NGF (5 ng/ml) or with GCM + NGF antibodies (500 ng/ml) or without addition (control). 10000 cells were plated per 35-mm dish. Neurons were counted after 48 h in culture. Results are expressed as the mean of triplicate determination ± SD. Where not shown, SD is smaller than the symbols.

neurons of different embryonic ages was very similar to that of glial conditioned medium (5). The assumption that the activity in brain extract originates from glial rather than neuronal cells is supported by two observations. The survival activity of brain extracts originating from animals of the first postnatal weeks increased in parallel with the rapid development of glial cells during this phase of ontogenesis (5). Recently, Lindsay reported that the survival of sensory neurons is also supported (independently of NGF) by conditioned medium produced by primary cultures of astroglial cells (47).

Although it was shown that astrocytes in culture also produce NGF (47), another instance where cells or tissues which do not contain measurable quantities of NGF in vivo start to produce NGF under culture conditions (29,30), no NGF can be detected in the brain with a specific, 2-site NGF radioimmunoassay (5).

COMBINATION OF NGF AND GLIAL CONDITIONED MEDIUM

At early embryological stages the combination of NGF and glial conditioned medium is more than additive (5), indicating that early in development, some sensory neurons require both NGF and the factor(s) contained in the glial conditioned medium for survival. With increasing age, the effect of the glial conditioned medium becomes greater and by day 16 of embryonic age, all the neurons which survive can be maintained by glial conditioned medium alone, NGF having no additional effect (5). All these results can be obtained if brain extract is substituted for glial conditioned medium (5), indicating again that that survival factors from these two sources are similar if not identical. It is tempting to speculate that in vivo the source for a potential glial factor could be the satellite cells which appear as early as day 5 in the embryonic chick dorsal root ganglia and by day 10 already completely surround the sensory neurons (58).

In summary, NGF is a specific neuronal survival factor: its ability to support the survival of dorsal root ganglionic sensory neurons is limited to a specific period during their development. An additional specificity is that only one sub-population of sensory neurons survives in response to NGF alone, although apparently all the neurons of the dorsal root ganglia can survive in culture if NGF is supplemented with factor(s) derived from glioma-conditioned media or brain extracts. These non-NGF factors may be clearly differentiated from NGF as they are able to support the survival of essentially all sensory neurons later during development, when NGF has no effect.

REFERENCES

1. Adler, R., Landa, K.B., Manthorpe, M., and Varon, S. (1979): Science, 204:1434-1436.
2. Aloe, L., and Levi-Montalcini, R. (1979): Proc. Natl. Acad. Sci. USA, 76:1246-1250.
3. Angeletti, P.U., Levi-Montalcini, R., Kettler, R., and Thoenen, H. (1972): Brain Res., 44:197-206.
4. Banks, B.E.C., Carstairs, J.R., and Vernon, C.A. (1979): Neuroscience, 4:1145-1155.
5. Barde, Y.A., Edgar, D., and Thoenen, H. (1980): Proc. Natl. Acad. Sci. USA, 77:1199-1203.
6. Barde, Y.A., Lindsay, R.M., Monard, D., and Thoenen, H. (1978): Nature, 274:818.
7. Bennett, M.R., and Nurcombe, V. (1979): Brain Res., 173:543-548.

8. Bjerre, B., Bjorklund, A., Mobley, W., and Rosengren, E. (1975): Brain Res., 94:263-277.
9. Bocchini, V., and Angeletti, P.U. (1969): Proc. Natl. Acad. Sci. USA, 64:787-794.
10. Bueker, E.D. (1948): Anat. Rec. 102:369-390.
11. Burnham, P.A., Raiborn, C., and Varon, S.(1972): Proc. Natl. Acad. Sci. USA., 69:3556-3560.
12. Campenot, R.B. (1977): Proc. Natl. Acad. Sci. USA, 74:4516-4519.
13. Chun, L.L.Y., and Patterson, P.H. (1977): J. Cell Biol., 75:694-704.
14. Chun, L.L.Y., and Patterson, P.H. (1977): J. Cell Biol., 75:705-711.
15. Cohen, S. (1960): Proc. Natl. Acad. Sci. USA, 46:302-311.
16. Cowan, W.M. (1973): In: Development and Aging in the Nervous System, edited by M. Rockstein, pp. 19-41. Academic Press, New York.
17. Detwiler, S.R. (1920): Proc. Natl. Acad. Sci. USA, 6:96-101.
18. Dumas, M., Schwab, M.E., and Thoenen, H. (1979): J. Neurobiol., 10:179-197.
19. Ebendal, T., and Jacobson, C.O. (1977): Exp. Cell Res., 105:379-387.
20. Ebendal, T., and Jacobson, C.O. (1977): Brain Res., 131:373-378.
21. Ebendal, T., Belew, M., Jacobson, C.O., and Porath, J. (1979): Neurosci. Lett., 14:91-95.
22. Ennis, M., Pearce, F.L., and Vernon, C.A. (1979): Neuroscience, 4:1391-1398.
23. Goedert, M., Otten, U., Schafer, T., Schwab, M.E., and Thoenen, H. (1980): Brain Res., (in press).
24. Goedert, M., Otten, U., and Thoenen, H. (1978): Brain Res., 148:264-268.
25. Greene, L.A. (1977): Develop. Biol., 58:96-105.
26. Greene, L.A. (1977): Develop. Biol., 58:106-113.
27. Hamburger, V. (1934): J. Exp. Zool., 68:449-494.
28. Hamburger, V. (1977): Neurosci. Res. Program Bull. 15: Suppl. 15:iii-37.
29. Harper, G.P., Al-Saffar, A.M., Pearce, F.L., and Vernon, C.A. (1980): Develop. Biol., 77:379-390.
30. Harper, G.P., Pearce, F.L., and Vernon, C.A. (1976): Nature, 261:251-253.
31. Harper, G.P., Pearce, F.L., and Vernon, C.A. (1980): Develop. Biol., 77:391-402.
32. Harper, G.P., and Thoenen, H. (1980): J. Neurochem., 34:5-16.
33. Hawrot, E., and Patterson, P.H. (1979): In: Methods in Enzymology, edited by W.B. Jakoby and I.H. Pastan, Vol. LVIII, pp. 574-584. Academic Press, New York.

34. Hefland, S.L., Riopelle, R.J., and Wessels, N.K. (1978): Exp. Cell Res., 113:39-45.
35. Hendry, I.A. (1975): Brain Res., 90:235-244.
36. Hendry, I.A. (1976): Rev. Neurosci. 2:149-194.
37. Hendry, I.A. Stockel, K., Thoenen, H., and Iversen, L.L. (1974): Brain Res., 68:103-121.
38. Herrup, K., and Shooter, E.M. (1975): J. Cell Biol., 67:118-125.
39. Jacobson, M. (1978): Developmental Neurobiology. Plenum Press, New York.
40. Johnson, Jr. E.M. (1978): Brain Res., 141:105-118.
41. Kessler, J.A., and Black, I.B. (1980): Proc. Natl. Acad. Sci. USA, 77:649-652.
42. Levi-Montalcini, R., Aloe, L., Mugnaini, E., Oesch, F., and Thoenen, H. (1975): Proc. Natl. Acad. Sci. USA, 72:595-599.
43. Levi-Montalcini, R., and Angeletti, P.U. (1966): Pharmacol. Rev., 18:619-628.
44. Levi-Montalcini, R., and Angeletti, P.U. (1968): Physiol. Rev., 48:534-569.
45. Levi-Montalcini, R. and Hamburger, V. (1951): J. Exp. Zool., 116:321-362.
46. Levi-Montalcini, R. and Hamburger, V. (1953): J. Exp. Zool., 123:233-278.
47. Lindsay, R.M. (1979): Nature, 282:80-82.
48. Lindsay, R.M., and Tarbit, J. (1979): Neurosci. Lett., 12:195-200.
49. Longo, A.M. (1978): Develop. Biol., 65:260-270.
50. MacDonnell, P.C., Nagaiah, K., Lakshmanan, J., and Guroff, G. (1977): Proc. Natl. Acad. Sci. USA, 74:4681-4684.
51. McLennan, I.S., and Hendry, I.A. (1978): Neurosci. Lett. 10:269-273.
52. Menesini Chen, M.G., Chen, J.S., Calissano, P., and Levi-Montalcini, R. (1977): Proc. Natl. Acad. Sci. USA, 74:5559-5563.
53. Mobley, W.C., Server, A.C., Ishii, D.N., Riopelle, R.J., and Shooter, E.M. (1977): N. Engl. J. Med., 297:1096-1104, 1149-1158, 1211-1218.
54. Murphy, R.A., Oger, J., Saide, J.D., Blanchard, M.H., Arnason, B.G.W., Hogan, C., Pantazis, N.J., and Young, M. (1977): J. Cell Biol., 72:769-773.
55. Oger, J., Arnason, B.G., Pantazia, N., Lehrich, J., and Young, M. (1974): Proc. Natl. Acad. Sci. USA, 71:1554-1558.
56. Otten, U., Katanaka, H., and Thoenen, H. (1978): Fourth International Catecholamine Symposium. pp. 115-117. Asilomar, Oxford.
57. Otten, U., Schwab, M.E., Gagnon, C., and Thoenen, H. (1977): Brain Res., 133:291-303.
58. Pannese, E. (1969): J. Comp. Neur., 135:381-422.
59. Paravicini, U., Stoeckel, K., and Thoenen, H. (1975): Brain Res., 84:279-291.
60. Purves, D. (1976): J. Physiol. London, 259:159-175.
61. Schwab, M.E. (1977): Brain Res., 130:190-196.

62. Schwab, M.E., Stockel, K., and Thoenen, H. (1976): <u>Cell Tiss. Res.</u>, 169:289-299.
63. Shorey, M.L. (1909): <u>J. Exp. Zool.</u>, 7:25-63.
64. Skaper, S.D., and Varon, S. (1979): <u>Brain Res.</u>, 163:89-100.
65. Skaper, S.D., and Varon, S. (1979): <u>Biochem. Biophys. Res. Commun.</u>, 88:563-568.
66. Stoeckel, K., Paravicini, U., and Thoenen, H. (1974): <u>Brain Res.</u>, 76:413-421.
67. Stoeckel, K., Gagnon, C., Guroff, G., and Thoenen, H. (1976): <u>J. Neurochem.</u>, 26:1207-1211.
68. Suda, K., Barde, Y.A., and Thoenen, H. (1978): <u>Proc. Natl. Acad. Sci. USA</u>, 75:4042-4046.
69. Thoenen, H., Barde, Y.A., Edgar, D., Hatanaka, H., Otten, U.,and Schwab, M.E. (1979): <u>Progr. Brain Res.</u>, 51:95-107.
70. Thoenen, H., Otten, U., and Schwab, M.E. (1979): In: <u>The Neurosciences, Forth Study Program</u>, edited by F.O. Schmitt and F.E. Worden, pp. 911-928. MIT-Press, Cambridge.
71. Thoenen, H., and Schwab, M.E. (1978): In: <u>Advances in Pharmacology and Therapeutics,</u> Vol. 5, edited by C. Dumont., pp. 37-59. Pergamon Press, New York.
72. Varon, S. (1975): <u>Exp. Neurol.</u>, 48:93-134.
73. Varon, S., and Bunge, R.P. (1978): <u>Ann. Rev. Neurosci.</u>, 1: 327-361.
74. Young, M., Oger, J., Blanchard, M.H., Asdourian, H., Amos, H., and Arnason, B.G.W. (1974): <u>Science</u>, 187:361-362.

Neurosecretion and Brain Peptides,
edited by J. B. Martin, S. Reichlin, and K. L. Bick.
Raven Press, New York © 1981.

Chemical Differentiation of Sympathetic Neurons

D. D. Potter, S. C. Landis, and E. J. Furshpan

Department of Neurobiology, Harvard Medical School, Boston, Massachusetts 02115

The development of the sympathetic nervous system has been studied for many years. One aspect which has recently attracted attention in several laboratories is the control of transmitter choice in the principal neurons. A majority of these neurons is adrenergic in adults (in higher vertebrates they secrete norepinephrine (NE)). A minority, apparently a few percent in some sympathetic ganglia, is cholinergic; these neurons secrete acetylcholine (ACh) and innervate certain sweat glands and blood vessels. Some principal neurons also contain peptides; immunocytochemical evidence has been obtained for the presence of somatostatin-like, vasoactive intestinal polypeptide (VIP)-like and enkephalin-like substances (for a review see ref. 6). Moreover, Hokfelt and his colleagues have reported that in sympathetic ganglia of adult rats some neurons contain not only somatostatin-like or enkephalin-like immunoreactivity, but also dopamine β-hydroxylase, an enzyme of the catecholamine pathway. This led them to postulate that these neurons secrete both NE and a peptide. Since all sympathetic principal neurons are apparently derived from a common embryological source, the neural crest, an obvious question is how their precursors were directed during development into these different transmitter states.

This question has been investigated <u>in vivo</u> (8,1,5) and <u>in vitro</u> (12,2,9). There is strong evidence both <u>in vivo</u> and <u>in vitro</u> that the choice between NE and ACh is influenced by non-neuronal cells. For example, it has been shown in culture that non-neuronal cells from a variety of tissues cause many rat sympathetic neurons, placed in culture on the day of birth when they appear morphologically to be adrenergic, to develop cholinergic properties over a period of one to several weeks. These properties include the ability to synthesize and store ACh and to form functional cholinergic synapses with each other, with cardiac myocytes and with skeletal myotubes. The non-neuronal cells can exert this effect by way of the medium (conditioned medium, CM), and the effect is graded. In mass cultures containing several thousand neurons, the larger the number of non-neuronal cells co-cultured with the neurons or the higher the proportion of CM fed to the neurons, the greater the ratio of cholinergic metabolism to adrenergic metabolism and the higher the incidence of synapses and varicosities of cholinergic transmission between the neurons. The chemical identity of the factor(s) in CM is not yet known.

To study the non-neuronally induced transition from adrenergic to cholinergic status, it is useful to grow the neurons singly or in small numbers in microcultures and then to determine the transmitter status with biochemical, electron microscopical or physiological methods. The sensitivity of current biochemical assays for synthesis of NE or ACh permits study of single neurons only after the cells have reached appreciable size, rather late in the process of transmitter choice (4-5 weeks in culture). At this stage of development, Reichardt and Patterson (13) found that most single neurons synthesize and accumulate detectable amounts of only one transmitter: always NE when the neurons were grown in the absence of non-neuronal cells; almost always ACh when the neurons were grown in the presence of large numbers of cardiac cells (myocytes and fibroblasts); NE or ACh with about equal probability when the neurons were fed 50% CM. Only a few single neurons appeared to synthesize both transmitters. This result raised the possibility that the transmitter-control mechanism can operate in a flip-flop manner, and that, at least in certain culture conditions, the final state is exclusive.

At earlier times in the transition the transmitter status was studied in single neurons in microcultures by combining physiological and electron microscopical methods (3,7). Neurons were grown for 2-3 weeks on cardiac cells and the transmitter status determined by recording the effect of the neuron on the myocytes (cholinergic inhibition; adrenergic excitation); the fine structure of the synapses and varicosities made by that neuron on itself and on the myocytes was then observed. While some neurons were functionally and morphologically adrenergic and some cholinergic, there was also a population which was intermediate in status. These dual-function neurons produced both a cholinergic inhibition and an adrenergic excitation of the myocytes and possessed synapses which contained both "empty" synaptic vesicles, characteristic of cholinergic junctions, and small, granular vesicles, characteristic of adrenergic junctions.

Dual function neurons which store and release more than one transmitter appear to be a novelty, although there are recent studies consistent with dual function in certain adult neurons (6). The existence of dual function neurons seen in the course of development of the nervous system provides further evidence for the transition from adrenergic to cholinergic status. Microcultures provide a sensitive way to study this transition. In this chapter we report findings on dual-function sympathetic neurons in such cultures.

METHODS

Microcultures (about 0.5mm in diameter) with one or several developing rat sympathetic neurons and beating cardiac myocytes were produced as described previously (3). The neurons, dissociated from the superior cervical ganglion (SCG), were placed in culture on the day of birth. To determine the transmitter status of the neurons at various times thereafter (7-88 d), the culture dish was placed in an electrophysiological set-up, a neuron and cardiac myocytes were impaled with microelectrodes and the effect of action potentials evoked in the neuron on the electrical behavior of the myocytes was recorded. Hyperpolarization and inhibition of beating were shown to be cholinergic by block of the effect with $1-4 \times 10^{-7}$M atropine sulfate; depolarization and increased beat frequency were shown to be adrenergic by block with $1-3 \times 10^{-6}$M DL-propranolol.HCl, $2-5 \times 10^{-7}$M DL-alprenolol.HCl or $2-5 \times 10^{-5}$M DL-sotalol.HCl. A perfusion system similar to that described in O'Lague et al. (9) permitted continuous recording from the neuron and myocytes, often while blocking drugs were perfused into, and subsequently washed out of, the culture dish.

After the physiological assay, some microcultures were prepared for electron microscopy by fixation with glutaraldehyde followed by osmium, to examine general fine structure, or with potassium permanganate to reveal storage of NE in synaptic vesicles. Each culture dish contained an orderly array of 25 microcultures. By noting the position, in the array, of the microculture studied physiologically, it was possible to identify this microculture with assurance during preparation for electron microscopy. When only one neuron was present in the microculture, the fine structure and physiology of that neuron, including its endings, could be directly correlated.

In preliminary experiments, the transmitter status of individual neurons was tested repeatedly, the culture dish being returned to the incubator after each recording session. In spite of efforts to maintain sterility, many cultures became contaminated during the recording sessions. The longest observation period of the same neuron was 30 days during which 4 physiological assays were made. Again, by noting the position of the microculture in the culture dish, it was possible to return to the same neuron.

CHARACTERIZATION OF DUAL-FUNCTION NEURONS

The appearance of a microculture which contained a single sympathetic neuron and beating cardiac myocytes is shown in Fig. 1. In such microcultures, beating myocytes could easily be recognized with phase microscopy, and under visual control the neuronal cell body and a myocyte could be impaled with microelectrodes. Most dual-function neurons were studied at culture ages of 2-4 weeks. However, dual effects on cardiac myocytes were observed as early as 9 d and as late as 41 d.

To test the transmitter status of a neuron as shown in Fig. 1, the neuron was stimulated, usually with a train of brief current pulses through the recording microelectrode (<u>e.g.</u>, 5-50/s for 1-5 s) and the effect on the myocytes was simultaneously recorded. In cultures examined 2-4 wk after the neurons were plated, most neurons exhibited dual-function. An example is shown in Fig. 2c.

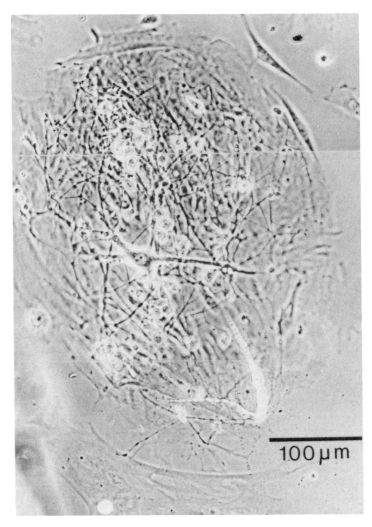

FIG. 1. A microculture containing a single neuron, 13d old. This neuron exhibited dual function as in Fig. 2.

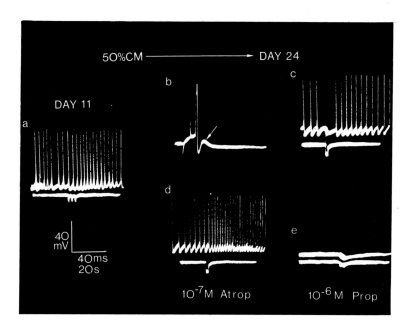

FIG. 2. A neuron whose transmitter status was assayed twice,
first on day 11 (a) and then on day 24. From day 11 to day 24
the neuron was fed with 50% CM. Further description in the text.
Calibrations: 40mV, 40ms for (b); 40mV, 20s for a,c,d,e.

     The neuron was stimulated at 5/s for about 3s (deflection of
lower trace). This produced a hyperpolarization of the myocytes
(upper trace) and a pause in the spontaneous cardiac activity,
followed by a slightly increased frequency of cardiac action
potentials. To check whether the inhibition of the myocytes was
cholinergic, the muscarinic blocker atropine sulfate (10^{-7}M) was
added (Fig. 2d); the inhibition disappeared, and neuronal activity now produced a pronounced positive chronotropic effect. When
the atropine was washed out, the initial inhibition returned (not
shown). At a later stage in the experiment, to check whether the
excitation of the myocytes was adrenergic, the β-blocker DL-prop-

ranolol (10^{-6}M) was added. This eliminated the spontaneous low-frequency beat present at this stage, and when the neuron was stimulated at 20/s for 3s only a hyperpolarization was produced in the myocytes (Fig. 2e). This sequence of effects on the myocytes, combined with evidence that in similar conditions neurons in mass cultures collectively synthesize and accumulate ACh and NE (11), leaves little doubt that the neuron under study secreted both ACh and NE (dual function). It is plausible that the two transmitters were secreted simultaneously and that in control solution (Fig. 2c) the action of ACh preceded a more prolonged action of NE. When the two substances are simultaneously applied locally to myocytes from a micropipette, cholinergic inhibition precedes adrenergic excitation (not shown). As in almost all dual-function neurons, each neuronal action potential elicited a cholinergic excitatory postsynaptic potential in the neuron (arrow, Fig. 2b) which arose at synapses made by this neuron on itself (autapses). No adrenergic effect was detected at such autapses because the neuronal resting potential is relatively insensitive to NE.

The fine structure of the varicosities of this neuron is shown in Fig. 3. As was typical of the varicosities of dual-function neurons, both "empty" synaptic vesicles and small, granular synaptic vesicles are present. Since most thin sections made at random through the varicosities contain both "empty" and granular vesicles, it is plausible that both transmitters can be secreted from the same varicosity (7).

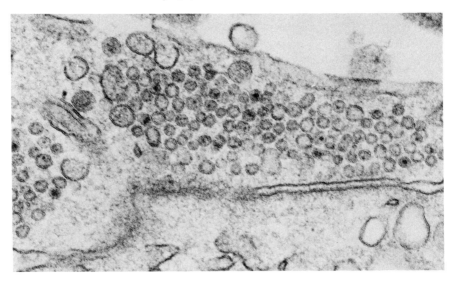

FIG. 3. Electron micrograph of an autapse in the microculture whose assay is shown in Fig. 2. After permanganate fixation to localize endogenous stores of norepinephrine, approximately one quarter of the synaptic vesicles contained granular precipitate. x 75,000.

The Relative Intensity of Cholinergic and Adrenergic Effects Varied from One Dual-Function Neuron to Another

Among the more than 100 dual-functon neurons studied electrophysiologically at various culture ages, the relative strength of the two effects on cardiac myocytes varied from predominantly adrenergic to predominantly cholinergic. Part of this range of variation is illustrated in Figs. 2 and 4 (2 different neurons). Substantial effects of both kinds are seen in Fig. 2c,d and e. In Fig. 4b,c and d the cholinergic effect is relatively stronger.

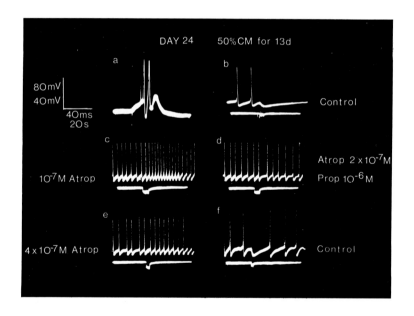

FIG. 4. A 24-day old neuron grown for 13d in 50% CM in a microculture adjoining the one assayed in Fig. 2. Description in the text. Calibrations: 40mV, 40ms for (a); 80mV, 20s for (b)-(e); 40mV, 8s for (f).

In this case, a single impulse in the neuron produced a visible hyperpolarization of the myocytes (not shown; amplitude depended on the level of the myocyte membrane potential) and sustained stimulation of the neuron at 1/s prevented the myocyte from beating. The effect of stimulation of the neuron at 10/s (Fig. 4b) appeared purely inhibitory and cholinergic. However, block of the cholinergic effect with atropine disclosed a previously-hidden, rather weak adrenergic effect (Fig. 4c). The cholinergic effect of the neuron on itself at "autapses"; (cf., refs. 19,3) was greater in the second case than in the first (Fig. 2b vs. Fig. 4a) in conformity with the relatively stronger cholinergic effect on the myocytes.

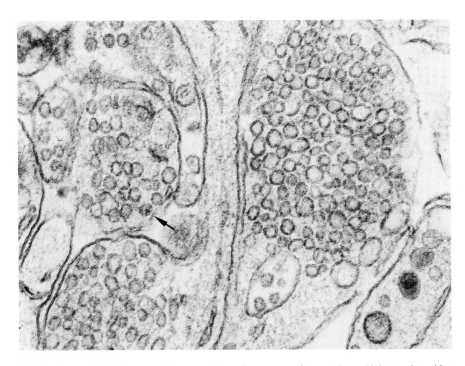

FIG. 5. Electron micrograph of several varicosities in the microculture whose records are shown in Fig. 4. After permanganate fixation to localize endogenous stores of norepinephrine, less than one percent of the synaptic vesicles contained granular precipitate. Arrow,, small granular vesicle. x75,000.

During a transition from an initial adrenergic state to a final cholinergic state one might expect, in a population of solitary neurons, to observe differences in the relative strengths of the two properties, as in Figs. 2-5. Thus it is plausible that the second neuron (Figs. 4,5) was in a more advanced stage of the transition than the first neuron (Figs. 2,3). However, the effect of the neuron on the myocytes depended not only on the amounts of the two transmitters secreted but also on the post-synaptic sensitivity to the transmitters. Is it possible that the two neurons had identical transmitter status but were assayed on myocytes with different sensitivities? It is unlikely that the difference between the two microcultures resided in the myocytes because 1) the two microcultures were adjacent in the same dish and received myocytes from the same cell suspension, and 2) there were clear fine structural differences on the presynaptic side (Figs. 3,5) consistent with the physiological differences (Figs. 2,4). A more satisfactory physiological assay of transmitter status would include a test of the relative sensitivity of the myocytes to the two transmitters. In the absence of such tests we tentatively conclude that the two neurons of Figs. 2-5 were in different stages of the transition in transmitter status.

Repeated Assays on the Same Neuron

In the experiments just described and in previous work on the properties of the cultured neurons, only single assays of transmitter status were made on a given neuron or culture at various times during development. Taken together, this evidence shows that many individual sympathetic neurons of the rat are still plastic at birth with respect to the choice between NE and ACh and, under the influence of non-neuronal cells, can shift from relatively or completely adrenergic status to cholinergic status. We have investigated this transition further by recording successively from specific neurons in microcultures containing cardiac cells.

In 16 experiments we have recorded at least twice (maximum: 4 times) from the same neuron; after a recording session, the culture dish was either returned to the incubator or fixed for electron microscopy. All the observed changes were from adrenergic to cholinergic. No single neuron was followed throughout the whole transition, but partial transitions, from apparently adrenergic to clear dual-function and from dual-function to relatively more cholinergic dual-function or apparently cholinergic were observed.

An example of a partial transition is shown in Fig. 2. This neuron was tested first on day 11; Fig. 2a shows that the effect of 3 trains of impulses in the neuron (each 40/s for 1s) was weakly excitatory. No attempt was made to see if a hidden cholinergic effect was present. The neuron was returned to the incubator and for the next 13d was fed medium half of which had been conditioned for 2d by cardiac cells (50% CM). On day 24 the neuron was tested again, and a substantial cholinergic effect was present (dual function) as described above (Fig. 2b,c,d,e). In the 13d between assays, there was a change from no obvious cholinergic effect to a clear one.

In most dishes studied in this series, several neurons in adjoining microcultures were assayed. We usually found that neurons in the same dish did not undergo the transition synchronously. This was true even of pairs of neurons whose cell bodies were side-by-side in the same microculture. For example, in one such microculture, one of the two neurons changed from apparently

purely adrenergic to dual-function over a 30d period and the
other neuron from dual-function to apparently purely cholinergic.
Cases were also seen in which there was no obvious change in
transmitter status on two successive assays separated by 2-15d in
dishes in which another neuron underwent a partial transition.

The significance of this asynchrony is not known, but several
factors might have contributed. The terminal divisions of the
neuroblasts of the embryonic rat SCG occur over a period of about
7 days (4); neurons with later "birthdays" may lag behind those
with earlier "birthdays". Moreover, there was considerable
variation in the number of heart cells to which the neurons were
exposed from one microculture to the next. Finally, the observed
heterogeneity may reflect differences between neurons originally
destined to innervate different targets in vivo (e.g., heart,
iris, blood vessels, salivary glands, brown fat.)

DISCUSSION

In this chapter we report several new findings on dual-function, sympathetic principal neurons in culture. Adrenergic and
cholinergic properties can be expressed in different proportion
(as would be expected, there is a correspondence between physiological and fine structural observations on the same neuron).
Partial transitions in the direction adrenergic-to-cholinergic
can be observed by following particular neurons over time. The
neurons display wide variability in the time course of the transition, under the culture conditions used. These points are all
compatible with previous findings on "mass cultures". The transitions we observed occurred more slowly than those reported by
Patterson and Chun (12), but this may have been due to a relatively weak cholinergic induction in the microcultures; the
cardiac cells in the array of 25 microcultures per dish collectively covered only about 0.5% of the bottom of the dish, while in
the experiments reported by Patterson and Chun (12), the neurons
were grown in the presence of 62% CM.

The transmitter status of individual neurons of the rat SCG
before birth and during the first two months in vivo has so far
been investigated only with microscopical methods. The presence
of small, granular vesicles is a sensitive assay for the synthesis and storage of NE, but no comparable fine structural assay
is at hand for the synthesis or storage of ACh. A sensitive
immunocytochemical assay for choline acetyltransferase would
markedly improve understanding of the differentiation of sympathetic neurons at all stages of development. A recent report by
Wakshull, Johnson and Burton (15) that principal neurons can be
isolated from the SCG of rats up to 12 weeks old also offers new
promise for studying the control of transmitter choice in vivo.

The recent demonstrations by Hokfelt and his collegues (6) of
somatostatin-like, VIP-like and enkephalin-like immunoreactivity
in rat sympathetic neurons make it highly likely that transmitter
choice in sympathetic principal neurons is wider than previously

recognized and that multiple-function is expressed in adult neurons, as it is during development in culture.

ACKNOWLEDGEMENT

The research reported here has been supported by USPHS research grants NS11576, NS03273, NS02253 and a Grant-In-Aid from the American Heart Association with funds contributed in part by the Massachusetts Affiliate.

REFERENCES

1. Black, I.B. (1978): Ann. Rev. Neurosci., 1:183-214.
2. Bunge, R., Johnson, M., and Ross, C.O. (1978): Science, 199:1409-1416.
3. Furshpan, E.J., MacLeish, P.R., O'Lague, P.H., and Potter, D.D. (1976): Proc. Natl. Acad. Sci. USA, 73:4225-4229.
4. Hendry, I.A. (1977): J. Neurocytol., 6:299-309.
5. Hill, C.E., and Hendry, I.A. (1979): Neurosci. Lett., 12: 133-139.
6. Hokfelt, T., Johansson, O., Ljungdahl, A., Lundberg, J.M., and Schultzberg, M. (1980): Nature, 284:515-521.
7. Landis, S.C. (1976): Proc. Natl. Acad. Sci. USA, 73:4220-4224.
8. LeDouarin,N.M., Teillet, M.A., Ziller, C., and Smith, J. (1978): Proc. Natl. Acad. Sci. USA, 75:2030-2034.
9. O'Lague, P.H., Potter, D.D., and Furshpan, E.J. (1978): Dev. Biol., 67:424-443.
10. Patterson, P.H. (1978): Ann. Rev. Neurosci. 1:1-17.
11. Patterson, P.H., and Chun, L.L.Y. (1974): Proc. Natl. Acad. Sci. USA, 71:3607-3610.
12. Patterson, P.H., and Chun, L.L.Y. (1977): Dev. Biol., 60: 473-481.
13. Reichardt, L.F., and Patterson, P.H. (1977): Nature, 270:147-151.
14. Van der Loos, H., and Glaser, E.M. (1972): Brain Res., 48:355-360.
15. Wakshull, E., Johnson, M.I., and Burton, H. (1979): J. Neurophysiol., 42:1426-1436.

Neurosecretion and Brain Peptides,
edited by J. B. Martin, S. Reichlin, and K. L. Bick.
Raven Press, New York © 1981.

Regulation of Noradrenergic and Petidergic Development: A Search for Common Mechanisms

Ira B. Black and John A. Kessler

Division of Developmental Neurology, Department of Neurology, Cornell University Medical College, New York, New York 10021

Study of the relatively simple, well-defined autonomic nervous system has elucidated a number of principles governing neuronal ontogeny and plasticity. In particular, analysis of sympathetic neurons has indicated that intercellular interactions, occurring throughout development, regulate neuronal survival, phenotypic expression and maturation (for review ref. see 1). It is now apparent, for example, that phenotypic expression and development of transmitter characters are influenced by environmental cues. Sympathetic neurons may alter transmitter functions qualitatively as well as quantitatively, in response to appropriate stimuli in vivo (11,12,31,35) or in vitro (42). The relatively well-characterized nature of the cholinergic and noradrenergic transmitter systems allows detailed analysis of the events underlying these aspects of plasticity.

Specific molecular markers are available for the study of sympathetic ontogeny. Choline acetyltransferase (ChAc), the enzyme that catalyzes the synthesis of acetylcholine (21), is highly localized to presynaptic sympathetic terminals in ganglia (26) and may be used to monitor maturation of these elements (2). On the other hand, tyrosine hydroxylase (T-OH), the rate-limiting enzyme in catecholamine biosynthesis (39), is localized to post-synaptic adrenergic neurons (6) and may be employed as an index of development of these cells (2). T-OH catalyzes the conversion of tyrosine to L-dopa, the first step in catecholamine synthesis (39). Dopa decarboxylase (DDC), which converts L-dopa to dopamine, and dopamine-β-hydroxylase (DBH), which converts dopamine to norepinephrine (NE), are also highly restricted to postsynaptic neurons and may serve as indices of adrenergic ontogeny (6). Precise biochemical-morphological correlation is made possible by

the availability of histofluorescence techniques for the visualization of catecholamines in situ (20); immunocytochemical methods may be employed to visualize noradrenergic enzymes (11,12).
Study of ontogenetic mechanisms in other neuronal subsystems has
been hampered by lack of suitable molecular indices of development and plasticity. Consequently, it has been difficult to
determine whether the principles governing autonomic ontogeny
also regulate development of other classes of neurons. Recent
work, however, indicates that the putative sensory transmitter,
Substance P (SP), is a useful marker for development of dorsal
root ganglia (DRG), and suggests that similar mechanisms may
govern peptidergic and noradrenergic maturation. In this paper,
previous extensive studies of sympathetic development are reviewed, with emphasis on the postnatal period. These observations are compared with sensory maturation in the embryonic and
prenatal period in an effort to define common mechanisms and
processes.

TRANS-SYNAPTIC REGULATION OF NORADRENERGIC DEVELOPMENT

Cellular interactions at multiple levels of the autonomic
neuraxis regulate postnatal development of sympathetic neurons.
Orthograde and retrograde trans-synaptic factors are necessary
for normal maturation of presynaptic (preganglionic) cholinergic
nerves as well as postsynaptic (postganglionic) cholinergic noradrenergic neurons in sympathetic ganglia. Descending information
within the spinal cord regulates normal development of the proxi-
mate cholinergic perikarya as well as the second order noradrenergic neurons in the peripheral sympathetic ganglia (4,9,25).
Conversely, target organs influence survival and maturation of
innervating noradrenergic neurons as well as the second order
afferent cholinergic neurons (17-19). Finally, postsynaptic noradrenergic neurons are necessary for normal presynaptic cholinergic development (7). Some of these interactions are considered
in greater detail, since emerging evidence suggests that the
mechanisms involved are probably not restricted to sympathetic
development, but may govern sensory peptidergic ontogeny as well.

During perinatal ontogeny, there is a marked increase in
synapse numbers within the sympathetic ganglion, a developmental
increase which immediately precedes the 6- to 10-fold rise in
postsynaptic T-OH activity (5). These observations suggested that
the development of T-OH activity might be dependent on formation
of intraganglionic synaptic contacts. To examine this possibility, ganglia were unilaterally decentralized in neonatal animals.
The presynaptic cholinergic trunk innervating the superior cervical sympathetic ganglion (SCG) was transected in 3- to 4-day-old
mice and rats, and in each animal the contralateral intact ganglion served as a control. Decentralization of the ganglion
prevented the normal developmental increase in postsynaptic T-OH
activity (5), suggesting that presynaptic cholinergic neurons
regulate the development of postsynaptic T-OH activity through a

trans-synaptic process.   In addition, the other postsynaptic
catecholamine biosynthetic enzymes, DDC and DBH, also failed to
develop normally in decentralized ganglia.

What are the trans-synaptic messages involved in this regulation? To approach this question, neonatal animals were treated
with long-acting ganglionic blocking agents. These drugs prevent
post-synaptic depolarization by competing with acetylcholine for
postsynaptic receptor sites. In fact, treatment of neronates
with chlorisondamine or pempidine prevented the normal development of postsynaptic T-OH activity, reproducing the effects of
ganglion decentralization (2,3). Moreover, as was the case with
decent ralization, postsynaptic DDC and DBH also failed to
develop normally.   These effects appeared to be specific for
nicotinic receptor blockade, since treatment of animals with
atropine, the classic muscarinic antagonist, had no effect on
development (2,3).   Consequently, presynaptic neurons regulate
postsynaptic development through the mediation of acetylcholine
and its interaction with nicotinic receptors on the postsynaptic
membrane. On this basis, it is not necessary to postulate the
existence of some as yet unidentified presynaptic growth factor,
since the normal presynaptic transmitter also subserves this
growth function.

Acetylcholine may regulate postsynaptic maturation through the
conventional process of depolarization or may utilize unique
mechanisms. The recent work of Potter and colleagues, described
elsewhere in this volume, supports the former contention.
In summary, postnatal sympathetic neurons maintained in cell
culture increase the ratio of catecholamine to acetylcholine
biosynthesis in response to depolarizing stimuli. Consequently,
in vivo and in vitro studies suggest that acetylcholine influences transmitter phenotypic characters by depolarizing
sympathetic neurons.

TARGET REGULATION OF NORADRENERGIC DEVELOPMENT

In addition to orthograde trans-synaptic regulation, survival
and development of innervating sympathetic neurons are regulated
by target organs through retrograde trans-synaptic mechanisms.
Target extirpation in neonatal rats or mice prevents the normal
survival of the noradrenergic neurons destined to innervate the
targets.   The loss of the noradrenergic neurons can be prevented
by administration of the trophic agent, nerve growth factor (NGF)
(17,19,27). Moreover, end organs continue to regulate noradrenergic enzymes in the adult.   Inhibition of retrograde axonal
transport with colchicine in postganglionic noradrenergic axons
reduces ganglion T-OH activity, and this effect is partially
reversed by NGF treatment (33). NGF itself is a protein, present
in a number of organs which causes massive sympathetic overgrowth
when injected into the neonatal animal (for review see 22).
Moreover, treatment of neonates with anti-NGF antiserum prevents

sympathetic development, suggesting that the protein plays a role in normal sympathetic ontogeny (22). However, the precise function of NGF in development remains to be elucidated.

In addition to regulating neuronal survival and enzyme activity, the characteristic pattern of innervation of a target is determined by the target itself and not by the innervating neurons. In the iris, for example, the noradrenergic innervation rconsists of a dense plexus in which thick preterminal axons undergo extensive branching to form a dense network of varicose terminals, a pattern radically different from that in other peripheral structures and in the cerebral cortex. Since NGF molecules from different targets are identical, and since the same neurons innervate different NGF-containing targets differently (40) interactions between targets and neurons probably involve other unidentified factors, as well as NGF.

To analyze target-neuron interactions in greater detail, we developed a tissue culture system using the 14 gestational day (E14) mouse superior cervical ganglion (SCG) (14). This system was particularly advantageous because, at this early developmental stage, sympathetic neurons survive and grow in the absence of added NGF, in contrast to all previously described sympathetic systems. Consequently, we were able to define some of the mechanisms underlying target actions without the complicating presence of NGF in the medium. We initiated studies by culturing ganglia alone, or in the presence of target submaxillary salivary glands to study biochemical and morphologic development. After 2 days of incubation, T-OH activity in ganglia grown with the target had increased 10-fold compared with zero-time controls, and was twice that of ganglia grown alone. After five days in culture, ganglia grown with target exhibited T-OH activity 7-fold higher than that in ganglia grown alone (15). The target, consequently, exerted both a stimulatory and maintenance effect on biochemical development. Moreover, ganglia grown with target elaborated neurites into and around the salivary gland, whereas very few neurites were elaborated in other directions (15).

To determine whether the target effects were mediated by NGF, ganglia were grown with and without targets, in the presence or absence of anti-NGF antiserum. Anti-NGF did not alter the stimulatory influence of target tissue on biochemical or morphologic ontogeny, suggesting that the target effects on the embryonic ganglion were not mediated by NGF (15). In fact, we are now isolating and characterizing a diffusible target factor which increases T-OH activity, enhances neurite outgrowth and is different from NGF by a number of criteria (8). Regardless of the mediating molecules involved, it is apparent that the peripheral field of innervation exerts a profound effect on noradrenergic development during prenatal and postnatal ontogeny. Recent work suggests that peripheral targets also influence peptidergic maturation.

DEVELOPMENT OF SUBSTANCE P IN THE DORSAL ROOT GANGLION

Although suitable biochemical markers of sensory ontogeny have become available only recently, morphologic development of the DRG has been well-studied. It is known, for example, that the peripheral field of innervation influences sensory ontogeny; limb amputation prevents normal morphologic growth of the innervating

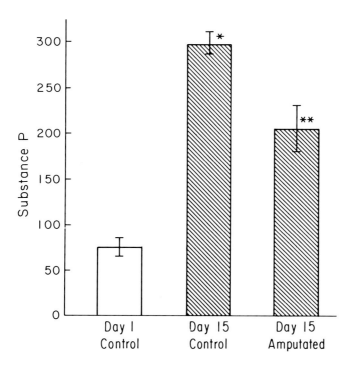

FIG. 1. Effects of forelimb amputation on SP in the C_6DRG. Unilateral forelimb amputation was performed on 8 animals on the first day of life. Two weeks later the C_6DRG on the side of amputation and on the contralateral control side were examined for SP content. SP is expressed as mean pg/ganglion $\pm$ SEM (vertical bars).
  *Differs from day 1 control at p < .001
  **Differs from day 1 control at p < .001 and from day 15
    control at p < .01
(Reprinted with permission from Kessler and Black, <u>Proc. Natl. Acad. Sci.</u>, in press.)

ganglion (23,45). Although the factors mediating limb-DRG interactions are undefined, NGF may be involved, since the protein
enhances growth of the embryonic DRG (13,38), and is subjected to
retrograde transport from limb to ganglion (10). Using SP as an
index of sensory maturation, we have begun characterizing sensory
development in greater detail, and have found that limb-DRG
interactions parallel those already defined for target-sympathet-
ic regulation.

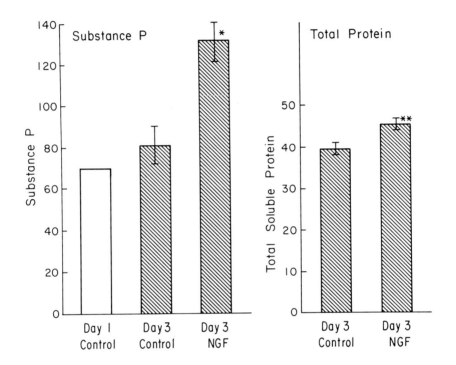

FIG. 2.   Effects of NGF on SP and total protein in the C_6DRG.
Neonates were injected subcutaneously with 100 µl of either NGF
(10^3 units) in saline (8 animals) or saline (8 animals).   Two
days later the C_6DRG was examined for SP content and total pro-
tein. SP is expressed as mean pg/ganglion $\pm$ SEM (vertical bars).
Protein is expressed as mean µg/ganglion $\pm$ SEM (vertical bars).
   *Differs from respective control at p < .001
   **Differs from respective control at p < .025
(Reprinted with permission of Kessler and Black, <u>Proc</u>. <u>Natl</u>.
<u>Acad</u>. <u>Sci</u>., in press).

Recent work suggested that SP might be an excellent index of DRG ontogeny. SP is an undecapeptide, which is heterogeneously distributed in the nervous system (30,32,43,46), and highly localized to certain perikarya in the DRG, with neurites in the spinal cord (28,29,43,46). This distribution, and recent physio-logic and biochemical experiments (28,29,36,41,46), strongly suggested that SP may be a sensory neurotransmitter (see Jessell *et al.*, this volume).

The sixth cervical (C_6) DRG was chosen for study, since this is a relatively large ganglion which innervates the rat forelimb, allowing convenient target manipulation. The normal postnatal development of SP was defined initially. The peptide increased more than 5-fold during the first 5 weeks of life from 50-70 pg/ganglion at birth (34). Total DRG protein increased only 3-fold during this period, rendering the change in specific SP content highly significant.

To begin to characterize the relationship between ganglion SP development and its field of innervation, unilateral forelimb amputation was performed in neonates. Two weeks thereafter SP was measured in the ipsilateral C_6DRG, the contralateral ganglion with a normal target complement, and in ganglia of sham-operated controls. In fact, limb amputation prevented the normal develop-ment of ipsilateral ganglion SP, whereas contralateral ganglion SP and that in sham-operated rats developed normally (Fig. 1).

What factor(s) mediate the apparent limb-ipsilateral DRG interaction? NGF was examined initially, since it has long been known that it alters DRG morphologic development in the embryo (10,13,38). In pilot studies we simply sought to determine whether the postnatal DRG can respond to NGF by treating newborn rats with the protein and examining SP 2 days later. NGF admin-istration resulted in nearly a 2-fold increase of C_6DRG SP, indi-cating that postnatal ganglia do respond to the protein (Fig. 2).

To determine whether NGF can prevent the effects of amputation on ganglion SP development, rats were treated with the factor subsequent to amputation. After unilateral forelimb extirpation, one group of rats received saline, while another group received NGF daily for 9 days. The four sets of ganglia were assayed for SP, the ganglia contralateral to surgery serving as controls in each group. As expected, amputation prevented the normal devel-opment of ipsilateral ganglion SP in saline-treated controls. However, NGF prevented the failure of SP development consequent to amputation. In the NGF-treated animals there was no signifi-cant difference between ipsilateral and contralateral ganglion SP (Fig. 3). Consequently, NGF blocked the effects of amputation on SP development and, in a sense, was capable of substituting for the limb.

A number of generalizations may be warranted. First, it is apparent that SP represents a specific and convenient index of DRG development, and that the peptide may be used to monitor normal and abnormal maturation. Consequently, development of SP in DRG may be analogous to development of transmitter enzymes in sympathetic ganglia. Nevertheless, it is useful to recall that

SP is probably not present in all DRG neurons (28), and most probably reflects maturation of a sub-population in the ganglion.

It is quite clear, however, that the DRG peptidergic neurons are critically dependent on the peripheral field of innervation for normal maturation. The peripheral target may regulate DRG maturation in a manner analogous to the role of targets, or afferent nerves, in sympathetic development. The parallel is

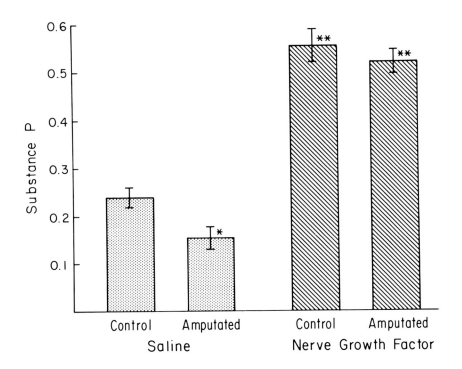

FIG. 3. Effects of combined amputation and NGF treatment. Unilateral forelimb amputation was performed in neonates (16 animals). Animals were then injected subcutaneously with 100 μl of either NGF (10^3 units) in saline (8 animals) or saline (8 animals) daily for 9 days. The C_6DRG on both the amputated side and the contralateral control side were examined for content of SP. SP is expressed as mean pg/ganglion ± SEM (vertical bars).

   *Differs from saline control at p<.01
   **Differs from saline control at p<.001
(Reprinted with permission from Kessler and Black, <u>Proc. Natl. Acad. Sci.</u>, in press).

somewhat complicated, however, since the DRG is an afferent structure, whereas the sympathetic ganglion is efferent. Consequently, sympathetic impulse flow is directed towards the target, while sensory impulses emanate from the target. These considerations notwithstanding, it is quite clear that the target may regulate both sympathetic and sensory development through the elaboration of diffusible signals. Moreover, the present work, viewed in conjunction with previous studies on postnatal sympathetic maturation (17,19,27), suggests that end organs may influence sensory and sympathetic maturation through the mediation of NGF in the postnatal period. Of course, it is entirely possible that targets also elaborate other factors, as described above. Our conclusions, however, must be regarded as tentative, since the studies cited do not indicate whether targets normally elaborate NGF, and whether sensory neurons are normally dependent on NGF during the postnatal period. We are presently evaluating the effects of anti-NGF to approach this question.

The apparent commonality of regulatory mechanisms in sympathetic and senory neurons may be attributable to their common origin from the embryonic neural crest (47). However, target regulation has been observed in a number of other peripheral and central systems (16,24,37,44), suggesting that this is a general neuronal phenomenon. We are presently characterizing additional processes in peptidergic ontogeny to determine whether some of the other mechanisms governing sympathetic development play a role in the sensory system as well.

REFERENCES

1. Black, I.B. (1978): Ann. Rev. Neurosci., 1:183-214.
2. Black, I.B. and Geen, S.C. (1973): Brain Res., 63:291-302.
3. Black, I.B. and Geen, S.C. (1974): J. Neurochem., 22:301-306.
4. Black, I.B., Bloom, E.M., and Hamill, R.W. (1976): Proc. Natl. Acad. Sci. USA, 73:3575-3578.
5. Black, I.B., Hendry, I.A., and Iversen, L.L. (1971): Brain Res., 34:229-240.
6. Black, I.B., Hendry, I.A., and Iversen,L.L. (1971): Nature, 231:27-29.
7. Black, I.B., Hendry, I.A., and Iversen, L.L. (1972): J. Physiol. Lond., 221:149-159.
8. Bloom, E.M., Coughlin,M.D., and Black, I.B. (1979): Soc. for Neurosci. Abstr.
9. Bloom, E.M., Hamill, R.W., and Black, I.B. (1976): Brain Res., 115:525-528.
10. Bunso-Bechtold, J. and Hamburger, V. (1979): Proc. Natl. Acad. Sci. USA, 76:1494-1496.
11. Cochard, P., Goldstein, M., and Black, I.B. (1978): Proc. Natl. Acad. Sci. USA, 75:2986-2990.
12. Cochard, P., Goldstein, M., and Black, I.B. (1979): Develop. Biol., 71:100-114.
13. Cohen, S. (1960): Proc. Natl. Acad. Sci. USA, 46:302-311.

14. Coughlin, M.D., Boyer, D.M., and Black, I.B. (1977): Proc. Natl. Acad. Sci. USA, 74:3438-3442.
15. Coughlin, M.D., Dibner, M.D., Boyer, D.M., and Black, I.B. (1978): Develop. Biol., 66:513-528.
16. Cowan, W.M. (1970): In Contemporary Research Methods in Neuroanatomy, edited by W.J.H. Nauta and S.O.E. Ebbeson, pp. 217-251. Springer, New York.
17. Dibner, M.D. and Black, I.B. (1976): Brain Res., 103:93-102.
18. Dibner, M.D., and Black,I.B. (1976): J. Neurochem., 27:323-324.
19. Dibner, M.D., Mytilineou, C., and Black, I.B. (1977): Brain Res., 123:301-310.
20. Falck, B., Hillarp, N.A., Thieme, G., and Torp, A. (1962): J. Histochem. Cytochem., 10:348-354.
21. Fonnum, F. (1970): Norw. Def. Res. Establ. Rep., 58.
22. Green, L.A. and Shooter, E.M.: Ann. Rev. Neurosci., (in press).
23. Hamburger, V (1934): J. Exp. Zool., 68:449-494.
24. Hamburger, V. (1975): J. Comp. Neurol., 160:535-546.
25. Hamill, R.W., Bloom, E.M., and Black, I.B. (1977): Brain Res., 134:269-278.
26. Hebb, C.O. and Waites, G.M.H. (1956): J. Physiol. Lond., 132:667-671.
27. Hendry, I.A. and Iversen, L.L. (1973): Nature, 243: 550-554.
28. Hokfelt, T., Kellerth, J.O., Nilsson, G., and Perrow, B. (1975): Science, 190:889-890.
29. Hokfelt, T., Kellerth, J.O., Nilsson, G., and Perrow, B. (1975): Brain Res., 100:235-252.
30. Hokfelt, T., Myerson, B., Nilsson, J., Perrow, B., and Sachs, C. (1976): Brain Res., 104:181-186.
31. Jonakait, G.M., Wolf, J., Cochard, P., Goldstein, M. and Black, I.B (1979): Proc. Natl. Acad. Sci. USA, 76:4683-4686.
32. Kanazawa, I. and Jessell, T. (1976): Brain Res., 117:362-367.
33. Kessler, J.A. and Black, I.B. (1979): Brain Res., 171:415-424.
34. Kessler, J.A. and Black, I.B. (1980): Proc. Natl. Acad. Sci. USA, 77:649-652.
35. Kessler, J.A., Cochard, P., and Black, I.B. (1979): Nature, 280:141-142.
36. Konishi, S. and Otsuka, M. (1974): Nature, 252:734-735.
37. Landmesser, L. and Pilar, G. (1974): J. Physiol. London, 241:715-736.
38. Levi-Montalcini, R., Myer, H., and Hamburger, V. (1954): Cancer Res., 14:49-57.
39. Levitt, M., Spector, S., Sjoerdsma, A.,and Udenfriend, S. (1965): J. Pharmacol. Exp. Ther., 148:1-8.
40. Olson, L. and Malmfors, T. (1970): Acta Physiol. Scand. Suppl., 348:1-111.
41. Otsuka, M. and Konishi, S. (1975): Cold Spring Harbor Symp. Quart. Biol., 40:135-143.

42. Patterson, P.H. (1978): <u>Ann. Rev. Neurosci.</u>, 1:1-17.
43. Perrow, B. (1953): <u>Acta Physiol. Scand. Suppl.</u>, 29:105:1-90.
44. Prestige, M.C. (1967): <u>J. Embryol. Exp. Morphol.</u>, 18:359-387.
45. Shorey, M.L. (1909): <u>J. Exp. Zool.</u>, 7:25-63.
46. Takahashi, T. and Otsuka, M. (1975): <u>Brain Res.</u>, 87:1-11.
47. Tennyson, V. (1965): <u>J. Comp. Neurol.</u>, 124:267-317.

Neurosecretion and Brain Peptides,
edited by J. B. Martin, S. Reichlin, and K. L. Bick.
Raven Press, New York © 1981.

Introduction to Section V:
Blood–Brain Barrier, Cerebrospinal Fluid, and Cerebral Blood Flow

Earl A. Zimmerman

*Department of Neurology, College of Physicians and Surgeons, Columbia University,
New York, New York 10032*

This section comprises four chapters dealing with the entry of peptides in the brain and cerebrospinal fluid (CSF). Poorly understood at the present, the subject is important to future diagnostic and therapeutic use of the neuropeptides. Studies of hypothalamic function based on measurements of peptides in circulating blood have been disappointing. Even when they are measurable, such values may not reveal what is happening in brain since we now know that many peptides, including somatostatin, thyrotropin releasing hormone (TRH), vasoactive intestinal peptide (VIP) and cholecystokinin, are synthesized in peripheral tissues as well as in brain. CSF assay is therefore being explored as a possible means of elucidating the role of peptides in brain in relation to normal and pathological alterations in man. However, CSF assay may not reflect focal disease because most peptides are produced in many different areas of the brain (affected differently in disease), and they may enter or be removed from the CSF at different sites. Despite the recent exciting reports that a number of peptides injected systemically into experimental animals influence memory, sexual activity or other functions, their entry into brain in "significant" amounts has not been established with certainty. How parenterally administered peptides might enter the brain, what amounts are "significant", and how they act are basic issues about which little is currently known. The chapters bring us up to date on traditional thinking about these issues, and offer some new directions for further research.

Weindl and Sofroniew discuss in detail the circumventricular organs. Most have fenestrated capillaries lacking a blood-brain barrier that could permit passage of neuropeptides from brain into the general circulation: for example, neurohypophysial peptides from the neurohypophysis and hypophysiotropic hormones from the median eminence. A number of different peptidergic fibers end on fenestrated capillaries in the organum vasculosum of the lamina terminalis (OVLT), but their function is unknown. The relatively larger numbers of gonadotropin releasing hormone (GnRH)-containing fibers to OVLT has attracted considerable attention since this structure is located near the preoptic area where it may regulate ovulation, at least in subprimate species. The authors doubt that GnRH transported by venous drainage from OVLT to preoptic neurons is important since capillaries at the latter site are not fenestrated. Jackson points out in his chapter that OVLT drains to the systemic circulation, but definite roles for GnRH in the periphery at levels encountered are not known. He also suggests that GnRH-containing nerve terminals may secrete into the third ventricle at this site. An earlier ultrastructural study by David Scott and colleagues (4) showed numerous free nerve endings penetrating the floor of the third ventricle containing granules which could possibly contain peptides or catecholamines. By light microscopy, our own immunocytochemical studies of a number of peptides often reveals processes, particularly in the 80 mm-thick sections, which appear to traverse the third ventricular wall. My own expectation is that further immunoelectron microscopic and physiologic studies of CSF will reveal intraventricular secretion to be a common phenomenon.

The circumventricular organs also permit entry of substances into brain at specific sites that may be important for sensing the peripheral environment. One example is angiotensin II that is involved in regulation of drinking behavior. Peptidergic innervation of the areas near the area postrema -- particularly by vasopressin and oxytocin fibers to the solitary and dorsal motor vagal nuclei -- should be emphasized. As discussed later by Raichle, these peptides and angiotensin II may interact at this site in the floor of the fourth ventricle to modify brain capillary permeability and to regulate blood pressure.

A clear discussion of the blood-CSF and blood-brain barriers is provided by Pardridge and co-authors. They point out that the surface area of the blood-brain barrier is 5000 times greater than the blood-CSF barrier. Rapid distribution of cirulating peptides would require lipid solubility or transport-mediation for which there is little evidence. Earlier reports of rapid entry of radiolabeled enkephalin into brain were not confirmed. Like enkephalin, permeability of TRH is also low. Molecules with a relatively high plasma/CSF ratio (compared to inulin) are transported by a blood-CSF barrier which is selective and limited. Some substances demonstrated to be transported by this route are: prealbumin, prolactin, α-fetoprotein, some immunoglobulins, and possibly insulin. The insulin story is complicated because it may also be made in brain. Furthermore, the authors suggest that saturable specific receptors for insulin are present in brain endothelial cells. They proposed a new interesting mechanism whereby circulating peptides, i.e., insulin, may modulate brain without entering it, by acting on receptors on the endothelial cells forming the blood-brain barrier. Insulin may be transported into nervous tissue by additional mechanisms. van Houten and co-authors (5) found radiolabeled insulin in neurons of the ventromedial nucleus of the hypothalamus and the dorsal motor vagal nucleus in the medulla. Although they suggest that vasopressin might be acting through a similar mechanism in altering the blood-brain barrier, in the next chapter Raichle suggests that vasopressin may act indirectly in this regard through norepinephrine pathways.

Raichle presents an important new idea that hormonal factors may be capable of modifying the blood-brain barrier. Vasopressin and angiotensin II increase capillary permeability but, unlike norepinephrine, do not stimulate endothelial cell cyclic AMP. Raichle reviews the compelling evidence that norepinephrine originating in brainstem innervates brain capillaries and suggests that vasopressin pathways originating in hypothalamus may act on this system by connections reviewed in this volume (see Zimmerman, this volume). Angiotensin II may, in turn, act on the vasopressin system. Vasopressin fibers originating in hypothalamus also appear to innervate choroid plexus, and the hormone may regulate CSF production at this site. Finally, Raichle extends the thought to regulation of brain volume by speculating that vasopressin might act on astroglia via a norepinephrine-sensitive (Na^+-K^+) ATPase. One wonders how the homozygous Brattleboro rat, which cannot produce vasopressin, survives at all! Still, it seems likely that at least the first two parts of the vasopressin hypothesis are likely to be important from the data at hand.

Yet another dimension concerning peptides and blood vessels are the reports of nerve fibers containing VIP on cerebral arteries mentioned in the chapter by Jackson and of oxytocin found in our laboratory (Nilaver and Zimmerman, unpublished) and by Swanson. These and other peptides may directly regulate cerebral blood flow.

Jackson extensively reviews previous hypotheses concerning the role of peptides in CSF and recent efforts to measure them in human CSF by radioimmunoassay. The "alternate route" hypothesis whereby brain peptides active in the anterior pituitary are first secreted into third ventricular CSF and transported by tanycytes to hypophysial portal blood remains plausible, but unproven, after many years of hard work. All the data are either circumstantial or negative. There is no question that injected radiolabeled peptides are transported from third ventricle to portal blood, and relatively small amouonts of GnRH or TRH given by that route release anterior pituitary hormones. But do they normally so act? Reports of GnRH in tanycytes by light microscopic immunocytochemistry from my laboratory and others were not subsequently substantiated or interpreted to be GnRH-containing nerve endings by electron microscopy. Jackson also reviews the recent negative data of Cramer and Barraclough (2) that third ventricular GnRH did not rise after electrical stimulation of the medial preoptic area despite a marked rise in serum luteinizing hormone (LH) in rats. It is also not known if anterior pituitary hormones are taken back into CSF from portal blood by tanycytes. The source of anterior pituitary hormones in CSF is also uncertain. In the case of ACTH the story is even more complicated since it is now known that neurons of the arcuate nucleus produce this peptide. To my knowledge, CSF ACTH has not been measured after total hypophysectomy, but it is clear that it remains in brains of experimental animals after the operation. The contention that all brain ACTH comes from the pituitary is untenable, although it is still possible that some may yet arise from this site. The idea that the pituitary secretes to the brain put forth by Bergland and Page (1) based on important new anatomical studies and supported by evidence of possible reverse flow in the hypophysial portal system by Oliver et al. (3) remains hard to prove. There is some evidence that the brain, as well as the pituitary, may also produce LH and growth hormone (GH). Very high concentrations of anterior pituitary hormones found in the CSF of patients with hypersecretory tumors and suprasellar extension are probably due to disruption of the "pituitary-CSF barrier". Improved CT scans have generally obviated the usefulness of CSF assay of ACTH, GH or prolactin in patients with hypersecreting pituitary tumors. Jackson points out that CSF prolactin may be elevated without suprasellar extension, presumably due to transfer of the hormone from the system circulation.

Data concerning the hypothesis that brain peptides produced in one area of the brain affect another region via a CSF pathway is also reviewed. Currently available evidence is insufficient to determine whether it is an important, albeit relatively slow, means of communication within the CNS.

Most peptides that have been assessed in human CSF are detectable by radioimmunoassay. An exception in both humans and monkeys is GnRH. Reports of angiotensin II levels are also suspect in view of recent evidence of non-specific interfering substances. There is a need for better biochemcial characterization of peptides in CSF. Recent reports of changes in CSF peptide levels in certain neurological diseases as discussed by Jackson raise the exciting possibility that assay of neuropeptides in the CSF may have clinical applicability. These early reports justify the need for further clinical studies.

REFERENCES

1. Bergland, R.M., and Page, R.B. (1979): Science, 204: 18-24.
2. Cramer, O.M., and Barraclough, C.A. (1975): Endocrinology, 96: 913-921.
3. Oliver, C., Mical, R.S., and Porter, J.C. (1977): Endocrinology, 101: 598-604.
4. Scott, D.E., Kozlowski, G.P., and Sheridan, M.D. (1974): Int. Rev. Cytol., 37: 349-388.
5. van Houten, M., Posner, B.I., Kopriwa, B.M., and Brawer, J.R. (1979): Endocrinology, 105: 666-673.

Neurosecretion and Brain Peptides,
edited by J. B. Martin, S. Reichlin, and K. L. Bick.
Raven Press, New York © 1981.

Relation of Neuropeptides to Mammalian Circumventricular Organs

A. Weindl and *M. V. Sofroniew

*Department of Neurology, Technical University, and *Department of Anatomy, Ludwig-Maximilians University, 8000 Munich, West Germany*

The circumventricular organs (CVO) of the mammalian brain are areas of specialized tissue located at strategic positions of the midline ventricular system (Fig. 1). With the exception of the subcommissural organ, they are highly vascularized, containing fenestrated capillary loops surrounded by perivascular connective tissue spaces and are permeable to proteins and peptides (45,48). This lack of the blood-brain barrier, as demonstrated by increased permeability to horseradish peroxidase, is essential for the neurohemal secretion of peptide neurohormones from the neural lobe, median eminence and the organum vasculosum of the lamina terminalis (OVLT), as well as for the secretion of indolalkylamines from the pineal. The lack of the blood-brain barrier may be equally important for hemo-neural interactions. Certain constituents of the blood, including peptides, which have no access to the central nervous system elsewhere, may thus be sensed by receptor neurons located in or abutting a CVO. This may be of importance in the subfornical organ, area postrema and OVLT.

Vascular injection studies were carried out as previously described (46). The endogeneous neuropeptides vasopressin, oxytocin, neurophysin, somatostatin, luteinizing hormone release hormone (LHRH), β-endorphin and adrenocorticotrophin (ACTH) were localized in the brains of rhesus monkey, squirrel monkey, tree shrew, guinea pig, rat, and mouse, using the unlabelled antibody-enzyme peroxidase-antiperoxidase method of Sternberger (40). Our modifications of this procedure, the production of the antisera used, and the tests conducted to verify the specificity of the staining obtained have been described elsewhere (34,37).

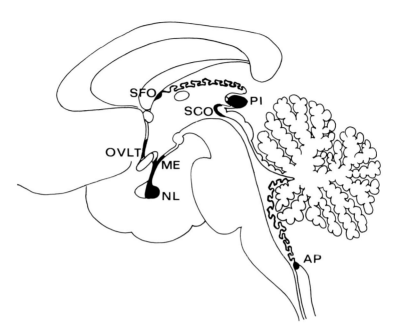

FIG. 1. Drawing of the mid-sagittally sectioned human brain. Outlined in black are the circumventricular organs: AP, area postrema; ME, median eminence; NL, neural lobe; OVLT, organum vasculosum of the lamina terminalis; PI, pineal body; SCO, subcommissural organ; SFO, subfornical organ.

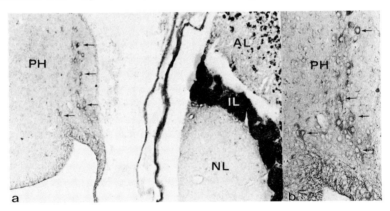

FIG. 2. Rat. Horizontal section through the pituitary and posterior hypothalamus (PH). β-endorphin immunoperoxidase reaction. The neural lobe (NL) does not contain β-endorphin immunoreactivity in contrast to the surrounding intermediate (IL) and anterior lobe (AL). Immunoreactive neurons (←) in the PH appear to send no fibers to the NL. a. Survey 64x. b. Detail of a. 120x.

NEUROHYPOPHYSIS

Neural Lobe

Fibers of magnocellular neurons of the supraoptic and paraventricular nuclei project through the internal zone of the
median eminence to the permeable capillaries of the neural lobe,
where terminals containing neurophysin, vasopressin or oxytocin
are found around the fenestrated vessels (12). The neural lobe
is the organ of storage and release of vasopressin, oxytocin and
their related neurophysins, and contains the highest quantities
and concentrations of a neuropeptide within the nervous system.
This immense reservoir of neurohormones is vital for certain
regulatory mechanisms such as antidiuresis and blood pressure
control as well as for milk-ejection and uterine contraction. On
the other hand, the location of the neural lobe "outside the
brain" may also be a means of protecting central neurons from an
undirected overflow of these peptides. Morphologic studies show
an extensive network of projections from hypothalamic vasopressin
and oxytocin neurons to various levels of the central nervous
system (9,36). Central administration of these peptides appears
to have behavioral effects (11), as well.

Somatostatin immunoreactivity has also been shown in fibers in
the neural lobe of a number of species including rat (18), oppo-
sum (51), tree shrew, rhesus monkey and human (7). The function
of these immunoreactive somatostatin and of LRH fibers (18) in
the neural lobe is still unclear.

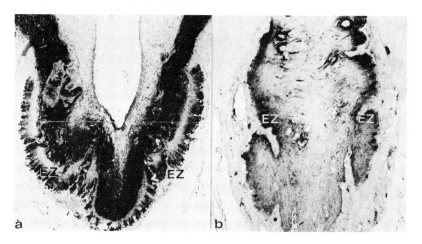

FIG. 3. Rhesus monkey. Infundibulum. Frontal sections at
different levels. Immunoperoxidase reaction for neurophysin (a)
and somatostatin (b). a. Branches of neurophysin fibers of the
hypothalamo-hypophyseal tract terminate in large numbers at
portal vessels of the external zone (EZ). 32x. b. Somatostatin
fibers contact portal vessels in the dorso-lateral EZ. 42x.

Although cells producing both β-endorphin and ACTH (23,44) are present in the intermediate and anterior lobes surrounding the neural lobe, we have at present found no β-endorphin or ACTH fibers in the neural lobe (Fig. 2a). Neurons producing both β-endorphin and ACTH (6,33) are present in the medio-basal hypothalamus (Fig. 2). Fibers from the β-endorphin/ACTH neurons project rostrally to the anterior hypothalamic-preoptic area, including the lamina terminalis, bend around the anterior commissure to the dorsal thalamus, and project caudad to the mesencephalic central grey (3,43).

Median Eminence

The median eminence or infundibulum forms the proximal part of the neurohypophysis. It is commonly divided into an internal and external zone. Haymaker (17) reports that this division does not exist in the primate; however, immunohistochemical staining clearly shows both an internal and external zone in the rhesus monkey infundibulum (Fig. 3). The median eminence of many species presents a mosaic distribution of peptide hormones and amines. In the internal zone, vasopressin, oxytocin and neurophysin fibers (Fig. 4) pass to the neural lobe. A number of vasopressin and neurophysin fibers branch off from the internal zone and terminate at portal vessels of the external zone (Fig. 4a). Their increase after adrenalectomy, and the suppression of this increase following glucocorticoid, but not mineralocorticoid substitution (38,41), supports observations that vasopressin may be involved in ACTH release (53).

In the tree shrew and rhesus monkey, somatostatin fibers contacting portal vessels are distributed in a more central division (Figs. 3,4) of the external zone. A similar distribution is found in the rat (18). Thyrotropin releasing hormone (TRH) terminals appear to be located primarily in the medial part of the rat median eminence (18). Noteworthy is also the relationship of amines in the median eminence to this regional distribution of the various peptides. The juxtaposition of dopamine and LHRH terminals in the lateral parts of the median eminence suggests a functional interaction between dopamine and LHRH release, and appears to support the findings of an inhibitory effect of dopamine on gonadotropin secretion (18). Fibers immunoreactive with enkephalin (18), but not with β-endorphin/ACTH antiserum are present in the external zone. The role of enkephalin and other peptides such as Substance P and angiotensin II (18) in the median eminence has not been clarified. The proposition of an alternate route of hormone transport from the cerebrospinal fluid to the portal blood (21) is not supported by immunohistochemical findings. Early reports of neurophysin and LHRH immunoreactivity within tanycyte processes projecting from the ventricular to the portal surface could not be confirmed in various laboratories, and is regarded as non-specific staining. Evidence of hormone transport

either from the ventricle to the portal vessels or in the opposite direction is lacking. A hypothesis of retrograde flow in the
portal vessels has been given (2). This postulate has been used
to explain the presence of radioimmunoassayable ACTH in the
medio-basal hypothalamus and other periventricular regions. However, ACTH immunoreactivity is still present within neurons of

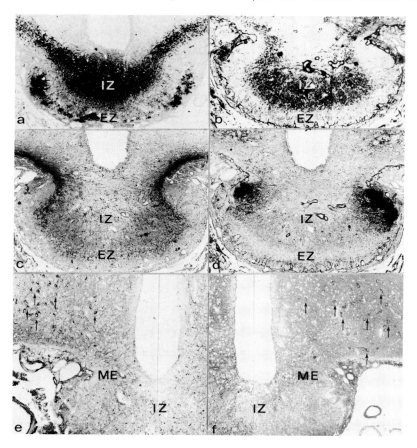

FIG. 4a-f.   Tree shrew. Median eminence (ME). Frontal sections.
Immunoperoxidase reaction for vasopressin (a), oxytocin (b), LHRH
(c), somatostatin (d), β-endorphin (e) and ACTH (f).   The peptides are regionally distributed.   a,b.   Vasopressin (a) and
oxytocin (b) fibers are located in the internal zone (IZ).
Branches of vasopressin, but not oxytocin, fibers terminate at
portal capillaries of the external zone (EZ).   c,d.   LHRH fibers
(c) terminate more laterally, somatostatin fibers (d)more
medially in the EZ. e,f.   Fibers of β-endorphin (e) and ACTH (f)
neurons (↑) in the infundibular nucleus do not terminate in the
EZ.   a-f.

the medio-basal hypothalamus after hypophysectomy (43).   The
existence of significant retrograde flow under physiologic circumstances has yet to be shown.

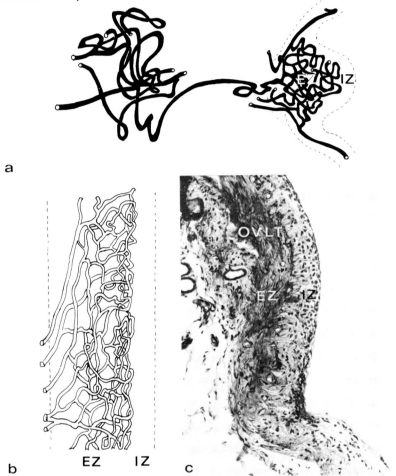

FIG. 5 a-c. Squirrel monkey.   Organum vasculosum of the lamina
terminalis (OVLT).   a,b.   Drawings of the vascularization.
Microfil-injection. Thick horizontal (a) and sagittal (b) sections. A tortuous intrapial plexus supplies the complex capillary
network of the cone-shaped vascular or external zone (EZ).   The
plexus of the EZ consists of a dense network of convoluted capillary loops. a,b. 130x.   c.   Sagittal paraffin section.   PAS-
hematoxylin staining.   The EZ contains blood vessels and connective tissue, and can clearly be differentiated from the parenchyma of the internal zone (IZ) which is covered by non-ciliated
ependyma. 140x.

ORGANUM VASCULOSUM OF THE LAMINA TERMINALIS

Although the OVLT resembles the median eminence closely (45), its function has not yet been clarified. The blood vessels form an internal and external network of capillary loops, and the organ can be divided into an external vascular and internal parenchymal zone (Fig. 5). The parenchymal zone is covered by oligociliated ependymal cells with basal processes (tanycytes). Numerous neuronal processes containing dense core vesicles are present in the internal zone, and in contact with the fenestrated capillaries of the external zone (45).

Small caliber vasopressin and neurophysin fibers originating from paravocellular neurons in the suprachiasmatic nucleus, pass through the internal zone of the organ (Fig. 6c,d) en route to target neurons present in the nucleus of the diagonal band of the lateral septum (35). Most of these fibers appear to pass through the OVLT. Very few fibers terminate at capillaries in the external zone (Fig. 6c,d). Thus our findings do not show a projection from the suprachiasmatic nucleus to the OVLT as described by others (9). These fibers are not present in the Brattleboro rat (Fig. 6e), which lacks vasopressin due to a genetic defect. The presence of radioimmunoassayable vasopressin in the dissected OVLT and surrounding area (26) is not sufficient evidence that it is released there, particularly since morphological evidence indicates that these fibers continue elsewhere.

LHRH fibers originating from perikarya in the preoptic/anterior hypothalamic and precommissural region terminate in large numbers at vessels of the OVLT in the rat (Fig. 6a). In the guinea pig, only a part of the fibers in this region appear to terminate at the OVLT vessels, and more laterally running fibers seem to contact neuronal perikarya in the lamina terminalis. Several experimental studies showing alterations of the LHRH content during the menstrual cycle (32), or morphological changes after castration or hypophysectomy (52), were contradicted by others (16,22), and have helped little to clarify the role of release of LHRH. A vascular transport of LHRH from the OVLT to the perikarya of origin in the median preoptic region proposed by Palkovits et al. (27) is not likely, because fenestrated capillaries are not present in the preoptic region for the reentrance of peptides into neural tissue.

Somatostatin fibers possibly deriving from periventricular perikarya of the anterior hypothalamic nucleus also terminate at blood vessels of the rat OVLT (Fig. 6b). In the rat, a number of β-endorphin/ACTH fibers also pass through the preoptic/anterior hypothalamic area, and a few appear to be in contact with the dorsal part of the OVLT. In the guinea pig, magnocellular neurosecretory perikarya containing primarily oxytocin and neurophysin surround the OVLT (Fig. 7). Dendrites of these perikarya appear to enter the OVLT (Fig. 7e). Transmitters such as serotonin and GABA (5) present in fibers of the OVLT may interact with peptides.

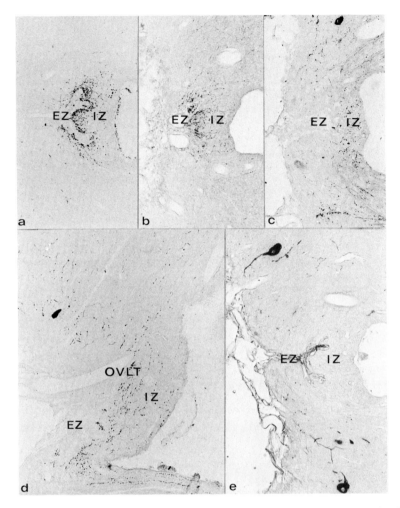

FIG. 6 a-e.   Rat.   Organum vasculosum of the lamina terminalis
(OVLT). Horizontal (a-c,e) and sagittal (d) sections.   Immunoperoxidase reaction for LHRH (a), somatostatin (b), and neurophysin (c-e).   While LHRH (a) and somatostatin (b) fibers
terminate at vessels of the external zone (EZ), small caliber
neurophysin fibers pass ventro-dorsally through the internal zone
(IZ).   e.   These fibers are absent in the Brattleboro rat.   a-e.
120x.

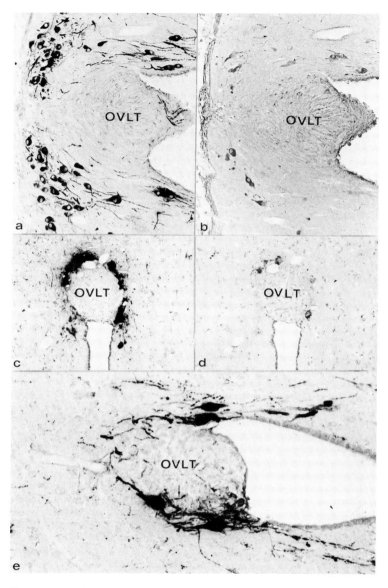

FIG. 7 a-e. Guinea pig. Organum vasculosum of the lamina terminalis (OVLT). Horizontal (a,b,e) and frontal (c,d) sections.
Immunoperoxidase reaction for neurophysin (a,c,e), vasopressin
(b) and oxytocin (d). a,b. Magnocellular neurophysin and some
vasopressin neurons are located in the vicinity of the ventral
part of the OVLT. 100x. c,d. More dorsally, neurophysin and
oxytocin neurons encircle the OVLT. 120x. e. Dendritic processes of neurophysin neurons seem to extend into the OVLT (↓).
256x.

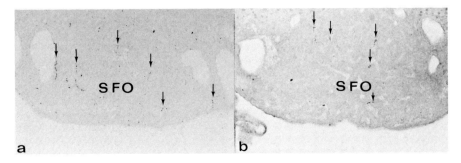

FIG. 8 a,b. Rat. Subfornical organ (SFO). Frontal sections. Immunoperoxidase reaction for neurophysin (a) and vasopressin (b). Few fibers (↓) are located in the SFO. 80x.

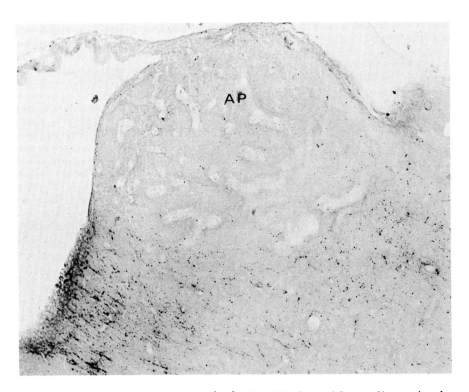

FIG. 9. Rat. Area postrema (AP). Sagittal section. Neurophysin immunoperoxidase reaction. Neurophysin fibers are concentrated in the nucleus of the solitary tract bordering the AP, but not entering it. 160x.

The OVLT may function not only as a neuro-hemal outlet of hypothalamic peptides into the bloodstream. The intimate relationship of neurophysin and mainly oxytocin producing magnocellrular perikarya to the OVLT, suggests that these neurons and their dendritic processes are exposed to constituents of the blood entering the tissue of the OVLT, but not having access to neurons in areas protected by the blood-brain barrier. This structural relationship may be of importance for a hemo-neural function whereby certain peptides, proteins, amines or other constituents of the blood are sensed by neurons having specific receptor properties. Electrophysiological observations (14) indicate that neurons sensitive to angiotensin II are located in the vicinity of the OVLT in the rat. This supports the observation that lesions of the anterior wall of the third venticle including the OVLT abolish the dipsogenic effect of this peptide (19).

SUBFORNICAL ORGAN

The ventricular surface of the subfornical organ (SFO) is covered by non- or oligociliated ependymal cells. Numerous small neurons (parenchymal cells) are present in the SFO. Often the endoplasmic reticulum of these cells undergoes vacuolization to form giant vacuoles, whose role is unclear. No neurosecretory granules are found within parenchymal cells (10). While no hypothalamic hormones are present in perikarya within the SFO, scattered single vasopressin, oxytocin and neurophysin fibers

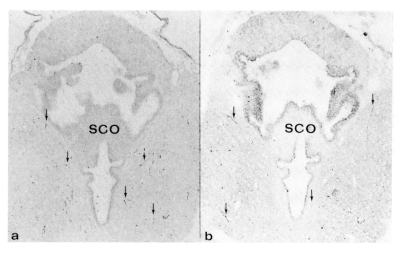

FIG. 10 a,b. Mouse. Subcommissural organ (SCO). Frontal sections. Immunoperoxidase reaction for neurophysin(a) and β-endorphin (b). Rostro-caudally directed neurophysin or β -endorphin fibers (↓) in the central grey do not contact secretory cells of the SCO. 80x.

(Fig. 8) as well as occasional LHRH (51) and somatostatin (50) fibers were found. We have not observed β-endorphin/ACTH fibers entering the SFO. Several studies including electrophysiological data suggest that the SFO may have neurons with receptor properties for the dipsogenic effect of angiotensin II (29). The location of the SFO between the interventricular foramina, its connection with the choroid plexus, and its vascular permeability may be important for a function in the regulation of body fluids.

AREA POSTREMA

The area postrema (AP) located at the transition from the fourth ventricle to the central canal has a structure similar to the SFO (45). However, vacuolated cells are not found. Although the AP is immediately surrounded by very dense fields of neurophysin, oxytocin and some vasopressin fibers teminating in the nucleus of the solitary tract and dorsal nucleus of the vagus (36), no fibers containing neurohypophyseal peptides could be

TABLE 1. Relation of neuropeptides to CVOs of the rat

		AVP	OT	ST	LHRH	β-END/ACTH
Neural lobe	a	++++	++++	+	•	−
median eminence						
internal zone	a	++++	++++	•	•	−
external zone	b	++	•	++++	++++	−
OVLT	a	•	•	+	++	•
	b	++	•	+	++	++
Subfornical	a	•	•	•	•	−
organ	b	•	•	•	•	−
area postrema	a	−	−	−	−	−
	b	+	++	+	−	−
pineal	a	•	•	−	−	−
subcommissural	a	−	−	−	−	−
organ	b	+	•	−	•	•

a fibers terminating at capillaries of CVO
b fibers close to, but not in CVO
AVP, vasopressin; β-end, β-endorphin; OT, oxytocin ST, somatostatin
• occasional presence of a single fiber
+ presence of a number of fibers
− no fibers present

observed within the AP. Furthermore, neither LHRH, somatostatin,
nor β-endorphin/ACTH fibers have been found within the AP, although somatostatin fibers also form a dense field of axon terminals in the region around the AP (49). This field includes the
nucleus of the solitary tract. Catecholamine fibers (15) appear
to play a role in the neural activity of the AP.

Various receptor functions have been proposed for the AP,
e.g., as a chemo-trigger zone for circulating emetic substances
in the vomiting reflex (4), or for the pressor effect of circu-

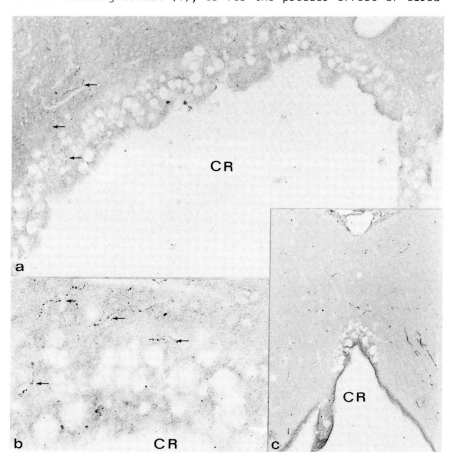

FIG. 11 a-c.   Rat.   Collicular recess (CR). Sagittal (a,b) and
frontal (c) sections. Immunoperoxidase reaction for β-endorphin
(a,b) and neurophysin (c). Some β-endorphin fibers (←) seem to
pass between the subependymal vacuoles of the CR (a,b).   In
contrast, neurophysin fibers in this area are not in contact with
the vacuoles (c). a. Survey 132x. b. Detail of a.   320x.   c.
80x.

lating angiotensin II (20).   Morest (25) has reported that
neurons of the nucleus of the solitary tract have collaterals
connected with the AP. The immediate proximity of the AP to this
nucleus which contains neurons regulating cardiovascular reflexes
and blood pressure should be considered in future experiments
aimed at elucidating the function of the AP.

PINEAL BODY

The pineal body contains specialized neurons with secretory
activity (pinealocytes) and supporting cells. Melatonin is produced under the influence of light deprivation (45). Pavel (28)
has proposed that arginine-vasotocin is produced by pinealocytes
as the pineal peptide. We have previously reported that no
pineal cells contain immunoreactive arginine-vasotocin and
neurophysin (50). However, occasionally a neurophysin fiber was
observed in the ventral part of the pineal of the guinea pig
(51), and single vasopressin and oxytocin fibers in the rat (8).
Radioimmunoassay measurements revealed no arginine-vasotocin, but
small quantities of vasopressin and oxytocin within the pineal
(13).

SUBCOMMISURAL ORGAN

The subcommissural organ (SCO), although located in the immediate vicinity of the pineal (48), has no functional connections
with it. The secretory ependymal cells discharge a mucopolysach-
aride into the ventricular fluid, which is transformed into
single filaments which converge to form Reissner's fiber (31,47).
This fiber can be traced through the aqueduct and fourth ventricle to the caudal end of the central canal (39).   Although
secretory material of the SCO and Reissner's fiber can be stained
due to their large number of disulfide bridges reacting with
neurosecretory stains, they are not immunoreactive with antisera
to hypothalamic hormones. Rostro-caudally directed neurophysin/
vasopressin (Fig. 10a) as well as β-endorphin/ACTH fibers (Fig.
10b) in the central canal are found in the vicinity, but not in
contact with SCO cells. The report of radioimmunoassayable
arginine-vasotocin in the SCO (30) is not supported by immunohistochemical observations (50).   The SCO is innervated by
serotonin fibers (24). Scanning electron microscopy revealed that
particles and cells of the cerebrospinal fluid attach to the
surface of the Reissner's fiber (47).
Table 1 provides a summary of the relation of several
neuropeptides to the CVO's of the rat.

COLLICULAR RECESS

The cerebral aqueduct of the mesencephalon widens to form a
dorsal recess between the inferior colliculi, the collicular
recess. The walls of the collicular recess consist of ependymal

folds and subependymal vacuoles. In the rat, some neurophysin fibers are found in the vicinity of the vacuoles (Fig. 11c), and a few β-endorphin/ACTH fibers intermingle with the vacuolated cells (Fig. 11a,b). The function of these vacuoles remains unclear. The lack of a special vascularization and of features similar to other CVOs, makes it unlikely that the "collicular recess organ" (42) has a function comparable to that of other CVOs.

SUMMARY

In summary, highly vascularized CVOs of the mammalian brain are the site of increased vascular permeability for peptides and other molecules which generally do not cross the blood-brain barrier. In the CVOs the blood-brain barrier is shifted from the level of the capillaries to the tight junctions of the oligociliated ependymal cells. The neurohypophysis is the well known target of various peptidergic neuroendocrine neurons. In the neural lobe, peptide hormones from magnocellular neurons are stored and released into the general circulation in the median eminence, releasing and inhibiting hormones enter the hypothalamo-adenohypophyseal portal circulation. The OVLT appears to be an additional vascular outlet for LHRH and somatostatin. In the pineal, no pinealocytes stain positively for arginine-vasotocin; however, occasionally a single neurophysin (vasopressin or oxytocin) fiber has been observed. In the subfornical organ and area postrema which do not appear to have a primary neuroendocrine function, hemo-neural interactions may be important for effects of circulating peptides and other molecules on specific receptors. In the subcommissural organ, which does not have a special vascular permeabilty, ependymal cells secrete Reissner's fiber, a mucopolysaccharide, whose function in unclear.

ACKNOWLEDGEMENTS

The authors thank R. Kopp-Eckmann, I. Wild, H. Asam, A. Nekic, for technical support and P. Campbell for editorial assistance. This work was supported by DFG grant We 608/6.

REFERENCES

1. Barker, J.L. (1976): Physiol. Rev., 56:435-452.
2. Bergland, R.M., and Page, R.B. (1979): Science, 204:18-24.
3. Bloom, F., Battenberg, E., Rossier, J., Ling, N., and Guillemin, R. (1978): Proc. Natl. Acad. Sci. USA, 75:1591-1595.
4. Borison, H.L., and Brizzee, K.R. (1951): Proc. Soc. Exp. Biol. Med., 77:38-42.
5. Bosler, O. (1977): Cell Tiss. Res., 182:383-399.
6. Bugnon, C., Bloch, B., Lenys, D., and Fellmann, D. (1979): Cell Tiss. Res., 199:177-196.

7. Bugnon, C., Fellmann, D., and Bloch, B.(1977): <u>Cell Tiss.</u> <u>Res.</u>, 183:319-328.
8. Buijs, R.M., and Pevet, P. (1980): <u>Cell Tiss Res.</u>, 205: 11-17.
9. Buijs, R.M., Swaab, D.F., Dogterom, J., and Van Leeuwen, F.W. (1978): <u>Cell Tiss. Res.</u>, 186:423-433.
10. Dellmann, H.D., and Simpson, J. (1979): In: <u>International</u> <u>Review of Cytology</u>, Vol. 58, edited by G.H. Bourne, and J.F. <u>Danielli</u>, pp. 333-421. Academic Press, New York.
11. De Wied, D. (1976): <u>Life Sci.</u>, 19:685-690.
12. Dierickx, K. (1980): <u>In: International Review of Cytology</u>, Vol. 62, edited by G.H. Bourne, and J.F. Danielli, pp. 119-185. Academic Press, New York.
13. Dogterom, J., Pevet, P., Buijs, R.M., Snijdewint, F.G.M., and Swaab, D.F. (1979): <u>Acta Endocrinol, Suppl.</u> 225:413.
14. Felix, D., and Phillips, M.I., (1979): <u>Brain Res.</u>, 169:204-208.
15. Fuxe, K., and Owman, C.(1965): <u>J. Comp. Neurol.</u>, 125:337-354.
16. Gautron, J.P., Pattou, E., and Kordon, C. (1977): <u>Molec.</u> <u>Cellul. Endocrinol.</u>, 8:81-92.
17. Haymaker, W. (1969): In: <u>The Hypothalamus</u>, edited by W. Haymaker, E. Anderson, and W.J.H. Nauta, pp. 219-250. Charles C. Thomas, Springfield.
18. Hokfelt, T., Elde, R., Fuxe, K., Johansson, O., Ljungdahl, A., Goldstein, M., Luft, R., Efendic, S., Nilsson, G., Terenius, L., Ganten, D.,Jeffcoate, S.L., Rehfeld, J., Said, S., Perez de la Mora, M., Possani, L., Tapia, R., Teran, L., and Palacios, R. (1978): In: <u>The Hypothalamus</u>, edited by S. Reichlin, R.J. Baldessarini, and J.B. Martin, pp. 69-135. Raven Press, New York.
19. Johnson, A.K. and Buggy, J. (1978): <u>Am. J. Physiol.</u>, 234: R122-R129.
20. Joy, M.D., and Lowe, R.D.(1970): <u>Nature</u>, 228:1303-1304.
21. Knigge, K.M., and Scott, D.E. (1970): <u>Am. J. Anat.</u>, 129: 223-244.
22. Kobayashi, R.M., Lu, K.H., Moore, R.Y., and Yen, S.S.C. (1978): <u>Endocrinology</u>, 102:98-105.
23. Mains, R.E., Eipper, B.A., and Ling, N. (1977): <u>Proc. Natl.</u> <u>Acad. Sci. USA.</u>, 74:3014-3018.
24. Mollgard, K., and Wiklund, L. (1979): <u>J. Neurocytol.</u>, 8: 445-467.
25. Morest, D.K. (1967): <u>J. Comp. Neurol.</u>, 130:277-300.
26. Negro-Vilar, A., and Samson, W.K. (1979): <u>Brain Res.</u>, 169: 585-589.
27. Palkovits, M., Mezey, E., Ambach, G., and Kivovics, P. (1978): In: <u>Brain-Endocrine Interaction III. Neural Hormones</u> <u>and Reproduction</u>, edited by D.E. Scott, G.P. Kozlowski, and <u>A. Weindl</u>, pp. 302-312. Karger, Basel.
28. Pavel, S. (1971): <u>Endocrinology</u>, 89:613-614.
29. Phillips, M.I. (1978): <u>Neuroendocrinology</u>, 25:354-377.

30. Rosenbloom, A.A., and Fisher, D.A. (1975): Endocrinology, 96: 1038-1039.
31. Reissner, E. (1860): Arch. Anat. Physiol., 545-588.
32. Setalo, G., Vigh, S., Schally, A.V., Arimura, A., and Flerko, B. (1976): Acta Biol. Acad. Sci. Hung., 27:75-77.
33. Sofroniew, M.V. (1979): Am. J. Anat., 154:283-289.
34. Sofroniew, M.V.,Madler, M., Muller, O.A., and Scriba, P.C. (1978): Fresenius Z. Anal. Chem., 290:163.
35. Sofroniew, M.V., and Weindl, A. (1978): Am. J. Anat., 153: 391-430.
36. Sofroniew, M.V., and Weindl, A. (1981): In: Endogenous Peptides and Learning and Memory Processes, edited by J.L. Martinez, R.A. Jensen, R.B. Messing, H. Rigter, and J.L. McGaugh. Academic Press, New York. (In press).
37. Sofroniew, M.V., Weindl, A., Schinko, I., and Wetzstein, R. (1979): Cell Tiss. Res., 196:367-384.
38. Sofroniew, M.V., Weindl, A., and Wetzstein, R. (1977): Acta Endocrinol. Suppl. 212:72.
39. Sterba, G. (1969): In: Zirkumventrikulare Organe und Liquor, edited by G. Sterba, pp. 17-32. Fischer, Jena.
40. Sternberger, L.A. (1974): Immunocytochemistry, Prentice Hall, Englewood Cliffs.
41. Stillman, M.A., Recht, L.D., Rosario, S.L., Seif, S.M., Robison, A.G., and Zimmerman, E.A. (1977): Endocrinology, 101: 42-49.
42. Stumpf, W.E., Hellreich, M.A., Aumuller, G., Lamb, J.C. IV, and Sar, M. (1977): Cell Tiss. Res., 184:29-44.
43. Watson, S.J., Richards, C.W., and Barchas, J.D. (1978): Science, 200:1180-1182.
44. Weber, E., Voigt, K.H., and Martin, R. (1978): Brain Res., 157:385-390.
45. Weindl, A. (1973): In: Frontiers in Neuroendocrinology, edited by W.F. Ganong and L. Martini, pp. 3-32. Oxford University Press, New York.
46. Weindl, A., and Joynt, R.J. (1972): In: Brain-Endocrine Interaction. Median Eminence: Structure and Function, edited by K.M. Knigge, D.E. Scott, and A. Weindl, pp. 281-297. Karger, Basel.
47. Weindl, A., and Schinko, I. (1975): Brain Res., 88: 319-324.
48. Weindl, A., and Schinko,I. (1980): In: Cerebrospinalflussigkeit-CSF, edited by D. Dommasch, and H. G. Mertens,pp. 66-78. Thieme, Stuggart.
49. Weindl, A. and Sofroniew, M.V. (1978): Drug Res., 28: 1264-1268.
50. Weindl, A., and Sofroniew, M.V. (1978): In: Brain-Endocrine Interaction III. Neural Hormones and Reproduction, edited by D.E. Scott, G.P. Kozlowski, and A. Weindl, pp. 117-137. Karger, Basel.
51. Weindl, A., and Sofroniew, M.V. (1980): In: Ferring Symposium on Brain and Pituitary Peptides, edited by W. Wuttke, A. Weindl, K.H. Voight, and R.R. Dries, pp. 97-109. Karger, Basel.

52. Wenger, T. (1976): Brain Res., 101:95-102.
53. Yates, F.E. and Maran, J.W. (1974): In: Handbook of Physio-
     logy, Section 7: Endocrinology, Vol. IV, Part 2, edited by
     R.O. Greep, and E.B. Astwood, pp. 367-404. American
     Physiological Society, Washington, D.C.

Neurosecretion and Brain Peptides,
edited by J. B. Martin, S. Reichlin, and K. L. Bick.
Raven Press, New York © 1981.

Neuropeptides and the Blood–Brain Barrier

William M. Pardridge, Harrison J. L. Frank, *Eain M. Cornford,
*Leon D. Braun, *Paul D. Crane, and *William H. Oldendorf

*Department of Medicine, Division of Endocrinology and Metabolism, UCLA School of
Medicine; and *Department of Neurology, UCLA School of Medicine, Los Angeles,
California 90024, and Research Service, Veterans Administration Medical Center,
Brentwood, Los Angeles, California 90073*

An understanding of the mechanisms by which the neuropeptides
enter the brain from blood is important for at least two reasons.
Firstly, many, if not all, of the neuropeptides are synthesized
in peripheral organs, principally the gut (10), and may therefore
potentially function in a feedback system with the central nervous system (CNS). Secondly, there is evidence that the systemic
administration of large doses of peptides results in a pharmacologic action on the brain (6, 16); thus, the neuropeptides may
have clinical efficacy in brain disorders in man. The transport
of circulating neuropeptides either into the brain interstitial
space or into the cerebrospinal fluid (CSF) involves the trans-
cellular movement of the peptide through one of two barrier
systems (Fig. 1): (i) the brain capillary endothelia, i.e., the
blood-brain barrier (BBB), or (ii) the epithelia of the choroid
plexus or other circumventricular organs (CVO), i.e., the blood-
CSF barrier. This chapter will discuss the possible ways in which
blood-borne neuropeptides may traverse these two barrier systems.

THE BLOOD-CSF BARRIER

The surface area of the blood-CSF barrier is only about 0.02%
of the surface area of the BBB (5). This gross quantitative
difference between the two barrier systems is often overlooked,
and the BBB and blood-CSF barrier are frequently regarded as one
system. The two membrane systems are segregated both anatomically and functionally. Despite the very small surface area of
the blood-CSF barrier, this membrane is of great importance to
blood-brain transport, since it is probable that some circulating
substances enter brain not via transport through the BBB but
through the blood-CSF barrier. For example, available evidence

indicates some circulating peptides (e.g., insulin) and plasma
proteins (e.g., prealbumin) are selectively transported into CSF
via the blood-CSF barrier.

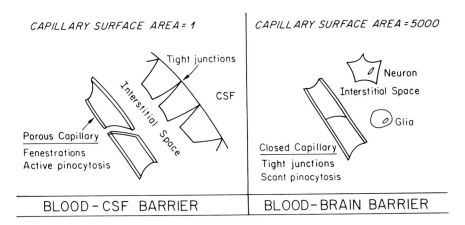

FIG. 1.   The blood-brain barrier (BBB) and blood-CSF barrier
comprise the two major membrane systems segregating brain extracellular space (ECS) from the systemic ECS. The blood-CSF barrier is found at the circumventricular organs (CVO): choroid
plexus, median eminence, organum vasculosum of the lamina terminalis, subfornical organ, subcommissural organ, and area postrema.
The capillaries in the CVO's are porous and allow the rapid distribution of protein and small molecules into the immediate
interstitial space. The presence of low resistance tight junctions on the ventricular side of the ependymal cells prevents the
further distribution of circulating substances into the CSF. The
ependymal junctions are porous in non-CVO areas, wherein brain
capillaries possess tight junctions; therefore, a reciprocal
relationship exists between the anatomical distribution of the
BBB and the blood-CSF barrier. Since the surface area of the BBB
is 5,000-fold that of the blood-CSF barrier, the rapid distribution of circulating compounds into the brain interstitial space
is dependent on the diffusibility of the substance through the
BBB via either lipid-mediation or carrier-mediation.

One index that is useful in determining whether circulating
peptides are selectively transported into the CSF is the CSF/
plasma ratio. The studies of Saunders and co-workers (7) have
shown that the CSF/plasma ratios for non-metabolizable compounds
such as sucrose (Molecular Weight [MW $\sim$ 350] and inulin (MW $\sim$ 5,000)
are 0.08 and 0.03, respectively. The CSF/plasma ratio for albumin
(MW $\sim$ 68,000) is approximately 0.007 (11). Moreover, virtually all
plasma proteins with a MW $\leq$ 160,000 are found in the CSF (30),
and the CSF/plasma ratio is inversely related to the MW

of the protein (7). Since most neuropeptides have a MW between that of sucrose (350) and inulin (5,000), a CSF/plasma ratio of 0.03-0.08 would be expected if the peptide gained access to the CSF only via the non-specific routes available to compounds such as sucrose or inulin. A CSF/plasma ratio greater than 0.03-0.08 is indicative of either (i) selective transport of the peptide from blood to brain through either the blood-CSF barrier or the BBB, or (ii) direct neurosecretion of the peptide into the CSF. For example, the CSF/plasma ratio of prealbumin, a plasma protein which binds both thyroid hormones and retinol-binding protein (24), is 0.07 (11), or about tenfold the ratio for a protein (e.g., albumin) of similar MW. Since it is unlikely that brain synthesizes and secretes prealbumin into the CSF, it may be concluded that this protein is selectively transported into the CSF, probably through the blood-CSF barrier (7). The CSF/plasma ratio for insulin also exceeds the ratio expected for the MW (5,000) of this peptide; the CSF/plasma ratio for insulin is 0.25 (33) vs. that for inulin, 0.03 (7), which also has a MW$\sim$5,000. It is known that circulating insulin is not transported through the BBB (9) but is transported into the CVO (9,31) and into CSF (14,33). Therefore, the high concentration of insulin in the CSF may be a result of the operation of selective transport mechanisms, probably at the blood-CSF barrier. However, since there is evidence that insulin may be synthesized in brain (12), a second but seemingly less likely explanation for the high level of insulin in CSF may be that the peptide is secreted directly into CSF by brain. Prolactin is another peptide which, like insulin, is characterized by a relatively high CSF/plasma ratio, 0.17 (29). This suggests that selectve transport mechanisms may exist at the blood-CSF barrier for the uptake of circulating prolactin. Evidence has also been reported which suggests that selective uptake mechanisms mediate the entry into CSF of α-fetoprotein (28) and certain immunoglobulins (13).

THE BLOOD-BRAIN BARRIER

More than 99% of the interstitial space of brain is segregated from the circulating extracellular space by an epithelial-like capillary endothelia barrier (2). The presence of high-resistance tight junctions between all adjacent brain endothelia and of a paucity of pinocytosis or fenestrations means that circulating substances enter brain via only (i) carrier-mediation or (ii) lipid-mediation (20). Given the low lipid solubility of the peptide molecules (26), specific peptide transport systems, similar in nature to those described for a number of metabolic substrates (22) and for thyroid hormones (23), would have to be present in the BBB in order for circulating peptides to readily gain access to the brain interstitial space. Initial reports with the carotid injection technique (15) indicated that labeled enkephalins readily cleared the BBB with extractions of unidirectional influx of approximately 12% (Table 1). Cornford and co-workers (3), however, also used the carotid artery injection technique and showed that the extraction of unidirectional influx of radiochemically pure enkephalins was only 2-3%, as opposed to the extraction for a non-diffusible reference such as dextran, 1% (Table 1). The low but measurable transport of enkephalin was non-saturable up to pharmacologic enkephalin levels, and no regional differences in peptide transport were observed (3); these data indicated the absence of a specific BBB transport

TABLE 1. Restrictive blood-brain barrier transport of enkephalins and TRH

Compound	Extraction (%)	Reference
^{3}H-methionine-enkephalin	12.7[a]	Kastin et al. (15)
^{3}H-methionine-(D-ALA^2)-enkephalinamide	2.6[b]	Rapoport et al. (26)
^{3}H-methionine-enkephalin	2.7 $\pm$ 0.3[c]	Cornford et al. (3)
^{3}H-leucine-enkephalin	2.7 $\pm$ 0.3	Cornford et al. (3)
^{3}H-TRH	1.0 $\pm$ 0.1	Cornford et al. (3)
^{14}C-dextran	0.9 $\pm$ 0.2	Cornford et al. (3)

[a]Calculated from BUI = E/E_{HOH}, where BUI = brain uptake index, 15% (15), E = extraction of peptide, and E_{HOH} = water reference extraction at 5 sec after carotid injection, 85% (3). The high BUI originally reported for enkephalin (15) has apparently not been confirmed (see p. 408 of ref. 16).

[b]Calculated from E = P x S/F, where E = extraction, P = permeability constant, 2.5 x 10^{-6} cm/sec (26), S = brain capillary surface area, 240 cm^2/g (26), and F = average cerebral blood flow, 1.4 cm^3/min/g (27), for the regions studied by Rapoport et al. (26).

[c]mean $\pm$ S.D. (n = 3-6 rats).

mechanisms for the enkephalins (3). The finding of a low but measurable transport of labeled enkephalins through the BBB has recently been confirmed (26) with a single venous injection technique (Table 1).

The low permeability of the BBB to neuropeptides such as the enkephalins or TRH (Table 1) is quantitatively similar to barrier permeability to other compounds, e.g., the catecholamines, which serve neurotransmitter or neurotransducer roles in both the CNS and the periphery. The restricted permeability of the BBB to the catecholamines or to the enkephalins prevents a ready exchange of these substances between the CNS and the periphery and thereby allows these chemical messengers to function independently in both compartments. Until a specific transport system for the enkephalins or for any of the neuropeptides is identified within the BBB, a practical approach to the evaluation of the role of the BBB in blood-brain exchange of the peptides is the use of the catecholamine model. The enhancement of catecholamine activity in brain via the systemic administration of catecholaminergic compounds is achieved not by giving unmodified catecholamines, which traverse the BBB poorly (19), but rather by administering one of two types of catecholamine derivatives (i) latentiated catecholamines which by virtue of added substituents are highly lipid soluble compounds (4), or (ii) catecholamine precursors, e.g., tyrosine (8) or DOPA (18), compounds which traverse the BBB by carrier-mediation (19,21). Given the absence of BBB peptide transport systems (3), the synthesis of highly lipid soluble peptides would appear to be the most practical approach to circumventing the low permeability of the BBB to the neuropeptides.

BLOOD-BRAIN BARRIER PEPTIDE RECEPTORS

The previous two sections have emphasized our present view that circulating petides may slowly gain access to the CSF by transport through the blood-CSF barrier, but that the rapid distribution of peptides into the brain interstitial space is prevented by the restrictive permeability properites of the BBB to the peptides. Although there is at present no reported evidence showing specific carrier-mediated mechanisms for peptide transport through the BBB, there are data which demonstrate the operation of BBB receptors for peptides. That is, circulating peptides may transmit information to the brain side of the BBB by binding and activating specific peptide receptors on the blood side of the BBB. Van Houten and Posner (32) have recently reported in vivo evidence showing the presence of saturable, specific receptors for circulating insulin on the endothelia of brain capillaries. Two of the present authors (Frank and Pardridge) have recently confirmed the presence of specific saturable receptors for ^{125}I-insulin on brain endothelia using an in vitro radioreceptor assay with isolated bovine brain capillaries (Fig. 2). Approximately 70% of insulin binding to isolated capillaries is saturable (dissociation constant = 0.9 nM), and this binding is not inhibited by growth hormone, prolactin, or thyrotropin. The brain capillary radio-receptor assay (Fig. 2) may be extended to other labeled peptides and to the search for other BBB peptide receptors. For example, reported data indicate that vasopressin modulates BBB function (17,25; Raichle, this volume), presumably by first activating brain endothelial receptors. The presence of peptide receptors at the BBB provides a mechanism by which circulating peptides may modulate brain function without entering brain interstitial and synaptic spaces. Should BBB peptide receptors prove to have a physiologic function, an important area

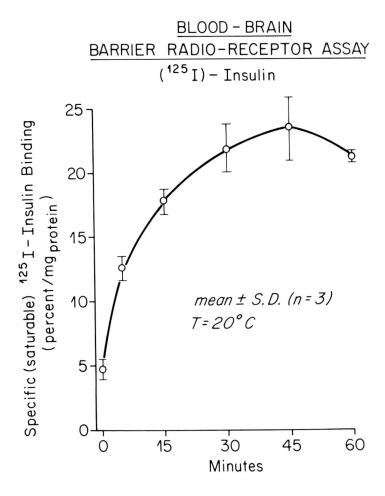

FIG. 2. (^{125}I)-insulin binds a saturable and specific receptor on the BBB membrane, i.e., the brain capillary endothelial cell. The saturable component of insulin binding reaches equilibrium at 20o in by 45 min. Saturation studies (data not shown) indicated the dissociation constant of inusulin binding to the capillary is 0.9 nM. Insulin binding was weakly inhibited by proinsulin, but not by growth hromone, prolactin, or thyerotropin. Capillaries were isolated from fresh bovine brains, obtained at a local slaughterhouse, by method of Brendel et al. (1).

in need of further investigation would be the elucidation of the second messengers which transmit signals from the putative BBB peptide receptors to brain cells.

SUMMARY

1. No evidence has been reported to date which indicates that peptides such as insulin, the enkephalins, or TRH traverse the BBB by specific transport systems. Therefore, the use of latent-iated (lipid-soluble) derivatives of peptides provides the most practical approach to circumvent the restricted permeability of the BBB to peptides. In contrast to the BBB, the blood-CSF barrier appears to selectively transport certain peptides (e.g., insulin) or plasma proteins (e.g., prealbumin) from blood into CSF. However, since the surface area of the BBB is 5,000-fold greater than the surface area of the blood-CSF barrier, it is unlikely that transport through the blood-CSF barrier permits rapid distribution of circulating peptides into brain intersti-tial space.

2. The presence of BBB peptide receptors such as those which have recently been demonstrated for insulin (ref. 32 and Fig. 2), provides a mechanism by which neuropeptides may transmit signals to the <u>brain</u> side of the BBB via binding and activate receptors on the <u>blood</u> side of the BBB. In this way, circulating neuro-peptides may potentially rapidly influence brain activity without traversing the BBB or entering brain interstitial or synaptic spaces.

ACKNOWLEDGEMENTS

These studies were supported by NIH grant AM-25744 (WMP), the American Diabetes Association, Southern California Affiliate (HJLF), the Veterans Administration (WHO), and NIH grant AI-15692 (EMC). Lawrence J. Mietus provided outstanding technical assist-ance in the isolation of bovine brain capillaries. Charlotte Limberg and Anita Anderson prepared the manuscript.

REFERENCES

1. Brendel,K., Meezen, E., Carlson, E.C. (1974): <u>Science</u>, 185: 953-955.
2. Brightman, M.W., Prescott, L., Reese, T.S. (1975): In: <u>Brain-Endocrine Interaction II,</u> edited by K.M. Knigge, D.E. Scott, H. Kobayashi, and S. Ishii, pp. 146-165. S. Karger AG, Basel.
3. Cornford, E.M., Braun, L.D., Crane, P.D., Oldendorf, W.H., (1978): <u>Endocrinology,</u> 103:1297-1303.
4. Creveling, R.C., Daly, J.W., Tokuyama, T., Witkop, B. (1969): <u>Experientia</u>, 25:26-27.
5. Crone, C. (1971): In: <u>Ion homeostasis of the Brain</u>, edited by B.K. Siesjo and S.C. Sorensen, pp. 52-62. Munksgaard, Copenhagen.

6. DeWied, D. (1977): Acta Endocrin., (suppl. 214) 85:9-18.
7. Dziegielewska, K.M., Evans, C.A.N. Malinowska, D.H., Mollgard, K., Reynolds, J.M., Reynolds, M.L., Saunders, N.R. (1979): J. Physiol., 292:207-231.
8. Gibson, C.J., Wurtman, R.J. (1978): Life Sci., 22:1399-1406.
9. Goodner, C.J., Berrie, M.A. (1977): Endocrinology, 101:605-612.
10. Grossman, M.I. (1979): Fed. Proc., 38:2341-2343.
11. Hagen, G.A., Elliott, W.J. (1973): J. Clin. Endocrinol. Metab., 37:415-422.
12. Havrankova, J., Roth, J., Brownstein, M.J. (1979): J. Clin. Invest., 64:636-642.
13. Hemmings, W.A. (1978): Proc. R. Soc. Lond., 200:175-192.
14. Hill, D.E., Schedewie, H.K., Chalhub, L., Sziszak, T., Boughter, M., Owen, J. (1978): Clin. Res., 26:71A.
15. Kastin, A.J., Nissen, C., Schally, A., Coy, D.H. (1976): Brain Res. Bull., 1:583-589.
16. Kastin, A.J., Olson, R.D., Schally, A.V., Coy, D.H. (1979): Life Sci., 25:401-414.
17. Landgraf, R., Hess, J., Ermisch, A. (1978): Acta Biol. Med. Ger., 37:655-658.
18. Mena, I., Cotzias, G.C. (1975): N. Engl. J. Med., 292:181-184.
19. Oldendorf, W.H. (1971): Am. J. Physiol., 221:1629-1639.
20. Oldendorf, W.H. (1975): In: The Nervous System, edited by D.B. Tower, pp. 279-289. Raven Press, New York.
21. Pardridge, W.M. (1977): J. Neurochem., 28:103-108.
22. Pardridge, W.M., Oldendorf, W.H. (1977): J. Neurochem., 28:5-12.
23. Pardridge, W.M. (1979): Endocrinology, 105:605-612.
24. Peterson, P.A., Rask, L., Ostberg, L., Andersson, L., Kamwendo, F., Pertoft, H. (1973): J. Biol. Chem. 218:4009-4022.
25. Raichle, M.E., Grubb, Jr., R.L. (1978): Brain Res., 143:191-194.
26. Rapoport, S.I., Klee, W.A., Pettigrew, K.D., Ohno, K. (1979): Science, 207:84-86.
27. Sakurada, O., Kennedy, C., Jehle, J., Brown, J.D., Carbin, G.L., Sokoloff, L. (1978): Am. J. Physiol. 234:H59-H66.
28. Saunders, N.R. (1977): Exp. Eye Res. Suppl., 25:523-550.
29. Schroeder, L.L., Johnson, J.C., Malarkey, W.B. (1976): J. Clin. Endocrinol. Metab., 43:1255-1260.
30. Tourtellotte, W. (1970): J. Neurol. Sci., 10:279-304.
31. van Houten, M., Posner, B.I., Kopriwa, B.M., Brawer, J.R. (1979): Endocrinology, 105:666-673.
32. van Houten, M., Posner, B.I. (1979): Nature, 282:623-625.
33. Woods, S.C., Porte, Jr., D. (1977): Am. J. Physiol., 233:331-E334.

Neurosecretion and Brain Peptides,
edited by J. B. Martin, S. Reichlin, and K. L. Bick.
Raven Press, New York © 1981.

Hypothesis: A Central Neuroendocrine System Regulates Brain Ion Homeostasis and Volume

M. E. Raichle

Department of Neurology and Neurological Surgery, Mallinckrodt Institute of Radiology,
Washington University School of Medicine, St. Louis, Missouri 63110

Precise control of brain volume, through adjustment of cell water and electrolyte content, is important for the normal function of the brain not only because it is confined to the rigid environment of the skull (13), but also because changes in cell volume may affect important functional relationships between cells (15,33,60). This volume homeostasis must be achieved in the face of fluctuating osmotic and hydrostatic forces imposed by the incoming blood supply while respecting functionally critical ionic gradients within the brain.

Considerable evidence indicates that the brain is capable of controlling its fluid and electrolyte balance under a variety of circumstances (for reviews see 1,14,25). The precise mechanism involved in this type of regulation, as well as the cells participating are, at best, poorly characterized.

It is the purpose of this paper to summarize briefly extant information in support of a working hypothesis that three cell groups (brain capillary endothelial cells, secretory cells of the choroid plexus, astroglia) can be expected to perform in a coordinated manner, the regulation of the internal ionic environment of the brain. A unique element of this hypothesis is that the activity of regulating the ionic environment of the brain is orchestrated by a central neuroendocrine system capable of affecting all three cell types.

CAPILLARY ENDOTHELIUM

It is generally accepted that the blood brain barrier (BBB) is very important in the regulation of the internal environment of the brain. Because of its very large surface area relative to the choroid plexus (5000:1) and its intimate contact with the neuropil it must be viewed as a major component in such a regulatory process. Despite the potential importance of the BBB in the regulation of brain fluid and electrolyte balance, its role has been viewed as largely static (43) in contrast to the choroid plexus, where dynamic regulation of fluid and electrolyte exchange is well recognized although incompletely characterized. Thus, the BBB is often viewed as a membrane system that, once developed, very effectively excludes a wide variety of substances from the central nervous system (e.g., various ions, proteins), while through the mechanism of facilitated diffusion, admits and discharges a limited number of essential compounds, (e.g., glucose, specific amino acids (8)).

Several pieces of evidence now suggest that this view of the BBB as a static participant in the normal regulation of brain fluid and electrolyte homeostasis (43) is probably incorrect. In fact, these data (summarized in Table 1) suggest that the brain capillaries have the potential to play a very dynamic role in the regulation of brain water and electrolyte permeability in a manner analogous to the mammalian kidney and other membranes known to regulate fluid and electrolyte permeability (e.g., various urinary bladders, frog skin, intestine). These data are briefly reviewed below.

TABLE   1.   Membranes that regulate water and electrolyte permeabilities

FEATURE	RENAL TUBULE (intestine, toad bladder, frog skin)	BRAIN CAPILLARY
Tight Junctions	4*	4
Mitochondrial Rich Cells	53	42
Specific Cellular Enzymes	16	11,16
Relatively Impermeable to H$_2$0	6	44
Innervation by Adrenergic Neurons	2,56	19,20,22-24 34,37,49,57
ΔPermeability by:		
Δ osmolality	5	45
Δperfusion pressure	40	47
sympathetic stimulation	3,21	17,48
vasopressin (ADH)	5	41,49
angiotensin II	7	+

*Numbers indicate appropriate references to representative publication.

+This communication.

Brain capillaries exhibit a number of <u>anatomical</u> and <u>biochem-</u><u>ical</u> features unique to membranes known to regulate water and electrolyte permeability. These include tight junctions between constituent endothelial cells (4); a high mitochondrial content in endothelial cells (42); and a similar complement of intra- cellular enzymes (11,16). Of special importance in this regard is the recent demonstration of a membrane bound (Na^+ - K^+) ATPase on isolated brain capillaries (11).

Brain capillaries appear to be functionally innervated by adrenergic neurons originating from within brain (i.e., brain- stem). This was first suggested by anatomical studies showing a close association between noradrenergic nerve varicosities and capillaries at the light microscopic level using immunohistochem- ical techniques for the demonstration of the enzyme dopamine-β hydroxylase (20). Electron microscopic studies have since confirmed the actual presence of true synapses on a limited num- ber of capillaries (23,24,34,49,57). Considerable controversy, however, has surrounded these anatomical observations because of the infrequent occurrence of true synapses as seen at the elec- tron microscopic level and the apparent distance between many capillaries and noradrenergic varicosities as seen in the light microscope. This controversy has been largely resolved by studies demonstrating the presence of adrenergic receptors on brain capillaries (19,22,37). We examined the effects of various neurotransmitters and arginine vasopressin on adenosine 3',5'- monophosphate (cyclic AMP) in the isolated microvessels of the rat brain (22). We chose to examine the cyclic AMP system be- cause of its close link with certain neurotransmitter receptors and the abundant evidence relating this substance to hormone- induced changes in water and electrolyte permeability in other tissues. Our experiments (22) revealed that norepinephrine increases the concentration of cyclic AMP in incubated suspen- sions of brain capillaries. This response was matched by other drugs that stimulate β-receptors but the α-agonist phenylephrine was without effect. β-adrenergic blockade abolished the response while α-adrenergic blockade produced no change. Similar obser- vations were made and reported almost simultaneously by Nathanson and Glaser (37). Taken together, our observations as well as Nathanson's strongly suggested the presence of β-adrenergic receptors on the brain capillaries. Their presence has now been confirmed by Harik <u>et al</u>. (19) who reported actual radioligand binding to β (type 2) receptors on isolated brain capillaries. Thus a functional innervation of brain capillaries by at least the central noradrenergic system seems very likely. Because of the limited, if nonexistent, role played by capillaries in overall cerebrovascular resistance, and their dominant role in exchange processes, it seems a reasonable hypothesis that this innervation is concerned with the regulation of exchange of substances between the blood and the brain.

Brain capillaries also exhibit several <u>functional</u> character- istics of membranes which regulate water and <u>electrolyte</u> permea- bilities. These include a restricted permeability to water (44);

a prompt and reversible increase in water permeability when subjected to transient hyperosmolarity (45); a prompt and reversible
decrease in permeability to water when subjected to an increase
in perfusion pressure (47); a change in permeability associated
with central (locus coeruleus) as well as peripheral (superior
cervical gangion) stimulation (17,48); and an increase in permeability to centrally administered vasopressin (41,46) as well as
angiotensin II (unpublished).

The role of brain peptides such as vasopressin and angiotensin
II in the regulation of brain vascular permeability remains to be
clarified. Of special note in this regard is the fact that neither of these peptides affect cyclic AMP levels in isolated
microvessels (22) despite the fact that both alter brain water
permeability when injected into the cerebral ventricles of monkeys. In the case of vasopressin it is possible that it acts at
the capillary by some mechanism not involving cyclic AMP (10).
However, the demonstration by Tanaka et al. (59) that intraventricular vasopressin alters brain norepinephrine turnover in a
number of areas, the discovery by Swanson (58) of an anatomical
connection between the noradrenergic system and the vasopressin
and oxytocin systems in brain and the recent demonstration that
brain catecholamines play an important role in mediating vasopressin release in response to both osmotic and hypovolemic
stimuli (39) argue for the noradrenergic mediation of the vasopressin effect in vivo. Angiotensin II likewise probably produces its effect indirectly. It is of interest in this regard
that angiotensin II causes the release of vasopressin (54).

If, in fact, brain norepinephrine, vasopressin and angiotensin
II form part of a central neuroendocrine system concerned with
brain fluid and electrolyte homeostasis and cell volume regulation, it is to be expected that their influence extends to cells
in the brain other than just the endothelium of the brain capillaries. Two likely candidates in this regard are the secretory
cells of the choroid plexus and the astroglia.

CHOROID PLEXUS

A fairly convincing case can be made for the hypothesis that
the activity of the secretory cells of the choroid plexus is
under neuroendocrine control. Several studies have now clearly
documented the presence of nerve fibers on the secretory cells as
well as the capillaries of the choroid plexus (9,29,30,36). These
fibers originate not only from the peripheral sympathetic chain
but also from the paraventricular and supraoptic nuclear fields.
The functional significance of these anatomical observations is
strongly supported, first, by data suggesting the presence of
specific neurotransmitter and hormone receptors on the secretory
cells of the choroid plexus (38,50); and, second, by data showing
that sympathetic nerve stimulation (32) as well as exposure of
the choroid plexus to vasopressin (12) and acetylcholine analo-

gues (31) not only alters cerebrospinal fluid formation, but also produces ultrastructural changes in the affected cells (52).

ASTROGLIA

The role of the glia cell in brain extracellular fluid and ionic homeostasis has been the subject of extensive investigations and speculation (for review see ref. 55). It is quite clear from an anatomical standpoint that the glia cell is so situated between neurons and capillaries that it could play a major role in the ionic homeostasis of the brain extracellular fluid. These cells not only surround the neurons but also cover approximately 80% of the surface area of the capillaries. Furthermore, substantial physiological data suggest that these cells have the capacity to alter the extracellular ionic environment of the brain by the uptake (active as well as passive) of potassium, in exchange for other ions. Is this activity of the glia cell under any type of neurohormonal control? Several recent observations at the least suggest that glia cell activity may be under neurohormonal control. First, Kimelberg et al. (26,27) have shown that the activity of glia $(Na^+-K^+)ATPase$ as well as carbonic anhydrase is increased in the presence of norepinephrine. Their observations further suggest that the effect of norepinephrine is mediated through the β-adrenergic stimulation of glia cyclic AMP. This observation is of particular importance because of the key role these two enzymes play in the regulation of ionic movements between the glia cell and its environment. Second, Harik et al. (18) were able to show that cerebral norepinephrine depletion slows the recovery of the redox ratio of cytochrome a,a_3 during increased metabolic demands induced by local cortical stimulation. One explanation for this observation is that the clearance of the extracellular potassium, accumulated as the result of the increased metabolic activity, is greatly slowed in the absence of norepinephrine. Unfortunately, extracellular potassium concentration was not monitored in these experiments. Finally, MacKenzie et al. (35) report that acute increases in brain norepinephrine concentration cause a marked increase in cerebral oxidative metabolism and blood flow despite depressant effect on neuronal activity (30,61). One must consider the possibility that norepinephrine is stimulating glia cell $(Na^+-K^+)ATPase$ and effecting the movement of potassium from the extracellular fluid into the cell. An analogy to this is the well described control of avian erythrocyte plasma potassium balance by plasma norepinephrine (for review see ref. 53) through stimulation of β-adrenergic receptors on these cells. Although mammalian erythrocytes do not possess β-receptors or responses of this type to norepinephrine, presumably because of the increased sophistication of the mammalian kidney, it seems reasonable to suggest that in the special case of the brain such responsiveness may be retained by the glia cell.

The above observations suggest but do not prove that norepinephrine stimulated glia cells may be actively involved in brain extracellular fluid ionic homeostasis and/or cell volume regulation. Further experiments including actual measurements of brain extracellular fluid ionic concentrations during manipulations of brain norepinephrine, vasopressin and angiotensin II; better characterization of specific receptors for these substances on glia and more quantitative information on specific enzymatic activation are obviously needed.

In summary, it seems reasonable to suggest, on the basis of preliminary experimental evidence, that norepinephrine, vasopressin and angiotensin II form part of a central neuroendocrine system designed to regulate the fluid and ionic environment and, possibly, cell volume of the brain. It seems reasonable to suggest, further, that three cell types are primarily involved in a coordinated fashion as effectors in this type of regulation, brain capillary endothelial cells, glia cells and the secretory cells of the choroid plexus. Elucidation of such a system should aid considerably in our understanding of the cellular and biochemical basis for various functional states attributed to brain vasopressin and norepinephrine as well as a more rational basis for the treatment of brain edema.

REFERENCES

1. Andreoli, T.E., Grantham, J.J., Rector, F.C. Jr., editors (1977): Disturbances in Body Fluid Osmolality. American Physiological Society, Bethesda.
2. Barajas, L. (1978): Fed. Proc., 37:1192-1201.
3. Bello-Reuss, E., Trevino, D.L., Gottschalk, C.W. (1976): J. Clin. Invest., 57:1104-1107.
4. Bennett, H.S., Luft, J.H., and Hampton, J.C. (1959): Am. J. Physiol., 196:381-390.
5. Bindslev, N., Tormey, J. McD., Pietras, R.J., and Wright, E.M. (1974): Biochi. Biophys. Acta (Amst), 332:286-297.
6. Bindslev, N., Wright, E.M. (1976): J. Membrane Biol., 29:265-288.
7. Bolton, J.E., Munday, K.A., Parson, B.J., York, G.B. (1975): J. Physiol., 253:411-428.
8. Bradbury, M. (1979): The Concept of a Blood Brain. John Wiley and Sons, New York.
9. Brownfield, M.S., Kozlowski, G.P. (1977): Cell Tiss. Res., 178:111-127.
10. Christensen, S. (1978): Pflugers Arch., 374:229-234.
11. Eisenberg, H.M., Suddith, R.L. (1979): Science, 206:1083-1085.
12. Fishman, R.A. (1959): J. Clin. Invest., 38:1698-1708.
13. Fishman, R.A. (1974): Res. Publ. Assoc. Res. Nerv. Ment. Dis., 53:159-171.
14. Fishman, R.A. (1974): In: Brain Dysfunction in Metabolic Disorders, edited by F. Plum, pp. 159-171. Raven Press, New York.

15. Geinismann, Y.Y., Larina, V.N., and Mats, V.N. (1971): Brain Res., 26:247-257.
16. Goldstein, W.G. (1977): New Engl. J. Med., 296:632.
17. Grubb, R.L., Jr., Raichle, M.E., and Eichling, J.O. (1978): Brain Res., 144:204-207.
18. Harik, S.I., LaManna, J.C., Light, A.I., Rosenthal, M. (1979): Science, 206:69-71.
19. Harik, S.I., Sharma, V.K., Weatherbe, J.R., Warren, R.H., Banergee, S.P. (1980): Europ. J. Pharm., 61:207-208.
20. Hartman, B.K., Zide, S., and Udenfriend, S. (1972): Proc. Natl. Acad. Sci. USA, 69:2722-2726.
21. Haywood, G.P., Isaia, J., Maetz, J. (1977): Am. J. Physiol., 232 R110-R115.
22. Herbst, T.J., Raichle, M.E., and Ferrendelli, J.A. (1979): Science, 204:330-332.
23. Iijuma, T., Wasano, T., Tagawa, T., and Ando, K. (1977): Cell Tiss. Res., 179:143-155.
24. Itakura, T., Yamamoto, K., Tobyama, M., and Shimizu, N. (1977): Stroke, 8:360-365.
25. Katzman, R. (1976): Fed. Proc. (Symposium), 35:1244-1247.
26. Kimelberg, H.K., Naunri, S. Biddlecome, S., Bourke, R.S. (1978): In: Dynamic Properties of Glia Cells, edited by Schoffeniels, E., Frank, G., Tower, D.B., Herty, L., pp. 347-357. Pergamon Press, New York.
27. Kimelberg, H.K., Biddlecome, S., Bourke, R.S. (1979): Brain Res., 173:111-124.
28. Krnjevic, K. (1974): Physiol. Rev., 54:418-540.
29. Lindvall, M., Edvinsson, L., Owman, C. (1977): Exp. Neurol., 55:152-159.
30. Lindvall, M., Owman, C. (1978): Cell Tiss. Res., 192:195-203.
31. Lindvall, M., Edvinsson, L., Owman, C. (1978): Neurosci. Lett., 10:311-316.
32. Lindvall, M., Edvinsson, L., Owman, C. (1978): Science, 201:176-178.
33. Lipton, P. (1973): J. Physiol. (London), 231:365-383.
34. McDonald, D.M. and Rasmussen, G.L. (1977): J. Comp. Neurol., 173:475-496.
35. MacKenzie, E.T., McCulloch, J., Harper, M.A. (1976): Am. J. Physiol., 231:489-494.
36. Nakamura,S., Milhorat, T.H. (1977): Brain Res., 153:285-293.
37. Nathanson, J.A., Glaser, G.H. (1979): Nature, 278:567-569.
38. Nathanson, J.A., (1979): Science, 204:843-844.
39. Miller, T.R., Handelman, W.A., Arnold, P.E., McDonald, K.M., Molinoff, P.B., Schrier, R.W. (1979): J. Clin. Invest., 64:1599-1607.
40. Mortillaro, N.A., Taylor, A.E. (1976): Circ. Res., 39:348-358.
41. Noto, T., Nakajima, T., Saji, Y., Nagawa, Y. (1978): Endocrinol. Japan, 25:591-596.

42. Oldendorf, W.J., Cornford, M.E., and Brown, W.J. (1977): Ann. Neurol., 1:409–417.
43. Rapaport, S.I. (1978): J. Theor. Biol., 74:439–467.
44. Raichle, M.E., Eichling, J.O., Straatmann, M.G., Welch, M.J., Larson, K.B., and Ter-Pogossian, M.M. (1976): Am. J. Physiol., 230:543–552.
45. Raichle, M.E., Grubb, R.L., Jr., and Eichling, J.O. (1977): Fed. Proc., 36:470.
46. Raichle, M.E., and Grubb, R.L., Jr. (1978): Brain Res., 143:191–194.
47. Raichle, M.E. and Grubb, R.L., Jr. (1978): Fed. Proc., 37:242.
48. Raichle, M.E., Hartman, B.K., Eichling, J.O., and Sharpe, L.G. (1975): Proc. Natl. Acad. Sci. USA,72:3726–3730.
49. Rennels, M.L. and Nelson, E. (1975): Am. J. Anat., 144:233–241.
50. Rudman, D., Hollins, B.M., Lewis, N.C., Scott, J.W. (1977): Am. J. Physiol., 232:E353–E357.
51. Rudolph, S.A., Lefkowitz, R.J. (1978): In: Membrane Transport in Biology, Vol. 1, edited by G. Giebisck, D.C. Tosteson, and H.H. Ussing, pp. 349–360. Springer-Verlag, New York.
52. Schultz, W.J., Brownfield, M.S., Kozlowski, G.P. (1977): Cell Tiss. Res., 178:129–141.
53. Scott, W.N., Sapirstein, V.S., and Voder, M.Y. (1974): Science, 184:797–799.
54. Simonnet, G., Rodriquez, F., Fumoux, F., Czernichow, P. and Vincent, J.D. (1979): Am. J. Physiol., 237:R20–R25.
55. Somjen, G.G. (1979): Ann. Rev. Physiol., 41:159–177.
56. Strum, J.M. and Danon, D. (1974): Anat. Rec., 178:15–40.
57. Swanson, L.W., Connelly, M.A., and Hartman, B.K. (1977): Brain Res., 136:166–173.
58. Swanson, L.W. (1977): Brain Res., 128:346–353.
59. Tanaka, M., de Kloet, E.R., de Wied, D., and Versteeg, G. (1977): Life Sci., 20:1799–1808.
60. Van Harreveld, A. and Fifkova, E. (1975): Exp. Neurol., 49:736–749.
61. Woodward, D.J., Moises, H.C., Waterhouse, B.D., Hoffer, B.J., Freedman, R. (1979): Fed. Proc., 38:2109–2116.

Neurosecretion and Brain Peptides,
edited by J. B. Martin, S. Reichlin, and K. L. Bick.
Raven Press, New York © 1981.

Neural Peptides in the Cerebrospinal Fluid

Ivor M. D. Jackson

*Tufts University School of Medicine, New England Medical Center,
Boston, Massachusetts 02111*

It is now well established that the central nervous system
(CNS) exercises control over the secretion of each adenohypophysial hormone through "releasing factors" synthesized and secreted
by peptidergic neurons in the hypothalamus, although the mammalian anterior pituitary gland itself lacks a direct nerve supply
from the brain (5). In accordance with the "portal vessel chemotransmitter hypothesis" (22) the hypothalamus secretes specific
pituitary regulatory substances which are released at nerve endings in close apposition to the portal vessel capillaries in the
median eminence (ME) for subsequent vascular transport via the
pituitary stalk to the adenohypophysis.

The isolation and synthesis of three of these hypothalamic
hypophysiotropic hormones, thyrotropin-releasing hormone (TRH),
luteinizing hormone-releasing hormone (LH-RH) and growth hormone
release inhibiting hormone (somatostatin) have provided powerful
tools for the investigation of hypothalamic-pituitary function
and have permitted the development of specific radioimmunoassays
for the measurement of these substances at low concentrations
(59). Following these methodologic developments it was recognized
that most of neural TRH and somatostatin is located in regions of
the CNS outside the hypothalamus. A number of other neural peptides have also been shown to be present in the hypothalamus as
well as in extrahypothalamic brain locations and extraneural
sites. Some of these peptides have the ability to influence pituitary function, but their physiologic role and mode of action,
both in the hypothalamus and elsewhere, have not been fully
elucidated at this time.

In addition to secretion of releasing factors by the tuberoinfundibular system into the portal vessel circulation, the possible role of the cerebrospinal fluid (CSF) in neuroendocrine
function has received much attention (34). The ventral medial
hypothalamus contains specialized cells (ependymal tanycytes)
that extend from the floor of the third ventricle through the

337

interstitial space of the ME to come into intimate contact with
the capillaries of the primary portal plexus. It has been postulated that the tanycytes may function as a bidirectional system -
conveying hypophysiotropic hormones from the third ventricle to
the ME (34) while at the same time permitting the adenohypophysial hormone to gain access to the CNS (4).

This review includes consideration of: 1) the significance of
the CNS as a route for the hypothalamic regulation of pituitary
function, 2) the role of the CSF as a conduit of communication
between hypothalamus and brain, pituitary and brain, and between
different regions of the CNS, 3) the pharmacologic and therapeutic effects of altering peptide levels in the CSF, and 4) the
potential clinical significance of measurements of endogenous CSF
peptides.

HYPOPHYSIOTROPIC FUNCTION

TRH, LH-RH and somatostatin have all been detected by radioimmunoassay (RIA) in extracts of mammalian hypothalamus. Their
concentrations are especially high in the ME and immunohistochemistry has demonstrated separate and distinct populations of nerve
terminals staining for each of these peptides. However, all are
found in the vicinity of portal capillaries (23).

The distribution of releasing hormones within the hypothalamus
has been utilized to impute a functional role to several anatomic
regions. Thus, the "thyrotrophic area", a region which includes
the paraventricualr nucleus has classically been considered the
center for TRH secretion from the hypothalamus. Utilizing a
microdissection technique which allows discrete nuclei to be
dissected from neural tissue, the highest concentration of TRH
in the hypothalamus is found in this region (8). The physiologic
significance of this anatomic area is demonstrated by the effects
of ablation (29) which causes hypothyroidism in the rat and
reduces the hypothalamic content of TRH by two-thirds (Table 1).

TABLE 1.   Effect of a lesion of the "thyrotrophic area" of the
hypothalamus on brain distribution of TRH* in the rat (29).

	Lesion	Control	Significance
Hypothalamus	3.6 ± 0.3	9.3 ± 0.5	p<0.001
Extra-hypothalamic brain	17 ± 0.9	18.9 ± 0.7	NS

* TRH in ng/organ (mean ± SEM)

The location of the LH-RH perikarya in the rat hypothalamus
appears to be the ventral portion of the medial preoptic area
(26). As these nerve fibers project from the POA to the ME they
come into intimate contact with the ependymal lining of the third

ventricle (Fig. 1) and may even decussate. [Of note, Ibata et al. (26) from lesion studies have provided evidence of a contralateral projection of LH-RH neurons to the ME.] Clearly the close relationship of these LH-RH fibers to the ventricular system provides a means for LH-RH to be secreted into the CSF. Likewise the cellular source of somatostatin nerve terminals in the ME is the anterior periventricular nucleus (2) where perikarya are located close to the ependymal border of the third ventricle and are strategically located to secrete directly into the CSF. However, immunoreactivity terminals for LH-RH or somatostatin on the surface of the third ventricle have not been demonstrated.

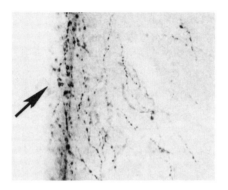

FIG. 1. Immunoperoxidase staining for LH-RH in female rat showing streams of beaded axons coursing through the ependymal layer in close contact with the surface of the third ventricle (IIIV). The arrow is located within the IIIV and points to the ependymal wall. Frontal section at level of ME x 250. (Courtesy of Dr. R. Lechan.)

Hypophysiotropic Effects of Peptides Placed in the CSF

TRH.
In studies performed in this laboratory, Gordon et al. (20) reported that the plasma TSH response to TRH was much less when the TRH was given directly into the third ventricle compared with that produced when the same dose was given intravenously or directly into the ME or pituitary gland (Fig. 2). These findings were interpreted to suggest that TRH transport from the CSF was not of major importance in the regulation of pituitary TSH secretion. Oliver et al. (48) subsequently showed that the administration of TRH into a lateral ventricle results in a significant release of TSH. There was a delay in reaching maximal TSH concentration following intravenous injection although the TSH levels following ventricular injection were maintained for a longer time than were those induced by intravenous injection of TRH. There was a good correlation between the timed course of TSH release

and the appearance of radioactivity in hypophysial portal blood following either intravenous or intraventricular injection of ³H-TRH. These workers reported that 289 pg of ³H-TRH was recovered in portal blood after intraventricular injection of 50 ng (³H)-TRH. Overall the data reported support the view that TRH is able to cross the ME from CSF into hypophysial portal blood, and that it is capable of stimulating the pituitary gland to release TSH.

In the human intra-spinal TRH administration, via lumbar puncture, evokes a rapid serum TSH and PRL rise in 60-70% subjects comparable to that achieved following intravenous administration (69). Kinetic studies with ¹²⁵I-TRH suggested that the TRH response did not reflect passage from the CSF to the systemic circulation.

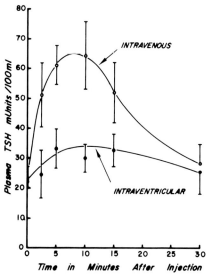

FIG. 2.   Plasma TSH levels following TRH injection in the rat (20).

LH-RH.
A number of studies have been reported on the effect of intraventricular administration of LH-RH on the release of LH. Ondo et al. (50) injected LH-RH into the third ventricle and cisterna magna and demonstrated a rise in plasma LH. Weiner et al. (72) also found a rise in LH but concluded from their studies in ovariectomized estrogen-primed rats that intraventricular administration of LH-RH was less potent in releasing LH than intravenously injected LH-RH. In further studies by Ben-Jonathan et al. (3), LH-RH was found in the portal blood after injection of 125 ng LH-RH intraventricularly and in some rats injected with as little as 5 ng of the decapeptide; the recovery in portal blood was approximately 5% of the quantity injected into the ventricular system. During the first 20-30 minutes following injec-

tion, intravenously-administered LH-RH was slightly more effective in stimulating LH release than was intraventricularly administered LH-RH. After this time, the plasma LH levels in rats given LH-RH intravenously fell steadily while the plasma LH levels in rats given LH-RH intraventricularly remained elevated. These workers concluded that LH-RH is transported from CSF to hypophysial portal blood in significant quantity, and that the decapeptide given intraventricularly is more effective on a prolonged basis in stimulating LH release than when given intravenously.

Other neural peptides.

A number of peptides widely distributed throughout the CNS, including hypothalamus, can be shown to influence pituitary function with different effects produced by intraventricular compared with systemic administration. Substance P (13) and neurotensin (40), for example, stimulate GH and PRL secretion following intravenous injection but inhibit GH and PRL release after intraventricular injection. In the case of Substance P, it appears that intraventricular administration stimulates somatostatin release. Such findings are consistent with the view that neural peptides (including opioid peptides, vasoactive intestinal peptide (VIP), and bombesin) may modulate pituitary function following secretion into the CSF.

Ependymal tanycyte hypothesis.

The hypothesis that the specialized ependymal cells (tanycytes) located at the floor of the third ventricle play a role in endocrine regulation was originally proposed on a purely morphological basis by Lofgren (39). Tanycytes are most abundant in those regions of the ME which correspond to the "hypophysiotropic area", and their structural features are compatible with the suggestion that they take up substances from the ventricular fluid and transport them for local release into the adjacent hypothalamus neuropil (65). Zimmerman et al. (74,75) have reported remarkable amounts of immunoreactive (IR)-LH-RH throughout the course of numerous tanycytes and in some arcuate perikarya of the mouse hypothalamus and suggested that the presence of endogenous IR-LH-RH in tanycytes is evidence that the ventricular system provides a pathway that is of importance in the neural regulation of gonadal function. (These workers also reported neurophysin in tanycytes of the ME). Studies by Naik (43) who reported that the intensity of LH-RH immunofluorescence changes in tanycyte cells during the estrous cycle of the rat provide support for this view, but no staining of the tanycytes was demonstrated by other workers (38,63) in the rat or mouse hypothalamus.

Cramer and Barraclough (17) have reported an LH-RH concentration of 0.33 pg/ml in third ventricular fluid taken from rats. Following electrical stimulation of the medial preoptic area (MPOA), which caused a marked elevation in serum LH, there was no rise in the CSF levels of LH-RH. These authors concluded that CSF does not serve as a vehicle for transport of LH-RH to the ME under physiologic conditions.

On ultrastructural study nerve terminals have been found to be closely apposed to the plasmalemmata of tanycytes in the rat ME (62), and these "axo-tanycytic" endings frequently give the impression of structural continuity across both membrane systems. Despite the fact that structural variations are demonstrable in ependymal cells during the estrous cycle (15), destruction of the tanycytes of the ME by electric cautery or the intraventricular injection of picric acid solution does not affect hypophysial gonadotropin secretion in the rat. These findings suggest that the tanycyte transport of LH-RH from the ventricle to the ME is not of physiologic importance in gonadotropin regulation and this is the generally accepted view of most investigators in this field.   It seems likely that the tanycyte pathway is at best a minor route by which the hypothalamic releasing hormones reach the adenohypophysis.

CSF AS A TRANSPORT SYSTEM

Hypothalamo-Brain Interaction

Significant concentrations of TRH are found in the rat extra-hypothalamic brain (28). Although such concentrations are small when compared with the levels in the hypothalamus, quantitatively over 70% of total brain TRH is found outside the strict anatomic confines of the hypothalamus. To determine whether extrahypo-thalamic brain TRH might be derived from the hypothalamus, pos-sibly via the CSF, we studied the effects of classical hypothal-amic "thyrotrophic area" lesions which brought about a reduction in hypothalamic TRH by two-thirds (29). The extrahypothalamic brain TRH content in these animals was unaffected suggesting that synthesis in such areas occurs in situ (Table 1). Following hypothalamic deafferentation not only were the levels of TRH in the extrahypothalamic brain unaltered but there was a marked reduction in hypothalamic content (9). Surgical isolation of the hypothalamus produces a similar depletion in the LH-RH and som-atostatin content without altering the levels in the extrahypo-thalamic brain (10,11). These findings suggest that much of hypothalamic TRH and other hypothalamic releasing hormones may be synthesized by cells outside this area or be dependent on extra-hypothalamic neuronal connections. In addition, they provide evidence against the view that the hypophysiotropic hormones within the hypothalamus are transported from extrahypothalamic brain sites via the CSF.

Adenohypophysial hormones in the CSF: pituitary-brain vascular connections.
Anterior pituitary hormones, until recently considered sys-temic hormones and essentially excluded from the CNS by the blood brain barrier (BBB), have been demonstrated to be present within brain tissue by a combination of techniques including RIA, immunohistochemistry and bioassay. Their source is controversial (36) but their presence in extrapituitary brain areas suggests a

role for these peptides in neuronal function, analogous to that speculated for the hypophysiotrophic hormones.

Much evidence (effects of hypophysectomy, studies with in vitro culture systems) supports the view that adenohypophysial hormones in brain tissue are synthesized in situ, although it has been suggested by one laboratory that adenohypophysial hormones in brain are derived solely from the pituitary (42).

Pituitary hormones secreted from the adenohypophysis can cross the BBB (56), but the presence of pituitary tumors especially with suprasellar extension leads to the highest CSF concentrations, the measurement of which may be helpful diagnostically (Fig. 3). The suprasellar extension of a pituitary adenoma appears to cause breakdown of the BBB and by a process of "misplaced exocytosis" the pituitary gland releases hormones directly into the CSF. There is, however, some evidence that prolactin (PRL) levels may be increased in patients with hyperprolactinemia due to pituitary adenomas without suprasellar extension (Fig. 3).

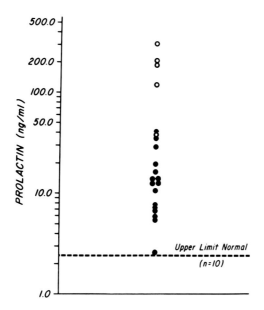

FIG. 3. CSF prolactin levels in patients with prolactinomas (56). o = suprasellar extension of tumor. ● = intrasellar tumor.

The initial suggestion by Popa and Fielding (55) that blood flowed in the portal vessels from pituitary to hypothalamus has recently gained support from the studies of Bergland and his colleagues (4) who have provided evidence that the pituitary secretes to the brain. Studies favoring retrograde transportion

in the portal vessels have been further provided by Oliver et al.
(49) raising the possibility that pituitary hormones may reach
the hypothalamus and influence its function.

There is evidence to support the concept of bidirectional
transport of substances from the ME to the CSF (73). Nakai and
Naito (44) have demonstrated that the ependymal cells in the frog
ME have intracellular bidirectional transport capabilities for
substances from both hypophysial blood and CSF. It is suggested
that once the pituitary hormones reach the ME by retrograde
portal vessel transport they could be carried by either retrograde neuronal pathways (cf. horseradish peroxidase) or ependymal
tanycytes to the CSF for distribution to the brain. Such
ascending transport could allow pituitary hormones to reach
hypothalamic nuclei ("short feedback loop"), and hypothalamic
hormones to either feedback on their own ("ultra-short feedback
loop") or other hypothalamic nuclei. In addition, this mechanism
might permit the distribution of hypothalamic peptides and pituitary hormones to distant parts of the brain, although such a
process remains to be demonstrated.

The CSF as a route for the regional distribution of neural
peptides: behavioral effects of intraventricular
administration.
TRH interacts with other neural peptides with respect to
temperature regulation and other behavioral effects. Most of
these responses have been demonstrated following intraventricular
injection and may not occur with systemic administration. Following intraventricular injection TRH produces hyperthermia and
behavioral excitation in the rabbit (25) and antagonizes the
hypothermic effect of morphine (25) and of β-endorphin in the rat
(24). Intracisternal administration of TRH to the rat also
produces hyperthermia and the elevation of temperature is antagonized by intracisternal injection of bombesin (7). Other effects
of TRH that can be recognized following intraventricular, but not
intravenous administration, include suppression of feeding and
drinking activity (70) and stimulation of colonic activity probably due to stimulation of central cholinergic receptors (67).

Intraventricular, but not intravenous, administration of
eledoisin, a peptide related to Substance P, in concentration of
10^{-11}M stimulates drinking behavior in the pigeon (18). Substance
P itself was also shown to stimulate dipsogenic behavior.
Opposite effects of these substances have been described in the
rat. The central administration of neurotensin produces a marked
hypothermic effect in rats and mice (45), a response which is
unaffected by passive immunization intraventricularly (7). Decreases in locomotor activity in rats and a marked dose-related
enhancement in pentobarbital-induced mortality, sedation and
hypothermia also occur. None of these effects were observed
after peripheral administration of neurotensin.

Evidence Favoring the Specificity of Behavioral Responses

Passive immunization studies.
The administration of sheep antisomatostatin $\check{Y}$-globulin, intraventricularly to the rat, significantly decreases the duration of strychnine-induced seizures and increases the pentobarbital LD_{50} as compared to controls (14). These findings support the view that endogenous somatostatin in the CSF and/or periventricular tissue modulates the response of the CNS to strychnine and pentobarbital and that the CSF may be a conduit important in the physiologic regulation of brain function.

Heroin self-administration in the rat is enhanced following intraventricular administration of vasopressin or PRL antiserum (68). These findings suggest that pituitary hormones are "physiologically" involved in heroin self-administration.

Regional turnover of a neurotransmitter.
Following intraventricular injection of TRH, somatostatin, neurotensin and angiotensin II into rats, evidence was obtained that these peptides modulate the turnover rate of acetylcholine (ACh) in the brain (41). However, whereas TRH increased turnover of ACh in the parietal cortex, the other peptides were ineffective. Somatostatin and neurotensin increased ACh in the diencephalon, whereas angiotensin II and TRH did not, and only somatostatin affected ACh turnover in the brainstem. The selective changes in ACh produced by each individual neural peptide in specific brain regions (12), supports the postulated role of these substances as specific agents in brain function.

It is probably that peptides placed in the CSF reach a specific neuronal region by diffusion from the CSF, since there is no brain CSF barrier. The presence of high affinity receptors, as reported for certain peptides in brain tissue, may well be of functional importance in peptide uptake from CSF and distribution in brain tissue.

BLOOD-BRAIN BARRIER: CIRCUMVENTRICULAR ORGANS

The ability of a molecule to penetrate the BBB is thought to reflect its degree of lipid solubility or the presence of specific transport systems. Kastin et al. (69) have reported that peptides and enkephalins readily penetrate the BBB in vivo following intracarotid injection. The studies reported by these workers were not confirmed by Cornford et al. (16), who measured brain uptake index (BUI) of 2-3% for met-enkephalin compared with 15% by Kastin and his colleagues, whose methodology was criticized by the former group. The BUI of TRH was reported to be 1% as was that of carnosine β-alanyl-histidine (16), a putative neurotransmitter. More recent studies by Rapaport et al. (58) have suggested that the BUI method which measures brain peptide uptake 15 seconds after injection may be relatively insensitive. These workers report that significant permeabilities of modified opioid peptides occur, (see chapter by Partridge, this volume).

In this laboratroy Engler and Jackson (unpublished observations) have found that a continuous infusion of "cold" TRH significantly depresses the entry of ^{125}I-TRH from the systemic circulation to the brain, a finding which raises the possibility of a specific transport system for TRH across the BBB.

Circumventricular Organs

The circumventricular organs are a group of specialized midline structures which differ from typical brain tissue with regard to the ultrastructure of their vascular, ependymal, glial and neuronal components (71) (see chapter by Weindl). They include the ME, neural lobe of the hypophysis, the organum vasculosum of the lamina terminalis (OVLT), the subfornical organ (SFO), subcommissural organ (SCO) and area postrema (AP). All these structures, except the SCO, are areas where the BBB is absent. Increased interest in the circumventricular organs has recently been generated by reports that LH-RH, TRH and somatostatin are located in the OVLT, SFO, and AP in high concentration (33,51).

Immunohistochemical studies in the rodent have localized LH-RH to the OVLT (Fig. 4) where 80% of the LH-RH present in the preoptic region occurs (33). The concentration of LH-RH in the OVLT is over 50% of that found in the rat ME with concentrations of LH-RH in the SFO, SCO, and AP being somewhat lower (33). In the

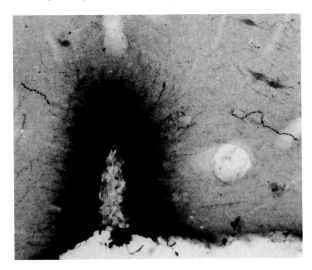

FIG. 4. Frontal section through the brain of the female rat at the level of the medial preoptic region showing dense staining of the OVLT (center) with LH-RH reaction product (PAP technique). Fusiform perikarya (? source of LH-RH in ME and OVLT) and beaded axons staining for LH-RH are evident in the preoptic nucleus (PON) laterally and superiorly to the OVLT. x 100 (Courtesy of Dr. R. Lechan.)

human, the LH-RH concentration in the OVLT is higher than in the
ME (35). There is no evidence to suggest that LH-RH in the OVLT
is present in cell bodies whose axons pass to the ME, for a "portal" relationship between the OVLT capillary vasculature and
sinusoids of the adenohypophysis does not occur, no LH-RH staining perikarya occur in OVLT, and terminal degeneration has not
been observed in the ME after rostral knife cuts (33). The
release of hormones such as LH-RH into the capillaries of the
OVLT is thought to reach the systemic circulation, yet a vascular
connection between the OVLT and the medial preoptic nucleus has
been reported (52). The architecture of the OVLT (and other
circumventricular organs) may allow LH-RH (and other neuropeptides present there) to be released into the CSF; it is also
possible that peptides may be taken up from the CSF by the circumventricular organs.

Function of circumventricular organs.
The function of TRH, LH-RH and somatostatin in the circumventricular organs other than ME is unknown. It is possible that
neural peptides concentrated there (either from CSF or by neuronal connections via axoplasmic flow) may be released into the CSF
to affect brain or hypothalamic-pituitary function. The LH-RH
released into the systemic circulation from the OVLT could constitute an additional effect on gonadotropin secretion from the
adenohypophysis.
Although the role of the classical hypophysiotrophic hormones
in the circumventricular organs is unknown, there is evidence
suggesting their importance in mediating the central effects of
angiotensin II (54). The SFO is a thirst center of the brain and
contains angiotensin II receptors. Angiotensin II is the most
potent dipsogenic substance known, and as little as 0.1 pg
applied to the SFO can induce a dipsogenic response (see chapter
by Epstein, this volume). The OVLT, which like the SFO, is
located in the anterior wall of the third ventricle is also a
dipsogenic receptor site for AII (66). The AP appears to be an
important area in blood pressure regulation, and a pressure
response is produced by the application of AII to this structure.
The SCO is involved in the secretion of a mucopolysaccharide substance termed Reissner's fiber, the function of which is unknown.
However, it has been proposed that the SCO might be involved in
regulating the composition of the CSF, particularly with respect
to its catecholamine concentration (49). Since not only biogenic
amines and related synthesizing hormones (61) but also hypophysiotropic hormones are to be found in circumventricular organs,
including the SCO (33), it is possible that neural peptides could
be involved in release of catecholamines from the SCO into the
CSF.

THERAPEUTIC EFFECTS OF PROCEDURES DESIGNED TO ALTER CSF (?
NEURONAL) CONCENTRATIONS OF PEPTIDES

Patients suffering from chronic intractable pain derive

alleviation of their symptoms following electrical stimulation of the medial thalamic and brainstem periaqueductal regions. This analgesia is associated with a 13-20 fold increase in β-endorphin-like immunoreactivity in human ventricular CSF (1). It can be hypothesized that β-endorphin-like material released into the CSF is transported to locations of the brain where analgesic effects are produced. Successful acupuncture in heroin addicts is associated with an increase in CSF, but not blood, met-enkephalin levels (15).

Whether the changes in opioid peptides found in the CSF represent a "spill over" from the brain or constitute a means of transport to specific brain areas is unknown at this time.

ENDOGENOUS NEURAL PEPTIDES IN THE CSF

The importance of the CSF in relation to the action of neural peptides at hypothalamic, pituitary and brain locations has been discussed. Clearly, however, direct evidence that endogenous neural peptides are present in the CSF, and that the levels of these substances show appropriate changes in response to physiologic stimuli, are necessary pieces of evidence to support the role of the CSF as a conduit for brain regulation. However, since the neural peptides occur not only in the hypothalamus and brain but also in the spinal cord, caution must be exercised in the interpretation of peptide levels measured in lumbar CSF. It is possible that significant changes could occur in the concentration of a peptide in the third ventricle without producing alterations in the levels elsewhere in the ventricular system. Further, even if lumbar CSF peptide concentrations reflect central levels, it is not clear if the peptide comes from the hypothalamus, brain or both. This issue is of particular importance since it is uncertain whether hypothalamic hormones located in extrahypothalamic sites are subject to the same feedback regulation as those present in the hypothalamus. A list of peptides reported present in the CSF is provided in Table 2.

TABLE 2. Peptides in CSF

TRH
LH-RH
Somatostatin
Opioid Peptides
Gastrin/CCK
VIP
Angiotensin II
Substance P
Adenohypophysial Hormones
Neurohypophysial Hormones

Measurement of CSF Neural Peptides

TRH.

Biologically active TRH has been detected in human cadaver CSF drawn from the third ventricle 2-5 hours after death. This material showed TSH releasing activity when incubated with human anterior pituitary tissue in vitro (27). In subsequent studies by Shambaugh et al. (64) TRH was quantitated in human lumbar CSF by radioimmunoassay. TRH was stable in CSF stored at 4C, TRH levels in AM or PM samples obtained from 15 women and 12 men showed levels of 40 pg/ml with no diurnal variation. No sex difference was evident. By contrast, CSF cortisol levels obtained concurrently were two fold higher in AM than PM. Immunoreactive TRH has also been reported in CSF obtained by cisternal puncture from patients undergoing neuroradiologic examination (47); the concentrations ranged from 60 to 290 pg/ml.

LH-RH.

Gunn et al. (21) reported that LH-RH was absent (<1 pg/ml) in 23 out of 26 samples obtained from patients for the purpose of venereal disease serology. In the other 3 samples levels of 22, 25 and 120 pg/ml were obtained but dilutions showed non-parallelism, thus creating doubts about identity. Rolandi et al. (60) however, reported LH-RH in the third ventricle of 5 hydrocephalic patients, the level ranging from 50 to 150 pg/ml.

Other neural peptides.

The presence in human CSF of the opioid peptides β-endorphin and met-enkephalin has been discussed earlier. Rehfeld and Kruse-Larsen (62) have found both gastrin and cholecystokinin (CCK) in human CSF. These workers, using specific radioimmunoassays reported the concentration of gastrin in CSF from 10 neurologically normal persons to range from 1.5 to 3.0 p mol/l whereas the concentration of CCK ranged from 4 to 55 p mol/l. On chromatography both peptides displayed a molecular heterogeneity similar to that found in extracts of brain tissue. Vasoactive intestinal peptide (VIP) which is present in neurons of both the central and peripheral nervous systems, is richly represented immunohistochemically in fibers innervating brain arteries. Such findings led Fahrenkurg et al. (19) to examine the CSF for VIP in patients undergoing myelography or pneumoencephalography. Mean values obtained were 50 p mol/l in "normal" CSF.

Angiotensin II levels of about 100 pg/ml have been reported in the rat cisterna magna (66). High levels of the octapeptide have also been reported in CSF from hypertensive patients and rats while other studies have been unable to confirm the presence of angiotensin II in CSF (see ref.54 for review).

Immunoreactive Substance P has been reported in the CSF of normal individuals at concentrations of 2.9-11.1 fmol/ml (mean 7.6 ± 0.6) in one study (48) and 25-45 pg/ml in another (26).

The presence of adenohypophysial hormones in CSF has been already discussed in this chapter and has been reviewed elsewhere (59). The neurohypophysial hormone, arginine vasopressin, in concentrations of 2.4 $\pm$ 0.7 pg/ml was reported in the CSF of 12 patients without endocrine or brain disease (31). The related nonapeptide arginine vasotocin, thought to be derived from the human fetal pineal gland and SCO, has been reported in relatively high concentrations in the CSF of newborns and infants (56).

DIAGNOSTIC SIGNIFICANCE OF CSF PEPTIDE LEVELS

The concentration of neural peptides in the CSF may be altered in neurologic disease. In a group of patients with cord or cerebral disease the somatostatin concentrations were increased in 20 out of 24 subjects suggesting that somatostatin elevations might indicate release from damaged brain tissue (55). However, low somatostatin content in CSF from patients with multiple sclerosis has recently been reported (72). Patients with brain atrophy have been reported to have reduced levels of VIP in the CSF (19).

Recently, Substance P levels in the CSF have been reported altered by neuronal disease. In patients with peripheral neuropathy and autonomic dysfunction (Shy-Drager Syndrome) there was a marked reduction in CSF Substance P concentration (48). These investigations suggested that lumbar CSF Substance P arises largely from spinal cord, nerve roots or dorsal root ganglia and that pathological processes affecting these structures may be reflected by a reduced CSF Substance P concentration. In contrast, patients with lumbar arachnoiditis have been reported to have CSF Substance P levels 6-10 times normal (26). The associated pain was reversed by morphine administration which concomitantly was associated with a fall of CSF Substance P levels to almost normal. These workers have proposed that the chronic pain of lumbar arachnoiditis may be caused by the release of Substance P from nociceptive afferent fibers. In other studies, patients with subarachnoid hemorrhage have been shown to have increased CSF vasopressin levels (31), while patients with depression have been reported to have increased CSF TRH levels (33). The diagnostic significance of anterior pituitary hormone levels in CSF has been discussed earlier.

The discrete and separate distribution of each neural peptide within the nervous system raises the possibility that alteration in the levels of individual peptides in the CSF might result from injury to specific regions of the brain or spinal cord.

SUMMARY

The widespread distribution of peptides throughout the CNS necessitates a reappraisal of the CSF as a conduit for these substances in the modulation of neuronal function in different brain regions. Some of the endogenous brain peptides have been detected in CSF, but their physiological significance has not been determined. Hypothalamic deafferentation in the rat leads to a reduction of the hypothalamic content of TRH and LH-RH with concomitant hypothyroidism and hypogonadism, respectively, without affecting the extrahypothalamic brain content of these peptides. Such findings would suggest that neuronal rather than CSF connections were important in the brain regulation of hypothalamic hypophysiotrophic hormone secretion and function. However, the physiologic significance of the CSF in neuronal function is supported by the effect of passive immunization studies with anti-somatostatin injected into the CSF which leads to changes in response to the subsequent administration of strychnine or barbiturate.

The neural peptides have limited access across the BBB, but in those areas (circumventricular organs) where the BBB is altered, hypothalamic hormones have been detected in high concentration (especially in the OVLT). Their function in these locations is not clearly understood.

Physiologic implications derived from measurement of neural peptides in the CSF is fraught with danger since it is unclear whether lumbar CSF peptide concentration is representative of changes only in the spinal cord or is indicative of central function. Further studies are required to clarify and characterize changes in peptide hormone concentration in different parts of the ventricular system in order to determine whether the CSF regulates brain function through its distribution of neural peptides or whether the CSF is merely a biologic fluid that provides the CNS a means of disposing of its waste products. There is some evidence that CSF levels of peptides can be altered in response to CNS pathology. Further evaluation is necessary, but the unique distribution of each of the neural peptides raises the exciting possibility that measurements of brain peptides may provide a sensitive "marker" for the anatomic diagnosis of specific areas of CNS damage.

ACKNOWLEDGEMENTS

I thank Dr. Ronald Lechan for the immunohstochemistry, and Ms. Amy Kassierer for typing the manuscript. This work was supported in part by NIH Grant AM 21863 to the author.

REFERENCES

1. Akil, H., Richardson, D.E., Barchas, J.D., and Li, C.H. (1978): Proc. Natl. Acad. Sci. USA, 75:5170-5172.
2. Alpert, L.C., Brawer, J.R., Patel, Y.C., and Reichlin, S.(1976): Endocrinology, 98:255-258.
3. Ben-Jonathan, N., Mical, R.S., and Porter, J.C. (1974): Endocrinology, 95:18-25.
4. Bergland, R.M., and Page, R.B. (1979): Science, 204:18-24.
5. Blackwell, R.E., and Guillemin, R. (1973): Ann. Rev. Physiol., 35:357-390.

6. Brawer, J.R., Sun Lin, P.S., and Sonnenschein, C. (1974): Anat. Rec. 179:481-490.
7. Brown, M., Rivier, J., Kobayashi, R., and Vale, W. (1978): In: Gut Hormones, edited by S.R. Bloom, pp. 500-558. Churchill Livingstone, Edinburgh.
8. Brownstein, M.J., Palkovits, M., Saavedra, J.M., Bassiri, R.M., and Utiger, R.D. (1974): Science, 185:267-269.
9. Brownstein, M.J., Utiger, R.D., Palkovits, M. and Kizer,J.S. (1975): Proc. Natl. Acad. Sci. USA, 72:4172-4179.
10. Brownstein, M.J., Arimura, A., Schally, A.V., Palkovits, M., and Kizer, J.S. (1976): Endocrinology, 98:662-665.
11. Brownstein, M.J., Arimura, A., Fernandez-Durango, R., Schally, A.V., Palkovits, M., and Kizer, J.S. (1977): Endocrinology, 100:246-249.
12. Burt, D.R., and Snyder, S.H. (1975): Brain Res., 93:309-328.
13. Chihara, K., Arimura, A., Coy, D.H., and Schally, A.V. (1978): Endocrinology, 102:281-290.
14. Chihara, K., Arimura, A., Chihara, M., and Schally, A.V. (1978): Endocrinology, 103:912-916.
15. Clement-Jones, V., McLoughlin, L., Lowry, P.J., Besser, G.M., Rees, L.H., and Wen, H.L. (1979): Lancet, 2:380-383.
16. Cornford, E.M., Braun, L.D., Crane, P.D., and Oldendorf, W.H. (1978): Endocrinology, 103:1297-1303.
17. Cramer, O.M., and Barraclough, C.A. (1975): Endocrinology, 96: 913-921.
18. Evered, M.D., Fitzsimmons, J.T., and DeCaro, G. (1977): Nature, 267:332-333.
19. Fahrenkrug, J., Schaffalitzky de Muckadell, O.B., and FahrenKrug, A. (1977): Brain Res., 124:581-584.
20. Gordon, J.H., Bollinger, J., and Reichlin, S. (1972): Endocrinology, 91:696-701.
21. Gunn, A., Fraser, H.M., Jeffcoate, S.L., Holland, D.T., and Jeffcoate, W.J. (1974): Lancet, 1:1057.
22. Harris, G.W. (1948): Physiol. Rev., 28:139-179.
23. Hokfelt, T., Elde, R., Fuxe, K., Johansson, O., et al. (1978): In: The Hypothalamus, edited by S. Reichlin, R.J. Baldessarini, and J.B. Martin, pp. 69-135. Raven Press, New York.
24. Holaday, J.W., Tseng, L.F.,Loh, H.H., and Li, C.H. (1978): Life Sci., 22:1537-1544.
25. Horita, A., and Carino, M.A. (1975): Psychopharmacol. Commun., 1:403-415.
26. Hosobuchi, Y., Emson, P., and Iverson, L. (1980): American Association of Neuro-Surgeons, Annual Meeting, New York City, April 20-24, p. 25.
27. Ibata, Y., Watanabe, K., Konoshita, H., Kubo, S. and Sano, Y. (1979): Cell Tiss. Res., 198:381-395.
28. Ishikawa, H. (1973): Biochem. Biophys. Res. Comm., 54: 1203-1209.
29. Jackson, I.M.D., and Reichlin, S. (1974): Endocrinology, 96:854-862.

30. Jackson, I.M.D., and Reichlin, S. (1977): Nature, 267:853-854.
31. Jenkins, J.S., Mather, H.M., and Ang, V. (1980): J. Clin. Endocrinol. and Metab., 50:364-367.
32. Kastin, A.J., Nissen, C., Schally, A.V., and Coy, D.H. (1976): Brain Res. Bull. 1:583-589.
33. Kirkegaard, C., Farber, J., Hummer, L., and Rogowski, P. (1979): Psychoneuroendocrinology, 4:227-235.
34. Kizer, J.S., Palkovits, M., and Brownstein, M.J. (1976): Endocrinology, 98:311-317.
35. Knigge, K.M., and Silverman, A.J.. (1972): Median Eminence Structure and Function, edited by K.M. Knigge, D.E. Scott, and A. Weindl, pp. 350-363. Karger, Basel.
36. Koch, Y., and Okon, E. (1979): Int. Rev. Exp. Path., 19:45-62.
37. Krieger, D.T., and Liotta, A.S. (1979): Science, 205: 366-372.
38. Kronheim, S., Berelowitz, M., and Pimstone, B.L. (1977): Clin. Endocrinol., 6:411-415.
39. Lechan, R.M., Alpert, L.C., and Jackson, I.M.D. (1976): Nature, 264:463-465.
40. Lofgren, F. (1959): Acta-Morph. Neerl. Scand., 2:220-229.
41. Maeda, K., and Frohman, L.A. (1978): Endocrinology, 103: 1903-1909.
42. Malthe-Sorenssen, D., Wood, P.L., Cheney, D.L., and Costa, E. (1978): J. Neurochem., 31:685-691.
43. Moldow, R.L., and Yalow, R.S. (1978): Life Sci., 22:1859-1864.
44. Naik, D.V. (1976): Cell Tiss. Res., 173:143-166.
45. Nakai, Y., and Naito, N. (1974): J. Electron Microscope, 23: 19-32.
46. Nemeroff, C.B., Bissette, G., Prange, Jr., A.J., Loosen, P.T., Barlow, T.S., and Lipton, M.A. (1977): Brain Res., 128:485-496.
47. Nozaki, M., Taketani, Y., Minaguchi, H., Kigawa, T., and Kobayashi, H. (1979): Cell Tiss. Res., 197:195-212.
48. Nutt, J.G., Mroz, E.A., Leeman, S.E., Williams, A.C., Engel, W.K., and Chase, T.N. (1980): Neurology, (in press).
49. Oliver, C., Charvet, J.P., Codaccioni, J.L., Vague, J., and Porter, J.C. (1974): Lancet, 1:873.
50. Oliver, C., Ben-Jonathan, N., Mical, R.S., and Porter, J.C. (1975): Endocrinology, 97:1138-1143.
51. Oliver, C., Mical, R.S., and Porter, J.C. (1977): Endocrinology, 101:598-604.
52. Ondo, J.G., Eskay, R.L., Mical, R.S., and Porter, J.C. (1973): Endocrinology, 93:231-237.
53. Palkovits, M., Brownstein, M.J., Arimura, A., Sato, H., Schally, A.V., and Kizer, J.S. (1976): Brain Res. 109:430-434.

54. Palkovits, M., Mezey, E., Ambach, G., and Kivovics, P. (1977): Brain-Endocrine Interaction III. Neural Hormones and Reproduction, 3rd Int. Symp. Wurzburg, pp. 302-312 S. Karger, Basel.
55. Patel, Y.C., Rao, K., and Reichlin, S. (1977): N. Engl. J. Med., 269:529-533.
56. Pavel, S. (1980): J. Clin. Endocrinol. Metab., 50:271-273.
57. Phillips, M.I. (1978): Neuroendocrinology, 25:354-377.
58. Popa, G., and Fielding, U. (1930): J. Anat., 65:88-91.
59. Post, K.D., Biller, B.J., and Jackson, I.M.D. (1980): In: Neurobiology of Cerebrospinal Fluid, edited by J.H. Wood. Plenum Press, New York. (In Press.)
60. Randall, R.V. (1979): Arch. Int. Med., 139:1092.
61. Rapaport, S.I., Klee, W.A., Pettigrew, K.D., and Ohno, K. (1979): Science, 207:84-86.
62. Rehfeld, J.F., and Kruse-Larsen, C. (1978): Brain Res., 155: 19-26.
63. Reichlin, S., Saperstein, R., Jackson, I.M.D., Boyd, A.E., III, and Patcl, Y. (1976): Ann. Rev. Physiol., 38:389-424.
64. Rolandi, E., Barreca, T., Masturzo, P., Gianrossi., R., Polleri, A., Perria, C. (1976): Lancet, 1:1080.
65. Saavedra, J.M., Browstein,M.J., Kizer, J.S., and Palkovits, M. (1976): Brain Res., 178:412-417.
66. Scott, D.E., and Paull, W.K. (1979): Cell Tiss. Res., 200:329-334.
67. Setalo, G., Vigh, S., Schally, A.V., Arimura, A., and Flerko, B. (1976): Brain Res., 103:597-602.
68. Shambaugh, G.E. III, Wilber, J.F., Montoya, E., Ruder, H., and Blonsky, E.R. (1975): J. Clin. Endocrinol. Metab., 41: 131-134.
69. Silverman, A.J., Knigge, K.M., Ribas, J.L., and Sheridan, M.N., (1973): Neuroendocrinology, 11:107-118.
70. Simpson, J.B., Saad, W.A., and Epstein, A.N. (1976): In: Regulation of Blood Pressure by the Central Nervous System, edited by G. Onesti, M. Fernandes, and K.E. Kim., pp. 191-202. Grune and Stratton, New York.
71. Smith, J.R., LaHann, T.R., Chestnut, R.M., Carino, M.A., and Horita, A. (1976): Science, 196:660-662.
72. Sorenson, K.V., Christensen, S.E., Dupont, E., Hansen, A.P., Pedereson, E., and Orskov, H. (1980): Acta. Neurol. Scand., 61:186-191.
73. Van Ree, J.M., and de Wied, D. (1977): Life Sci., 21:315-320.
74. Vigneri, R., Pezzino, V., Filetti, S., Squatrito, S., Corso, A., Maricchiolo, M., Polosa, P., and Scapagnini, U. (1977): Neuroendocrinology, 23:171-180.
75. Vijayan, E., and McCann, S.M. (1977): Endocrinology, 100:1727-1730.
76. Weindl, A. (1973): Frontiers in Neuroendocrinology, edited by W.F. Ganong, and L. Martini, pp. 1-32. Oxford University Press, London.

77. Weiner, R.I., Terkel, J., Blake, C.A., Schally, A.V., and Sawyer, C.H. (1972): Neuroendocrinology, 10:261-272.
78. Wittkowski, W. (1968): Electron Microscopical Studies on the Neurohypophysis of the Rat, 86:111-128.
79. Zimmerman, E.A., Hsu, K.D., Ferris, M., and Kozlowski, G.P. (1974): Endocrinology, 95:1-8.
80. Zimmerman, E.A., Kozlowski, G.P., and Scott, D.E. (1975): Brain-Endocrine Interaction II. The Ventricular System, edited by K.M. Knigge, D.E. Scott, H. Kobayashi, and S. Ishii. pp. 123-134. Karger, Basel.

Neurosecretion and Brain Peptides,
edited by J. B. Martin, S. Reichlin, and K. L. Bick.
Raven Press, New York © 1981.

Introduction to Section VI:
Functions of Neuropeptides in Homeostasis

Joseph B. Martin

Department of Neurology, Massachusetts General Hospital, Boston, Massachusetts 02114

Virtually all homeostatic functions are mediated by the brain. Feeding and drinking behavior are mainly integrated within the hypothalamus and involve an interaction of a number of pharmacologically defined neural pathways, including bioaminergic, cholinergic and peptidergic. Control of body temperature also is organized at the level of the hypothalamus and involves activation of both the autonomic nervous system and the hypothalamic-pituitary axis. Cardiovascular functions are less clearly determined by the hypothalamus; blood pressure control is mediated at several levels of the neuraxis, in particular by centers in the medulla. The chapters of this section review recent developments in the analysis of the role of neuropeptides in these functions.

The brain renin-angiotensin system has an ancient history insofar as neuropeptide interactions are concerned. The clarification of the biosynthetic pathway for angiotensin II, first established in peripheral tissues, provided a manageable system for experimental investigation of a similar pathway in the central nervous system (CNS). Angiotensin-induced thirst remains one of the most potent of known CNS peptide effects even though the physiologic significance of the mechanism remains in dispute. The chapters by Ganten and Epstein address these developments and controversies.

The importance of hypothalamic control of feeding behavior is well established, at least insofar as information gained by classical techniques of lesions and electrical stimulation are concerned. Elucidation of the neuropharmacology of this control has been more difficult. Several biogenic amine systems have been implicated, but definition of individual neurotransmitter effects has thus far failed to provide a unified hypothesis. Peptides have now entered the scene and Smith carefully describes the difficulties that arise and which must be addressed in assessment of CNS versus peripheral tissue effects. The results of this analysis indicate that a considerable amount of work remains to be done before a satisfactory working hypothesis can be developed.

The chapter by Brown examines the role of peptides in body temperature regulation. Somatostatin, bombesin, the opiod peptides and neurotensin have each been shown to have central thermoregulatory effects, inducing either hypothermia or hyperthermia after administration into the brain or cerebrospinal fluid. The remarkable potency of these peptides in thermoregulation is reminiscent of the effect of angiotensin in thirst. These observations offer a number of important physiological problems for future investigation, and the results can be expected to have clinical implications.

The central control of blood pressure and the abnormalities found in hypertension are considered in the chapter by Reis. Models that closely mimic essential hypertension have been extensively investigated with particular attention to the function of the medullary nucleus of the tractus solitarius. This region of the brainstem receives abundant afferent input from cranial nerves V, VII, IX and X, and functions to integrate these signals for regulation of heart rate and blood pressure. Only recently have neuropeptides begun to be incorporated into a general scheme of medullary cardiovascular control. Substance P is present in baroreceptor afferent nerves and in the nucleus of the tractus solitarius. Other peptides, including vasopressin and neurophysin(s), enkephalins and neurotensin, are also found in this region. The functions of these peptides in cardiovascular control remain to be defined.

Holaday and Faden have shown prescience in their approach to the investigation of the role of endogenous opioids in shock. Their data provide evidence that naloxone, an opiate antagonist, can reverse the morbidity and mortality of shock caused by hemorrhage (hypovolemia), endotoxins and spinal cord injury. These observations point to a potentially harmful effect of endogenous opioid peptide release in shock states and have already sparked interest on the part of internists to initiate clinical trials to examine the efficacy of naloxone administration.

The analysis of brain peptidergic systems in neuroendocrine regulation together with examination of effects mediated by the autonomic nervous system provide a fertile ground for future speculation and experimentation. Considering that more than thirty peptides have already been found in the hypothalamus and brainstem, it becomes evident that a great amount of work remains to be done before a full understanding is achieved. The results of this research already hold forth new promises for an understanding of the pathophysiology and treatment of certain disorders of central nervous system function.

Neurosecretion and Brain Peptides,
edited by J. B. Martin, S. Reichlin, and K. L. Bick.
Raven Press, New York © 1981.

The Brain Renin–Angiotensin System

D. Ganten, G. Speck, P. Schelling, and Th. Unger

*Department of Pharmacology, University of Heidelberg, and German High Blood Pressure
Research Institute, 6900 Heidelberg, West Germany*

The first approach to the discovery and function of peptides
in the brain has commonly been the extraction and identification
of the peptide and, subsequently, the administration of synthetic
peptides into the brain to test their pharmacology.

The discovery of the brain renin-angiotensin system was quite
different. Due to the intensive investigations of the kidney
renin-angiotensin system (RAS), its components are well known
(Fig. 1).

Angiotensin II (ANG II) is generated through a reaction
cascade involving different enzymes and peptide precursors.
Briefly, the decapeptide angiotensin I (ANG I) is cleaved from
the high molecular weight precursor angiotensinogen by the
protease renin (E.C.3.4.99.19) which splits the leu–leu bond in
positions 10 and 11 of the angiotensinogen molecule. Angiotensin
I is transformed into the effector peptide ANG II by the converting enzyme (CE) (E.C.3.4.15.1) which cleaves the C-terminal his-
leu dipeptide. The octapeptide ANG II finally reacts with specific receptors in target organs and is rapidly degraded by angiotensinases (E.C.3.4.99.3).

Due to the clinical interest in this enzyme system in hypertensive disease, pharmacological intervention in the RAS was
investigated early. The most important approaches were the development of ANG II receptor antagonists, inhibitors of the CE and
renin inhibitors such as pepstatin (15,34,36,47).

Taking advantage of this detailed knowledge of the plasma and
kidney RAS, and given the known high concentrations of renin in
the uterus and the submaxillary gland, which may even exceed that
of kidney (13), the existence of a brain RAS separate from the
kidney plasma has been investigated.

RENIN (E.C.3.4.15.1)

In the brain, enzymes which generate ANG I from natural angiotensin under <u>in vitro</u> conditions were first described by Fischer-

Ferraro et. al. (10) and Ganten et. al. (14) in 1971. These results were confirmed in different animal species and by several groups independently (6,20,24,26,36,39,41,44). The brain enzyme was consistently shown to generate ANG I upon incubation with homologous, heterogous, and synthetic substrates; it had no non-specific peptidase activity and did not hydrolyse ANG 1.

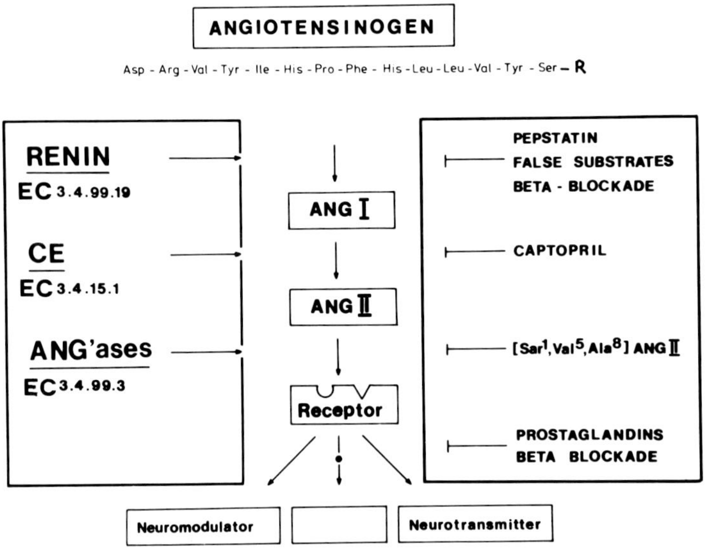

FIG. 1.   General outline of the renin-angiotensin system. The enzymes involved in the synthesis of the effector peptide angiotensin II from its high molecular weight precursor angioten-sinogen via the inactive intermediate decapeptide precursor angiotensin I are shown on the left; inhibitors of the enzymes and receptor blockers are shown on the right.

The physico-chemical properties of the brain enzyme show several similarities to kidney renin (4-6,10-17,20,22-24,31-33,35). It was suggested that brain renin could be identical with cathepsin D, an acid protease known to be present in brain tissue (39), since renin activity and acid protease activity had a similar distribution in dog brain, and since a semipurified brain enzyme preparation process exhibited both renin activity and acid protease activity (39). The same conclusion was drawn for the rat brain enzyme which was purified by pepstatin affinity chromatography as the main purification step (18). More recently, renin was separated from cathepsin D in brains of rats (24), humans (15,17), dogs (35), sheep (7) and mice (44) (Fig. 2). Some workers used affinity chromatography for purification with casein, pepstatin or hemoglobin as ligands (15,17,20). Other methods involving salt precipitation or gel filtration followed by procedures such as ion-exchange chromatography and isoelectric focusing have also been used successfully to purify renin from brain, sub-maxillary gland and kidney (4,7,35,37). Recently, high renin concentrations were also observed in mouse brain. Cathepsin D-

SEPARATION OF MOUSE BRAIN RENIN AND CATHEPSIN ON BLUE AGAROSE COLUMN

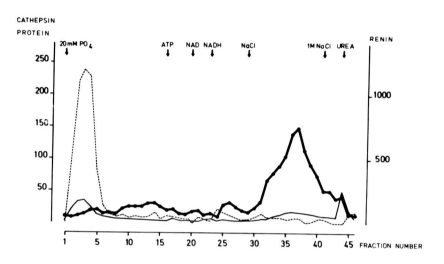

```
━━●RENIN pg ANG I /100 μl / h

------ CATHEPSIN μg BSA eq. /100 μl / h

━━━ PROTEIN    μg BSA / 100 μl
```

FIG. 2. Final purification step of mouse brain renin. Sepa-ration from cathepsin D-like acid protease activity was accomp-lished on Cibachron blue chromatography. Cathepsin D-like enzymes were not retained on the column. Brain renin eluted with a 1 M NaCl gradient. Mouse brain renin had no angiotensinase or non-specific protease activity. The optimum of enzyme activity was in the neutral pH range. Mouse kidney renin was separated from cathepsin D-like enzyme activity in the same way.

like activity could clearly be separated from true renin activity (44). The brain renin was inhibited by specific renin antibodies.
    Renin purified from rat brain exhibited the highest yield of ANG I in the neutral pH range (24). The pH curve for dog brain enzyme revealed three optima, a first at a very low pH, perhaps due to acid denaturation of the substrate, a second between pH 4 and pH 5, and a third at pH 6.5 (35). Human brain renin had its pH optimum of enzyme activity between 5.5. and 6.5. (15,17,44).

The ANG I generation from rat plasma substrate could be inhibited when rat brain renin was pre-incubated with a specific antibody against pure hog kidney renin (20,24,44). The same antibody had no influence on the reaction between purified dog brain renin and dog plasma angiotensinogen (35). Immunoreactive renin was measured in the pituitary of a hog nephrectomized 40 hours previously, by a direct radioimmunoassay (RIA) for hog kidney renin. The ANG I formation by the pituitary enzyme with hog and rat angiotensinogen was partly inhibited by this renin antibody (20). Renin-like immunofluorescence was demonstrated in mouse and rat brain using a specific antibody produced against pure mouse submaxillary gland renin (45). Human brain renin, which was purified from cathepsin D-like activity, has been shown to be biologically active, since injections of this preparation into brain ventricles of rats caused blood pressure increases which were completely reversed when the ANG II antagonist, saralasin, was applied. At the same time, a stimulation of angiotensin generation within the brain was measured by RIA (15,17,40).

It seems to be clear from the purification studies that different enzymes in the brain are able to split ANG I from several substrates, but also from homologous local brain angiotensinogen. One of these enzymes is closely related to kidney renin by its pH optimum, its lack of acid protease activity, its inhibition by antibodies raised against kidney renin, and its in vivo activity on brain angiotensinogen. Other enzyme fractions purified from brain are ascribed to acid proteases by their ability to degrade an acid denatured hemoglobin substrate. Some of these enzymes specifically cleave ANG I from heterologous and homologous plasma substrate at a lower pH and are not inhibited by kidney renin antibodies. It is possible that some of these enzymes contribute to angiotensin formation (17) or to the biosynthesis of other neuropeptides (3) in brain tissue.

CONVERTING ENZYME (E.C.3.4.15.1) AND ANGIOTENSINASES (E.C.3.4.99.3)

Converting enzyme is a rather non-specific dipeptide carboxypeptidase which cleaves C-terminal dipeptides from various substrates (9). This enzyme also is reported to degrade Substance P, kinins, and opioid peptides. Converting enzyme is located in synaptosomes and the choroid plexus also is a rich source. The regulatory role of the enzyme for the generation of ANG II in vivo remains to be studied. The availability of potent inhibitors of CE marks an important step forward in the study of brain peptide function (34). It may well be that CE is the key to our understanding of brain peptide synthesis and metabolism.

The term "angiotensinases" summarizes a group of amino peptidases, carboxypeptidases and endopeptidases. Their activity in brain tissue is high and leads to rapid inactivation of ANG II in vitro. For detailed information on these enzymes the reader is referred to recent reviews (12,15,36,39).

ANGIOTENSIN

Plasma angiotensinogen is a glycoprotein of hepatic origin. The molecular weight is estimated to be 56,000 daltons by disc electrophoresis of pure substrate (8). The leu–leu bond in positions 10 and 11, which is hydrolysed by renin, was recently reported to consist of leu–val in man (46). Thus, differences in the precursor molecules may exist and the diversity of angiotensinogens, renins and angiotensin peptides may be broader than previously assumed.

The presence of angiotensinogen in brain tissue was first reported in 1971 (14). It was extracted from dog brain and contamination with plasma angiotensinogen in dog brain was later confirmed by others, and even higher levels than in the original description were reported (38,39). Angiotensinogen has now been measured in brain tissue of dog, rabbit, sheep, and rat (13,28, 38,39,48). Angiotensinogen is also present in CSF of all species tested so far. Expressed as specific concentrations on a protein basis, the levels are 10 to 50 times higher than in plasma. The prohormone concentrations measured in brain tissue, CSF and plasma of different species are summarized in Table 1.

Partially purified brain and CSF angiotensinogens were compared with the plasma prohormone in order to obtain information about their origin. No biochemical differences were found between plasma and CSF angiotensinogen in dogs after gel filtration or after kinetic studies with hog renin (39). Rabbit angiotensinogens purified from plasma and brain were both shown to be glycoproteins. Both angiotensinogens resembled each other in their isoelectric-focusing profiles (38). Using a similar extraction method for sheep angiotensinogen, it was found that the prohormones from brain, CSF and plasma, respectively, varied from each other in their isoelectric-focusing profiles (28,48). Neuraminidase digestion, which cleaves terminal sialic acids from the angiotensinogen molecule, reduced the isoelectric points to 6.7 but did not abolish the differences between brain, plasma and CSF substrates. In vivo experiments showed that the angiotensinogen levels in plasma and brain were regulated in an independent way, since a striking rise of plasma substrate levels provoked by dexamethasone or by nephrectomy failed to change the CSF angiotensinogen concentrations (39). Furthermore, ^{125}I-angiotensinogen does not seem to cross the blood-CSF barrier, since no radioactivity entered the intraventricular space within 2 hours after the intravenous injection of the tracer. Hence, the penetration of plasma prohormone into the CSF appears to occur very slowly, if at all. This finding and the physico-chemical differences between angiotensinogen from brain, CSF and plasma, and the high specific concentrations in CSF favor the idea that angiotensinogen can be synthesized locally within the brain (for review see: 15,28,36,39,41,48).

The distribution of angiotensinogen agrees in general with that of receptor sites for ANG II (28). The presence of substrate within the brain has been convincingly documented in vivo, since intracranial renin injections trigger central effects which can be ascribed to ANG II and since effects are blocked by agents that interfere either with the synthesis of ANG II or with its receptors (15,42).

TABLE 1.   Angiotensinogen concentrations in brain tissue CSF and plasma of different species

	man	sheep	rabbit	rat	dog
plasma	3000[a]	1468[a]	1200[a]	602[b]	1053[a]
		15[b]		7[b]	16[b]
					861[a]
					12[b]
CSF	266[a]	328[a]		23[a]	205[a]
	468[b]	744[b]		52[b]	247[b]
					127[a]
					315[b]
brain		191[a]	261[a]	54[b]	
		19[b]			

Mean values are indicated in ng ANG I equivalents per ml (a) or in ng ANG I equivalents per mg protein (b). (For references see: 15,28,38,39,41,48).

BRAIN ANGIOTENSIN, ANGIOTENSIN RECEPTORS, AND POSSIBLE FUNCTIONS OF BRAIN ANGIOTENSIN

Angiotensin was first reported to be present in brain tissue of extracts of dogs and rats in 1971 (10,14). This material largely corresponded to synthetic ANG II by the classical criteria, but the identity of brain ANG II with plasma ANG II and its amino acid sequence remain to be determined.

The distribution of ANG II-like immunoreactivity in brain has been published (for review see: 12,36). The peptide is clearly localized in neuronal elements and rat brain cells that have been maintained in tissue culture.

Angiotensin receptors in brain have been studied by several groups and largely correspond to those in classical target organs such as the adrenal gland or arterial smooth muscle (2,43).

If we then accept that there is a complete endogenous RAS in the brain, the next question has to be asked: "What is its function?". This problem has been approached by different methods such as activity measurements of the brain RAS in various pathophysiological conditions and by pharmacological interference with the RAS. The latter approach will be exemplified in this paper. The converting enzyme inhibitor SQ 14225 (captopril) was used to inhibit the conversion of ANG I to the effector peptide ANG II. The ANG II analogue lsar-8ala-ANG II, saralasin, was employed to inhibit the action of ANG II at the receptor level.

Due to the higher concentrations of angiotensinogen in the CSF, ANG I and ANG II are formed when renin is injected into the brain ventricles. Angiotensinogen decreases over the same time, suggesting that brain renin cleaves ANG I from CSF angiotensinogen, which is then converted to ANG II (Fig. 3).

This experimental system, i.e., stimulation of the biosynthesis of endogenous brain angiotensin, we believe, is unique for the brain RAS and sets it apart from other brain peptide systems (21). It can be used to test the in vivo activity of purified enzymes, which has been done to confirm the in vivo activity of brain renin (15, 40). It can be pharmacologically blocked at all stages of the enzymatic cascade by inhibiting renin and CE activity or angiotensin receptors. Perhaps even more important, the biological role of endogenous brain peptides can be tested in this manner without having to resort to the injection of exogenous synthetic peptides which has several drawbacks. Of particular importance are the problems of unphysiologically high dosage and different receptor availability of the exogenous versus the endogenous peptide.

One biological effect of stimulated endogenous angiotensin biosynthesis is an increase of blood pressure which remains elevated for approximately 1 hour after renin administration, while ANG II is continuously being generated in the brain ventricles. Logically, this blood pressure increase can be inhibited by angiotensin receptor blockade with the analogue saralasin (36, 40,42). The renin-induced blood pressure increases are followed by a characteristic pattern of circulating hormones which are summarized in Fig. 4.

It is now well documented that brain angiotensin leads to an elevation of blood pressure via the stimulation of sympathetic tone, plasma norepinephrine and epinephrine, antidiuretic hormone (ADH) and corticosterone (15,16,36,42). Plasma corticosterone has been shown to increase blood vessel sensitivity to circulating pressor hormones. Interestingly, Wallis et al. (48) have shown that corticosterone increases brain angiotensinogen con

RENIN - ANGIOTENSIN COMPONENTS IN CSF OF DOGS FOLLOWING

RENIN INJECTION INTO THE BRAIN VENTRICLES

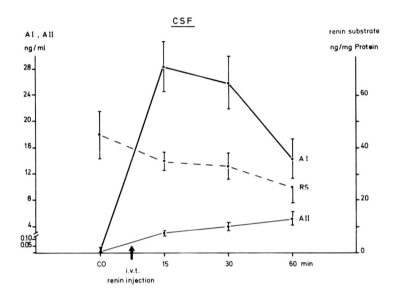

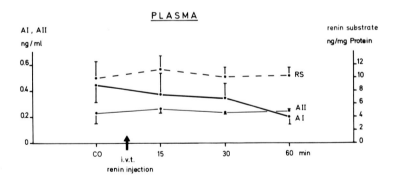

FIG. 3. In vivo angiotensin formation in CSF of dogs after renin injection (1.0 Goldblatt units) into the lateral brain ventricle. Renin substrate (angiotensinogen), angiotensin I (AI), and angiotensin II (AII) were measured in CSF and in plasma simultaneously 15 minutes before (Co) and 15, 30, and 60 minutes after renin administration. Changes in plasma were not significant before and after renin injection.

centration which, in turn, may activate the brain RAS and, thus, centrally and peripherally potentiate the mechanism just outlined.

The question whether central peptidgergic stimulation by the brain RAS could contribute to the elevation of blood pressure in hypertensive disease was studied in spontaneously hypertensive

CENTRAL PEPTIDERGIC STIMULATION

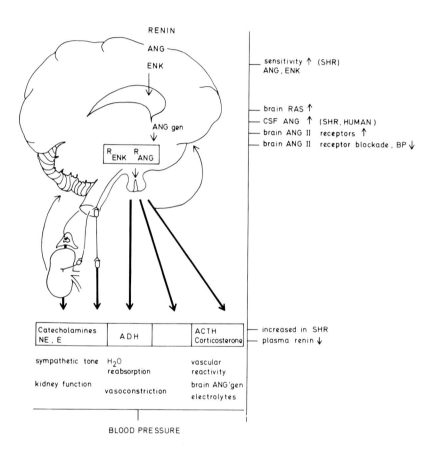

FIG. 4. Summary of possible mechanisms leading to an elevation of blood pressure by stimulation of central angiotensin receptors (R ANG): Typically, this leads to an increase of sympathethic tone, an elevation of plasma norepinephrine (NE) and epinephrine (E), of antidiuretic hormone (ADH), and corticosterone. A similar mechanism may be activated by stimulation of Leu-enkephalin receptors (R ENK) (16). Data supporting a role for angiotensin in the maintenance of high blood pressure in spontaneously hypertensive rats (SHR) are summarized on the right (↑ increased; ↓decreased.

rats, one model for human essential hypertension (15,16,26,29). These rats exhibit increased sympathetic tone and plasma catecholamines and elevated levels of adrenocorticotrophic hormone (ACTH) and corticosterone (16). In these rats, the plasma RAS is suppressed and they have an increased activity of the brain RAS.

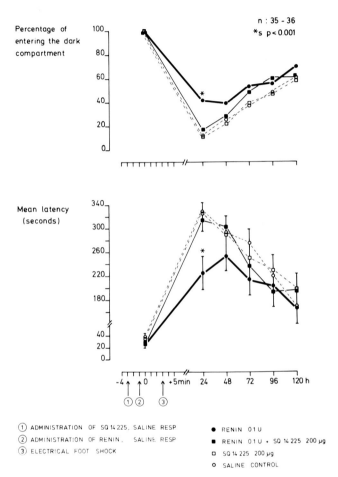

FIG. 5. Effects of endogenous brain ANG II on passive avoidance behavior in rats. Animals received an electrical foot shock after entering a black box on the first experimental day ③. Avoidance was tested every 24 hours for 120 hours. Renin, renin plus CE inhibitor SQ 14225, SQ 14225 alone or saline were injected 2 minutes before the first test at ① ② into the lateral brain ventricles. A disruption of avoidance learning was observed in the renin-treated rats. The frequency of rat entry into the dark compartment increased (upper panel) and the mean latency to enter the dark compartment decreased (lower panel) (27).

Also, CSF angiotensin levels have been reported to be elevated and the number of brain angiotensin receptors appears to be augmented. In addition, spontaneously hypertensive rats show a supersensitivity to central angiotensin and other brain peptides (Fig. 4).

Thus, if the brain RAS contributed to the elevation and maintenance of high blood pressure in the spontaneously hypertensive rat, pharmacological inhibition of the enzyme-peptide system and central blockade of the angiotensin receptors should result in a lowering of blood pressure. Indeed, this has been shown by several groups, including ourselves (15,16,29,36). Not only does this confirm the role of brain ANG II in blood pressure regulation, but also indicates that the peptide is synthesized centrally, because saralasin had no effect when administered peripherally, and the blood pressure-lowering effect persisted after nephrectomy when the kidney-plasma RAS was totally absent (29).

We conclude that one of the rules of the evolution of peptide hormones, namely that "everything can be made everywhere" is true for the RAS in the brain, but rule number two, namely that "nature has a tendency to conserve structure, not necessarily function" also needs to be considered. Consequently, the brain RAS may have completely different functions in addition to blood pressure regulation.

This is in fact indicated by a series of experiments with passive avoidance tests in rats (27,30). The animals received an electrical shock after entering a black box on the first experimental day. Avoidance was tested every 24 hours for 5 consecutive days. As expected, 100% of the rats entered the compartment before and less than 10% of the control rats after they had received the foot shock. However, if renin was injected into the brain ventricles, more rats forgot the painful foot shock experience and entered the dark compartment. Renin-treated rats also entered the compartment more quickly (27) (Fig. 5).

The specificity of this effect was idicated by the fact that if ANG II formation in the brain was prevented by the CE inhibitor SQ 14225, the renin-treated animals behaved exactly like the saline-treated control rats (27). This implies that apart from blood pressure regulation, brain ANG II may have completely different, hitherto unexpected, functions such as interference with memory processes but, to close on an optimistic note, this may be so only in rats.

ACKNOWLEDGEMENTS

Supported by the Deutsche Forschungsgemeinschaft (DFG) within the SFB 90 "Cardiovakulares System". We thank Mrs. Funke for competent secretarial help.

REFERENCES

1. Arregui, A., and Iversen, L.K. (1979): Biochem. Pharmacol., 28:2693–2696.
2. Bennett, J.B., and Snyder, S.H. (1976): J. Biol. Chem., 251: 7423–7430.
3. Benuck, M., Grynbaum, A., Cooper, T.B., and Marks, N. (1978): Neurosci. Lett., 10:3–9.
4. Corvol, P., Devaux, C., Ito, T., Sicard, P., Ducloux, J., and Menard, J. (1977): Circ. Res., 41 (Suppl. II): 616–622.
5. Corvol, P., Devaux, C., and Menard, J. (1973): FEBS Lett., 34:189–192.
6. Daul, C.B., Heath, D.G., and Garey, R.E. (1975): Neuropharmacology, 14:75–80.
7. Dworschakc, R.T., Gregory, T.J., and Printz, M.P. (1978): Fed. Proc., 37:1385.
8. Eggena, P., Chu, C.L., Barret, J.D., and Smabhi, M.P. (1976): Biochim. Biophsy. Acta, 427:208–217.
9. Erdos, E.G. (1975): Circ. Res., 36:247–255.
10. Fischer-Ferraro, C., Nahmod, V.E., Goldstein, D.J., and Finkielman, S. (1971): J. Exp. Med., 133:353–361.
11. Ganten, D. (1978): Circ. Res., 42:732–735.
12. Ganten, D., Fuxe, K., Phillips, M.I., Mann, J.F.E., and Ganten, U. (1978): In: Frontiers in Neuroendocrinology, Vol. 5, edited by W.F. Ganong, and L. Martini, pp. 61–99. Raven Press, New York.
13. Ganten, D., Hutchinson, J.S., Schelling, P., Ganten, U., and Fischer, H. (1976): Clin. Exp. Pharmacol. Physiol., 3:103–126.
14. Ganten, D., Minnich, J.L., Granger, P., Hyaduk, K., Brecht, H.M., Barbeau, A., Boucher, R., and Genest, J. (1971): Science, 173:64–65.
15. Ganten, D., and Speck, G. (1978): Biochem. Pharmacol., 27 2379–2389.
16. Ganten, D., Unger, Th., Rockhold, R., Schaz, K., and Speck, G. (1979): In: Radioimmunoassay of Drugs and Hormones in Cardiovascular Medicine, edited by A. Albertini, M. DePrada, and B.A. Peskar, pp. 33–43. Elsevier, Amsterdam.
17. Ganten, D., Speck, G., Meyer, D. Loos, H.-E., Schelling, P., Rettig, R., and Ungar, Th. (1980): In: Enzymatic Release of Vasoactive Peptides, edited by F. Gross and H.G. Vogel. Raven Press, New York. (In press).
18. Hackenthal, E., Hackenthal, R., and Hilgenfeldt, U. (1978): Biochim. Biophys. Acta, 522:574–588.
19. Hilgenfeldt, U., and Hackenthal, E. (1979): Biochim. Biophys. Acta, 579:375–385.
20. Hirose, S., Workman, R.J., and Inagami, T. (1979): Circ. Res., 45:275–281.

21. Hokfelt, T., Elde, R., Johannson, O., Ljungdahl, A., Schultzberg, M., Fuxe, K., Goldstein, M., Nilsson, G., Pernow, B., Terenius, L, Ganten, D., Jeffcoate, S.L., Rehfeld, I., and Said, S. (1978): In: Psychopharmacology: A Generation of Progress, edited by M.A. Lipton, A. DiMascio, and K.F. Fillam, pp. 39-66. Raven Press, New York.

22. Inagami, T., Hirose, S., Murakami, K., and Matoba, T. (1977): J. Biol. Chem., 252:7733-7737.

23. Inagami, T., Hirose, S., and Yokosawa, H. (1979): Fed. Proc., 38:636.

24. Inagami, T., Yokosawa, H., and Hirose, S. (1978): Clin. Sci. Mol. Med., 55:121s-123s.

25. Johnson, A.K., and Buggy, J. (1977): In: Central Actions of Angiotensin and Related Studies, edited by J.P. Buckley, and C.M. Ferrario, pp. 357-386. Pergamon Press, New York.

26. Kiprov, D., and Dimitrov, T. (1976): Comptes rendus de l'Acad. Sci. bulgare, 29:1543-1546.

27. Koiller, M., Krause, H.P., Hoffmeister, F., and Ganten, D. (1979): Neurosci. Lett., 14:71-75.

28. Lewicki, J.A., Fallon, J.H., and Printz, M.P. (1978): Brain Res., 158:359-371.

29. Mann, J.F.E., Phillips, M.I., Dietz, R., Haebara, H., and Ganten, D. (1978): Am. J. Physiol., 224:H629-H637.

30. Morgan, J.H., and Routtenberg, A. (1977): Science, 196:87-89.

31. Murakami, K., and Inagami, T. (1975): Biochim. Biophys. Res. Commun., 62:757-763.

32. Murakami, K., Inagami, T., and Haas, E. (1977): Circ. Res., 41 (Suppl. II): 4-7.

33. Murakmi, K., Inagami, T., Michelakis, A.M., and Cohen, S. (1973): Biochim. Biophys. Res. Commun., 54:482-487.

34. Ondetti, M.a., Rubin,B., and Cushman, D.W. (1977): Science, 196:441-443.

35. Osman, M.Y., Smeby, R.R., and Sen, S. (1979): Hypertension, 1:53-60.

36. Phillips, M.I., Weyhenmeyer, J., Felix, D., Ganten, D., and Hoffman, W.E. (1979): Fed. Proc., 38:2260-2266.

37. Poulsen, K., Vuust, J., Lykkegaard, S., Nielsen, A.H.O.J., and Lund, T. (1979): FEBS Lett., 98:135-138.

38. Printz, M.P., and Lewicki, J.A. (1977): In: Central Actions of Angiotensin and Related Studies, edited by J.P. Buckley, and C.M. Ferrario, pp. 57-64. Pergamon Press, New York.

39. Reid, I.A. (1977): Circ. Res., 41:147,153.

40. Rettig, R., Speck, G., Simon, W., Schelling, P., Fahrer, A., and Ganten, D. (1978): Klin. Wochenschr., 56 (Suppl. I): 43-45.

41. Schelling, P., Speck, G., Unger, Th., and Ganten, D. (1980): In: Advances in Experimental Medicine: A Centenary Tribute to Claude Bernard, edited by S. Parvez, and H. Parvez. Elsevier, Amsterdam. (In press).

42. Scholkens, B.A., Steinbach, R., and Jung, W. (1980): In: Enzymatic Release of Vasoactive Peptides, edited by F. Gross, and H.G. Vogel. Raven Press, New York. (In press).
43. Sirett, N.E., McLean, A.S., Bray, J.J., and Hubbard, J.I. (1977): Brain Res., 122:299-312.
44. Speck, G. and Ganten D.: (In press).
45. Taugner, R., Fuxe, K., Ganten, D., Hackenthal, E., and Rix, W. (1980); Neurosci. Lett., (in press).
46. Tewksbury, D.A., Dart, R.A., and Travis, J. (1979): Circulation, 59/60 (Suppl. II): 132.
47. Umezawa, H., Aoyagi, T., Morishima, H., Matsuzaki, M., Hamada, M., and Takeuchi, T. (1970); J. Antibiot. (Tokyo), 23:259-262.
48. Wallis, C.J., and Printz, M.P. (1980): Endocrinology, 106: 337-342.

Neurosecretion and Brain Peptides,
edited by J. B. Martin, S. Reichlin, and K. L. Bick.
Raven Press, New York © 1981.

Angiotensin-Induced Thirst and Sodium Appetite

Alan N. Epstein

Leidy Laboratory of Biology, University of Pennsylvania, Philadelphia, Pennsylvania 19104

In several respects angiotensin is an ordinary humoral agent. It, like the pituitary peptides, is the product of a synthetic cascade that reduces a protein to a polypeptide. Like the steroid and peptide hormones, it is a blood-borne agent that acts on target tissues distant from its site of synthesis in the periphery and in the brain (22). In fact, the analogy with the catecholamines goes further. Renin is found in several perhiperal organs (rodent salivary glands, muscular arteries, uterus) in addition to the kidney, but the kidney is the only source of circulating renin. The similarity of the catecholamines and the adrenal medulla is evident.

But angiotensin is extraordinary among humoral agents in the diversity of its actions and is unique in its behavioral effects. Considering only those actions that are most well-established (36) and that are produced by physiologically reasonable amounts of the hormone, angiotensin acts in the periphery to contract vascular smooth muscle, to promote the synthesis and release of aldosterone from the adrenal, to inhibit the release of renal renin, and to promote sodium reabsorption by the kidneys and the gut. It acts on the brain to release ADH (and possibly ACTH), to raise arterial pressure by sympathetic activation and, as described below, to arouse a thirst for water and an avidity for salt. The kaleidoscopic impression of this catalogue of effects is reduced by the idea that angiotensin defends the blood volume. Renin is released and angiotensin is generated in the plasma when blood volume or blood sodium is reduced, when blood pressure falls, and when there is sympathetic activation of the kidneys. Its actions gain and retain water and salt, and add to the force by which that mixture is moved about the body.

The arousal of thirst is surely one of its most extraordinary effects. Interest in it began 16 years ago as the result of a modest publication of Fitzsimons suggesting that the kidney produces a thirst hormone, but widespread interest in the possibility was not aroused until 1969 when he published a paper entitled "The role of a renal thirst factor in drinking induced by extracellular stimuli (16)." In this work Fitzsimons showed that the kidneys must be in the circulation, but they need not be producing urine, in order for the hypovolemia of acute vena caval ligation to be fully dipsogenic, and he demonstrated the inseparability of the dipsogenic and pressor activities of renal extracts. Interest in the phenomenon accelerated when we reported in 1970 (13) that angiotension is a specific and potent elicitor of drinking behavior when it is injected into the brain of the rat. This was confirmed (39) and is now known to be a phenomenon of all vertebrate groups from fish to primates (18).

THE POTENCY AND SPECIFICTY OF ANGIOTENSIN-INDUCED THIRST

Among the many effects of hormones on behavior, angiotensin's dipsogenic effect appears to be unique. I know of no other instance in animals like ourselves in which a complex motivated behavior is aroused by a single hormone acting on the brain in physiological amounts. That is, when applied directly to sensitive tissue in the brain, 1 picogram (or 10^{-15} mol) of angiotensin II in a concentration of 10^{-9} M can arouse a sleeping rat to seek, select and ingest water (see Fig. 2, below), and the amounts of water ingested are typical of those that are drunk when the animal drinks spontaneously (27). The specificity of the phenomenon is remarkable. The hormone elicits only drinking behavior. The animals do only what rats do when they drink (44). And the potency of the hormone for the arousal of drinking behavior is equally remarkable when it is recalled that in our experiments, both those employing hormone injected into the brain (44) and those in which the hormone is infused intravenously (25), it arouses thirst while acting outside of its usual context. Thirst does not have a single cause. It has many determinants. Hydrational, circadian, gustatory, and experiental factors as well as hormonal controls combine to generate the phenomenon of drinking behavior, and these nonendocrine determinants are not operating in the animals treated with angiotensin in our experiments. By design, we make our injections when the animal does not drink (in the daytimes, during an intermeal interval, and while the animal is resting or asleep in the rear of its cage). The hormone, therefore, acts alone to arouse thirst against a background of satiation. The effective dipsogenic doses are nevertheless comparable to those for the other actions of the hormone. For example, the same doses of angiotensin (10-20 ng/min) when given by continuous intravenous infusion to the unanesthetized rat produce physiological rises in plasma aldosterone (9), and blood pressure (23), and elicit drinking behavior (25).

The Precocity of Angiotensin-Induced Drinking

The phenomenon appears early in ontogeny. Louis Misantone, Susan Ellis and I have recently found (31) the same innate precocity in the development of angiotensin-induced drinking that Wirth and I (48) had found for other dipsogens such as cellular dehydration, hypovolemia, and β-adrenergic activation. Using Wirth's "water fountain" technqiue that allows rats of even the earliest ages to ingest water by direct consummatory responding, we obtained clear elicitation of water intake by injection of angiotensin into the cerebral ventricles of suckling rats at 4 to 5 days of age and thereafter. Moreover, at this age, which is fully two weeks prior to weaning, the rat pup is as sensitive to angiotensin as is the adult. Drinking is elicited by as little as 1 ng of hormone injected into the cerebral ventricles. Much needs to be done before we can understand the meaning of these facts, but they clearly suggest a role for the hormone in the development of ingestive behavior as well as in their expression in adulthood.

The Neurological Mechanism of Angiotensin-Induced Thirst

It is clear from autographic studies using isotopes of angiotensin that have unimpaired biological activity (43) and from those employing radioimmunoassay (42) that the hormones do not normally cross either the blood-brain or the blood-CSF barriers. Angiotensin is not carried from the blood into the parenchyma of the brain. It exerts its effects on the brain by utilizing the circumventricular organs (CVOs), the family of midline structures that lie on the surface of the ventricular walls outside the blood-brain barrier (see chapter by Weindl, this volume). They are all within the brain but outside the barrier, and angiotensin reaches them from the blood without special impediment. Simpson and Routtenberg (45) found, in 1973, that ablation of the subfornical organ (SFO) abolishes the drinking produced by intracranial injection of angiotensin. They suggested that the subfornical organ is a receptor organ within the brain for the dipsogenic action of the hormone. Simpson, Camardo and I (44) have now confirmed this idea in a series of experiments that showed, first, that the drinking induced by blood-borne hormone is eliminated by prior removal of the SFO and that the deficit so induced is permanent (Fig. 1). Three months after the ablation of the SFO, rats do not drink to intravenous angiotensin even when hypertensive doses are used. Moreover, the effect is specific. Animals with ablation of SFO drink normally in their home cages and in response to cellular dehydration, a thirst challenge that does not depend on mediation by angiotensin. The SFO lesion produces a selective and permanent loss of drinking induced by blood-borne angiotensin.

Secondly, as shown in Fig. 2, the SFO is exquisitely sensitive to the dipsogenic action of the hormone. In these same experi-

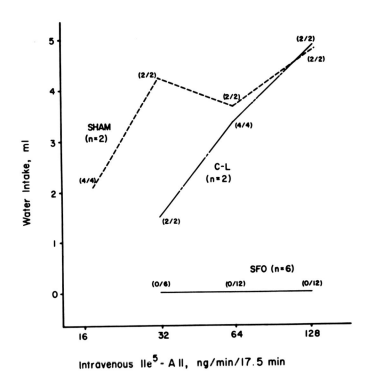

FIG. 1. The induction of thirst by intravenous angiotensin and its complete and permanent absence after ablation of subfornical organ (SFO). SHAM = rats without brain damage, C-L = rats with brain damage that did not remove the SFO, SFO = rats with 95% or more of the subfornical organ ablated. (Reprinted with permission from ref. 43.)

ments, injection of hormone into the nearby third ventricle or the overlying hippocampal commissure produced drinking, but only with doses that were 2 to 3 orders of magnitude greater. Lastly, these experiments demonstrated that infusion of the SFO, but not of the third ventricle, with Saralasin, a specific competitive inhibitor of angiotensin, selectively and reversibly blocks the drinking induced by intravenous angiotensin. This work, combined with electrophysiological evidence from two laboratories (8,37) demonstrating angiotensin-sensitive units in the SFO, makes its special role in thirst undeniable.

The agonist within the brain is the octapeptide and its structural requirements are similar in all respects to those for the myotropic and pressor responses in the periphery. Angiotensin I is not dipsogenic. It exerts its effect only after conversion to angiotensin (14).

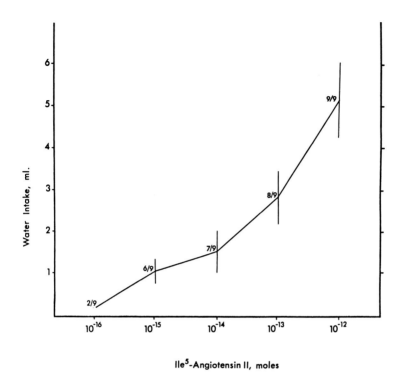

FIG. 2. Dose-response curve for thirst induced by injection of angiotensin directly into the SFO. Response fractions (at each concentration) give the number of animals tested (denominator) and the number of animals drinking (numerator). Note that threshold (50% of animals drinking) is between 10^{-16} and 10^{-15} moles of angiotensin II. (Reprinted with permission from ref. 43.)

To summarize: There are receptors for angiotensin within the SFO, and there is no special diffusion barrier between them and blood-borne angiotensin. The thirst aroused by blood-borne angiotensin is temporarily suppressed by treatment of the SFO with a competitive analog of angiotensin, and is permanently abolished by ablation of the organ. Clearly, the SFO is essential for the thirst that is aroused by blood-borne angiotensin of renal origin. The evidence parallels that for the relation between the area postrema and the central pressor response (36).

What I have just reviewed is the general consensus about angiotensin-induced drinking. That is, most of us concerned with

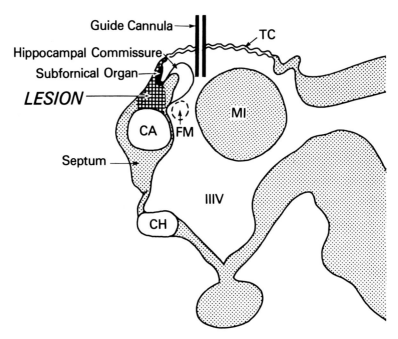

FIG. 3. Middle structures of the North American opossum's third
ventricle (IIIV). The intended lesion which includes the SFO is
shown (cross-hatching) as is the ventricular implant (guide cannula). CA = anterior commissure, CH = optic chiasm, FM = foramen
of Monro, MI = massa intermedia, TC = tela chloride. (Reprinted
with permission from ref. 1.)

TABLE 1. Drinking induced by intravenous hypertonic NaCl[*]

Experi-mental group (n)	No. of tests	No. (and %) of tests when drinking seen within 2 hours of onset of	Mean ($\pm$ SEM) intake in drinking animals	
			1 hour after infusion onset	2 hours after infusion onset
Intact (2)	3	3 (100%)	109.3 $\pm$ 13.3	128.7 $\pm$ 17.2
SFOX (4)	7	7 (100%)	68.6 $\pm$ 22.3	107.9 $\pm$ 20.0

[*]I.V. 2 M NaCl 0.816 ml/min for 25 min.

this problem agree that: 1) a role for the hormone in arousal of thirst is physiologically appropriate; 2) it arouses thirst with unparalleled potency; 3) it does so with unusual behavioral specificity; 4) it is effective in all major vertebrate groups without exception; 5) it utilizes the circumventricular organs for the brain, the SFO in particular, to produce its dipsogenic effect; and 6) it is not the sole cause of the complex phenomenon of drinking behavior.

Additional Receptor Sites

Although there is general agreement about the receptive role of the SFO for blood-borne angiotensin, additional receptor sites for the dipsogenic action of angiotensin that may originate within the brain have been proposed in the preoptic area (34) and in the region of the optic recess which includes the organum

TABLE 2.   Drinking induced by intravenous angiotensin in SFOX opossums

Experimental group (n)	No. of tests	No. (and %) of tests when drinking seen within 2 hours of onset of infusion	Mean ($\pm$ SEM) intake in drinking animals	
			1 hour after infusion onset	2 hours after infusion onset
A. Experiment 1*				
Intact (3)	17	13 (77%)	51.6 $\pm$ 9.8	52.6 $\pm$ 20.8
SFOX[***] (4)	30	5 (17%)	5.0 $\pm$ 1.2	10.4 $\pm$ 2.6
B. Experiment 2[**]				
Intact (3)	11	10 (91%)	48.7 $\pm$ 7.8	60.9 $\pm$ 10.8
SFOX[***] (4)	5	0 (00%)	NONE	NONE

[*]I.V. Angiotensin II 0.93 µg/min for 55 min
[**]I.V. Angiotensin II 1.86 µg/min for 55 min for two doses
[***]SFOX = SFO and adjacent tissue ablated

vasculosum of the lamina terminalis (OVLI) (7,26). These issues
have been reviewed elsewhere (12). It should be pointed out that
neither of these alternate receptive sites has met the stringent
criteria for a central neural receptive zone that have been successfully applied to the SFO. That is, ablation of them does not
produce a specific loss of angiotensin-induced drinking (26),
they do not have the minimal sensitivity that can be expected of
specific receptor organs, and effects, if any, of Saralasin have
not been tested.   Nevertheless, recent work by Findlay and
Epstein (15) demonstrates that sensitivity to the dipsogenic
action of angiotensin remains in brains from which the SFO has
been removed. For this work the American opossum, a marsupial,
was used. This species lacks a corpus callosum, making it possible to gain direct access to the contents of the third ventricle by surgical reflection of the cerebral hemispheres after
craniotomy. Exploiting this advantage, a group of possums was
implanted with third ventricular cannulas (as shown in Fig. 3);
in half of them, the SFO was ablated by suction (cross-hatched
area, Fig. 3). After recovery of spontaneous drinking (and eating) behavior, their responses were studied to 1) intravenous
dipsogens, hypertonic saline and angiotensin, administered by
infusion (49), and 2) injection of angiotensin directly into the
third ventricle.

The results of the intravenous infusions confirm those of
Simpson, Epstein and Camardo (44) in all important respects.
These experiments were done last, late in the animal's postoperative course, to assure that recovery of function was
complete. As shown in Table 1, all of the animals, both those
with intact brains and those with ablation of the SFO (SFOX),
drank in response to cellular dehydration produced by intravenous
hypertonic sodium chloride. There were not significant differences between the intact and SFOX groups in reliability and
volume of drinking or in thier latency to drink (not shown).
However, none of the SFOX animals responded to the intravenous
angiotensin on any occasion or at any dose, including the higher
dose, which was frankly hypertensive (Table 2).   These are, in
all cases, the same animals that drank intravenously after administration of sodium chloride. The small and unreliable drinking
associated with the lower dose of angiotensin is drinking that
occurs spontaneously when oppossums are wakened in the middle of
the day. It occurred with latencies that bore no consistent
relationship to the onset of angiotensin infusion, and was
eliminated by the higher dose, very likely because of the mild
distress produced by the pressor effect of the hormone. Therefore in the American opossum, just as in the rat, the SFO is
essential for the drinking aroused by blood-borne angiotensin.

The results of intracranial injection of angiotensin are shown
in Fig. 4. Angiotensin was injected directly into the anterior
cerebral ventricle in doses ranging from 0.1 to 4 µg. Intact animals (open bars) drink reliably to the 200 ng dose. The response
fraction at the base of each bar gives the number of tests at

that dose as denominator and the number of elicited drinking
episodes as numerator. The response increases and becomes more
rapid as the dose increases, and at the highest doses (2 and 4
μg) animals drink copiously, reliably and promptly. The outcome
is quite different for the SFOX animals. First, for some time
after surgery, they do not drink at all to intracranial angiotensin despite the recovery of spontaneous drinking. When they
recover, responsiveness to angiotensin injected into the third
ventricle (shaded bars in Fig. 4) is unreliable, attenuated, and
can be obtained only with the highest doses. Water consumption
in those animals that responded to high doses was markedly less
than in intact possums. Zero intake or a failure to respond is
not included in the average of the volumes consumed at each dose
(Fig. 4) and the columns of data are not unequal because of the
greater number of failures to respond among the SFOX animals.

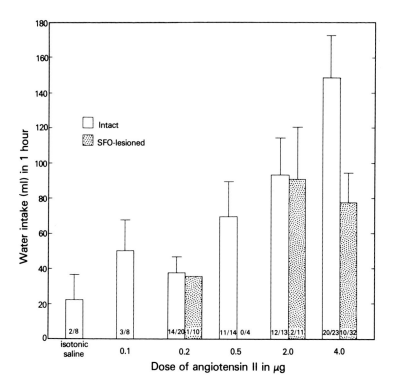

FIG. 4. Mean (+ SEM) of water intake in those animals which
drank within 60 mins of an intracerebroventricular injection of
angiotensin II. The response fraction (the number of tests in
which animals drank over the total number of tests) is shown for
intact animals, and for lesioned animals receiving doses of AII
between 0.2 and 4.0 μg.

One fact should be emphasized. Each of the SFOX animals responded on at least one occasion by drinking a considerable
amount of water to an intraventricular injection of either 2 or 4
µg of angiotensin. The possum brain can yield a dipsogenic response to angiotensin despite the loss of the SFO, and the same
seems to be the case for the rat (5). Responsiveness to angiotensin in SFOX-lesioned animals is unlikely to have physiological
significance, and may be accounted for by residual sensitivity to
angiotensin of structures in the third ventricular wall that are
normally associated with the SFO. The OVLT may be one such
structure. Recent anatomical work of Miselis, Shapiro and Hand
(32) demonstrates efferent connections from the SFO to the OVLT
as well as to the supraoptic nucleus (SON) and the medial nucleus
of the preoptic area suggesting that the circumventricular organs
of the third ventricle may function together in the control of
body water content. Damage to the optic recess and the lamina
terminalis, which includes the OVLT, would interrupt the connections described by Miselis and colleagues (32) and could produce
the transient adipsia, deficient ADH secretion, and unresponsiveness to angiotensin that have been reported to result from
lesions of the anterior-ventral third ventricular region (26).
The full meaning of these new facts is not clear but one appealing possibility is that the OVLT and the SON are controlled, in
part, by angiotensin-induced signals sent to them from the SFO.
In the absence of the SFO, they may still be capable of mediating
the arousal of thirst (and perhaps the release of antidiuretic
hor?r?rmone) but may only do so in response to the large doses of
the peptide.

Competition with Natural Antidipsogens

The possible existence within the brain of natural antidipso-
gens is suggested by the work of Kenney and Epstein (28) and of
Nicolaidis and Fitzsimons (35) who report suppression of angiotensin-induced thirst by intraventricular injection of prostaglandins. In addition, deCaro and his colleagues (11) report a
similar and potent effect of Substance P. Kenney and Epstein
(23) showed that: 1) prostaglandin Es (PGE) are responsible for
the suppression, 2) the effect can be obtained with doses of
PGEs as low as 10 ng which are not pyrexic, 3) the effect is not
secondary to the anorexia that others have reported with intracranial PGEs given at higher doses (3), and 4) the response is
specific at the low doses used for thirst induced by angiotensin.
Both the PGEs and Substance P are localized in the brain and
synthesis of the PGEs is stimulated by angiotensin in the periphery. We suggested that angiotensin could concurrently arouse
thirst and stimulate the synthesis and release of its own natural
dipsogenic antagonist, PGE. This idea is strengthened by two
recent reports. First, Fluharty has found that angiotensin-in

duced thirst is suppressed by prostaglandins produced in the brain (21). He showed that treatment with arachidonic acid, the precursor of the prostaglandins, prior to intraventricular injection of angiotensin suppressed the angiotensin response. In addition, Moe and Kenney (33) have confirmed a previous report (38) of enhancement of angiotensin's dipsogenic action during blockade of prostaglandin synthesis. Prolonged systemic treatment with indomethacin (in drinking water) augments the drinking produced by intravenous angiotensin.

AROUSAL OF SODIUM APPETITE

A direct role for angiotensin in sodium appetite is supported by recent studies of Avrith and Fitzsimons (2) and of Bryant,

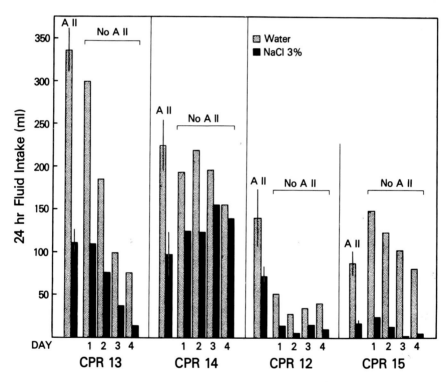

FIG. 5. Water and strong salt solution intake of 4 rats during intracranial infusion of angiotensin II (6 μ g/hr). The bars to the left in each panel represent the average of daily intakes during 4 days of continuous intracranial infusion. (Days 1, 2,3 and 4 in each panel labelled "No A II" indicate intakes after infusion.) Note the very large intakes of salt induced by intracranial angiotensin and their persistence after the infusions are terminated. (Reprinted with permission from ref. 5.)

Fitzsimons, Fluharty and Epstein (4) in which long-term infusions of the hormone into the anterior cerebral ventricles evoked impressive intakes of strong sodium chloride solutions. In our experiments 3% NaCl, which is ordinarily rejected by rats, is drunk in massive volumes during 4 days of continuous angiotensin infusion (4). Intakes frequently exceed 100 ml/day, and the appetite for salt persists after the infusions are terminated (Fig. 5). The animals drank little or no salt solution prior to infusion, but as shown by the pair of bars to the left of each panel, intake of both salt solutions and water rose during continuous angiotensin infusion to levels that, in three of the animals, are greater than any reported previously in the literature. Even rat CPR 15 drank volumes of 3% NaCl that are typical of the adrenalectomized rat. In the data shown here, each animal received 100 ng of angiotensin II/min. The appetite is aroused by doses as small as 100 pg/min. Although water intake is also excessive, it is not the cause of the salt ingestion. Animals drink salt excessively when it is the only fluid offered during infusion and they do not drink salt when, in the absence of angiotensin infusion, they are given 200-600 ml of water per day by automatic continuous gavage. Nor does sodium loss appear to be the cause of the excess salt ingestion. Large amounts of angiotensin produce a natriuresis both when given intravenously (30) and when injected into the brain (41). But low intravenous doses promote sodium reabsorption by the kidney (21), and our intracranial infusions may be similarly physiological. In the experiment of Avrith and Fitzsimons (2), sodium appetite is provoked by infusion of only 1 pmol/hr of angiotensin II. Moreover, the kidney is not necessary for the effect. They find that the appetite is produced in the nephrectomized rat by intracranial injection of renin (17). In recent experiments, we found that: 1) urinary salt losses in the first hour or two of infusion are not greater in animals infused with angiotensin (60 pmol/hr) and displaying the appetite than in those receiving saline and drinking little or no salt, and 2) the large salt losses produced by furosemide, which exceed those produced by intrancranial infusions, provoked sodium appetite unreliably and with latencies far greater than those produced by intracranial angiotensin. The persistence of the appetite after termination of angiotensin treatment is shown in Fig. 5. The duration and magnitude of the response are dose-dependent, and in some animals appear to continue indefinitely. We have also found that the appetite is specific. When infused with angiotensin, rats do not ingest NH_4 and show little interest in KCl. The persistence of the appetite and its specificity for sodium are also characteristic of the phenomenon when it is produced by other treatments such as dietary deficiency or peritoneal dialysis (14), both of which produce the effect as a consequence of reduced plasma sodium.

A role for angiotensin in the ingestion of salt is appropriate in the context of its control of salt excretion and is suggested by prior work (6,10,39). Plasma angiotensin levels are elevated by all of the treatments that produce sodium appetite except mineralocorticoid injection, an effect which may be mediated by aldosterone. Elevated angiotensin II levels that accompany sodium depletion raise plasma aldosterone, and both hormones cause sodium appetite when they are in pharmacological excess (20). The brain may be informed of the need for salt not by the salt deficiency itself but by the elevated levels of both angiotensin and aldosterone that are consequences of the deficiency.

Angiotensin is a potent elicitor of both water and salt intake and these behaviors are appropriate for a hormone whose overall function is defense of the blood volume. Keeping in mind that angiotensin cannot be the sole cause of either phenomenon, but is rather a participant in the complex determinants of both, future research will need to determine when and how much it contributes to the various kinds of spontaneous drinking behavior. This important information will <u>not</u> come from experiments that discourage the suggestion of a role for the hormone in normal drinking behavior (46). In the latter, nephrectomized rats, which have elevated renin substrate, were used. They were given sizeable doses of hog renin and were either sacrificed for determination of plasma activities or they were required to drink. Plasma renin levels were found to be abnormally high, as would be expected from the elevated substrate. Water intakes were found to be abnormally low, as would be expected from reduction in behavioral competence resulting from recent surgery and from the hypertension that must have occurred by combination of exogenous renin with elevated levels of its substrate. More useful information is likely to come from experiments like those of Malvin, Mouw and Vander (29) and of Hoffman and colleagues (24) who have reduced drinking behavior with intracranial infusions of Saralassin, used alone or in combination with other pharmacological blockers. In both of these reports, thirst was induced by water deprivation, an entirely natural thirst-provoking circumstance for the rat, and effects of endogenous angiotensin were examined directly by use of its specific competitive inhibitor.

ACKNOWLEDGEMENTS

The author's research is sponsored by grants from the NINCDS (03469) and NATO (RG 1532).

REFERENCES

1. Akert, K., Potter, H.D., and Anderson, J.W. (1961): *J. Comp. Neurol.*, 116:1-13.
2. Avrith, D., and Fitzsimons, J.T. (1978): *J. Physiol. (London)*, 282:40-41.

3. Baile, C.A., Simpson, C.W., Beam, S.M., McLaughlin, C.L., and Jacobs, H.L. (1973): Physiol. Behav., 10:1077-1085.
4. Bryant, R.W., Fluharty, S.J., and Epstein, A.N. (1978): Fed. Proc., 37:323.
5. Buggy, J., Fisher, A.E., Hoffman, W.E., Johnson, A.K., and Phillips, M.I. (1975): Science, 190:72-74.
6. Buggy, J., and Fisher, A.E. (1974): Nature (London), 250: 733-735.
7. Buggy, J., and Fisher, A.E. (1976): Pharmacol. Biochem. Behav., 4:651-660.
8. Buranarugsa, P., and Hubbard, J.I. (1979): J. Physiol. (London), 291:101-116.
9. Campbell, W.B., Brooks, S.M., and Pettinger, W.A. (1974): Science, 184:994-996.
10. Chiaraviglio, E. (1976): J. Physiol. (London), 255:57-66.
11. DeCaro, G., Massi, M., and Micossi, L.G. (1978): J. Physiol. (London), 279:133-140.
12. Epstein, A.N. (1978): In: Frontiers in Neuroendocrinology, Vol. 5, edited by W.F. Ganong, and L. Martini, 101-134. Raven Press, New York.
13. Epstein, A.N., Fitzsimons, J.T., and Rolls, B.J. (1970): J. Physiol. (London), 210:457-474.
14. Falk, J.L. (1961): In: Nebraska Symposium on Motivation, edited by M.R. Jones. University of Nebraska Press, Lincoln.
15. Findlay, A.L.R., Elfont, R.M., and Epstein, A.N. (1980): Brain Res. (in press).
16. Fitzsimons, J.T. (1969): J. Physiol. (London), 204:349-369.
17. Fitzsimons, J.T. (1978): Fed. Proc., 37:2669-2675.
18. Fitzsmons, J.T., (1979): The Physiology of Thirst and Sodium Appetite. Cambridge University Press, Cambridge.
19. Fitzsimons, J.T., Epstein, A.M., and Johnson, A.K. (1978): Brain Res., 153:319-331.
20. Fregly, M.J., and Waters, I.W. (1966); Physiol. Behav., 1:65-74.
21. Fluharty, S.J. (1980): Amer. J. Physiol., (in press).
22. Ganten, D.V. (1980): In. Neurosecretion and Brain Peptides, edited by J.B. Martin, S. Reichlin, and K.L. Bick. Raven Press, New York. (In press.)
23. Gross, F., Bock, K.D., and Turrian, J. (1961): Helv. Physiol. Acta, 19:42-47.
24. Hoffman, W.E, Ganten, V., Phillips, M.I., Schmid, P.S., Schelling, P., and Ganten, D. (1978): Amer. J. Physiol., 243:F41-47.
25. Hsiao, A., Epstein, A.M., and Camardo, J.S. (1977): Horm. Behav., 8:129-140.
26. Johnson, A.K., and Buggy, J. (1978): Amer. J. Physiol., 243: R122-R129.
27. Kissileff, H.R. (1969): J. Comp. Physiol. Psychol., 67: 284-300.
28. Kenney, N.J., and Epstein, A.N. (1978): J. Comp. Physiol. Psychol., 92:204-219.

29. Malvin, R.L., Mouw, D., and Vander, A.J. (1977): Science, 197:171-173.
30. Malvin, R.L., and Vander, A.J. (1967): Amer. J. Physiol., 213:1205-1208.
31. Misantone, L.J., Ellis, S., and Epstein, A.N. (1980): Brain Res., (In press.)
32. Miselis, R.R., Shaprio, R.E., and Hand, P.J. (1978): Soc. Neurosci. Abstr., 4:179.
33. Moe, K.E., and Kenney, N.J. (1980): paper presented to Eastern Psychological Association, Hartford.
34. Morgenson, G.J., and Kucharczyk, J. (1978): Fed. Proc., 37: 2683-2688.
35. Nicolaidis, S., and Fitzsimons, J.T. (1975): C.R. Acad. Sci. (D) Paris, 281:1417-1420.
36. Page, I.H., and Bumpus, F.M. (1974): Angiotensin. Springer-Verlag, New York.
37. Phillips, M.I., and Felix, D. (1976): Brain Res., 109:531-540.
38. Phillips, M.I., and Hoffman, W.E. (1977): In: Central Actions of Angiotensin and Related Hormones, edited by J.P. Buckley and C. Ferrario, pp. 325-356. Pergamon Press, New York.
39. Radio, G.J., Summy-Long, J., Daniels-Severs, A., and Severs, W.B. (1972): Amer. J. Physiol., 223:1221-1226.
40. Richter, C.P. (1936): Amer. J. Physiol., 115:155-161.
41. Severs, W.B., Daniels-Severs, A., Summy-Long, J., and Radio, G.J. (1971): Pharmacology, 6:242-252.
42. Shelling, P., Ganten, D., Heckl, R., Hutchinson, J.S., Spo-ner, G., and Ganten, U. (1976): In: Central Actions of Angiotensin and Related Hormones, edited by J.P. Buckley, and C. Ferrario, pp. 519-526. Pergamon Press, New York.
43. Shrager, E.E., Osborne, M.J., Johnson, A.K., and Epstein, A.N. (1975): In: Central Action of Drugs in Blood Pressure Regulation, edited by D.S. Davies, and J.L. Reid, 65-67. Universtiy Park Press, Baltimore.
44. Simpson, J.B., Epstein, A.N., and Camardo, J.S. (1977): J. Comp. Physiol. Psychol., 91:1220-1231.
45. Simpson, J.B., and Routtenberg, A. (1973): Science, 818: 1172-1174.
46. Stricker, E.M., Bradshaw, W.G., and McDonald, R.H. Jr. (1976): Science, 194:1169-1171.
47. Weindl, A. (1980): In: Neurosecretion and Brain Peptides, edited by J.B. Martin., S. Reichlin, and K.L.Bick. Raven Press, New York. (In press.)
48. Wirth, J.B., and Epstein, A.N. (1976): Amer. J. Physiol., 230:188-198.
49. Young, C.E., and McDonald, I.R. (1978): J. Physiol. (London), 280:77-85.

Neurosecretion and Brain Peptides,
edited by J. B. Martin, S. Reichlin, and K. L. Bick.
Raven Press, New York © 1981.

Brain–Gut Peptides and the Control of Food Intake

G. P. Smith and J. Gibbs

*Department of Psychiatry, Cornell University Medical College, New York, New York 10021;
and The Edward W. Bourne Behavioral Research Laboratory, New York Hospital–Cornell
Medical Center, Westchester Division, White Plains, New York 10605*

The function of gut peptides has been traditionally sought in the coordination of the cellular mechanisms for digestion and absorption of ingested food. The discovery of the potent effects of some of these peptides on the release of insulin and glucagon (the "entero-insular axis") has extended their function to the postabsorptive metabolism of ingested food as well.

In the past decade, a new function of gut peptides has been sought in the control of feeding behavior itself. The basis for this work has been two facts: First, feeding often stops before significant quantities of ingested food have been absorbed. Second, food stimuli release gut peptides by contacting the mucosal surface of the gut, i.e., prior to absorption.

These two facts have been combined to form the hypothesis that one or more gut peptides released by preabsorptive food stimuli act as a peripheral negative feedback signal(s) to stop feeding behavior (9). This hypothesis suggests that gut peptides are short-term satiety signals that are part of the mechanism(s) for postprandial satiety. This hypothesis is provocative because it proposes a testable, physiological mechanism(s) for the termination of a meal. The hypothesis is worth testing because our ignorance about mechanisms for postprandial satiety is total: No one knows what mechanism(s) terminates any meal in any organism at any time.

SATIETY EFFECT OF GUT PEPTIDES

To be considered a candidate for an endocrine satiety signal, a gut peptide must be released by preabsorptive food stimuli and enter the portal circulation before a meal has stopped. On the problematic basis of radioimmunoassay measurements, a number of gut peptides apparently satisfy this criterion. Most of these substances have now been tested for their effect on food intake. Ten of them inhibit intake significantly, at least in one species under one experimental condition (Table 1).

When a gut peptide is shown to inhibit food intake, further experiments are necessary (1) to demonstrate the potency of the inhibitory effect, (2) to investigate the possibility that the effect is a manifestation of toxicity, (3) to determine the site of action, and (4) to determine if the effect is a physiological function of the peptide. Of the ten peptides that inhibit food intake, only cholecystokinin (CCK) has been subjected to such extensive experimental scrutiny. For that reason, the rest of this chapter will deal primarily with the evidence concerning the satiety function of CCK.

TABLE 1. Effect of gut peptides on food intake

Gut Peptide	Inhibition of Feeding	Inhibition of Sham Feeding	Toxic Signs
Cholecystokinin	YES[9]	YES[10]	NO[9]
Bombesin	YES[6]	YES[7]	NO[6]
Gastrin	YES[11]	NO[15]	NO[11]
Secretin	YES[11]	NO[15]	NO[11]
Glucagon	YES[20]		NO[20]
Insulin	YES[17]		NO[17]
Somatostatin	YES[16]		NO[16]
Neurotensin	YES[8]		YES[8]
Substance P	YES[8]		YES[8]
Pancreatic Polypeptide	YES[19]		NO[19]
Gastric Inhibitory Peptide		NO[15]	

Potency of Satiety Effect

The best estimate of the potency of a gut peptide to produce satiety is to test it in the rat that sham feeds after overnight deprivation. Under these conditions, sham feeding never stops, the rats never satiate (31). Thus, if a gut peptide stops sham feeding, it is strong evidence for a satiety effect of that peptide. CCK (10) and bombesin (7) have satiety effects in the sham feeding rat. Gastrin, secretin and GIP do not (15). None of the other peptides that inhibit food intake have been tested.

The question of toxicity

The inhibition of food intake is necessary evidence of a satiety effect, but it is not sufficient, because the peptide could decrease food intake by sickness (nausea, malaise, visceral distress, etc.) or by the production of competing behaviors, e.g., running or drinking.

The inhibition of food intake by neurotensin and Substance P is associated with the frequent occurrence of toxic symptoms (Table 1). Substance P produced audible wheezing, dyspnea and frequent grooming behavior. Neurotensin produced increased drinking and unusual postures, such as treading of the paws and jerking of the head.

CCK (9,10) and bombesin (6,7), however, did not produce any such toxic or competing behaviors. Rats simply stopped eating sooner as if these peptides had accelerated the process of satiety.

It is possible, of course, that CCK or bombesin produce a subtle form of sickness that is manifested by decreased food intake. We have been concerned with this issue of distinguishing a satiety effect from a mild sickness. Most of the work has been done with CCK. The main results are:

1. Doses of CCK that inhibit liquid or solid food intake do not inhibit water intake (9).
2. Doses of CCK that inhibit food intake do not change body temperature (9).
3. Administration of CCK every other day for months in doses that inhibit food intake does not change the appearance of adult male rates or their body weights (9).
4. CCK does not change the initial rate of feeding; CCK simply makes rats (9), monkeys (5), and men stop (12) feeding sooner.
5. CCK can inhibit food intake in men without subjective reports of significant distress, discomfort or sickness (12,29).
6. When CCK inhibits sham feeding in rats (1) and monkeys (4), it also elicits a behavioral sequence that is characteristic of normal satiety.

We and others (3) have used a conditioned taste aversion test to determine if CCK was producing sickness. We abandoned this when it became clear that sickness was neither a necessary nor a sufficient condition for the production of a conditioned taste aversion (24).

Thus, we consider the convergence of the behavioral evidence in animals and men, and the subjective reports of men as compelling evidence that CCK produces satiety, not sickness.

Site of action.

Although it is common to assume that a behavioral effect of a peptide hormone indicates that the site of aciton is in the brain, current evidence suggests that the site of the satiety effect of CCK is in the abdomen. The evidence for a peripheral site is the fact that bilateral abdominal vagotomy abolishes or markedly reduces the satiety effect of CCK (14,23). In a preliminary attempt to identify which vagally innervated structure is important, we found that selective gastric vagotomy also abolished or markedly reduced the satiety effect (25), but selective hepatic or coeliac vagotomies did not.

We do not know if the effect of gastric vagotomy is the result of the loss of gastric motor or sensory fibers. Since the satiety effect of CCK was not blocked by atropine methyl nitrate (25), we hypothesized that removal of the vagal sensory fibers from the stomach is the critical lesion. Experiments are in progress to test that hypothesis and to determine what effect of CCK is monitored by vagal sensory fibers as a satiety signal.

Is Satiety a Physiological or Pharmacological Effect?

Given that exogenous CCK has a satiety effect, there remains the crucial question whether enough endogenous CCK is released by a meal to have a satiety effect. There is a short and a long answer to that question. The short answer is the Scottish verdict, "Not proven."

The long answer has two parts:

First, the threshold dose of CCK for significant inhibition of food intake is as low as 2.5 U/kg (about 0.1 µg/kg) in the rat (9) and is about 30 ng/kg in young men (12). Although these doses are very likely within the physiological range, they are being administered during real feeding when all of the endogenous satiety mechanisms, including endogenous CCK, are already activated.

Second, in the simpler case of sham feeding, where endogenous CCK is probably not released, the minimal dose required to inhibit sham feeding may be within the physiological range in rats (15), but it is clearly higher than the physiological range in the monkey (4). In considering these results, it should be remembered that the sham feeding procedure activates endogenous satiety mechanisms minimally. This is what makes it a reliable assay system. But if satiety is usually the result of synergism between a number of neural and endocrine mechanisms activated by food stimuli (and there is considerable evidence for this in the rat, see ref. 24), then the sham feeding rat or monkey will be a relatively insensitive assay because it requires the putative satiety signal to elicit satiety by itself instead of in concert with the other mechanisms. Thus, until the other mechanisms are identified and brought under experimental control, it is not possible to measure the threshold dose of CCK required for satiety. When that can be done and circulating CCK can be measured accurately, then we shall know if the satiety effect is physiological.

SATIETY EFFECT OF BRAIN PEPTIDES

The occurrence of these peptides in the brain and other parts of the central nervous system has raised the possibility that the peptides are involved in the central mechanisms for the control of food intake. This possibility has begun to be tested in the past few years and there are some promising results, particularly with CCK.

Maddison (18) reported that injections of impure CCK into the lateral ventricle of rats did not change meal size, but prolonged the intermeal interval. Nemeroff et al. (21) decreased food intake elicited by tail-pinch by intraventricular injection of CCK-8. The dose required, however, was larger than the threshold dose of peripherally administered CCK-8. Thus, Nemeroff et al. concluded that the inhibition was a peripheral effect of CCK-8.

Stern et al. (26,27), however, decreased food intake with intracerebral injections of very low doses of caerulein, a peptide that is structurally related to CCK and has all the known biological actions of CCK. Injections of caerulein into the lateral ventricle and into the ventromedial (VM) hypothalamic nucleus decreased food intake, but identical injections of caerulein into the lateral hypothalamus did not. Furthermore, they reported that the inhibition of food intake produced by caerulein administered intraperitoneally was abolished in rats with lesions of the VM area. This result apparently demonstrated that the site of action for peripherally administered caerulein was in the VM. Because of the close similarity of structure of caerulein and CCK-8, this result suggested that the VM is also the site of the satiety effect of peripherally administered CCK-8. But Kulkosky et al. (13) reported that peripherally administered CCK-8 produced its usual satiety effect in VM lesioned rats. And subsequent work with abdominal vagotomy (see above) supports a peripheral site of action for CCK. These discordant results with caerulein and CCK invite further experiments.

Since intraventricular injections of caerulein have inhibited food intake in doses that are clearly below the peripheral threshold, and, since caerulein and CCK bind to the same receptor in all known receptor systems, this is presumptive evidence for central CCK receptors involved in the inhibition of food intake. Despite this evidence, however, it has been extremely difficult to demonstrate inhibition of food intake by intraventricular injections of CCK-8 or by direct injection of CCK-8 into brain sites in the rat. The one convincing report of a central effect of CCK-8 was produced in sheep with infusions of CCK-8 (2). Identical infusions of CCK-8 in rats had no effect.

Woods et al. (30) have recently produced a significant inhibition of food intake and body weight by continuous ventricular infusions of insulin in baboons. They consider this to be evidence that insulin in the cerebrospinal fluid serves a major role in the coordination of food intake and body weight.

In addition to observing the effect of injections or infusions of peptides at central sites on food intake, an alternative strategy is to measure an abnormality in their synthesis, storage, release or receptor binding and activation in animals with characteristic disorders in the control of food intake. For example, Straus and Yalow (28) reported that genetically obese and hyperphagic mice (ob/ob) had significantly less immunoreactive brain CCK than lean littermates. But Schneider et al. (22) could not confirm this and they also could not measure any difference in immunoreactive CCK in the brains of a variety of rodents that had either neurological or genetic obesity and hyperphagia.

The demonstration that a number of gut peptides are released by food stimuli contacting the mucosal surface of the stomach and small intestine has led to the hypothesis that these peptides could serve as physiological mechanisms for postprandial satiety. A number of the gut peptides have been shown to inhibit food intake, but only CCK and bombesin have survived the critical test of inhibitng food intake and eliciting satiety in the sham feeding rat without demonstrable signs of sickness. But even in the case of CCK and bombesin, there is no direct evidence that their satiety effect is physiological, i.e., that endogenous CCK and bombesin are released in sufficient amounts to produce a satiety effect.

Many of the gut peptides are present in the brain and other parts of the central nervous system. This has led to the hypothesis that these neural peptides can also participate in the control of food intake. This "central" hypothesis has been investigated much less than the "peripheral" hypothesis. Despite a decade of experiments, it is clear that the investigation of the function of brain-gut peptides in the control of food intake has just begun.

ACKNOWLEDGEMENTS

Our work has been supported by research grants AM 17240 and MH 15455. We have been supported by Career Development Awards 5 K02 MH 70874 and 5 K02 MH00149.

REFERENCES

1. Antin, J., Gibbs, J., Young, R.C., and Smith, G.P. (1975): J. Comp. Physiol. Psychol., 89:784-790.
2. Della-Fera, M.A. and Baile, C.A. (1979): Science, 206:471-473.
3. Deutsch, J.A. and Gonzalez, M.F. (1978): Behav. Biol., 24:317-326.
4. Falasco, J.D., Smith, G.P., and Gibbs, J. (1979): Physiol. Behav., 23:887-890.
5. Gibbs,J., Falasco, J.D., and McHugh, P.R. (1976): Am. J. Physiol,. 230:15-18.

6. Gibbs, J., Fauser, D.J., Rowe, E.A., Rolls, B.J., Rolls, E.T., and Maddison, S.P. (1979): Nature, 282:208-210.
7. Gibbs,J. and Martin, C.F. (1979): Soc. for Neurosci. Abstr., 5:217.
8. Gibbs, J. and Martin, C., (1979): Unpublished data.
9. Gibbs, J., Young, R.C., and Smith, G.P. (1973a): J. Comp. Physiol. Psychol., 84:488-495.
10. Gibbs, J., Young, R.C., and Smith, G.P. (1973b): Nature, 245:323-325.
11. Grovum, W.L., (1977): Abstracts of Sixth International Conference on the Physiology of Food and Fluid Intake.
12. Kissileff, H.R., Pi-Sunyer,F.X., Thornton, J., and Smith, G.P. (1979): Am. J. Clin. Nutr., 32:939.
13. Kulkosky, P.J., Breckenridge, C., Krinsky, R., and Woods, S.C. (1976): Behav. Biol., 18:227-234.
14. Lorenz, D.N. and Goldman, S.A. (1978): Soc. for Neurosci. Abstr., 4:178.
15. Lorenz, D.N., Kreielsheimer, G., and Smith, G.P. (1979): Physiol. Behav., 23:1065-1072.
16. Lotter, E.C. and Woods, J.C. (1977): Abstracts of Sixth International Conference on the Physiology of Food and Fluid Intake.
17. Lovett, D. and Booth, D.A. (1970): Quart. J. Exp. Psychol., 22:406-419.
18. Maddison, S. (1977): Physiol.Behav., 19:819-824.
19. Malaisse-Lagae, F., Carpentier, J.L., Patel, Y.C., Malaisse, W.J., and Orci, L. (1977): Experientia, 33:915-917.
20. Martin, J.R. and Novin, D. (1977): Physiol. Behav., 19:461-466.
21. Nemeroff, C.B., Osbahr III, A.J., Bissette,G., Jahnke, G., Lipton, M.A., and Prange, Jr., A.J. (1978): Science, 200:793-794.
22. Schneider, B.S., Monahan, J.W., and Hirsch, J. (1979): J. Clin. Invest., 64:1348-1356.
23. Smith, G.P. and Cushin, B.J. (1978): Soc. for Neurosci. Abstr., 4:180.
24. Smith, G.P. and Gibbs, J. (1979): In: Progress in Psychobiology and Physiology Psychology, edited by J.M. Sprague and A.N. Epstein, pp. 179-242. Academic Press, New York.
25. Smith, G.P., Jerome, C., Eterno, R. and Clushin, B. (1979): Soc. for Neurosci. Abstr. 5:650.
26. Stern, J.J., Cudillo, C.A., and Kruper, J. (1976): J. Comp. Physiol. Psychol., 90:484-490.
27. Stern, J.J. and Page, P. (1977): Psychol. Rep., 40:3-8.
28. Straus, E. and Yalow, R.S. (1979): Science, 203:68-69.
29. Sturdevant, R.A.L. and Goetz, H. (1976): Nature, 261:713-715.
30. Woods, S.C., Lotter, E.C., McKay, L.D., and Porte, Jr., D. (1979): Nature, 282:503-505.
31. Young, R.C., Gibbs, J., Antin, J., Holt, J., and Smith, G.P. (1974): J. Comp. Physiol. Psychol., 87:795-800.

Neurosecretion and Brain Peptides,
edited by J. B. Martin, S. Reichlin, and K. L. Bick.
Raven Press, New York © 1981.

Peptides and Regulation of Body Temperature

Marvin Brown, Yvette Taché, Jean Rivier, and *Quentin Pittman

*Peptide Biology Laboratory and *Arthur Vining Davis Center for Behavioral Neurobiology,
The Salk Institute, San Diego, California 92138*

INTRODUCTION

Interest in the neurobiology of peptides has developed coin-
cident with their identification and chemical characterization in
mammalian brain. Studies on the central nervous system (CNS)
action of peptides are of considerable interest since several of
these peptides produce rather unique effects on the brain, which
are neither mimicked by nor dependent on other recognized putative neurotransmitters, e.g., changes in libido with luteinizing
hormone releasing hormone (21), analgesia with morphinomimetic
peptides (16), and elevation of plasma epinephrine with bombesin
(6,12). Several neuropeptides have profound influence on animal
thermoregulation.

Thyrotropin releasing hormone (TRH) placed intracerebroventricularly (icv) has been reported to elevate or reduce body
temperature in different animal species (5,14,19). Neurotensin
(2,4) and β-endorphin (3,8,17) given icv reduce body temperature.
It is possible that an endogenous opioid peptide may participate
in a thermoregulatory pathway, since naloxone, an opiate receptor
antagonist, has been reported to increase the colonic temperature
of acutely stressed animals (17). Bombesin, somatostatin and
analogs of somatostatin also have profound effects on regulation
of body temperature and will be discussed herein.

Bombesin

Bombesin is a tetradecapeptide originally isolated from the
skin of the European frog Bombina bombina (1). Recently the
structure of a 27 amino acid bombesin-like peptide isolated from
porcine gut, termed gastrin releasing peptide (GRP), has been
reported (18). GRP shares a common C-terminal decapeptide homology with bombesin (with the exception of a His/Gln interchange at
residue 8 from the C-terminus). We have previously reported the

presence in rat and ovine brain of a bombesin-like peptide [(determined immunologically circa 32 amino acids in length (15)].

Bombesin
pGlu-Gln-Arg-Leu-Gly-Asn-Gln-Trp-Ala-Val-Gly-His-Leu-Met-NH$_2$

Gastrin Releasing Peptide (GRP)
Ala-Pro-Val-Ser-Val-Gly-Gly-Gly-Thr-Val-Leu-Ala-Lys-Met-Tyr-Pro-

Arg-Gly-Asn-His-Trp-Ala-Val-Gly-His-Leu-Met-NH$_2$

While differences exist between GRP and our highly purified
extracts of mammalian brain bombesin-like peptide, it is likely
that GRP is a mammalian bombesin as suggested by homologies of
sequence and biological activities. Placement of bombesin or GRP
intracisternally (ic) (Fig. 1), icv or intrahypothalamically results in a prompt reduction of body temperature which is accentuated at ambient temperatures below thermoneutrality (4,5,13,
22,25). The hypothermia is dose dependent and reversible in
terms of magnitude and duration. A physiologic role of bombesin
in control of body temperature is suggested by the fact that the
peptide is most potent to produce hypothermia when placed into
the anterior hypothalamic-preoptic region (22), an area thought
to participate in regulation of body temperature (15). This
region contains bombesin-like immunoreactivity (9,10) and saturable, high affinity bombesin-binding sites (20). Specificity for
a bombesin site of action in the anterior hypothalamic-preoptic
area is suggested by the finding that placement of the peptide
into the amygdala, dorsal hippocampus, reticular formation,
periaqueductal grey, striatum, lateral hypothalamus, ventromedial

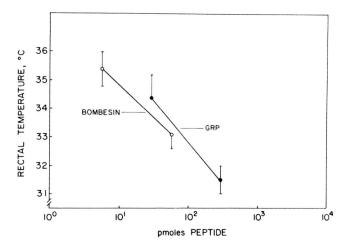

FIG. 1.   Bombesin and GRP given ic (10 µl) lower rectal
temperature of rats. Experiment performed at 4° C.

hypothalamus or posterior hypothalamus does not result in changes of body temperature (22).

Recent studies indicate that bombesin-like peptides not only prevent an animal's ability to thermoregulate against reduced ambient temperatures, but also impairs his ability to maintain euthermia when exposed to elevated ambient temperatures (25). These observations suggest that bombesin-like peptides produce a disruption of body temperature regulation such that an animal becomes poikilothermic (Fig. 2). Bombesin induced hypothermia is reversed by administration of substances which elevate body temperature, such as prostaglandins, somatostatin, somatostatin analogs or TRH (5,11). Animals receiving continuous icv administration of bombesin (15-20 hours) desensitize to its thermoregulatory disruptive effects (Fig. 3). While the mechanism by which this desensitization occurs is undetermined, we speculate that these effects may represent a receptor or cellular desensitization to the effects of bombesin.

While administration of dopaminergic, adrenergic or cholinergic agonists to animals results in a lower body temperature, it is unlikely that the hypothermic effects of bombesin are mediated by endogenous release of these neurotransmitters, since their respective antagonists do not interfere with bombesin effects (11). In addition, bombesin is about 10^3 times more potent than dopamine, apomorphine, carbachol and norepinephrine in lowering body temperature (11). Doses of bombesin that produce significant reduction of body temperature are not associated with any other obvious changes in behavior with the exception of a transient stereotypic scratching of the neck and head with the hind and forepaws. This is in marked contrast to the hypothermic effects of β-endorphin which in our hands only influences temper-

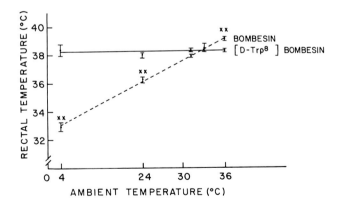

FIG. 2. Disruption of thermoregulation in rats by bombesin (1 μ g) in 10 μ l). An inactive analog of bombesin, [D-Trp^8] –bombesin does not affect thermoregulation.

ature regulation at doses that produce significant alterations of
motor activity.

Investigations into the mechanisms by which bombesin lowers
body temperature indicate that there is a slight, transient
increase of tail vein temperature following intracranial administration of bombesin (25); however, it is unlikely that this
effect could explain the degree of hypothermia produced. Current
investigations are underway to determine the effects of bombesin
and related peptides on heat production mechanisms. Hawkins and
coworkers (personal communication) have demonstrated that bombesin placed into the preoptic region significantly decreases O_2
consumption, data which support the concept that bombesin inhibits heat production.

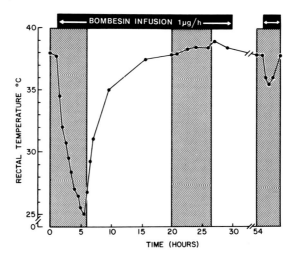

FIG. 3.   Effects of continuous icv infusion of bombesin into
rats.   Shaded areas represent exposure to 4° C, light areas are
23° C.

To study the neurobiology of endogenous bombesin we are developing peptide analogs which might act as receptor antagonists.
Efforts to develop a useful antagonist against bombesin have had
limited success.   Other analogs have been synthesized to study
the structure-activity relations of bombesin to produce hypothermia.   The most potent analogs of bombesin were those in which
positions 1 to 5 were altered, indicating that the C-terminal
decapeptide was sufficient for full biological activity. Gluta-
mine (Gln) at the 7th position and glycine at the 11th position
could be replaced with D-Gln and D-alanine (but not D-proline or
D-phenylalanine), respectively, without any change in potency.
Methionine at the 14th position could be replaced with a D-isomer
with retention of 10% of biological activity.   Any other altera-

tions at the C-terminus drastically reduced the biological potency of these peptides (23).

Among other naturally occurring peptides, alytesin was found
to have 100% of bombesin's potency, whereas litorin, neurotensin,
xenopsin, Substance P, physalaemin, and eledoisin were found to
be in the order of 10^4 times less potent. The shortest sequence
contained within bombesin that produces a full bombesin-like biologic effect is the following structure: Ac-Gln-Trp-Ala-Val-Gly-
His-Leu-Met-NH$_2$. Close agreement exists between biological
activity of bombesin analogs and their binding to synaptosomal
membranes. More detail on individual residue requirements for
biological activity has been discussed elsewhere (23).

TRH has been reported to produce a dose dependent elevation of
body temperature when administered ic or into the anterior hypothalamic-preoptic region of rats and rabbits (5,14). As mentioned above, TRH produces hyperthermia when placed into the
brain of some species (19). We have previously suggested a
possible interaction between bombesin and TRH in regulation of
body temperature and pituitary TSH secretion (11). Since bombesin does not prevent the effects of TRH in stimulating TSH
secretion, we concluded that the mechanism by which bombesin
inhibits cold-induced TSH secretion may be preventing the secretion or delivery of TRH to the pituitary following cold exposure
(11). These data together with the observation that TRH reverses
bombesin-induced hypothermia suggest the possibility that TRH may
be physiologically involved in maintenance of body temperature
and that bombesin hypothermia may result from the endogenous
inhibition of secretion of TRH.

Since our initial studies with bombesin, we have been intri-
gued with the possibility that endogenous bombesin-like peptides
might be involved in lowering of body temperature under physiologic circumstances such as adaptive hypothermia. The most
dramatic of these changes is that of hibernation. While a
substance has been isolated from plasma of hibernating animals,
which when injected into rats lowers body temperature, there is
no evidence that this substance is related to bombesin (24). We
are currently studying the possible role of bombesin in mediating
the lowering of body temperature associated with other more
subtle forms of adaptive hypothermia such as that associated with
fasting, insulin-induced hypoglycemia and 2-deoxyglucose treatment. Hibernation, or more subtle forms of adaptive hypothermia
such as daily torpor, may be best described as adaptive mechanisms to nutrient deprivation or mechanisms to conserve energy
producing substrate.

Somatostatin and Somatostatin Analogs

We have recently demonstrated that a group of analogs based on
the primary structure of somatostatin are potent when administered ic or icv to elevate body temperature and to reverse the
hypothermic effects of a variety of substances including bombesin, carbachol, and neurotensin (11). Native somatostatin shows

TABLE 1. Somatostatin (SS) and SS analogs: Relative potencies to prevent bombesin induced hypothermia and basal secretion of growth hormone (GH), insulin (I) and glucagon (G)

	Number of amino acids/peptide	Bombesin induced hypothermia	Relative Potency to inhibit[1]: Secretion of:		
			GH	I	G
1. (Tyr-Gly-Tyr)-Ala[1]-SS	17	< 0.1	100	284	93
2. $(Arg)_2$-Ala[1]-SS	16	< 0.1	100	100	100
3. SS	14	1	100	100	100
4. [D-Trp8]-SS	14	0.5	800	850	640
5. des-AA1,2[D-Trp8]-SS	12	0.2	400	-	-
6. des-AA1,2,4,5,12[d-Trp8]-SS	10	10	45	70	20
7. des-AA1,2,4,5,12,13-SS	8	100	0.2	< 1	< 1
8. des-AA1,2,4,5,12,13[D-Trp8]-SS	8	100	4	45	7
9. des-AA1,2,4,5,10,12,13[D-Trp8]-SS	7	< 0.1	< 0.01	< 1	< 1
10. des-AA1,2,4,5,6,10,12,13[D-Trp8]-SS	6	< 0.1	< 0.01	< 1	< 1
11. des-AA1,2,4,5,12[Pro3,D-Trp8,D-Pro14]-SS	9	100	-	-	-

[1] All potency estimates determined in 4 or 6 point bioassay comparing each analog to somatostatin, used as a reference standard.

low activity in elevating body temperature or reversing hypothermia induced by other substances (Table 1). Table 1 shows the
potencies of several analogs of somatostatin to reverse bombesin
induced hypothermia.

Analogs of somatostatin of reduced ring size appear to be more
potent to influence thermoregulation. The non-cysteine containing analog, desAA [1,2,4,5,12] [Pro[3],D-Trp ,D-Pro[14]] -somatostatin
is likewise very potent to reverse bombesin-induced hypothermia
(Table 1). Analogs which have been reduced in chain length to 7
or 8 residues are inactive to reverse bombesin hypothermia or to
mimic somatostatin effects on the secretion of growth hormone,
insulin or glucagon (26). It should be noted that most of the
somatostatin analogs actively influence thermoregulation as well
as pituitary and pancreatic secretion. DesAA[1,2,4,5,12,13] -som-
atostatin, however, shows low potency to inhibit growth hormone
and is completely inactive to inhibit insulin or glucagon secretion, but is extremely potent to reverse bombesin-induced hypothermia. N-terminal extension of somatostatin results in loss of
hyperthermic action while actions to inhibit secretion of growth
hormone, insulin and glucagon are retained. It is possible that
these modifications result in decreases in CNS distribution of
peptide while reduction of chain length enhances penetration to
site of action. Alternatively these results may suggest that CNS
somatostatin receptors are a subclass of somatostatin receptors
analogous to the subdivisions of adrenergic receptors.   We have
previously reported that receptors on pancreatic α and β cells
have somatostatin receptors with different specificities (7).

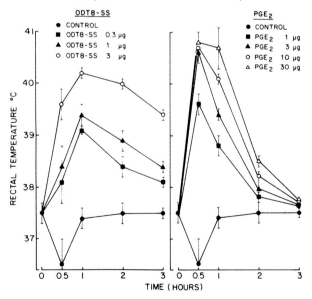

FIG. 4.   Somatostatin analog (ODT8-SS) and PGE_2 induced
hyperthermia. ODT8-SS and PGE_2 were given ic at time 0.

Icv administration of the somatostatin analog, Des-AA[1,2,4,5,12,13] [D-Trp[8]]-somatostatin results in prolonged hyperthermia (Fig. 4). Continuous administration of ODT8-SS results in eventual desensitization to its hyperthermic effects (Fig. 5). Similar to desensitization to bombesin, these results with a somatostatin analog may represent cell receptor desensitization.

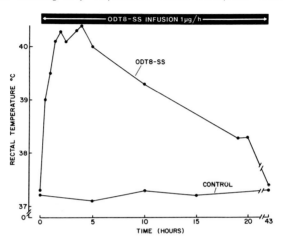

FIG. 5. Continuous infusion of somatostatin analog (ODT8-SS) results in desensitization to hyperthermia.

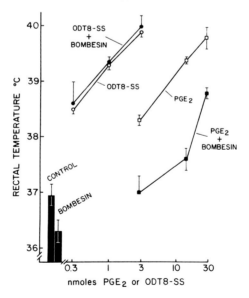

FIG. 6. Comparison of potencies of somatostatin analog (ODT8-SS) and PGE_2 to produce hyperthermia. PGE_2, but not ODT8-SS hyperthermia is reduced by bombesin (1 μg).

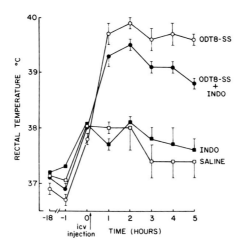

FIG. 7. Somatostatin analog (ODT8-SS, 10 μg) produces hyper-thermia which is not prevented by inhibition of prostaglandin synthesis by indomethacin. Indomethacin, 10 mg/kg, was admin-istered intraperitoneally 18 and 1 hours prior to ODT8-SS.

Cellular desensitization to somatostatin has been demonstrated <u>in vitro</u> by Vale <u>et al</u>. (27). Compared to PGE$_2$, ODT8-SS is more potent and longer acting to induce hyperthermia (Figs. 4 and 6). Of obvious interest is the question of whether the hyperthermic effects of somatostatin analogs are secondary to the release of endogenous prostaglandins. This possibility seems unlikely since administration of indomethacin does not prevent ODT8-SS induced hyperthermia (Fig. 7).

Further evidence for a fundamental difference between the hyperthermia induced by prostaglandins and that induced by ODT8-SS is that bombesin significantly reverses prostaglandin induced hypothermia but does not influence ODT8-SS induced hyperthermia (Fig. 7).

These results suggest that there are receptors and pathways within the brain capable of producing hyperthermia which do not appear to be prostaglandin mediated. Whether such a receptor sys-tem might be involved in the development of fever or maintenance of normal body temperature under conditions of reduced ambient temperature is clearly worthy of consideration.

CONCLUSION

The various actions of peptides on body temperature provide a potentially new neurochemical basis for thermoregulation. Future studies should explore the interactions of these peptides with one another and other non-peptide neurotransmitters in temperature regulation. Study of peptide participation in thermoregulation and other areas of cellular regulation will be enhanced by development of receptor antagonists and methods to inhibit endogenous peptide secretion and turnover rates. Use of endogenous passive immunization or desensitization methods may provide useful investigational tools pending development of receptor antagonists.

Identification of the neurochemical control of temperature regulation may lead to new methods of understanding thermoregulation; thermoregulation in terms of maintenance of a constant internal environment and in terms of adaptive thermoregulatory changes associated with nutrient metabolism.

ACKNOWLEDGEMENTS

Research supported by NINCDS grant NS-14263 and The Clayton Foundation for Research. The excellent technical assistance of Roberta Allen, Vicki Webb and Greta Berg and manuscript preparation by Susan McCall Garonski are appreciated.

REFERENCES

1. Anastasi, A., Erspamer, V., and Bucci, M. (1971): Experientia, 27:166-167.
2. Bissette, G., Nemeroff, C.B., Loosen, P.T., Prange, Jr., A.J., and Lipton, M.A. (1976): Nature, 262:607-609.
3. Bloom, F., Segal, D., Ling, N., and Guillemin, R. (1976): Science, 194:630-632.
4. Brown, M.R., Rivier, J., and Vale, W. (1977): Science, 196: 998-1000.
5. Brown, M.R., Rivier, J., and Vale, W. (1977): Life Sci., 20: 1681-1688.
6. Brown, M.R., Rivier, J., and Vale, W. (1977): Life Sci., 21: 1729-1734.
7. Brown, M., Rivier, J., and Vale, W. (1977): Endocrinology, 98:336-343.
8. Brown, M.R., Rivier, J., Kobayashi, R.M., and Vale, W. (1978): In: Gut Hormones, edited by S. Bloom, pp. 550-558. Churchill-Livingstone, Edinburgh.
9. Brown, M.R., Allen, R., Villarreal, J., Rivier, J., and Vale, W. (1978): Life Sci., 23:2721-2728.
10. Brown, M., and Vale, W. (1979): Trends in Neurosci., 2:95-97.
11. Brown, M., and Vale, W. (1980): Thermoregulatory Mechanisms and Their Therapeutic Implications, edited by B. Cox, P. Lomax, A.S. Milton, and E. Schonbaum, pp. 186-194. S. Karger, Basel.
12. Brown, M.R., Tache, Y., and Fischer, D. (1979): Endocrinology, 105:660-665.

13. Brown, M., Marki, W., and Rivier, J. (1980): Life Sci., submitted for publication.
14. Carino, M.A., Smith, J.R., Weick, B.G., and Horita, A. (1976): Life Sci., 19:1687-1692.
15. Hellon, R.F. (1975): Pharmacol. Rev., 26:289-321.
16. Goldstein, A., and Lowry, P.J. (1975): Life Sci., 17:927-932.
17. Holaday, J.W., Wei, E., Loh, H.H., and Li, C.H. (1978): Proc. Natl. Acad. Sci. USA, 75:2923-2927.
18. McDonald, T.J., Jornvall, H., Nilsson, G., Vagne, M., Ghatei, M., Bloom, S.R., and Mutt, V. (1979): Biochem. Biophys. Res. Commun., 90:227-233.
19. Metcalf, G. (1974): Nature, 252:310-311.
20. Moody, T.W., Pert, C.B., Rivier, J.E., and Brown, M.R. (1978): Proc. Natl. Acad. Sci. USA, 75:5372-5376.
21. Moss, R.L., and McCann, S.M. (1973): Science, 181:177-179.
22. Pittman, Q., Tache, Y., and Brown, M. (1980): Life Sci., 26:725-730.
23. Rivier, J.E., and Brown, M.R. (1978): Biochemistry, 17:1766-1771.
24. Swan, H., and Schatte, C. (1977): Science, 195:84-85.
25. Tache, Y., Pittman, Q., and Brown, M. (1980): Brain Res., 188: 525-530.
26. Vale, W., Rivier, J., Ling, N., and Brown, M. (1978): Metabolism, 27:1391-1401.
27. Vale, W., and Rivier, J. (1980): Proc. Sixth Intl. Cong. of Endocrinol., Melbourne, Australia (Feb. 10-16, 1980).
28. Villarreal, J.A., and Brown, M.R. (1978): Life Sci., 23:2729-2734.

Neurosecretion and Brain Peptides,
edited by J. B. Martin, S. Reichlin, and K. L. Bick.
Raven Press, New York © 1981.

The Nucleus Tractus Solitarius and Experimental Neurogenic Hypertension: Evidence for a Central Neural Imbalance Hypothesis of Hypertensive Disease

Donald J. Reis

*Laboratory of Neurobiology, Department of Neurology, Cornell University Medical College,
New York, New York 10021*

Over the past few years there has been increasing interest in the possibility that the sympathetic nervous system, in turn regulated by the central nervous system (CNS), is critical for the expression of essential hypertension in man. Disordered regulation of the arterial pressure by the brain may be the principal defect in the disease, leading inevitably to compensatory dysfunction in other organ systems including the kidneys, the renin, angiotensin-aldosterone systems, and in fluid and electrolyte balance (for reviews see refs. 3,7,16,25).

Any hypothesis that disordered brain function can, in fact, lead to the principal abnormalities of blood pressure control characteristic of the human disease requires the demonstration in experimental animals that impaired function of the brain can produce the constellation of cardiovascular abnormalities characteristic of the human disorder. The experimental model of hypertension should meet several criteria to show its similarity to the human disease. The hypertension should be: (a) neurogenic; (b) chronic and sustained; (c) the principal physiological abnormality and unassociated with substantial changes in behaviors which might themselves alter the arterial pressure; (d) reduced by administration of drugs which lower blood pressure in man (e.g., clonidine); (e) associated with other abnormalities of blood pressure control often seen in hypertensive patients, including lability (minute-to-minute variability of the arterial pressure); and (f) exaggerated reactivity during spontaneous behavior or in response to environmental stimuli; and should (g) result in cardiovascular pathology similar to the human disease.

NUCLEUS TRACTUS SOLITARIUS AND HYPERTENSION

During the past several years our laboratory has sought to establish that disorders of the CNS in animals can lead to some of these abnormalities of blood pressure regulation, thereby simulating the human disease. We have proposed that hypertension will result from an imbalance between the central neural networks which serve to excite sympathetic vasomotor neurons and those which inhibit them with the imbalance favoring a preponderance of sympathetic discharge. This concept has been termed a <u>central neural imbalance hypothesis of hypertension</u> (24). That such local imbalances in brain can produce chronic hypertension in animals had never been successfully established in the past (25).

As a target of inquiry, we have focused on one region of the brain, the intermediate third of the nucleus tractus solitarius (NTS). The NTS is the principal site of termination of visceral afferent fibers carried by the Vth, VIIth, IXth, and Xth cranial nerves. Of particular interest is the intermediate one third of the nucleus, largely in its medial portion, and its commissural subdivision. This region can be termed <u>the cardiovascular NTS</u> for several reasons: (a) It is the principal site of termination of primary afferent fibers arising from arterial baroreceptors (2,14,17,18,30) and as such it mediates the powerful inhibitory actions of baroreceptors on sympathetic discharge. (b) Electrical stimulation here will simulate baroreflexes producing hypotension, apnea and bradycardia (5). (c) In anesthetized animals lesions at this site will abolish baroreflexes (18). (d) It appears to be a site of action, at least in part, of several anti-hypertensive agents, including α-methyldopa (9). (e) In addition, the intermediate part of the NTS is richly innervated by fibers arising from numerous portions of the CNS, themselves known to have an important role in cardiovascular regulation, including the amygdala, parabrachial nucleus and medial hypothalamus (see 15). (f) Finally, it is richly innervated by a number of neurochemical transmitters and/or modulators which have been proposed to be important in cardiovascular control. These include a rich innervation of catecholamines, serotonin, and a variety of neuropeptides, including Substance P, enkephalin (14), and vasopressin (33). These considerations have led to the possibility that disordered transmission in the NTS could indeed result by impairing normal inhibitory output to sympathetic neurons, thereby resulting in hypertension. Impaired transmission to NTS also could involve disordered functions in one or several of the neurotransmitter systems so richly concentrated there (21).

In our studies we have examined the consequences of impaired function of neurotransmission in the NTS on blood pressure control. We have discovered that impaired transmission through this nucleus can result in abnormalities of pressure regulation simulating the human disease.

ACUTE FULMINATING NTS HYPERTENSION IN RAT

In 1973, Doba and Reis (10) first demonstrated that bilateral electrolytic lesions of the NTS in the unanesthetized rat result in the development of arterial hypertension within minutes after recovery from anesthesia (Fig. 1). The elevation of systolic pressure, usually greater than 200 mm Hg, is entirely neurogenic and due to a massive increase in peripheral vasoconstriction consequent to intense discharge of preganglionic sympathetic neurons, largely to blood vessels. The vasoconstriction is mediated by α-adrenergic receptors and is regionally differentiated, being most intense in skin, muscle, and portions of the gastrointestinal tract (31). The redistribution of blood flow is similar to that associated with interruption of baroreceptors and differs from other forms of experimental hypertension in the rat. As a consequence of the increased peripheral vasoconstriction there is ventricular overload thereby resulting in development of an increased left ventricular end-diastolic pressure, reduced stroke volume, diminished cardiac output and increased central venous pressure. Thus, following the NTS lesions, animals are almost immediately in cardiac failure. Over the ensuing hours, the cardiac failure evolves, leading ultimately to pulmonary edema and death.

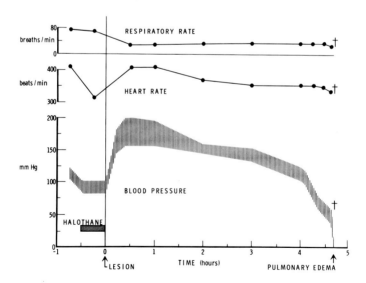

FIG. 1. Time course of changes in systemic arterial blood pressure, heart rate, and respiratory rate in a representative unanesthetized rat after production of bilateral lesions in the NTS. Just prior to death the rat developed pulmonary edema (from ref. 24).

The elevation in arterial pressure is unaltered by removal of the kidneys or the adrenal medulla. However, there is increased release of adrenal catecholamines during the syndrome (11). NTS lesions result in a disappearance of arterial baroreflexes. Acute NTS hypertension leads to profound changes in cyclic nucleotide metabolism in blood vessels similar to those seen in the spontaneously hypertensive rat and rats made hypertensive by DOCA-salt (1). The nucleotide changes are in the direction of reduced β-adrenergic sensitivity but also may be related to the initiating factors that ultimately lead to vascular hyperplasia.

The hypertension produced by NTS lesions is abolished by ether, halothane, anesthetics, α-blockade, ganglionic blocking agents, and reserpine (11). Of particular interest is the finding that this hypertension is also lowered by treatment with clonidine (27).

The expression of hypertension is dependent on the integrity of descending noradrenergic projections to the spinal cord (11). On the other hand, ascending noradrenergic and dopaminergic projections do not appear to influence the hypertension since intrahypothalamic injections of 6-hydroxydopamine (6-OHDA), producing massive degeneration of ascending catecholaminergic projections, do not influence the evolution of the syndrome.

Brain regions lying above the midbrain appear essential for the expression of the hypertension since midcollicular decerebration will abolish it (10). At the present time the location of the critical structures is unknown but recent studies (3) have suggested that NTS hypertension can be aborted by small lesions of ventral portions of the anterior third ventricle, thereby identifying this as the potential rostral site essential for the development of the elevated blood pressure.

These studies therefore demonstrate that impaired transmission through an inhibitory region of the brainstem of rat can lead to acute fulminating neurogenic hypertension. The mechanism of hypertension is related, at least in part, to a disturbance of baroreceptor reflex mechanisms centrally with release of sympathetic vasomotor neurons from inhibition, thereby leading to the profound vasoconstriction and hypertension.

CHRONIC LABILE HYPERTENSION IN CAT

The rapid development of malignant heart failure following NTS lesions in rat led us to determine whether or not a larger animal, cat, could survive the initial sympathetic overactivity following NTS lesions (19). Initially, the placement of bilateral electrolytic lesions in the cat resulted in an elevation of arterial pressure which most animals survived. As the blood pressure began to return to normal levels 8 to 16 hours later the characteristic features of the syndrome appeared. After recovery from a postoperative period, cats with NTS lesions developed a characteristic syndrome of chronic labile hypertension.

This syndrome of NTS hypertension in cat is characterized by five cardinal features:

(a) <u>Lability of arterial pressure.</u> Cats with NTS lesions exhibit an extreme second-to-second variability (lability) of blood pressure characterized by marked spontaneous fluctuations with both elevations and depressions often as great as 100 mm Hg (Fig. 2). The lability as influenced by environmental stimulation in a laboratory environment, is greater during the day.

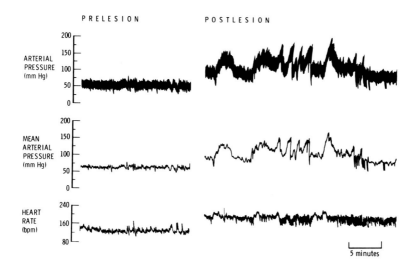

FIG. 2. Effect of lesions of NTS on the arterial pressure and heart rate of an individual cat. Prelesion tracing was taken two days before placement of the lesions when the cat was in quiet wakefulness and lying down. The postlesion result was taken 1 week after the lesion with the cat in the same behavioral state. Note the extreme lability of arterial pressure. (From ref. 24, with permission).

(b) <u>Sustained hypertension.</u> The arterial pressure of chronically lesioned animals is significantly elevated above controls. The elevation of pressure is modest (114 vs 80 mm Hg, experimental vs controls), is permanent and, as shown in other studies, dependent upon the sympathetic nervous system.

(c) <u>The exaggerated responsivity.</u> There is an exaggerated responsivity of the arterial pressure during spontaneous or evoked behaviors or in response to environmental stimulation (Fig. 3). In general, the direction of change of blood pressure is comparable to that seen when these behaviors are performed in unoperated controls: it is only the magnitude of the response which is altered by NTS lesions. The exaggeration affects both the reflex elevations and depressions of blood pressure.

(d) <u>Fixed tachycardia.</u>

(e) <u>Absence or marked attenuation of baroreceptor reflexes.</u>

(f) <u>Facilitated conditioning of arterial pressure.</u> The exaggerated responses of the arterial pressure to environmental stimuli following NTS lesions in the cat also result in facilitation of the conditioned elevation of arterial pressure which can be produced by classic conditioning procedures (20).   The elevation of arterial pressure conditioned by a tone followed by a mild

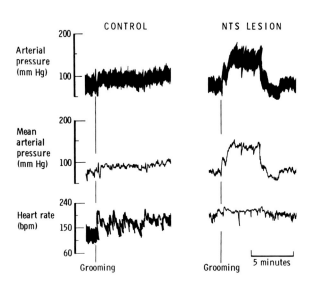

FIG. 3.   Effects of lesions of NTS on the changes in arterial pressure and heart rate associated with grooming. The light vertical lines indicate the onset of the behavior.   The responses were measured prior to placement of NTS lesions (control) and after placement of the lesions (NTS lesion) in the same cat. (From ref. 24, with permission).

cutaneous shock appears at earlier trials and with greater intensity in cats with NTS lesions than in their yoked controls. These findings demonstrate that central baroreceptor mechanisms serve to buffer the elicited pressor responses and, more importantly, that they may counterbalance environmental stimuli which may lead to exaggeration of the arterial pressure.

The observation that NTS lesions in cats facilitate conditioned elevation of blood pressure leads to a testable hypothesis; namely, that, for environmental stress to be of importance in the production of hypertension, it may be necessary for baroreflex integration to be impaired. Thus the failure of stress to produce sustained neurogenic hypertension in most animal models (25) may be explained by the efficacy of baroreceptor mechanisms to offset the emotional drive to sympathetic neurons. However, in a setting of impaired baroreflex function, emotional stress may act to produce disease. Conversely, baroreflex dysfunction alone may also fail to produce hypertension. Only when coupled with environmental or emotional stress will sustained hypertension appear.

SUSTAINED HYPERTENSION IN DOGS

Bilateral electrolytic lesions of the NTS in dogs also produce hypertension (4,29). Acutely there is a sympathetically mediated, marked elevation of blood pressure. However, after an initial phase the arterial pressure declines from the zenith but remains modestly elevated. Lability of arterial pressure appears and tachycardia persists. Baroreflex function is abolished. Hypertension is sensitive to clonidine and α-blockade (29). In the absence of any abnormality of circulating renin or aldosterone concentrations (4), the animals appear to simulate humans with the so-called low renin form of hypertension. Notable has been the observation (29) that after one to three months dogs with NTS lesions begin to develop sustained elevations of arterial pressure. Thus, in dogs, the NTS lesions may result in chronic sustained hypertension.

A LABILITY OF ARTERIAL PRESSURE PRODUCED BY IMPAIRMENT OF ADRENERGIC INNERVATION OF NTS

To determine whether impaired neurochemical mechanisms within the NTS can result in hypertension we next focused on the consequences of impairing the catecholaminergic innervation of the NTS in blood pressure control (32,37).

The NTS and adjacent regions of the dorsal medulla are richly innervated by catecholamine-containing neurons. The principal innervation is noradrenergic although dopamine and adrenalin are found in much smaller quantities in the region (9,28). The principal, but not exclusive, source of the noradrenergic innervation of NTS appears to arise from the so-called A2 group of catecholamine neurons whose cell bodies lie in medial and commissural portions of the NTS, the area postrema, and adjacent portions of the dorsal grey. The A2 neurons have been presumed, largely on pharmacological evidence, to exert a vasodepressor function possibly by facilitating baroreceptor reflexes (8).

In our studies we have analyzed in rat the effects of impairing the catecholaminergic innervation of NTS function by destroying the terminals within the NTS by the local injection of the neurotoxin 6-hydroxydopamine (6-OHDA) (32) or by electrolytic destruction of the cells of origin of much of this innervation, the A2 neuronal group in the medulla (37). 6-OHDA and A2 lesions, for the most part, produce similar lesions. Both treatments resulted in the development of chronic lability of arterial pressure without an elevation of arterial pressure. With 6-OHDA, baroreceptor reflexes remain although the gain of the reflex is somewhat depressed (32).

Our findings suggest that the catecholamine innervation of NTS (mostly noradrenergic) serves to stabilize the arterial pressure by acting in concert with baroreceptors. The fact that deficiencies in this system lead to lability of arterial pressure is of interest since lability is frequently an early sign of hypertension in man. However, if animals with A2 lesions are maintained for up to one year, despite the persistence of lability, there is no evidence of hypertension (34). Thus, at least in our animal model, lability of blood pressure per se is not an obligate antecedent of sustained hypertension.

The syndrome produced by interruption of the noradrenergic innervation of NTS demonstrates that interference with the normal function of a biochemically selective system in the rat brain can result in abnormalities of blood pressure control with many of the features similar to that of sustained hypertension in man. This animal model provides an opportunity to examine both the etiology and pathophysiology of sustained hypertension.

GLUTAMATE AND HYPERTENSION RELATIONSHIP TO THE PEPTIDES

We have recently proposed that the amino acid neurotransmitter glutamic acid may be the neurotransmitter of baroreceptor afferents (26,36). Our evidence includes the demonstration that: (a) the microinjection of small quantities (<0.1 ul) of l-glutamate into NTS simulates baroreceptor reflexes in the rat, the response being evoked by as little as 10^{-11} moles; (b) the response is anatomically specific (41,42); (c) a similar response is elicited by l-glutamate agonists, including kainic acid (KA), d-glutamate and aspartic acid; (d) the baroreceptor reflex is blocked and hypertension transiently produced by the local injection of the glutamate antagonist, glutamic acid diethyl ester (35); (e) removal of the nodose ganglion, the principal source of afferent fibers from baroreceptors in the aortic arch region, results in a 50% reduction in the high affinity uptake of l-glutamate restricted to NTS and a reduction in the content of endogenous glutamate (22,26,36).

Of particular interest is that the local administration of l-glutamate or its agonist KA into NTS in doses slightly higher than those required to elicit the baroreflex response results when anesthesia is terminated in fulminating neurogenic hypertension. The hypertension produced by KA is identical in all its components to that produced by electrolytic lesions of the nucleus and cannot be ascribed to neurotoxic action of the drug.

That blockade of l-glutamate receptors in NTS can lead to hypertension is of considerable theoretical importance with respect to the central neural imbalance hypothesis. It suggests that defects in the function of a single neurotransmitter system within NTS, that releasing glutamic acid, can produce hypertension. This raises the possibility that a biochemical defect resulting in deficient storage or release of glutamate or, conversely, a deficiency in the number or reduction in the affinity of glutamate receptors postsynaptically would be a sufficient condition within NTS to lead to hypertension.

These findings are also of some interest with respect to the biology of the neuropeptide Substance P. Substance P is present in baroreceptor afferent nerves, in neurons of the nodose ganglion and can be found in areas of the NTS in which baroreceptors terminate (12-14). However, microinjection of Substance P, even in large quantities, into NTS in the rat fails to produce any effects on blood pressure or heart rate. This is in sharp contrast to exquisite sensitivity in the region to the local administration of l-glutamate (Talman and Reis, unpublished). This raises the interesting question of the relationship of Substance P and glutamate fibers. Are they separate populations of neurons running in parallel from baroreceptors into the brainstem? Is it possible that Substance P and glutamate are contained in the same neuron?

SUMMARY AND CONCLUSIONS

1. Interference with neuronal transmission through the NTS can result, depending upon species and mode of perturbation, in a panoply of abnormalities of blood pressure control simulating many of the features of the human disease. These are summarized in Table 1.

TABLE 1. <u>Summary of abnormalities of arterial pressure elicited by impaired transmission in NTS</u>

---

1. FULMINATING NEUROGENIC HYPERTENSION (NTS lesions or blockade of l-glutamate in NTS by kainic acid in rat)
2. CHRONIC LABILE NEUROGENIC HYPERTENSION (NTS lesions in cat and dog)
3. CHRONIC SUSTAINED NEUROGENIC HYPERTENSION (NTS lesions in dog)
4. LABILE ARTERIAL PRESSURE WITHOUT HYPERTENSION (chemical lesions of noradrenergic innervation of NTS or destruction of A2 group in rat)
5. LABILE ARTERIAL PRESSURE <u>WITH</u> HYPERTENSION (NTS lesions in cats or dogs)
6. EXAGGERATED REACTIVITY OF ARTERIAL PRESSURE IN RESPONSE TO BEHAVIORAL, EMOTIONAL OR ENVIRONMENTAL STIMULI (NTS lesions in cat, dog; A2 lesions in rat)
7. HYPERTENSION LOWERED BY DRUGS EFFECTIVE IN TREATING HYPERTENSION IN MAN (<u>E.G.</u>, α-blockers, clonidine)

---

2. The abnormalities of pressure control resulting from abnormal transmission in NTS met most of the criteria of an animal model of central neurogenic hypertension. The only criterion yet to be met is that of pathology, a deficiency which may soon be overcome when animals, such as dogs, are maintained for prolonged periods of time.

3. The studies establish the possibility that subtle abnormalities involving neurochemical balances with NTS, resulting either from abnormal neurochemical transmission or variations of the organization of the nucleus with preponderance of one transmitter or deficiencies in others, can result in hypertension.

4. Impaired NTS function can produce an amplification of the action of the environmental stresses on blood pressure. Thus environmental stimuli or the expression of behaviors which normally result in trivial elevations of blood pressure will, after the NTS is perturbed, result in marked elevations.

5. <u>A neural or neurochemical imbalance in brain can produce hypertension.</u>

ACKNOWLEDGEMENTS

Thanks to Drs. Nobutaka Doba, Mark Nathan, Mark Perrone, David Snyder and William Talman for their contributions to these studies. The research was supported by grants from the NIH (HL 18974, NS 03346), and NASA (NSG 2259).

REFERENCES

1. Amer, M.S., Doba, N., and Reis, D.J. (1975): <u>Proc. Natl. Acad. Sci. USA</u>, 72:2135-2139.
2. Berger, A.J. (1979): <u>Neurosci. Lett.</u>, 14:153-158.
3. Brody, M.J., Haywood, J.R., and Touw, K.B. (1980): <u>Ann. Rev. Physiol.</u>, 42:441-453.
4. Carey, R.M., Dacey, R.G., Jane, J.A., Winn, H.R., Ayers, C.R., and Tyson, G.W. (1979): <u>Hypertension</u>, 1:246-254.
5. Crill, W.E., and Reis, D.J. (1968): <u>Am. J. Physiol.</u> 214:269-276.
6. Curtis, D.R. (1976): In: <u>Glutamic Acid: Advances in Biochemistry and Physiology</u>, edited by L.J. Filer, Jr., S. Garratini, M.R. Klare, W.A. Reynolds, and R.J. Wurtman, pp. 163-175. Raven Press, New York.
7. DeJong, W., Provost, A.P., and Shapiro, A.P. (1977): <u>Hypertension and Brain Mechanisms</u>. Elsevier, Amsterdam.
8. DeJong, W., Zandberg, P., and Bohus, B. (1975): <u>Prog. Brain Res.</u>, 42:285-298.
9. DeJong, W., Zanberg, P., Wijnen, H., and Nijkamp, F.P. (1979): In: <u>Nervous System and Hypertension</u>, edited by P. Meyer, and H. Schmitt, pp. 165-172. Wiley-Flammarion, New York.

10. Doba, N., and Reis, D.J. (1973): Circ. Res., 32:584-593.
11. Doba, N., and Reis, D.J. (1974): Circ. Res., 34:293-301.
12. Gillis, R.A., Heelke, C.J., Hamilton, B.L., Norman, W.P., and Jacobowitz, D.M. (1980): Brain Res., 181:476-481.
13. Helke, C.J., O'Donahue, T.L., and Jacobowitz, D.J. (1980): Peptides, 1:1-9.
14. Katz, D.J., and Karten, H.J. (1979): Brain Res., 171:187-195.
15. Loewy, A.D., and McKeller, S. (1980): Fed. Proc., 39:2495-2503.
16. Meyer, P., and Schmitt, H. (1979): Nervous System and Hypertension. Wiley-Flammarion, New York.
17. Miura, M., and Reis, D.J. (1969): Am. J. Physiol., 217:142-153.
18. Miura, M., and Reis, D.J. (1972): J. Physiol., 223:525-548.
19. Nathan, M.A., and Reis, D.J. (1977): Circ. Res., 40:72-81.
20. Nathan, M.A., Tucker, L.W., Severini, W.P., and Reis, D.J. (1978): Science, 201:71-73.
21. Palkovitz, M., Mezey, E., and Jaborsky, L. (1979): In: Nervous System and Hypertension, edited by P. Meyer, and H. Schmitt, pp. 18-30. Wiley-Flammarion, New York.
22. Perrone, M.H., Talman, W.T., and Reis, D.J. (1980): Fed. Proc., 39:362.
23. Pickel, V.M., Joh, T.H., Reis, D.J., Leeman, S.E., and Miller, R.J. (1979): Brain Res., 160:387-400.
24. Reis, D.J. (1980): Res. Publ. Assoc. Res. Nerv. Ment. Dis. (In press.)
25. Reis, D.J., and Doba, N. (1974): Prog. Cardiovasc. Dis., 17:51-71.
26. Reis, D.J., Talman, W.T., Perrrone, M., Doba, N., and Kumada, M. (1980): In: Baroreceptors and Hypertension, edited by P. Sleight. Oxford University Press, Oxford. (In press.)
27. Rockhold, R.W., and Caldwell, R.W. (1979): Neuropharmacology, 18:347-354.
28. Saavedra, J.M. (1979): In: Radioimmunoassay of Drugs and Hormones in Cardiovascular Medicine, edited by A. Albertini, M. Da Prada, and B.A. Peskar, pp. 199-215. Elsevier, Amsterdam.
29. Schmitt, H., and Laubie, M. (1979): In: Nervous System and Hypertension, edited by P. Meyer, and H. Schmitt, pp. 173-201. Wiley-Flammarion, New York.
30. Seller, H., and Illert, M. (1969): Pflugers Arch., 306:1-19.
31. Snyder, D.W., Doba, N., and Reis, D.J. (1978): Circ. Res., 42:87-91.
32. Snyder, D.W., Nathan, M.A., and Reis, D.J. (1978): Circ. Res., 43:662-671.
33. Swanson, L.W., and Hartman, B.K. (1975): J. Comp. Neurol., 163:567-506.
34. Talman, W.T., Alonso, D.R., and Reis, D.J. (1980): In: Baroreceptors and Hypertension, edited by P. Sleight. Oxford University Press, Oxford. (In press.)

35. Talman, W.T., Perrone, M.H., and Reis, D.J. (1980): Fed. Proc., 39:362.
36. Talman, W.T., Perrone, M.H., and Reis, D.J. (1980): Science, (In press.)
37. Talman, W.T., Snyder, D.W., and Reis, D.J. (1980): Circ. Res., (In press.)

Neurosecretion and Brain Peptides,
edited by J. B. Martin, S. Reichlin, and K. L. Bick.
Raven Press, New York © 1981.

Naloxone Reverses the Pathophysiology of Shock Through an Antagonism of Endorphin Systems

John W. Holaday and Alan I. Faden

Department of Medical Neurosciences, Division of Neuropsychiatry, Walter Reed Army Institute of Research, Washington, D.C. 20012

Since their discovery, the physiological roles for endogen-ous-opiate peptides (collectively termed endorphins) have been the subject of intense scientific investigation. Their involve-ment in pain modulation has been relatively well resolved (15,25); however, other physiological and behavioral functions ascribed to endorphins may be equally important (3,13,15,16). Perhaps because of the pain relieving and euphoric properties of opiates, the endorphins have been considered panacea peptides by scientists and lay-persons alike. Nonetheless, opiates can also produce deleterious effects such as respiratory depression and hypotension (9,10,23). In this report, we present the results of our investigations into the pathophysiological function of endor-phins as toxic factors contributing to the hypotension and mor-tality of shock.

It is known that endorphin systems are activated by stress (1,12) and that the pharmacological administration of endogenous and exogenous opiates can result in marked hypotension (9,10,23). Since shock states are characterized by profound physiological stress, we investigated the possibility that the stress of shock would activate endorphin systems and thus contribute to the hypo-tension which characterizes the shock syndrome. More important-ly, if endorphins contribute to the pathophysiological and lethal effects of shock, then blockade of endorphins by opiate antagon-ists should reverse the endorphin component of hypotension and improve survival.

Following our initial report on the therapeutic effects of naloxone in conscious rats subjected to endotoxic shock (14), we have completed further studies designed to evaluate the effects of naloxone in endotoxic, hemorrhagic, and spinal shock (see next chapter) in rats, dogs, and cats (4-8,15,17,21-24,27). Here we will discuss the role of endorphins in the pathophysiology of endotoxic and hemorrhagic shock in rats and dogs. Recent evidence which addresses the source of endorphins as well as their site and mechanisms of action in shock will also be presented.

ENDOTOXIC SHOCK

Rats

As many anesthetic agents cause release of endorphins or alter pituitary responses to stress, we elected to study unanesthetized rats. Twenty-four hours prior to study, catheters were placed in the external jugular vein and tail artery; both cannulae were threaded subcutaneously to emerge from the occipital area and extended outside the cage through a wire-spring shield. This permitted evaluation of conscious, free moving rats which remained in their home cages and were not subjected to handling stresses during the study (14). Blood pressure (BP) and heart rate (HR) were continously recorded using a micro-transducer connected to a polygraph.

To induce endotoxic shock, E. coli lipopolysaccharide endotoxin was administered intravenously (iv at a dose which produced 70% mortality in pilot studies (12-60 mg/kg, depending on lot number). When mean arterial pressure (MAP) fell to a pre-established level of 65-70 mm Hg, rats were injected iv with equal volumes of either saline vehicle or varying doses of naloxone HCl. Cannulae were flushed with 0.3 ml saline to ensure complete drug delivery.

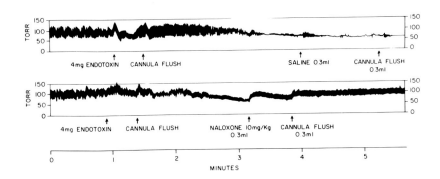

FIG. 1. Effects of intravenous saline (top) and naloxone (bottom) on the precipitious fall in blood pressure produced by 4 mg endotoxin in representative rats. Saline (0.3 ml) or 10 mg/kg naloxone was injected after blood pressure fell to 65-70 mm Hg (torr). (Reprinted with permission from ref. 14).

In Fig. 1, the responses of representative rats to saline (top) or 10 mg/kg naloxone HCL (bottom) are compared following endotoxin-induced hypotension. Saline injection produced no

significant change in BP over this interval, whereas naloxone
produced a rapid return of BP to control levels within seconds
following administration.

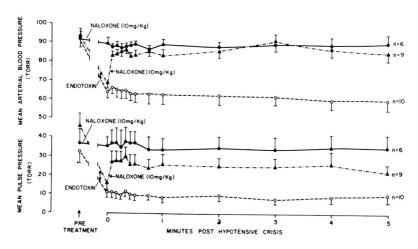

FIG. 2. The effects of naloxone or saline on mean arterial
pressure (MAP, top) or pulse pressure (bottom) are compared with
and without endotoxin-induced hypotension. Naloxone alone
(●——●) did not affect MAP of PP. Endotoxin produced a significant drop in both MAP and PP which was reversed by naloxone
(△—·—△) but unaffected by saline (0----0). Points represent
averages ± SEM. (Reprinted with permission from ref. 14).

A comparison of MAP and pulse pressure (PP, a crude index of
cardiac performance) responses for a group of rats is seen in
Fig. 2. After endotoxin hypotension, naloxone treatment resulted
in a rapid and significant improvement in MAP and PP, whereas
saline treatment was without effect. Additionally, naloxone
administration to control animals which were not subjected to
endotoxemia had no effect on MAP and PP, thus indicating a selective action of naloxone in reversing shock hypotension instead of
a direct cardiovascular effect of the drug by itself.
We observed a typical biphasic BP response following endotoxin
injection, with an initial nadir occurring within 10-15 min, followed by a partial recovery and a second fall in BP at about 45
min (14). The restoration of BP produced by naloxone was more
rapid and complete when administered during the initial hypotensive crisis; however, a 15-20 mm Hg improvement in MAP was still
produced by naloxone when injections were made during the second
hypotensive interval (6). When administered prophylactically
prior to endotoxin injection, naloxone significantly attenuated
both hypotensive intervals (14). However, the effects of naloxone on BP were less pronounced when given prior to shock than
when given following shock (14).

In order to determine the dose-response profile for these therapeutic effects of naloxone, we subjected conscious rats to endotoxemia as before. After MAP fell to 65-70 mm Hg, rats were treated with low-dose naloxone (0.1 mg/kg bolus, 0.05 mg/kg/hr infusion), middle dose naloxone (1.0 mg/kg bolus, 0.5 mg/kg/hr), high dose naloxone (10.0 mg/kg bolus, 5.0 mg/kg/hr), or equivalent volumes of saline. Physiological parameters were monitored for 12 hrs; drug infusions were maintained until 24 hr survival was determined (6,17).

As can be seen in Fig. 3, even the low dose (0.1 mg/kg) naloxone resulted in a significant increase in MAP. However, maximum responses were obtained following 1.0 mg/kg and 10.0 mg/kg bolus injections.

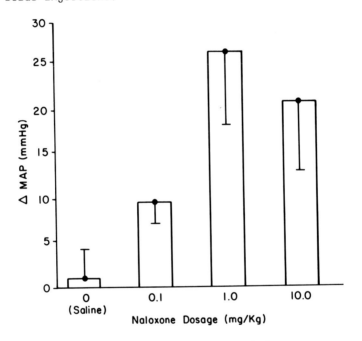

FIG. 3. The maximal change in mean arterial pressure responses were compared following injection of saline or varying doses of naloxone HCl in endotoxin-induced hypotensive rats. Vertical bars are SEM, n=10 per group. A dose-response relationship is indicated. (Reprinted with permission from ref. 6).

The fact that even a very low dose (0.1 mg/kg) of naloxone produced an improvement in MAP was suggestive of a specific interaction at the receptor level. However, Waterfield and Kosterlitz (29) recommend that the stereospecificity of opiate antagonist action be demonstrated in order to exclude pharmacological effects <u>not</u> due to opiate-receptor involvement.

Since (+) naloxone has no measurable opiate-receptor affinity or narcotic-antagonist effects (22), we compared injections of

(+) naloxone HCl (1.0 mg/kg) to an equivalent dose of (-) naloxone HCl known to be effective in this system.   Fig. 4 shows that
(+) naloxone was totally devoid of effects upon MAP or PP, whereas its stereoisomer, (-) naloxone, significantly improved both
measures. This stereospecificity of naloxone in endotoxic shock
provides evidence for its specific action upon opiate receptors,
probably by competitively antagonizing the agonistic effects of
endorphins.

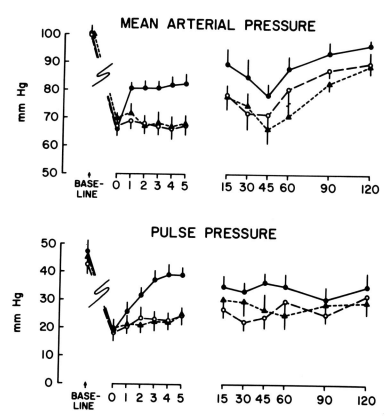

FIG. 4. The stereospecific effects of naloxone on cardiovascular
parameters were compared to vehicle injections following endotoxic hypotension.   After MAP fell to 65-70 mm Hg, 1 mg/kg (-)
naloxone HCl (●——●), 1 mg/kg (+) naloxone HCl, or an equivalent
volume of water (O——O) were injected intravenously.   Time in
minutes across horizontal axis following injections; vertical
bars are SEM; n=9 rats/group. (Reprinted with permissin from
ref. 6).

Despite the persistent improvement in BP produced by naloxone,
24 hr survival was not significantly improved in this rat model
of bacteremic shock (6).   This finding was not unexpected, since
endotoxemia in rats results in significant pathophysiology in

lungs, kidneys, liver, and the gastrointestinal system which does not necessarily involve hypotension (11,28,30). In other species where these effects are less pronounced, naloxone improves as well as survival BP (24).

Dogs

To investigate the question of species specificity as well as to evaluate cardiovascular mechanisms by which naloxone reverses shock, we studied endotoxemia in dogs. Studies were conducted on pentobarbital anesthetized animals because of surgical procedures required for monitoring complex cardiovascular parameters. Naloxone had no signficant effect on these parameters in the pentobarbital-anesthetized, unshocked dogs used as controls (24). A femoral arterial catheter was used to monitor MAP. A pigtail catheter was placed in the left ventricle (LV) and the first derivative of LV pressure with respect to time (LV dp/dt max) was used as an index of LV contractility. A triple lumen Swan-Ganz catheter was passed into the pulmonary artery to measure pulmonary arterial wedge pressure (PAw). This catheter was equipped with a thermistor tip to allow for estimates of cardiac output (CO) by the thermodilution technique. Total peripheral vascular resistance (TPVR) and stroke volume (SV) were calculated from CO, MAP, and HR.

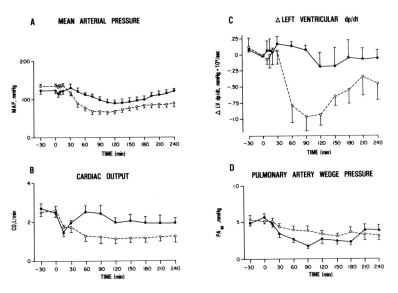

FIG. 5. Anesthetized dogs were administered endotoxin at t=0, either saline (n=14, o —o) or 2.0 mg/kg (-) naloxone HCl (n=6, ● —●) were administered at t=15 min. Naloxone significantly improved MAP, CO, and LV dp/dt, whereas no significant effect on PAw was observed. Vertical bars are SEM. (Reprinted with permission from ref. 4).

E. coli endotoxin was injected at a dose (0.1 mg/kg) which produced a 24 hr mortality of 80% in untreated dogs. Fifteen minutes following endotoxin, dogs were treated iv with saline (n=14) or naloxone (2 mg/kg, n=6) in an equivalent volume. Survival was monitored at 24 hrs.

In all animals, endotoxin produced a striking decline in MAP, LV dp/dt, and CO (Fig. 5). A 2 mg/kg dose of naloxone followed by an infusion at 2 mg/kg/hr significantly attenuated the drop in MAP and LV dp/dt and reversed the decline in CO. SV was also significantly improved by naloxone (data not shown). By contrast naloxone had no effect on PAw, HR, or TPVR when compared to saline controls. These findings suggest that naloxone exerts its therapeutic effects by improving myocardial contractility, not by altering vascular resistance or venous return (24).

In contrast with rats, naloxone significantly improved 24 hr survival following endotoxic shock in dogs. Only 3 of 14 (21%) of saline-treated dogs survived, whereas 5 of 6 (83%) of naloxone-injected animals were alive on the day after endotoxin injection.

HEMORRHAGIC SHOCK

Rats

Despite the therapeutic efficacy of naloxone in endotoxic shock, these effects of naloxone might have been unique to interaction with endotoxins and not generalizable to other forms of shock. To test our hypothesis that endorphins are more ubiquitously involved in the etiology of shock states, we studied the effects of naloxone in conscious rats subjected to hemorrhagic shock (4).

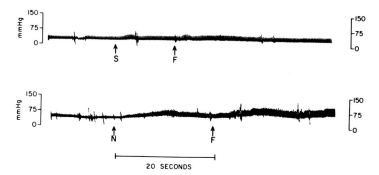

FIG. 6. Representative polygraph recording showing the effect of saline (S, top) or 1 mg/kg naloxone (N, bottom) on blood pressure following hemorrhagic hypothension. A 0.3 ml saline flush (F) was used to ensure drug delivery. (Reprinted with permission from ref. 4).

Rats were prepared with venous and arterial cannulae as described above; 24 hrs later, hemorrhagic shock was produced by withdrawing blood from the venous catheter. MAP was maintained at 40 mm Hg for a period of 20 min. This resulted in the withdrawal of approximately 50% of the total blood volume at the time of treatment and produced 50% mortality in untreated rats (4).

Fig. 6 shows the effects of naloxone on blood pressure following hemorrhagic shock in representative individual rats. Naloxone, at a dose of 1 mg/kg, significantly improved MAP and PP when compared to saline-injected, control rats (Fig. 7, n=15 rats/group). More importantly, in this model of hemorrhagic shock, naloxone significantly improved survival with 13 of 15 naloxone-treated and 8 of 15 saline-treated rats surviving 24 hrs (Fisher's exact probability test, p<.05).

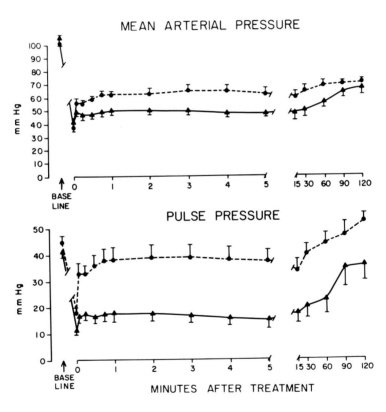

FIG. 7. The effects of 1.0 mg/kg naloxone (●- - -●) and saline (△——△) treatment on MAP (top) and PP (bottom) following hemorrhagic shock. Naloxone treatment significantly improved these cardiovascular parameters, whereas saline was without a significant effect. Fifteen rats were studied in each group. Vertical bars are ± SEM.

Dogs

For hemorrhagic shock studies, anesthetized dogs were prepared as described above for endotoxic shock. These animals were bled down to 45 mm Hg, and this MAP was maintained for 60 min. At this time, the dogs were treated with either naloxone (2 mg/kg iv bolus, 2 mg/kg/hr) or saline at equal volumes. One hour after treatment, shed blood was returned to surviving animals.

The data depicted in Fig. 8 demonstrate the significant improvement in MAP, CO, and LV dp/dt produced by naloxone injections at t=60 min. As seen in endotoxic shock studies with dogs, there was no significnat effect of naloxone on PAw, HR, TPVR, or portal venous pressure. Thus, in both endotoxic and hemorrhagic shock in dogs, naloxone appears to exert its therapeutic benefit through increases in cardiac contractility and not by means of altered vascular resistance or venous return (24,27).

In Fig. 8, data for saline-injected dogs end abruptly at t=75 min. This occurs because hypovolemic dogs treated with saline, unlike naloxone-treated dogs, do not survive long enough to receive reinfusion of shed blood at t=120 min. This important observation confirms our findings in dogs subjected to endotoxic shock (24) as well as rat hemorrhagic shock (4) which demonstrate that naloxone significantly improves survival in shocked animals.

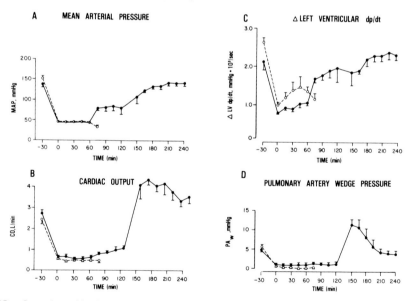

FIG. 8. Anesthetized dogs were bled to MAP of 45 mm Hg (t=0) and maintained at that pressure for 60 min. At that time, 2.0 mg/kg (-) naloxone HCl (n=5, ●——●) or equivolume saline (n=6, O----O) were administered intravenously. All saline-injected dogs died within the next 30 min. At 120 min., shed blood was returned to naloxone-treated dogs. Vertical bars are SEM.

MECHANISMS

Since naloxone is described as a pure-opiate antagonist, its effects in altering physiological and behavioral parameters have been used as evidence for a blockade of endorphin-mediated events (1,3,13,15,16). From this perspective, our results provide experimental support for the hypothesis that endorphin systems significantly contribute to the pathophysiology of shock produced by a variety of means in two different species.

After establishing the therapuetic effects of naloxone in rat models of endotoxic (6,8,14,24), hypovolemic (4,27), and spinal shock (5,7,19; see following chapter), the question as to the source of endorphins and their site of action became central. We chose the rat hemorrhagic shock model to investigate our hypothesis that pituitary endorphins were acting upon opiate receptors in the central nervous system (CNS) to depress myocardial contractility through autonomic pathways to the heart (18,20).

Hypophysectomized and sham-hypophysectomized rats (Zivic-Miller labs) were prepared with arterial and venous cannulae as before. Additionally, a guide tube was affixed to the cranium and placed extradurally 2 mm rostral from the bregma and 1 mm lateral to the mid-sagittal suture (19). This tube allowed injection of drugs into the right-lateral ventricle of conscious rats subjected to controlled hemorrhage on the day after surgical preparation.

Hemorrhagic shock was produced as before by withdrawing blood through the venous cannula until MAP reached about 35-40 mm Hg where it was maintained for 20 min. Hypophysectomized and sham-operated rats were subdivided into 2 groups, half receiving intraventricular (ivt) saline (20 ul over 20 sec.) and the other half receiving naloxone (10 μg) in the same volume ivt.

Within 2-3 min following ivt injection of naloxone in sham-operated, hypovolemic rats, MAP increased to 40 mm Hg above pre-treatment values (Fig. 9). Sham-operated rats which were injected with an equivalent volume of saline ivt had a 25 mm Hg increase in MAP. Nonetheless, the increase in MAP produced by ivt naloxone was significant (18,20).

In sharp contrast, neither ivt naloxone nor saline produced any change in MAP in hypophysectomized rats (Fig. 9). Pulse pressure data paralleled MAP since PP was only improved by ivt naloxone in rats with intact pituitary glands. Additionally, intravenous injection of naloxone is as effective as ivt naloxone in producing and sustaining an increase in MAP and PP in rats with intact pituitaries (20). The increased MAP produced by ivt saline injections was probably a Cushing effect due to increased intracranial pressure; this effect was blocked by hypophysectomy or spinal-cord transection (19,20).

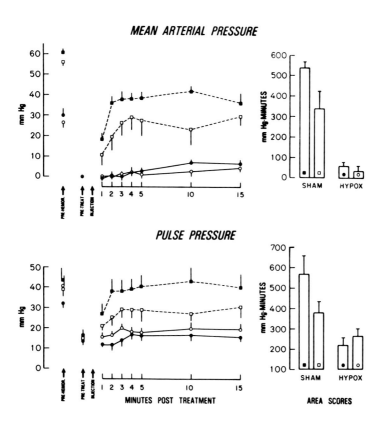

FIG. 9. The effects of intraventricular injections of 10 µg (-) naloxone HCl (solid squares or circles) or equivolume saline (open squares or circles) following induction of hypovolemic shock in sham-hypophysectomized (dashed lines) or hypophysectomized (solid lines) conscious rats. Mean arterial pressures were normalized to pre-treatment values. Areas under response-time data (left) were integrated for individual rats (right). Vertical bars represent SEM; sham=8-9 rats/group, hypox (hypophysectomized)=11 rats/group.

These studies support the view that pituitary endorphins are toxic pathophysiological factors in shock. The reversal of hypotension by microgram quantities of ivt naloxone in hypovolemic shock confirms the findings from spinal shock (5,7,19, see next chapter) which point to the importance of CNS sites in mediating the cardio-depressant effects of endorphins. It is possible that circulating endorphins gain access to the CNS in circumventricular sites such as the area postrema which lack a blood-brain barrier (13,15,16). Additionally, the area postrema is rich in opiate receptors and adjacent to medullary cardio-respiratory centers (16).

Recently, we have obtained further evidence which points to the importance of pituitary endorphins in shock pathophysiology. Adrenal dysfunction, such as occurs in Addison's disease or following extirpation of adrenal glands, potentiates shock susceptibility (11) and elevates pituitary β-endorphin levels (2,12). We studied the effects of ivt and iv naloxone in adrenalectomized rats subjected to endotoxic shock and found that adrenalectomy increased shock susceptibility at least 15 fold. Moreover, naloxone significantly improved cardiovascular parameters in adrenalectomized animals. Thus, the absence of adrenal steroid feedback in adrenalectomized rats results in elevated pituitary β-endorphin (2,12) and may also produce the concomitant enhancement of shock susceptibility (21).

Once in the CNS, endorphins appear to act upon specific opiate receptors since their effects are stereospecifically reversed by minute amounts of naloxone. We have evidence suggesting that endorphins act upon autonomic centers in the CNS which indirectly depress cardiovascular function via neuronal inputs into the heart (see next chapter).

The specific-opiate antagonist naloxone rapidly improved blood pressure and significantly decreased mortality associated with endotoxic and hemorrhagic shock in rats and dogs. Naloxone injections into control, non-shocked animals produced no change in cardiovascular parameters, indicating a selective action in reversing shock hypotension instead of a direct cardiovascular effect. As little as 0.1 mg/kg intravenous naloxone improved blood pressure in endotoxic rats; 10 μg naloxone administered intracerebroventricularly had therapeutic effects in rats subjected to hemorrhagic shock. The therapeutic efficacy of these low doses of naloxone, combined with its stereospecificity of action in reversing shock states, indicates the direct involvement of opiate receptors. Since intraventricularly administered naloxone was without effect in hypophysectomized animals at doses which significantly improved blood pressure in animals with intact pituitaries, it appears that pituitary endorphins released during the stress of shock act upon opiate receptors in the brain to depress cardiovascular function. Studies in dogs as well as rats revealed that naloxone produced the improvement in blood pressure by increasing cardiac contractility, not by altering vascular resistance or venous return. Collectively, our findings indicate the source as well as mechanism of endorphin involvement in the pathogenesis of a variety of shock states in a number of species. These results predict the therapeutic efficacy of opiate antagonists in human shock states as well.

ACKNOWLEDGEMENTS

We thank T. Jacobs and C. Johnson for their expert technical assistance and S. Danchik, P. Conners, A. Trees and V. LaGrange for preparation of the manuscript.

FOOTNOTES: This material has been reviewed by the Walter Reed Army Institute of Research, and there is no objection to its presentation and/or publication. The opinions or assertions contained herein are the private views of the authors and are not to be construed as official or as reflecting the views of the Department of the Army or the Department of Defense.

REFERENCES

1. Akil, H., Madden, J., Patrick, R.L., and Barchas, J.D. (1976): Opiates and Endogenous Opiate Peptides, edited by H.W. Kosterlitz, pp. 63-70. North Holland, Amsterdam.
2. Akil, H., Watson, S.J., Barchas, J.D., and Li, C.H. (1979): Life Sci., 24:1659-1666.
3. Belenky, G.L., and Holaday, J.W. (1979): Brain Res., 177: 414-417.
4. Faden, A.I., and Holaday, J.W. (1979): Science, 205:317-318.
5. Faden, A.I., and Holaday, J.W.(1979): In: Endogenous and Exogenous Opiate Agonists and Antagonists, edited by E.L. Way, pp. 483-486. Pergamon Press, New York.
6. Faden, A.I., and Holaday, J.W. (1980): J. Pharmacol. Exp. Ther., (in press).
7. Faden, A.I., Jacobs, T.P., and Holaday, J.W. (1980): Trans. Am. Neurol. Assn., (in press).
8. Faden, A.I., and Holaday, J.W. (1980): Rev. of Infectious Dis., (in press).
9. Fennessy, M.R., and Rattray, J.F. (1971): Eur. J. Pharma-col., 14:1-8.
10. Florez, J., and Mediavilla, A. (1977): Brain Res., 138: 585-590.
11. Gilbert, R.P. (1960): Physiol. Rev., 40:245-279.
12. Guillemin, R., Vargo, T., Rossier, J., Minick, S., Ling, N., Rivier, C., Vale, W., and Bloom, F. (1977): Science, 197: 1367-1369.
13. Holaday, J.W., Wei, E., Loh, H.H., and Li, C.H. (1978): Proc. Natl. Acad. Sci. USA, 75:2923-2927.
14. Holaday, J.W., and Faden, A.I. (1978): Nature, 275:450-451.
15. Holaday, J.W., Belenky, G.L., Faden, A.I., and Loh, H.H. (1979): In: Neuro-psychopharmacology, edited by B. Saletu, pp. 503-514. Pergamon Press, Oxford.

16. Holaday, J.W., and Loh, H.H. (1979): In: Neurochemical Mechanisms of Opiates and Endorphins, edited by H. Loh, and D. Ross, pp. 227-258. Raven Press, New York.

17. Holaday, J.W., and Faden, A.I. (1979): In: Endogenous and Exogenous Opiate Agonists and Antagonists, edited by E.L. Way, pp. 479-482. Pergamon Press, New York.

18. Holaday, J.W., and Faden, A.I.(1979): Physiologist, 22:57.

19. Holaday, J.W., and Faden, A.I. (1980): Brain Res., (in press).

20. Holaday, J.W., O'Hara, M., and Faden, A.I. (1980): (submitted).

21. Holaday, J.W., and Faden, A.I. (1980): Third Annual Conf. on Shock, Lake of the Ozarks, MO.

22. Iijima, I., Minamikawa, J., Jacobson, A.E., Brossi, A., Rice, K.C., and Klee, W.A. (1978): J. Med. Chem., 21: 398-400.

23. Lemaire, I., Tseng, R., and Lemaire, S. (1978): Proc. Natl. Acad. Sci. USA, 75:6240-6242.

24. Reynolds, D.G., Gurll, N.J., Vargish, T., Lechner, R., Faden, A.I., and Holaday, J.W. (1980): Circ. Shock, (in press).

25. Terenius, L. (1978): Annu. Rev. Pharmacol. Toxicol., 18: 189-204.

26. Thomas, L., (1979): The Medusa and the Snail. Viking Press, New York.

27. Vargish, T., Reynolds, D.G., Gurll, N.J., Lechner, R., Holaday, J.W., and Faden, A.I. (1980): Circ. Shock, (in press).

28. Waisbren, B.A. (1964): Am. J. Med., 36:819-824.

29. Waterfield, A.A., and Kosterlitz, H.W. (1975): In: The Opiate Narcotics, edited by A. Goldstein, pp. 35-38. Pergamon Press, New York.

30. Zweifach, B.W., and Thomas, L. (1957): J. Exp. Med., 106: 385-401.

Neurosecretion and Brain Peptides,
edited by J. B. Martin, S. Reichlin, and K. L. Bick.
Raven Press, New York © 1981.

A Role for Endorphins in the Pathophysiology of Spinal Cord Injury

Alan I. Faden and John W. Holaday

Department of Medical Neurosciences, Division of Neuropsychiatry, Walter Reed Army Institute of Research, Washington, D.C. 20012

Acute injuries of the cervical spinal cord produce changes in autonomic function in addition to the well recognized effects on sensory, motor and reflex activity (21,23). The autonomic changes can involve blood pressure (BP), heart rate (HR), respiration and temperature. Of these, alterations in BP are probably most important. Significant hypotension is often observed in the early post-injury period (23) and may contribute to the spinal cord damage by exacerbating the spinal cord ischemia which follows spinal injury (7). Since autoregulation of spinal blood flow is impaired after spinal trauma (5,19), blood flow to the cord becomes directly dependent upon systemic blood pressure. Restoration of blood pressure after spinal injury may therefore improve spinal cord blood flow and reduce the functional neurologic deficit by limiting ischemic damage. However, the decline in BP caused by spinal transection or spinal injury is minimally responsive to routine therapeutic measures such as the use of vasopressor agents (6,13).

We have suggested that activation of endorphin systems is in part responsible for the hypotension which occurs in endotoxic and hypovolemic shock and have shown that the specific opiate-antagonist naloxone significantly improves BP as well as survival in these shock models (8,9,16,20,22). To determine whether endorphin-release might contribute to the pathophysiologic changes of "spinal shock", we compared the effects of naloxone and saline following C-7 spinal transection in the rat and cat (10,11,17). In addition, we have evaluated the effects of naloxone treatment on BP and neurologic recovery in a cat spinal-injury model (12).

RAT SPINAL-TRANSECTION

Adult male Sprague-Dawley rats (500-800 g) were anesthetized with ketamine and pentobarbital. A tail-artery catheter, connected to a microtransducer allowed continuous BP measurements.   An external jugular vein catheter was used for intravenous (iv) drug administration.   Placement of a transcranial guide tube permitted intraventricular (ivt) drug administration into the right lateral ventricle.   The spinal cord was transected at C-7, using an extradural ligature as described elsewhere (10,11,17).

Spinal transection produced a transient rise in mean arterial pressure (MAP) and bradycardia, followed by a gradual decline of MAP to 19.3 ± 2.0 SEM mm Hg below pre-severance values.   Approximately 20 minutes post-transection, when MAP had stabilized, either drug or drug vehicle was administered.

In 22 animals, naloxone (10 mg/kg) or saline was administered iv. Naloxone treatment rapidly restored MAP to pre-transection levels, whereas equal-volume saline treatment was without effect (Fig. 1).   Significant differences in MAP between the groups were maintained over the entire two hour monitoring period [repeated measures analysis of variance (ANOVA), p <0.05].   Naloxone treatment also significantly attenuated (ANOVA, p<0.05) the decline in colonic temperature observed after spinal transection (Fig. 2). In contrast to its effects in spinal-transected animals, naloxone given iv in doses up to 10 mg/kg had no effect on MAP in anesthetized rats which were not injured (n=3, data not shown).

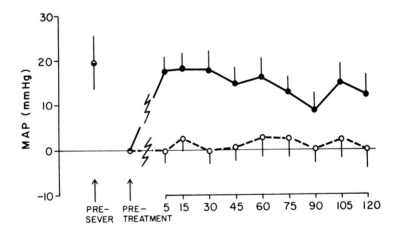

FIG. 1.   Following hypotension produced by spinal transection, rats received either iv naloxone HCl (n=11, ●———●) or equal-volume saline (n=11, O---O). Naloxone treatment significantly improved MAP, whereas saline treatment was without effect. Points represent mean values ± SEM.   MAP has been normalized to pre-treatment values.

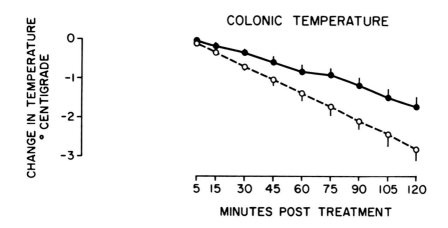

FIG. 2.   Effect of iv (-) naloxone HCl (n=7, ●————●) or saline
(n=7, O----O) on colonic temperature following cervical spinal
transection in rats. Naloxone attenuated the rate of temperature
decline. Points represent mean values ± SEM.

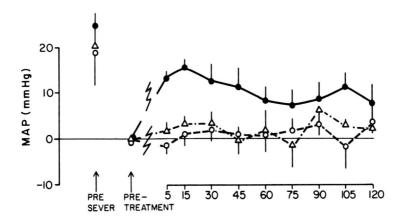

FIG. 3.   After spinal transection, rats received either 48 µg (-)
naloxone HCl (n=7, ●————●), 48 µg (+) naloxone HCl (n=6, △—●—△)
or saline (n=7, o----o).   Drugs were administered iv at equal
volumes.   (-) naloxone significantly improved MAP in contrast to
either (+) naloxone or saline which were without effect.   Points
represent mean values ± SEM.   MAP has been normalizd to pre-
treatment values.

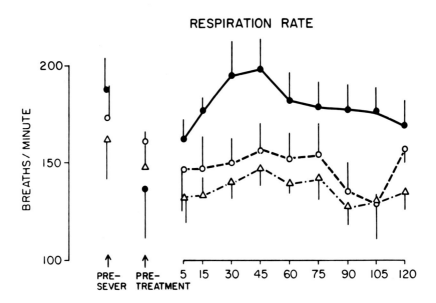

FIG. 4. Effect of ivt (-) naloxone (n=7, ●———●), (+) naloxone
(n=6, △ —•— △) or saline (n=7, O----O) on respiration rate after
spinal transection in rats. Naloxone effects were stereospecific. Points represent mean values ± SEM.

To determine whether these effects of naloxone were stereospecifically mediated at opiate receptors within the central
nervous system (CNS), drugs were administered ivt in another 20
animals. Following the hypotension produced by spinal transection, rats received either (-) naloxone (48 µg, n=7), (+)
naloxone (48 µg, n=6) or saline (n=7). Drugs were administered
into the right lateral ventricle at eoual volumes (20 µl), given
through a Hamilton syringe over 20 seconds. Whereas both (+)
naloxone and saline were without effect, (-) naloxone rapidly
improved MAP (Fig. 3). MAP in the (-) naloxone group remained
significantly higher than in the other treatment groups over the
first post-treatment hour (ANOVA, p <0.05). Respiratory rates
were also significantly and stereospecifically increased by ivt
(-) naloxone treatment (Fig. 4). However, this dose of (-)
naloxone (48 µg) had no effect on MAP or respiration when given
iv (n=4, data not shown), thereby confirming (-) naloxone's
central site of action.
     Since spinal transection isolates the spinal sympathetic
nervous system from supraspinal control, the parasympathetic
nervous system becomes the primary mechanism for the supraspinal
regulation of cardiovascular function. Because our methods also
isolated the spinal subarachnoid space from the ventricular

cerebrospinal fluid, the effect of (-) naloxone on MAP after
spinal transection appeared to be mediated by parasympathetic
cholinergic fibers. To test this, we repeated the ivt studies,
dividing the rats into four treatment groups: naloxone alone,

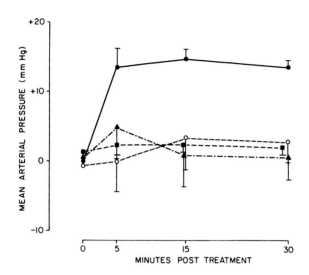

FIG. 5. Following hypotension caused by spinal transection, rats
received either ivt naloxone (48 µg, n=23) or saline (n=6,
O— —O). Naloxone animals were divided into those with preceding
vagotomy (n=9,▲—•—▲), those with preceding methylatropine treatment (n=8,■- - -■) and those without vagotomy or methylatropine
(n=6, ●——●). Naloxone significantly improved MAP in comparison
with saline controls. MAP has been normalized to the pre-treatment values. Points represent mean values ± SEM.

naloxone plus bilateral cervical vagotomy, naloxone plus methylatropine, and saline alone. Naloxone (48 µ g) and saline were
administered as before. Following spinal transection, naloxone
treatment significantly improved MAP in comparison to saline
controls, as previously demonstrated (Fig. 5). This effect of
naloxone on MAP was completely blocked by either vagotomy or
methylatropine administration.

CAT SPINAL TRANSECTION

To determine whether the effects of naloxone in spinal shock
were species-specific, we repeated these spinal-transection
studies in the cat. Adult female cats (2-3 Kg) were anesthetized
with pentobarbital. A femoral artery catheter connected to a
pressure transducer permitted continuous measurement of BP.
Spinal cord transection produced a brief rise in MAP followed
by a gradual decline in MAP to 52.0 ± 3.2 mm Hg below pre-sever-

ance values. When MAP had stabilized, either naloxone (1 mg/Kg) or saline were administered in equal volumes. Naloxone treatment significantly (ANOVA, p <0.05) improved MAP in comparison to saline controls (Fig. 6). As in the rat, this effect of (-) naloxone on MAP was completely blocked by bilateral cervical

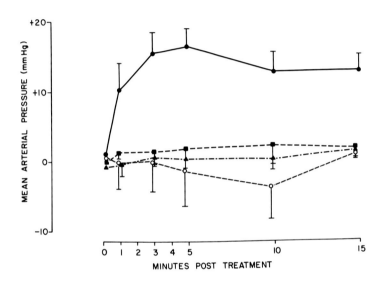

FIG. 6. Following hypotension caused by spinal transection, cats received either iv naloxone (1 mg/Kg, n=21) or saline (n=9, 0 ———0). Naloxone treated animals were divided into three groups, naloxone alone (n=9, ●——●), naloxone plus vagotomy (n=9, ▲–•–▲), or naloxone plus atropine (n=3, ■–-–-–■). Naloxone significantly improved MAP in comparison to saline controls. This effect of naloxone on MAP was blocked by either atropine or vagotomy. MAP has been normalized to pre-treatment values. Points represent mean values ± SEM.

vagotomy. It was also blocked by prior atropine (2 mg/Kg) administration. In contrast to naloxone's effects in spinal-transected animals, iv naloxone at a dose as high as 10 mg/Kg had no effect on MAP in intact animals (n=3, data not shown).

SUMMARY OF SPINAL TRANSECTION STUDIES

From these studies in the rat and cat we suggest that endor-phins play a pathophysiologic role in "spinal shock". Since ivt naloxone in the rat stereospecifically improved MAP at a dose which was ineffective when given iv, naloxone effects on BP appear to be mediated through actions at opiate receptors within the CNS. Moreover, vagotomy or muscarinic antagonists block these cardiovascular changes in both species indicating that naloxone effects on BP in this model are mediated by efferent cholinergic pathways. That naloxone reversed the hypotension after spinal transection, along with the role of the parasympathetic nervous system, appears to question the classic concept that this decline in BP results primarily from interruption of descending sympathetic pressor pathways. Finally, the ability of naloxone to restore BP in "spinal shock" suggested that opiate antagonists might be of therapeutic benefit in the management of spinal cord injury in man.

CAT SPINAL TRAUMA

To determine whether opiate antagonists might improve neurologic outcome after spinal injury, we compared the effects of naloxone and saline in a cat spinal-trauma model. Adult cats (2-3 Kg) were anesthetized with pentobarbital. BP was continuously monitored from a femoral artery catheter; a femoral vein catheter was used for iv drug administration. Laminectomy was performed to expose spinal segments C-6 to T-1. With the dura intact, the C-7 spinal segment was traumatized by dropping a 20 gram lead weight a distance of 25 centimeters to produce a force of 500 gram-cm; the weight was dropped through a glass guide tube onto a 10 mm^2 plastic foot plate which had been contoured to fit over the spinal cord. This method was initially developed by Allen (2) and has been widely employed as a spinal cord injury model (1,24). The injury parameters were chosen after pilot studies showed that they resulted in a reproducible spastic quadriparesis in untreated animals.

Forty-five minutes after injury, cats received either iv naloxone or saline. Two independent studies were performed. In the acute study (AST, n=12), drugs were infused for 24 hours; at this time animals were sacrificed. In the chronic study (CSI, n=12), drugs were infused for four hours, these animals were sacrificed three weeks post-injury. The naloxone dosage (2 mg/Kg bolus; 2 mg/Kg/hr)was established from previous experiments and was identical in both studies. BP was recorded for four hours in both the acute and chronic studies. Neurologic status was rated by two neurologists who were unaware of treatment group. BP findings in acute animals were scored at 24 hours and chronic animals at 24 hours, 1, 2 and 3 weeks. Function was rated using a modification of an established 5-point scale (3). This scale was based primarily on motor function: 0 = no voluntary movement; 1 = minimal voluntary movement; 2 = good voluntary movement, but unable to support; 3 = able to support weight, but unable to walk; 4 = able to walk, but with substantial spasticity and/or ataxia; 5 = normal function. Half-point scoring was permitted, and forelimb and hindlimb scores were determined separately.

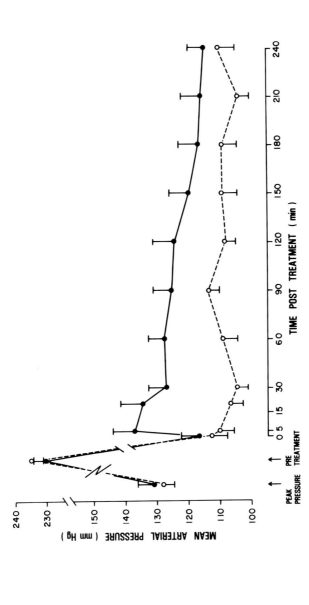

FIG. 7. Spinal trauma produced a transient rise in MAP followed by a gradual decline in MAP to approximately 20 mm Hg below baseline. Forty-five minutes following spinal injury, cats received either naloxone or saline. Naloxone treatment (n=12, ●——●) significantly improved MAP in contrast to saline controls (n=12, ○−−−−○). Data from AST and CSI studies have been pooled. Points represent mean values $\pm$ SEM.

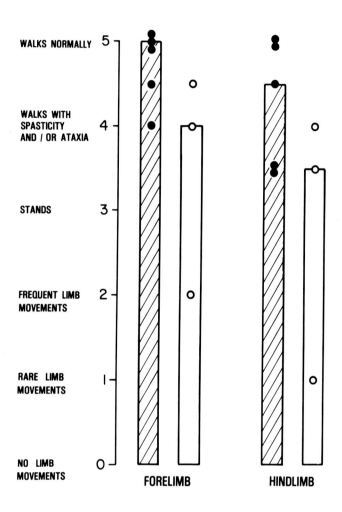

FIG. 8. Effects of naloxone (▨•▨) or saline (▭•▭) on neurologic recovery at three weeks in the CSI study. Neurologic function was rated separately for forelimbs and hindlimbs using a
5-point scale. Neurologic recovery was significantly improved by
naloxone treatment (see text). The different number of animals in
the two groups (n=5, naloxone; n=3, saline) reflects the higher
mortality in the saline animals. Histogram bars represent median
values.

Spinal trauma produced a transient pressor response followed by a decline of MAP to approximately 15 mm Hg below pre-trauma levels at 45 minutes. In both AST and CSI studies, naloxone treatment resulted in a substantial increase in MAP which persisted for several hours (Fig. 7). In contrast, MAP in saline controls continued to decline over the first post-treatment hour, with significant differences between naloxone and saline groups persisting over the first two post-treatment hours (ANOVA, p <0.05).

Most important, naloxone treated animals also had better neurologic scores than saline controls at 24 hours in the AST study and at 24 hours, 1,2, and 3 weeks in the CSI study. Whereas the differences between naloxone and saline animals failed to reach significance at 24 hours in either study, significant differences were observed at 1, 2, and 3 weeks (Wilcoxon signed-ranks test, p <0.05). These differences between naloxone and saline treated animals are more impressive when the neurologic scores are translated into functional equivalents. At three weeks the median naloxone animal was able to walk well, whereas the median saline animal was unable to walk without support (Fig. 8). Mortality rates were also higher in saline treated animals in CSI (3/6 saline vs 1/6 naloxone), pulmonary edema being the primary cause of death. Pulmonary edema is also a recognized complication in human spinal injury (18) and suggests a further possible therapeutic benefit for naloxone. In addition, since the cats that died had lower neurologic scores than those that survived, the difference in mortality between the groups served to select the higher neurologic scores in the saline animals.

Traumatic spinal injuries cause neurologic dysfunction either through direct damage to neurons or white matter tracts or through secondary progressive ischemic changes (4,19). Because there appears to be little functional spinal cord regeneration in man, therapeutic measures in the spinally-injured have aimed to reduce the secondary ischemia. Corticosteroids, in large pharmacologic doses, have been most consistently demonstrated to be effective (6,24). Whereas the beneficial effects of steroids in both spinal injury and shock have been attributed to their ability to stablilze membranes and/or to reduce edema (6,24), we have speculated (16) that the effectiveness of steroids in shock may be secondary to their well-established ability to inhibit the release of pituitary endorphins (15). From the present studies it appears plausible that the therapeutic effects of steroids in spinal injury relate in part to their ability to inhibit endorphin release.

SUMMARY OF SPINAL TRAUMA STUDIES

These studies have demonstrated that the opiate-antagonist naloxone improves blood pressure and functional neurologic recovery after spinal injury. From these findings we suggest that endorphins are released in response to spinal injury and contribute to the hypotension and to the ultimate neurologic deficit. Naloxone's ability to reverse the presumed endorphin-mediated hypotension in this model supports the hypothesis that its therapeutic effects may be secondary to its improvement of spinal cord blood flow, thereby reducing the ischemic damage caused by spinal cord trauma.

REFERENCES

1. Albin, M.S., White, R.J., Acosta-Rua, G., and Yashon, D. (1968): J. Neurosurg., 29:113-119.
2. Allen, A.R. (1911): JAMA, 57:878-880.
3. Ducker, T.B., and Hamit, H.F. (1969): J. Neurosurg., 30: 693-697.
4. Ducker, T.B., Kindt, G.W., and Kempe, L.G. (1971): J. Neurosurg., 35:700-708.
5. Ducker, T.B., and Perot, P.L. (1971): Trans. Am. Neurol. Assoc., 96:229-231.
6. Ducker, T.B., Salcman, M., and Daniell, H.B. (1978): Surg. Neurol., 10:71-76.
7. Faden, A.I., and Holaday, J.W. (1979): In: Endogenous and Exogenous Opiate Agonists and Antagonists, edited by E.L. Way, pp. 483-486. Pergamon Press, London.
8. Faden, A.I., and Holaday, J.W. (1979): Science, 205:317-318.
9. Faden, A.I., and Holaday, J.W. (1980): J. Pharm. Exper. Ther., 212:441-447.
10. Faden, A.I., and Holaday, J.W. (1980): Trans. Amer. Neurol. Assn., (in press).
11. Faden, A.I., Jacobs, T.P., and Holaday, J.W.: (Submitted).
12. Faden, A.I., Jacobs, T.P., Holaday, J.W., and Rigamonti, D.: (Submitted).
13. Faden, A.I., Jacobs, T.P., and Woods, M. (1978): Exp. Neurol., 61:301-310.
14. Freeman, L.W., and Wright, T.W. (1953): Ann. Surg., 137: 433-443.
15. Guillemin, R., Vargo, T., Rossier, J., Minick, S., Ling, N., Rivier, C., Vale, W., and Bloom, F. (1977): Science, 197: 1367-1369.
16. Holaday, J.W., and Faden, A.I. (1978): Nature, 275:450-451.
17. Holaday, J.W., and Faden, A.I. (1980): Brain Res., (in press).
18. Meyer, G.A., Berman, I.R., Doty, D.B., Moseby, R.V., and Gutienez, V.S.(1971): J. Neurosurg., 34:168-177.
19. Osterholm, J.L. (1974): J. Neurosurg., 40:5-33.
20. Reynolds, D.G., Guill, N.J., Vargish, T., Lechner, R.B., Faden, A.I., and Holaday, J.W. (1980): Circ. Shock, (in press).
21. Tibbs, P.A., Young, B., McAllister, R.G., Brooks, W.H., and Tackett, L. (1978): J. Neurosurg., 49:558-562.

22. Vargish, T., Reynolds, D.G., Guill, N.J., Lechner, R.B., Holaday,J.W., and Faden, A.I. (1980): Circ. Shock, (in press).
23. Yashon, D. (1978): Spinal Injury, PP. 283-286. Appleton-Century-Crofts, New York.
24. Yashon, D. (1978): SpinalInjury, pp. 248-254. Appleton-Century-Crofts, New York.

Neurosecretion and Brain Peptides,
edited by J. B. Martin, S. Reichlin, and K. L. Bick.
Raven Press, New York © 1981.

Introduction to Section VII:
Circadian Rhythms

Dorothy T. Kreiger

*Division of Endocrinology, Department of Medicine, Mt. Sinai School of Medicine,
New York, New York 10029*

One of the major adaptive mechanisms developed during the course of evolution in response to external environmental changes is the temporal organization of behavior and of physiological processes. The well known phenomena of hibernation, migration, seasonal reproductive activity, and seasonal hunting are a few examples of such temporal organization. Although most of these rhythms are determined by yearly cycles, the most common biological rhythms are circadian, that is with a period of approximately 24 hours. This is characteristic of many hormonal rhythms, as well as bodily responses to external agents, i.e., stress, toxins and drugs. Indeed it is now appreciated that virtually all bodily constituents have rhythmic functions. These exist on cell organ, and multi-system levels, all integrated by hierarchial control mechanisms.

Although the most common frequency is one of approximately 24 hours there are ultradian rhythms (periodicity less than 24 hours) and those with periods greater than 24 hours (i.e., months or years). Rhythms are further characterized as to whether they are endogenous or exogenous. This can be most readily determined by observing whether the given rhythm persists (as an endogenous rhythm would) in the absence of a periodic environmental or physiological input or disappears under such conditions. An endogenous rhythm can be entrained by a periodic environmental input. That is, a periodic environmental factor (usually referred to as a synchronizer or Zeitgeber) may determine the phase of a given endogenous rhythm. In the absence of this kind of synchronizing influence, the endogenous rhythm "free-runs" with a period slightly different than that imposed upon it by the Zeitgeber. For example, an endogenous circadian rhythm which has a period of 24 hours when entrained by a given environmental cue, will free-run with a period slightly less or greater than a 24-hour period in isolation from this cue.

The anatomical and biochemical nature of the "internal clock(s)" controlling rhythms of various frequencies, the inter-relationship of rhythms of different physiological variables as

447

well as the relationship of ultradian and circadian rhythms of the same physiological variable are matters of general interest. Also important are the delineation of the external inputs involved in the generation of exogenous rhythms and the functional significance of biological rhythms in health and disease. Much of the work in this area is at a relatively early stage, and currently involves a multiplicity of approaches at many levels. Information derived from such studies is applicable to an understanding of basic biological processes and has significant clinical implications and practical applications. As an example of the latter, it is necessary to be aware of the existence of rhythms in a blood hormone or chemical factors so that determinations of the concentration or physiological response of such a variable can be made at similar times of day. Such considerations are significant with regard to timing of drug administration, since the lethal dose of many drugs vary across the circadian frame.

In the following sections only certain aspects of biological rhythmicity can be considered. Dr. Moore's paper is a significant contribution to understanding of the role of the suprachiasmatic nucleus in the genesis of many rhythms. Other studies have demonstrated a unique circadian periodicity in the metabolic activity of this nucleus. This anatomical area appears to represent a major biological clock which may be coupled to other unspecified clocks for the expression of specific rhythmic functions. Still to be delineated are the neurotransmitters involved (both afferent and efferent). The nature of the efferent connections and the mechanisms of entrainment of the rhythm of this nucleus to environmental stimuli also still remain to be explored. Dr. Zucker's studies provide an elegant demonstration of the interrelationship of circadian rhythms to those that take place over a longer time period, (i.e., the estrus cycle in rodents), and may lead to an understanding of neural integration in such systems. Dr. Weitzman's studies present an overview of human circadian rhythms and their adaptability to cycles other than 24 hour ones, as well as illustrating the existence of episodic variation in such hormone concentrations. Until recently the existence of such ultradian rhythms was not even suspected. It now appears that the presence of such ultradian rhythms may well be involved with receptor regulation, providing a means whereby such receptors can recover from the phenomenon of "down-regulation" which normally occurs in response to increasing ligand concentrations.

Neurosecretion and Brain Peptides,
edited by J. B. Martin, S. Reichlin, and K. L. Bick.
Raven Press, New York © 1981.

The Suprachiasmatic Nucleus, Circadian Rhythms, and Regulation of Brain Peptides

Robert Y. Moore

Departments of Neurology and Neurobiology, Health Sciences Center, State University of New York at Stony Brook, Stony Brook, New York 11794

INTRODUCTION

Circadian rhythms are ubiquitous features of living organisms. In the context to be used here, the term "circadian" will be applied to those biological rhythms with a period that approximates 24 hours and whose cyclicity persists, or free-runs, in the absence of external timing cues. Because circadian rhythms are normally entrained to the solar cycle of light and dark, they usually exhibit a period of exactly 24 hours and will maintain a consistent phase relationship to the cycle of light and dark. For example, plasma cortisol levels in the human are highest at the beginning of the light portion of a light-dark cycle. It is now clearly recognized that circadian rhythms represent a major component of the regulation of homeostatic functions of the internal milieu of an organism which permits maximal adaptation of the organism to its environment. In this respect, then, circadian rhythms function as major components of regulatory systems which participate in promoting the survival of an organism. This presentation examines the neural mechanisms which underlie circadian phenomena in mammals and indicates some aspects of the importance of circadian rhythms in the regulation of central nervous system peptide producing neurons.

NEURAL MECHANISMS OF CIRCADIAN RHYTHMS

Circadian rhythms have two dominant features. First, under normal environmental conditions circadian rhythms are entrained to a light-dark cycle. In mammals, section of the optic nerve results in a loss of entrainment with circadian rhythms becoming free-running. This indicates that information concerning the timing of the light-dark cycle is transmitted to the central nervous system via the eye and central visual projections. Sec-

449

ond, this visual input must be coupled to endogenous central circadian oscillating mechanisms which are capable of maintaining circadian rhythmicity in the absence of external input. Consequently, there are two major problems in elucidating the central mechanisms underlying circadian rhythms. First, it is necessary to determine the visual pathways which participate in entrainment of circadian rhythms. Second, the location and organization of central circadian oscillating systems must be elucidated. Studies of circadian rhythms under a number of circumstances indicate that the central circadian oscillating system in mammals must be composed of at least two independent but mutually coupled oscillators (24).

Over the past decade considerable information has been accumulated concerning both the visual pathways participating in entrainment of circadian rhythms in mammals and on the location of central circadian oscillating mechanisms (cf. 14-16, 30, for reviews). Much of the work in this area has been carried out on the rat or hamster and most of the studies noted below were carried out on these species. The visual pathways in the rat had been extensively studied by Hayhow and his associates (6,7) in the early 1960s. These studies form the basis for a series of ablation studies which clearly demonstrated that the entrainment of circadian rhythms could be accomplished in the absence of all of the then known components of the primary and accessory optic projections (14,18,20). On this basis, studies were undertaken to demonstrate a direct projection from retina to hypothalamus, a projection which was known to occur in lower vertebrates and which would appear, if present, to be the most likely candidate for a pathway participating in entrainment of circadian rhythms. Studies using the autoradiographic tracing method demonstrated the existence of a direct projection from retina to the suprachiasmatic nucleus of the hypothalamus in the rat (8,19) and subsequent studies (Table 1) have demonstrated that this projection is a consistent feature of the mammalian visual system. It is present in prototherian, metatherian and eutherian mammals and shows little alteration in mammalian phylogeny. Indeed, it appears to be the single most consistent feature of the mammalian visual system. It is now evident that the retinohypothalamic projection to the suprachiasmatic nucleus is sufficient for the entrainment of circadian rhythms but it has not been established that it is essential (cf. Klein and Moore (11)). That is, no experiment has been carried out in which the retinal hypothalamic projection was transected leaving the remaining visual pathways and the suprachiasmatic nucleus intact. And, although the retino-hypothalamic projection probably plays a major role in mediation of entrainment of circadian rhythms, there is evidence that other visual pathways also participate in this function (28).

TABLE 1. The retinohypothalamic tract in mammals-species with known projection from retina to suprachiasmatic nucleus

	Selected References
Prototherian Mammals	
Platypus (Ornithorhynchus anatimus)	3
Metatherian Mammals	
Opossum (Didelphis virginiana)	13
Eutherian Mammals	
Rat (Rattus norvegicus)	8,12,19
Mouse (Mus musculus)	44
Guinea Pig (Cavis porcelius)	8
Rabbit (Oryctolagus cuniculus)	8
Ferret (Mustelo furo)	40
Hedgehog (Hemiechinus auratus)	13
Cat (Felis domesticus)	8,13
Tree Shrew (Tupaia glis)	13
Galago (Galago senegalensis)	13
Marmoset (Saguinus oedipus)	13
Squirrel Monkey (Saimiri sciurens)	41
Macaque Monkey (Macaca mulatta)	8,13
Chimpanzee (Pan troglodytes)	42

Discovery of the retinal projection to the suprachiasmatic nucleus obviously suggested that ablation of the nucleus should have some significant effect upon circadian rhythms. In experiments carried out independently by Moore and Eichler (18,35) and Stephan and Zucker (35), it was established that ablation of the suprachiasmatic nucleus in the rat resulted in loss of circadian rhythms in activity, drinking and adrenal corticosterone content. Subsequent studies carried out in a number of laboratories have confirmed and extended these observations. The results of a number of these studies are summarized in Table 2. These indicate that in the rodent, bilateral ablation of the suprachiasmatic nuclei results in a loss of all circadian function. There is some question as to whether ultradian rhythms in activity may be preserved in the hamster (27). In the rat it is clear that ultradian rhythms in components of the sleep-wake cycle persist in animals with suprachiasmatic nucleus lesions whereas no circadian component is evident following such lesions (9).

These studies raise the question whether the suprachiasmatic nucleus represents a primary circadian oscillator in the rodent brain. As a further test of this concept, the effects of suprachiasmatic nucleus lesions produced in the neonatal period were tested (21,22). In general, it has been a consistent observation that lesions produced in the neonatal period result in either preservation or recovery of function in a manner which would not be observed in the adult animal. Bilateral suprachiasmatic nucleus lesions were produced in two-day old rats and rhythms in drinking and activity were studied at 50-100 days of age and at approximately one year of age. At two days of age the

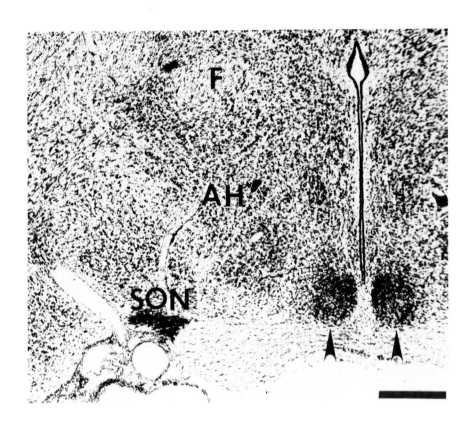

FIG. 1.   Photomicrograph of a Nissl-stained coronal section
through the rat hypothalamus showing the suprachiasmatic nuclei
composed of dense accumulations of small cells lying immediately
above the optic chiasm (arrows). The ventral portion of the
third ventricle extends between the nuclei. The supraoptic
nuclei (SON) lie lateral to the optic chiasm and ventral to the
anterior hypothalamic area (AH). The postcommissural fornix (F)
is present dorsally. Marker bar=0.5 mm.

retinohypothalamic tract has not developed and the rat does not
exhibit a circadian rhythm in any known function. Animals subjected to bilateral suprachiasmatic nucleus ablation at two days
of age have been studied at intervals from the late neonatal
period to one year of age and no evidence of circadian function
was noted at any postoperative time so that neonatal lesions are
equivalent to those made in the adult (21,22).

TABLE 2. Suprachiasmatic nucleus lesions: effects on circadian rhythms

Investi-gator(s)	Animal	Rhythm(s)	Post-lesion Rhythm Status	Post operative Surv. (days)
Stephan and Zucker (35)	Rat	Activity Drinking	Abolished	120
Moore and Eichler (18)	Rat	Adrenal Corticosterone	Abolished	21
Moore and Klein (20)	Rat	Pineal Serotonin N-Acetyltrans-ferase	Abolished	21
Rusak (27)	Hamster	Activity	Abolished	~120
Ibuka et al. (9)	Rat	Sleep-wake	Abolished	~100
Brown-Grant and Raisman (2)	Rat	Estrous cycle, Ovulation	Abolished	~100
Raisman and Brown-Grant (25)	Rat	Ovulation, Activity, Feed-ing, Drinking, Corticosterone Serum, Pineal Serotonin N-Acetyltransferase	Abolished	120-180
Moore (17)	Rat	Drinking	Abolished	54
Stephan and Nunez (37)	Rat	Drinking, Activity, Sleep-wake Temperature	Abolished	40
Saleh et al. (31)	Rat	Activity	Abolished	*
Saleh and Winget (32)	Rat	Heart Rate	Abolished	*
Stephan and Kovacecic (36)	Rat	Learning	Abolished	*

TABLE 2. (Con't.)

Investigator(s)	Animals	Rhythm(s)	Post-lesion Rhythm Status	Post-operative Surv. (days)
Mosko and Moore (22, 23)	Rat	Activity, Drinking, Estrous Cycle	Abolished	∿365
Willoughby and Martin (45)	Rat	Growth Hormone	Abolished	∿20
VanDen Pol and Powley (45)	Rat	Feeding, Drinking	Abolished	∿14

*Length of postoperative survival period not reported.

The results of the lesion studies present one major problem of interpretation. It is not possible to be certain that the uniform loss of a large variety of circadian rhythms following suprachiasmatic nucleus ablation necessarily implies that the suprachiasmatic nucleus is a primary oscillator. An alternative explanation would be that the nucleus is a necessary output for another primary oscillator. However, two recent studies lend further support to the view that the suprachiasmatic nucleus is a critical component of the central circadian system in the rat. First, Inouye and Kawamura (10) have shown that a circadian pattern of neuronal firing persists in suprachiasmatic nuclei isolated from all neuronal input. Second, Schwartz et al. (33) have demonstrated a circadian rhythm in metabolic activity in the suprachiasmatic nuclei but in no other brain region. Thus, the effect of suprachiasmatic nucleus ablation, the presence of a circadian rhythm in metabolic activity in the nucleus, and recording of intact circadian neuronal firing patterns in isolated suprachiasmatic nuclei strongly suggest that the suprachiasmatic nucleus represents a primary circadian oscillator in the rodent brain. This immediately raises the question as to whether the nucleus is sufficiently complex in organization to constitute a system of at least two, mutually coupled oscillators. Cell counts indicate that the nucleus contains approximately 10,000 neurons on each side. Golgi studies indicate that the nucleus is complex in organization and contains at least three major subdivisions. The rostral one-fourth of the nucleus receives no retinal input (24) and cannot be induced, in the rat, to receive retinal input even in the absence of the caudal three-fourths of the nucleus (20). Further studies using the horseradish peroxidase (HRP)-retrograde transport method (Moore and Riley, unpublished) have demonstrated that the suprachiasmatic nuclei have extensive association connections within each nucleus and extensive commissural connections between the nuclei that are topographically organized. That is, in a single nucleus there are connections from rostral to caudal, caudal to rostral and between components of the nuclei. In addition, one suprachiasmatic nucleus projects topographically upon the other through a commissural pathway running beneath the third ventricle. Both of these projections would be essential if the suprachiasmatic nuclei are to be viewed as constituting sets of mutually coupled circadian oscillators. If the suprachiasmatic nucleus is viewed as a primary oscillator, this model has the advantage that visual input is directly coupled into the oscillator via the retinohypothalamic projection. It should be noted, however, that this is an hypothesis. There is no evidence to exclude the possibility that other circadian oscillators do not exist in the rodent brain and, particularly, that the circadian systems of other mammals may not be differently organized.

THE SUPRACHIASMATIC NUCLEUS AND REGULATION OF BRAIN PEPTIDES

If the suprachiasmatic nucleus is accepted as an important component of central circadian oscillating mechanisms, it would appear highly likely that it participates in the regulation of peptide neuron systems as these systems regulate a number of functions which are circadian in nature. An example noted previously is the circadian rhythm in the secretion of adrenal corticoids. A number of similar examples could be cited but one deserves special attention. It has long been known that lesions in the suprachiasmatic region interfere with the estrous cycle in the female rat (cf. Rusak and Zucker (30) for review). Evidence accumulated over the past ten years indicates that the estrous cycle represents a four-day cycle in rodents such as the rat and hamster which is, in turn, a function of four successive circadian events culminating in the surge of luteinizing hormone (LH) on the day of proestrus (1,2,4,30,38). The suprachiasmatic nucleus appears to project predominantly caudally into the periventricular hypothalamus and to the region of the ventral tuberal area (39). Neurons producing the luteinizing hormone-releasing hormone (LHRH) are present extensively in the suprachiasmatic region and project to median eminence (34). Extensive studies now available indicate that very selective ablation of the suprachiasmatic nucleus results in persistent vaginal estrus (2,30), even when the lesions are produced prior to the onset of puberty (23). Thus, LHRH neurons require a circadian signal to produce a surge in LH and ovulation. In addition, the suprachiasmatic nuclei are essential for mediation of other regulatory influences, such as those from the pineal (29), in the control of reproductive function. Consequently, the circadian system can be viewed as a regulatory system which has important functions in

the regulation of brainpeptide neuron systems. The circadian
system, and the suprachiasmatic nuclei in particular, appear
necessary for normal homeostatic regulation of a series of
hormonal events and the integration of these with important
aspects of an animal's behavior. Thus, the circadian system
plays an important role not only in organization of the internal
milieu in order to permit maximal adaptation of the organism to
its environment but also is important in regulation of mechanisms
essential for the survival of the species.
   In conclusion, the observations reviewed above have analyzed
the function of the central circadian oscillating systems in the
regulation of brain peptides. In the rodent, the retinal hypothalamic projection to the suprachiasmatic nucleus appears to be
a visual pathway important to the mediation of entrainment of
circadian rhythms in the light-dark cycle. In addition, the
suprachiasmatic nuclei appear to be critical components of central circadian oscillating mechanisms.

REFERENCES

1. Alleva, J.J., Waleski, M.W., and Alleva, F.R. (1971): Endo-
    crinology, 88:1368-1379.
2. Brown-Grant, K., and Raisman, G. (1977): Proc. R. Soc.
    Lond. (Biol.), 198:279-296.
3. Campbell, C.B.G., and Hayhow, W.R. (1972): J. Comp.
    Neurol., 145:195-208.
4. Fitzgerald, K.M., and Zucker, I. (1976): Proc. Natl. Acad.
    Sci. USA, 73:2923-2927.
5. Guldner, F.H. (1976): Cell Tissue Res., 165:509-544.
6. Hayhow, W.R., Sefton, A., and Webb, C. (1962): J. Comp.
    Neurol., 118:295-322.
7. Hayhow, W.R., Webb, C., and Jervie, A. (1960): J. Comp.
    Neurol. 115:187-216.
8. Hendrickson, A.E., Wagoner, N., and Cowan, W.M. (1972): Z.
    Zellforsch, 135:1-26.
9. Ibuka, N., Inouye, S.-I>T>, and Kawamura, H. (1977): Brain
    Res., 122:33-47.
10. Inouye, S.-I.T., and Kawamura, H. (1979): Proc. Natl. Acad.
    Sci. USA, 76:5962-5966.
11. Klein, D.C., and Moore, R.Y. (1979): Brain Res., 174: 245-
    262.
12. Mai, J.K., and Junger, E. (1977): Cell Tiss. Res., 183:
    221-237.
13. Moore, R.Y. (1973): Brain Res., 49:403-409.
14. Moore, R.Y. (1978): In: The Pineal Gland and Reproduction,
    edited by R.J. Reiter, pp. 1-29. Karger Press, Basel.
15. Moore, R.Y. (1979): In: Endocrine Rhythms, edited by D.T.
    Krieger, pp. 63-87. Raven Press, New York.

16. Moore, R.Y. (1979): In: Biological Rhythms and their Central Mechanism, edited by M. Suda, O. Hayaishi, and H. Nakagawa, pp. 343-354. Elsevier/North-Holland Biochemical Press, Amsterdam.
17. Moore, R.Y. (1980): Brain Res., 183:13-28.
18. Moore, R.Y., and Eichler, V.B. (1972): Brain Res., 42:201-206.
19. Moore, R.Y., and Lenn, N.J. (1972): J. Comp. Neurol., 146: 1-14.
20. Moore, R.Y., and Klein, D.C. (1974): Brain Res., 71:17-34.
21. Mosko, S.S., and Moore, R.Y. (1979): Brain Res., 164:1-15.
22. Mosko, S.S., and Moore, R.Y. (1979): Brain Res., 164:17-38.
23. Mosko, S.S., and Moore, R.Y. (1979): Neuroendocrinol. 29: 350-361.
24. Pittendrigh, C.S. (1974): In: The Neurosciences-Third Study Program, edited by F.O. Schmitt, and F.G. Worden, pp. 437-458. MIT Press, Cambridge, Massachusetts.
25. Raisman, G., and Brown-Grant, K. (1977): Proc. R. Soc. Lond., (Biol.), 198:297-314.
26. Riley, J.N., and Moore, R.Y. (1977): Soc. Neurosci. Abstr., 3:355.
27. Rusak, B. (1977): J. Comp. Physiol., 118:145-164.
28. Rusak, B. (1977): J. Comp. Physiol., 118:165-172.
29. Rusak, B. (1980): Biol. Reproduction, 22:148-154.
30. Rusak, B., and Zucker, I. (1979): Physiol. Rev., 59: 449-526.
31. Saleh, M.A., Hard, P.J., and Winget, C.M. (1977): J. Interdiscipl. Cycle Res., 8:341-346.
32. Saleh, M.A., and Winget, C.M. (1977): Physiol. Behav., 19: 561-564.
33. Schwartz, W., Davidsen, L.C., and Smith, C.B. (1980): J. Comp. Neurol., 189:157-167.
34. Setalo, G., Vigh, S., Schally, A.V., Arimura, A., and Flerko, B. (1976): Brain Res., 103:597-602.
35. Stephan, F.K., and Zucker, I. (1972): Proc. Natl. Acad. Sci. USA, 69:1583-1586.
36. Stephan, F.K., and Kovacevic, N.S. (1978): Behav. Biol., 22:456-462.
37. Stephan, F.K., and Nunez, A.A. (1977): Behav. Biol., 20:1-16.
38. Stetson, M.H., and Gibson, J.T. (1977): J. Exp. Zool., 201: 289-294.
39. Swanson, L.W., and Cowan, W.M. (1975): J. Comp. Neurol., 160:1-12.
40. Thorpe, P.A. (1975): Brain Res., 85:343-346.
41. Tigges, J., and O'Steen, W.K. (1974): Brain Res., 79:489-495.
42. Tigges, J., Bos, J., and Tigges, M. (1977): J. Comp. Neurol., 172:367-380.
43. Van Den Pol, A.N., and Powley, T. (1979): Brain Res., 160: 307-326.
44. Wenisch, H.J.C. (1976): Cell Tiss. Res., 167:547-561.

45. Willoughby, J.O., and Martin, J.B. (1978): Brain Res., 151: 413-417.

Neurosecretion and Brain Peptides,
edited by J. B. Martin, S. Reichlin, and K. L. Bick.
Raven Press, New York © 1981.

Circadian Rhythms, Brain Peptides, and Reproduction

Irving Zucker and Marie S. Carmichael

Department of Psychology, University of California, Berkeley, Berkeley, California 94720

The recognition of the endogenous nature and widespread occurrence of rhythms in cellular and behavioral function fostered development of the discipline of biochronometry (biological time measurement). Rhythms with periods of approximately 24 hours have been described for hundreds of morphological, cellular and behavioral processes, including blood concentrations of the hormones estradiol, progesterone, testosterone, corticosterone, LH, FSH, prolactin (e.g., 29). Under constant environmental conditions the majority of daily rhythms tested persist for at least several cycles with periods that differ from 24 hours and thereby qualify as endogenous circadian rhythms.

SUPRACHIASMATIC NUCLEI AS CIRCADIAN CLOCKS

Cellular mechanisms underlying mammalian circadian rhythm generation are not understood. Nor is there definitive information on the number of circadian clocks or their interrelation (clocks are inferred from the existence of observable rhythms). Localization and physiological study of mammalian clocks were facilitated by the identification of the suprachiasmatic nuclei (SCN) of the rat hypothalamus as putative circadian oscillators (37,60). Almost all circadian rhythms investigated so far in several mammals are disrupted when the SCN are bilaterally ablated (see ref. 49 for review). The SCN are not the sole sources of circadian rhythmicity since circadian rhythms persist in tissue culture preparations of avian pineals (30) and in hamster adrenal glands (2,55) and presumably in other structures isolated from nervous system influences (12). Nevertheless, the SCN may play a critical role as overall integrators of circadian rhythmicity (47) or as hierarchal oscillators that impose phase information on other circadian clocks. If the SCN were master circadian clocks they could facilitate temporal integration of the internal milieu as well as promote the organism's adaptations to the external environment.

Neural pathways by which light synchronizes circadian rhythms are reviewed elsewhere in this volume; a direct retinohypothalamic tract seems sufficient to entrain circadian rhythms linked to the SCN (see 35,49,74 for further discussion).

RHYTHMS IN BRAIN PEPTIDES

The rhythmic nature of brain peptide biosynthesis and release has scarcely been investigated. Bioassay procedures have been used to document a diurnal rhythm in rat hypothalamic corticotropin-releasing hormone content (26). Collu et al. (15) presented evidence for diurnal fluctuations in thyrotropin releasing hormone (TRH) content of rat hypothalamus and amygdala; TRH rhythms also were detected in rat retina although these did not appear to be endogenous (52). Whole brain enkephalin content reportedly does not undergo diurnal fluctuations (25) but only two time points were sampled and prominent rhythms may have gone undetected. Clear diurnal rhythmicity in luteinizing hormone releasing hormone (LHRH) content is established for several rat hypothalamic nuclei (57) and for the organum vasculosum lamina terminalis and median eminence circumventricular organs (51). In no instance has the endogenous nature of a diurnal peptide rhythm been established. The commonplace occurrence of diurnal rhythmicity in chemicals in the central nervous system [e.g., norepinephrine (NE), serotonin (5-HT), dopamine (DA), γ-aminobutyric acid (GABA)] as well as in numerous other parameters of neural activity (49) suggests that brain peptides too are generally subject to circadian organization. Whether each of these rhythms is controlled by an SCN-dependent oscillatory network is an open question.

NEUROTRANSMITTERS AND THE SCN

Neurotransmitters involved in generation and entrainment of mammalian circadian rhythms have yet to be identified. The SCN are a primary projection site for serotonergic fibers from the midbrain raphe nuclei (21) and have a high concentration of 5-HT (50). However, marked reduction of brain 5-HT content induced by destruction of the raphe nuclei, or by treatment with parachlorophenylalanine did not eliminate several behavioral circadian rhythms (8).

Moderately high levels of NE have been reported for nerve terminals in the SCN; levels of DA also are moderate in these nuclei (10); the enzyme choline acetyltransferase, which correlates well with the amount of acetylcholine in nervous tissue, is present in moderate concentrations in the SCN (10).

Attempts to influence rhythms by injecting putative neurotransmitters in the vicinity of the SCN have been largely unsuccessful; for example, infusion of 5-HT, DA, TRH, Substance P, glycine and GABA each were ineffective in altering pineal enzyme rhythms (71). However carbachol treatment did mimic the effects of light on the SCN (71). Electrophysiological studies

indicate that 80% of SCN cells are excited by acetylcholine, further implicating a cholinergic mechanism of synaptic transmission
(38). The responsiveness of the SCN to putative transmitters may
depend on the phase in the circadian cycle at which they are
applied and also could be contingent on the use of physiologically meaningful concentrations. These factors should be considered in future studies.

BRAIN PEPTIDES AND THE SCN

Some of the brain peptides found in the rat SCN are listed in
Table 1. With the exception of LHRH, only one time point in the
circadian cycle was sampled in each study. Since negative
results with immunohistochemical and radioimmunoassay studies do
not definitively establish absence of peptides (18), the list of
SCN peptides probably will grow. On the other hand, some of the
substances currently identified may be immunologically related to
the stipulated peptide and not the peptide itself (17).

TABLE 1. Localization of peptides in the rat SCN

Peptide	Localization	References[a]
Vasopressin	Cell bodies in medial, dorsal and rostral parts of SCN; major projections to septum, thalamus and habenula; projection to median eminence in dispute.	58,70
Somatostatin	Perikarya in dorsal and and medial portion of SCN; cell types distinct from those containing vasopressin.	17,27
Vasoactive Intestinal Peptide	In nerve terminals, especially in basal part of nucleus.	22
Substance P	Substantial concentration, presumably in nerve terminals.	9
LHRH	Localization to cell bodies is controversial; found in trace amounts.	10,27,54
TRH	In scattered nerve terminals.	10,18,27

[a] Only a partial list of relevant citations

Vasopressin-neurophysin pathways originating in the SCN have
recently been confirmed and their projections traced to the septum, medial dorsal thalamus and lateral habenula (58). In con-

trast to an earlier report (70), no fibers were traced to the median eminence and it was concluded that the vasopressin-SCN projections make axosomatic contact with neurons and do not terminate on capillaries (58). The projection to the habenula is a potentially short route for SCN innervation of the pineal (cf. 46).

Since the retinohypothalamic projection does not terminate in the vicinity of vasopressin-containing SCN neurons, "it is questionable whether these neurons participate in SCN mediation of circadian rhythms" (58). This prediction is consistent with the persistence of circadian rhythmicity in Brattleboro rats with absence of vasopessin-containing neurons in the SCN (40,61).

Each brain peptide within the SCN is a potential neurotransmitter or neuromodulator for entrainment of SCN neurons by light acting through the retinohypothalamic tract. Similarly, each SCN brain peptide is a possible mediator of the circadian rhythm generating process or the coupling mechanism by which the SCN may drive oscillators or effectors involved in rhythm expression. Experimental data relevant to these issues are presently unavailable.

PINEAL GLAND: MODEL SYSTEM FOR PEPTIDE RHYTHM STUDIES

Rhythmic processes in the pineal gland may be useful as a general model for neuroendocrine integration of biological clocks (cf. 13,31).

1. Neural innervation. The SCN are perceived as controlling pineal function via a multi-synaptic pathway projecting into periventricular hypothalamus, lateral hypothalamus, interomedial lateral columns of upper thoracic cord and superior cervical ganglia. Post-ganglionic noradrenergic fibers provide the final link in this pathway (36). Lesions of the SCN eliminate pineal circadian rhythms (32).

2. Structural rhythms. There are distinct diurnal rhythms in the number of granulated vesicles within pinealocytes (4). The number of adrenergic binding sites in the gland also undergoes diurnal variations which are not accompanied by changes in receptor binding affinity (45).

3. Melatonin rhythms. Pineal melatonin content fluctuates diurnally; e.g., marked increases occur during the dark phase in hamsters maintained in a 14 h light: 10 h dark photoperiod (LD 14:10)(65). Melatonin determinations in blood and urine of several other species confirm the rhythmic nature of its secretion and excretion (33).

4. Target tissue rhythms. In hamsters melatonin-binding receptors in brain were saturable at two phase points in the circadian cycle with a diurnal rhythm in binding capacity and no change in receptor affinity for the radiolabeled ligand (68). The number of binding sites in brain was greatest near the end of the daily light phase.

5. <u>Melatonin and reproduction.</u> Rhythms in pineal hormone secretion may play a significant role in regulating the reproductive cycles of photoperiodic rodents; in these species the reproductive system regresses when day lengths fall below a critical minimum usually attained at the end of the summer (reviewed in 73). This process is mediated by the pineal gland as pinealectomy prevents or reverses short-day induced gonadal regression (3). Many of the effects of short-day exposure on reproduction can be simulated in hamsters by appropriately timed injections of melatonin (5); <u>e.g.</u>, the gonads of males and females maintained in long days can be regressed by once daily melatonin injections

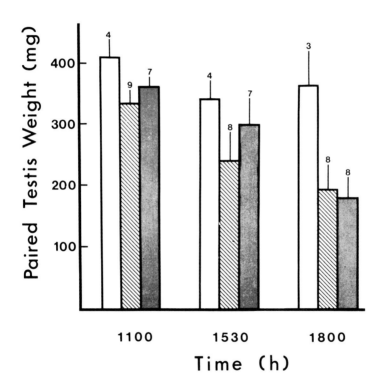

FIG. 1. Testicular weights (M+SEM) of <u>Peromyscus leucopus</u> injected subcutaneously once daily at times indicated with melatonin (0, 5, 25, μg per injection, respectively for open, cross-hatched and gray bars). Injections were given for 10 weeks. The 14 h light phase began daily at 0700 h. Sample sizes are indicated above each bar. (From Johnston and Zucker, submitted for publication).

given at a particular phase of the circadian cycle. Injections
at other times are ineffective (44,64,66). This pattern of
results has been confirmed with the white-footed mouse, Peromys-
cus leucopus (Fig. 1) and may be a general feature of the neuroendocrine axis of photoperiodic rodents (73). Rhythms of responsiveness in melatonin target tissues may mediate these effects;
thus, the greatest number of brain binding sites for melatonin
occurs at the time of day that the reproductive system is most
responsive to the effects of exogenous melatonin (68).

Circadian patterns of melatonin secretion may influence the
availability of receptors for binding melatonin. Vacas and
Cardinali (68) speculate that exposure to high levels of melatonin at inappropriate times of day (as occurs in animals treated
with constant release Silastic capsules) may produce desensitization or down-regulation of the number of brain receptor binding
sites (cf. 14). Well separated peaks and troughs in melatonin
secretion may be instrumental in assuring an appropriate number
of binding sites during times of peak melatonin secretion. The
failure of constant release melatonin capsules (67) to mimic the
effects of timed injections in hamsters is consistent with this
formulation, although we note that in P. leucopus sustained release of melatonin is effective in regressing the testes of mice
maintained in long days (Johnston and Zucker, submitted for publication).

The respective roles of the SCN and of circadian organization
in determining responsiveness to melatonin are presently unclear.
Ablation of the SCN eliminates the effects of photoperiod on the
reproductive axis of male hamsters (48). This could reflect obli-
teration of unspecified circadian rhythms important for day
length measurement or a disruption in secretory function attributable to denervation of the pineal gland. Alternatively, the
SCN, which appears to concentrate melatonin selectively (11)
might constitute an essential melatonin target tissue. Recently,
hamsters with SCN lesions were found to be similar to pinealectomized animals and different from intact controls in that their
gonads regressed during a regimen of three times daily melatonin
treatment (7) (Fig. 2). Thus the SCN are not a critical target
tissue for the anti-gonadal effects of melatonin nor is the circadian disorganization produced by SCN ablation inconsistent with
the full anti-gonadal response to melatonin. However, the extent
to which the SCN and circadian rhythmicity can be dispensed with
under physiological conditions is not addressed by these experiments. We consider it not unlikely that SCN-synchronization of
pineal secretory rhythms with rhythms in brain sensitivity to
pineal products is functionally important to the pineal anti-
gonadal response.

These findings emphasize the importance of temporal patterns
of hormone secretion for normal neuroendocrine integration. This
theme is echoed in work on primate menstrual cycles in which
pulsatile rather than tonic patterns of LHRH release are essen-

tial for episodic LH release.   The temporal pattern of LHRH
secretion may prevent desensitization of the pituitary to LHRH
(cf. 3).
      In hamsters entrained to a 24 h day (e.g., LD 16:8) the
estrous cycle has a period of 96 h (1).   The endogenous nature of
this cycle becomes apparent under constant lighting conditions
(LL) wherein the rhythm free-runs with a period significantly
different from 96 h (1,20). A timing mechanism entrained by LD
cycles controls the precise phase relations among gonadotropin
release, behavioral receptivity and ovulation in hamsters
(1,20,62,63) and in rats (34).

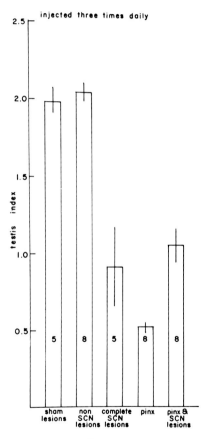

FIG. 2.   Mean testis indices (gonadal length x width divided by
body weight) of hamsters after 7 weeks of melatonin treatment
administered 3 times daily at 1000, 1300 and 1600 h.   The 14 h
light phase began at 0800 h daily.   Sample size and SEM are
indicated on each bar. Pinx = pinealectomized.   (From ref. 7.)

Several lines of evidence suggest that the underlying timer for the estrous cycle has a circadian rather than a circaquadridian (approximately 4 day) period: i) Spontaneous or drug-induced changes in the period of the estrous cycle are almost always in intervals of 24 h and seldom in integral multiples of 96 h (1,19,24,72). ii) Estrogen-primed ovariectomized (ovx) hamsters show daily LH surges as do intact non-cycling females with regressed gonads, or ovx and ovx-adrenalectomized females under short-day conditions (6,39,53). iii) Under constant dim illumination, the hamster estrous cycle free-runs with a period that is a quadruple multiple of the concurrently recorded circadian wheel running rhythm (20). iv) Short term treatment with deuterium oxide lengthens the period of the circadian and estrous cycles but does not affect the phase relation between the two rhythms (20).

Although these data suggest that a circadian mechanism underlies the temporal organization of the estrous cycle, both rhythms could be generated by a single oscillator or by separate but mutually coupled oscillators. Since the estrous and circadian rhythms have yet to be uncoupled either hypothesis remains tenable. Yet another type of explanation dispenses with the circadian framework and emphasizes the time course inherent in ovarian follicle maturation as a major determinant of the pattern of estradiol secretion, the timing of ovulation and the onset of behavioral receptivity (56). If complete follicular development requires approximately 96 h then the estrous cycle period will be four times the period of the circadian cycle but the relation between the two rhythms may be coincidental rather than causal.

Ongoing work in our laboratory attempts to clarify these issues. We reasoned that, if a common circadian pacemaker generates both the activity and estrous rhythms, then the period of the estrous cycle should remain a quadruple multiple of the activity period. To test this hypothesis we entrained hamsters to artificial days that differed considerably from T=24 h (T is the sum of the L and D phases of the entraining photoperiod). We anticipated that as the limits of entrainment were approached (i.e., the point at which the LD cycle no longer drives the circadian rhythm) either the activity or the estrous rhythm might remain entrained to the LD cycle while the other would free-run; alternatively, both rhythms might simultaneously break away from entrainment and free-run with periods not related in the usual 1:4 ratio. In contrast, if the period of the estrous cycle is determined by the time required for follicular maturation and the latter process is independent of circadian rhythms, then estrus should recur at approximately 96 h intervals independently of the status of the circadian system.

Results obtained thus far from hamsters entrained to photoperiod cycles with T substantially less than 24 h support the circadian organization of the estrous cycle (Table 2). In every instance where a hamster showed regular estrous cycles their period was a quadruple multiple of the entrained activity rhythm. There were no statistically significant differences between the

periods of the daily activity onsets and the estrous cycle
periods divided by 4. Since several hamsters had estrous cycles
of approximately 88 h at T=22 h and two females maintained 4 consecutive cycles of approximately 87 h when entrained at T=21.75 h
(Fig. 3), a 96 h interval for follicular maturation does not
limit the period of the hamster estrous cycle.

TABLE   2.   Relation of activity and estrous rhythms under
             entrainment to T-cycles less than 24 h

T-cycle Duration	N	Period of Activity rhythm		Period of Estrous rhythm	
		M + SD	Range	M + SD	Range
23.32	8	23.34 + .04	23.26–23.36	93.33 + .09	93.20–93.40
23.00	8	23.00 + .03	22.95–23.04	92.02 + .09	91.88–92.21
22.00	7	22.09 + .10	21.97–22.29	88.17[b]+ .42	87.68–88.88
21.75	2	21.86 + .16	21.75–21.97	87.16 + .23	87.00–87.32

[a]Hamsters displayed 4 consecutive estrous cycles unless
otherwise indicated; no hamster showed an estrous period of 96 h.
The period of estrous cycles was computed from successive onsets
of behavioral estrus. Periods of the activity rhythm were calculated from daily activity onsets spanning the 4 consecutive
estrous cycles.
    Only two consecutive estrous cycles were established for three
of these hamsters.
    (From Carmichael, Nelson, and Zucker, unpublished observa-
tions).

    One circadian model of the estrous cycle (49) presumes that a
neurogenic stimulus from the SCN stimulates LHRH release from
the median eminence into the primary capillary plexus of the hypophyseal portal system. The temporal coordination of increased
LHRH release, increased pituitary sensitivity to LHRH (under the
influence of estrogenic stimulation) and LHRH sensitization of
the pituitary to subsequent LHRH release are thought to depend on
periodic influences from the SCN (49). These events are major
factors in production of the LH surge. Retriction of this surge
to one day of the estrous cycle may reflect also the availability
of estrogens and progesterone. For detailed elaboration on these
points the reader may consult ref. 49.

THE SCN AND FEMALE REPRODUCTIVE CYCLES

One implication of the circadian model of estrous cycles is that disruption of circadian rhythmicity should eliminate or distort normal estrous cycles (20). Disturbances in estrous cycles often have been reported after destruction of the rat SCN. Critchlow (16) was among the first to establish acyclicity in rats after SCN ablation and to propose that entrainment of estrus to the LD cycle was mediated by a retinohypothalamic tract terminating in the SCN. Many subsequent studies, some with concurrent confirmation of circadian disorganizations, reported an elimination of ovulatory cycles subsequent to SCN destruction (e.g., 42). The positive-feedback response of LH to estrogen-progesterone treatment also was eliminated by SCN ablation (23). Discussion of the relative contributions of the SCN and medial preoptic nuclei to estrous cycles is beyond our present scope (but see ref. 49). We do emphasize, however, the high correlation between treatments that disrupt circadian organization and elimination of estrous cyclicity (20).

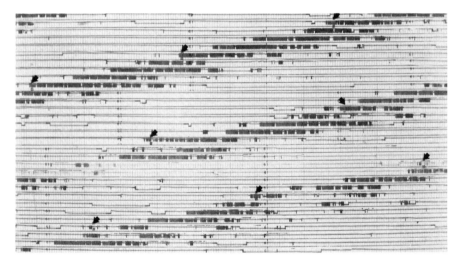

FIG. 3. Wheel running activity and estrus onset of a hamster entrained to non-24 h days. Each horizontal strip represents 24 h with records from successive days pasted one beneath the other. The solid black lines represent sustained wheel running. Arrows mark the onset of behavioral estrus. The first four estrous cycles occurred while the animal's wheel running was entrained to a 22 h day and the last four during entrainment to T=21.75 h. One arrow is slanted to the right to indicate that the precise time of heat onset was not determined on this test. (From Carmichael, Nelson and Zucker, unpublished observations).

The integrity of the SCN is <u>not</u> essential to normal reproductive cycles in some primate species. Complete SCN ablation did not eliminate the menstrual cycle of a rhesus monkey, nor was increased secretion of LH in response to repetitive injections of estradiol benzoate interfered with (41). Note, however, that circadian rhythms have not been implicated in the organization of the primate menstrual rhythms (49) so this result is not particularly damaging to the circadian hypothesis for estrous cycle disorganization.

An interesting observation that merits further study is the persistence of robust diurnal rhythms in serum cortisol and prolactin after complete destruction of the SCN (41,59). In both studies monkeys were tested while exposed to LD cycles and presumably during maintenance of normal laboratory routines of daytime feeding and animal care. The persistence of hormone rhythms under these conditions is not inconsistent with circadian disorganization; studies of hamsters with SCN lesions document "entrainment" of circadian activity rhythms during exposure to LD cycles even when no circadian organization is evident under constant conditions (47). Studies of SCN-lesioned monkeys housed in aperiodic environments are needed to resolve this issue.

PROSPECTUS

We anticipate that future research will document circadian rhythms in the number of receptor binding sites for numerous brain peptides. Furthermore, we predict that circadian rhythms in the biosynthesis and release of brain peptides will be detected and a functionally significant role for these rhythms elaborated in maintaining responsiveness of the receptor-binding apparatus. The alternation between peaks and troughs of secretory activity may be a critical event in maintaining brain receptor sensitivity. Catt and Dufau 914) have remarked that prolonged exposure of target tissues to high circulating concentrations of hormones decreases sensitivity to the biological effects of these hormones, presumably from a loss of some of the receptor population in target cells. This phenomenon of desensitization or down-regulation could account for the resistance to hormones that develops in states characterized by high circulating hormone levels (14). The converse situation in which the presence of a peptide hormone increases receptors for that hormone also has been reported (69) and as noted may occur during action of trophic hormones on their target cells (14). Within this framework, we speculate that circadian rhythms in neurotransmitter, neuromodulator and hormone secretions play an important role in temporal organization of pulsatile release of peptidergic hormones necessary to the maintenance of target tissue sensitivity to agonists. The study of such rhythms should add an exciting dimension to endocrine investigations.

ACKNOWLEDGMENTS

We are grateful to Rosemary Hendrick for typing the manuscript and to Alyssa Zucker for bibliographic assistance. Preparation of the manuscript was supported by funds from the Committee on Research, University of California, and by Grant HD-02982.

REFERENCES

1. Alleva, J.J., Waleski, M.V., Alleva, F.R., and Umberger, E. J. (1968): Endocrinology, 82:1227-1235.
2. Andrews, R.V. (1968): Comp. Biochem. Physiol. 26:179-193.
3. Belchetz, P.E., Plant, T.M., Nakai, Y., Keogh, E.J., and Knobil, E. (1978): Science, 202:631-633.
4. Benson, B., and Krasovich, M. (1977): Cell Tiss. Res., 184: 499-506.
5. Bittman, E.L. (1978): Science, 202:648-650.
6. Bittman, E.L., and Goldman, B.D. (1979): J. Endocrinol. 83: 113-118.
7. Bittman, E.L., Goldman, B.D., and Zucker, I. (1979): Biol. Reprod., 21:647-656.
8. Block, M., and Zucker, I. (1976): J. Comp. Physiol. 109: 235-247.
9. Brownstein, M.J., Mroz, E.A., Kizer, J.S., Palkovits, M., and Lecman, S.E.(1976): Brain Res., 116:299-305.
10. Brownstein, M.J., Palkovits, M., Saavedra, J.M., and Kizer, J.S. (1976): In: Frontiers in Neuroendocrinology, Vol. 4, edited by L. Martini, and W.F. Ganong, pp. 1-23. Raven Press, New York.
11. Bubenik, G.A., Brown, G.M., and Grota, L.J. (1976): Brain Res., 118:417-427.
12. Bunning, E.(1973): The Physiological Clock. Springer-Verlag, Berlin.
13. Cardinali, D.P. (1979): Trends Neurosci., Oct. 250-253.
14. Catt, K.J. and Dufau, M.L.(1977): Ann. Rev. Physiol., 39: 529-557.
15. Collu, R., DuRuisseau, P., Tache, Y., and Ducharme, J.R. (1977): Endocrinology, 100:1391-1393.
16. Critchlow, V. (1963): In: Advances in Neuroendocrinology, edited by A.V. Nalbandov, pp. 377-402. Univ. of Illinois Press, Urbana.
17. Dierickx, K., and Vandesande, F. (1979): Cell Tiss. Res., 2- 201:349-359.
18. Elde, R., and Hokfelt, T. (1978): In: Frontiers in Neuro-endocrinolgy, edited by W.F. Ganong, and L. Martini, pp. 1-33. Raven Press, New York.
19. Everett, J.W., and Sawyer, C.H. (1950): Endocrinology, 47: 198:218.
20. Fitzgerald, K.M., and Zucker, I. (1976): Proc. Natl. Acad. Sci. USA., 73:2923-2927.

21. Fuxe, K. (1965): <u>Acta Physiol. Scand.</u>, 64: Suppl. 247: 37-85.
22. Fuxe, K., Hokfelt, T., Eneroth, P., Gustafsson, J.A., and Skett, P. (1977): <u>Science</u>, 196:899-900.
23. Gray, G.D., Sodersten, P., Tallentire, D., and Davidson, J.M. (1978): <u>Neuroendocrinology</u>, 25: 174-191.
24. Greenwald, G.S. (1971): <u>Endocrinology</u>, 88:671-677.
25. Gwynn, G., Frederickson, R.C., and Domino, E.F. (1979): <u>Neurosci. Soc. Abstr.</u>, 5:527.
26. Hiroshige, T. (1974): In: <u>Biological Rhythms in Neuroendocrine Activity</u>, edited by M. Kawakami, pp. 267-280. Igoku Shoin, Tokyo.
27. Hokfelt, T., Elde, R., Fuxe, K., Johansson, O., Ljungdahl, H., Goldstein, M., Luft, R., Efendic, S., Nilsson, G., Terenius, L., Ganten, D., Jeffcoate, S.S., Rehfeld, J., Said, S., Perez de la Mora, M., Possani, L., Tapia, R., Teran, L., and Placios, R. (1978): In: <u>The Hypothalamus</u>, edited by S. Reichlin, R.J. Baldessarini, and J.B. Martin, pp. 69-136. Raven Press, New York.
28. Kafka, M.S., Wirz-Justice, A., Naber, D., and Lewy, A.J. (1979): <u>Neurosci. Soc. Abstr.</u>, 5:338.
29. Kalra, P.S., and Kalra, S.P. (1979): <u>J. Ster. Biochem.</u>, 11: 981-987.
30. Kasal, C.A., Menaker, M., and Perez-Polo, J.R. (1979): <u>Science</u>, 203:656-658.
31. Klein, D.C., (1978): In: <u>The Hypothalamus</u>, edited by S. Reichlin,, R.J. Baldessarini, and J.B. Martin, pp. 303-328. Raven Press, New York.
32. Klein, D.C. and Moore, R.Y. (1979): <u>Brain Res.</u>, 174: 245-262.
33. Lynch, H.J., Ozaki, Y., and Wurtman, R.J. (1978): <u>J. Neur. Trans.</u>, (Suppl), 13:251-264.
34. McCormack, C.E. and Sridaran, R.(1978): <u>J. Endocrinology</u>, 76: 135-144.
35. Moore, R.Y. (1978): In: <u>Frontiers in Neuroendocrinology</u>, edited by W.F. Ganong and L. Martini, pp. 185-206. Raven Press, New York.
36. Moore, R.Y. (1978): In: <u>The Pineal and Reproduction,</u> edited by R.J. Reiter, pp. 1-30. Karger, Basel.
37. Moore, R.Y., and Eichler, V.B. (1972): <u>Brain Res.</u>, 42:201-206.
38. Nishino, H., and Koizumi, K. (1977): <u>Brain Res.</u>, 120: 167-172.
39. Norman, R.L., Blake, C.A., and Sawyer, C.H. (1973): <u>Endocrinology</u>, 93:965-970.
40. Peterson, G.M., Watkins, W.B., and Moore, R.Y. (1980): <u>Behav. Neur. Biol.</u>, (in press).
41. Plant, T.M., Moossy, J., Hess, D.L., Nakai, Y., McCormack, J.T., and Knobil, E. (1979): <u>Endocrinology</u>, 105:465-473.
42. Raisman, G., and Brown-Grant, K. (1977): <u>Proc. R. Soc. Lond. Ser. B.</u>, 198:297-314.

43. Reiter, R.J. (1973): Ann. Rev. Physiol., 34:305-328.
44. Reiter, R.J., Blask, D.E., Johnson, L.Y., Rudeen, P.K., Vaughan, M.K., and Waring, P.J. (1976): Neuroendocrinology, 22:107-116.
45. Romero, J.A., Zatz, M., Kebabian, J.W., and Axelrod, J. (1975): Nature, 258:435-436.
46. Ronnekleiv, O.K., and Moller, M. (1979): Exper. Brain Res., 37:551-562.
47. Rusak, B. (1977): J. Comp. Physiol., 118:145-164.
48. Rusak, B., and Morin, L.P. (1976): Biol. Reprod., 15:366-374.
49. Rusak, B., and Zucker, I. (1979): Physiol. Rev., 59:449-526.
50. Saavedra, J.M., Palkovits, M., Brownstein, M.J., and Axelrod, J. (1974): Brain Res., 77:157-165.
51. Samson, W.K., and McCann, S.M. (1979): Neurosci. Soc. Abstr., 5:457.
52. Schaeffer, J.M., Brownstein, M.J., and Axelrod, J.(1977): Proc. Natl. Acad. Sci. USA, 74:3579-3581.
53. Seegal, R.F., and Goldman, B.D. (1975): Biol. Reprod., 12:223-231.
54. Setalo, G., Vigh, S., Schally, A.V., Arimura, A., and Flerko, B. (1976): Brain Res., 103:597-602
55. Shiotsuka, R., Jovonovich, J., and Jovonovich, J. (1974): In: Chronobiolgoical Aspects of Endocrinology, edited by J. Aschoff, F. Ceresa, and F. Halberg, pp. 255-267. Schattauer, Stuttgart.
56. Short, R.V. (1974): In: Chronobiological Aspects of Endocrinology, edited by J. Aschoff, F. Ceresa, and F. Halberg, pp. 221-228. Schattauer, Stuttgart.
57. Snabes, M.C., Kelch, R.P., and Karsch, F.J. (1977): Endocrinology, 100:1521-1525.
58. Sofroniew, M.V., and Weindl, A. (1978): Am. J. Anat., 153:391-430.
59. Spies, H.G., Norman, R.L., and Buhl, A.E. (1979): Endocrinology, 105:1361-1368.
60. Stephan, F.K., and Zucker, I. (1972): Proc. Natl. Acad. Sci. USA, 69:1583-1586.
61. Stephan, F.K., and Zucker, I. (1974): Neuroendocrinology, 14:44-66.
62. Stetson, M.H., and Anderson, P.J. (1980): Am. J. Physiol., 238:R23-R27.
63. Stetson, M.H., and Gibson, J.T. (1977): J. Exp. Zool., 201:289-294.
64. Tamarkin, L., Lefebvre, N.G., Hollister, C.W., and Goldman, B.D. (1977): Endocrinology, 101:631-634.
65. Tamarkin, L., Reppert, S.M., and Klein, D.C. (1979): Endocrinology, 104:385-389.
66. Tamarkin, L., Westrom, W.K., Hamill, A.I., and Goldman, B.D., (1976): Endocrinology, 99:1534-1541.
67. Turek, F.W. (1977): Proc. Soc. Exp. Biol. Med., 155:31-34.
68. Vacas, M.I., and Cardinali, D.P. (1979): Neurosci. Lett. 15:259-263.

69. Vale, W., Rivier, C., and Brown, M. (1977): Ann. Rev. Physiol., 39:473-527.
70. Vandesande, F., Dierickx, K., and Demey, J. (1975): Cell Tissue Res., 156:377-380.
71. Zatz, M., and Brownstein, M.J. (1979): Science, 203:358-361.
72. Zucker, I. (1967): J. Endocrinology, 38:269-277.
73. Zucker, I., Johnston, P.G., and Frost, D.(1980): In: Prog. Reprod. Biol., 5: (in press).
74. Zucker, I., Rusak, B., and King, Jr., R.G. (1976): In: Advances in Psychobiology, edited by A.H. Riesen and R.F. Thompson, pp. 35-74. Wiley-Interscience, New York.

Neurosecretion and Brain Peptides,
edited by J. B. Martin, S. Reichlin, and K. L. Bick.
Raven Press, New York © 1981.

Biological Rhythms in Man: Relationship of Sleep–Wake, Cortisol, Growth Hormone, and Temperature During Temporal Isolation

Elliot D. Weitzman, Charles A. Czeisler, Janet C. Zimmerman, and *Martin C. Moore-Ede

*Laboratory of Human Chronophysiology, Department of Neurology, Montefiore Hospital and Albert Einstein College of Medicine, Bronx, New York 10467; and *Department of Physiology, Harvard Medical School, Boston, Massachusetts 02115*

In 1729 de Mairan (17) first reported to the French Royal Academy of Sciences in Paris that a biological organism (a "sensitive plant") would continue to have a rhythm of activity when light-dark entraining stimuli were absent. De Candolle (16) in 1832 first described a free-running rhythm in constant darkness in the same species of plant with a progressive phase advance of 1.5 to 2 hours per day. Subsequent studies have demonstrated that organisms not entrained by 24-hour "zeitgebers" (time cues) develop daily cycles with periods greater or less than 24 hours (27). Extensive research in animals utilizing a rest-activity measurement has demonstrated that these "free-running" period lengths are species-specific and genetically influenced. When an animal is maintained in constant conditions, the new cycle length can be remarkably constant for months to several years (15,28-30). In 1962, Aschoff and Wever (1) studied three men and three women living for 8 to 19 days in a "deep cellar" in Munich. They first demonstrated that normal humans also maintain a "circadian" rest-activity cycle which would "free-run" with a non-24 hour period when the subjects were isolated from all time cues. Many subsequent studies have repeatedly confirmed those observations in man (22-25), and have extended the measurements to include body temperature, urinary electrolytes and certain hormonal metabolic products (2,4,5,8,9). In almost all instances of several hundred such studies now performed (10,11,21,33,34,-37-39,45), including cave and controlled laboratory environments, the period lengths of such rest-activity rhythms have been greater than 24 hours, the majority occurring at approximately 25 hours.

Important conclusions have been reached. These include the change of phase angle relationship between body temperature and rest time (9,47), the ability of different variables to develop different cycle lengths during free-running (3,7), and the concept of multiple oscillators normally synchronized with each other but which can become desynchronized under free-running conditions (48, 49). The importance of "social" entraining rather than light-dark cues for man has been emphasized (6), and in some subjects the ability to develop and sustain very long rest-activity rhythms (between 30 and 50 hours in length) has been recognized (2,4,10, 19,21).

It had been assumed that the "rest" segment was a sleep episode in these studies determined either by "lights out", absence of activity, or "bed-rest" time. Except for the cave studies of Jouvet et al. (21) and the short "isolation" studies of Webb et al. (37-39), systematic studies of the temporal complexity of polygraphically defined sleep stages have not been reported. Analysis of the previously reported detailed sequential durations of the "rest" (selected lights out) time indicates that a significant day-to-day variability is present in almost all subjects, suggesting that interval sleep stage amounts and timing may be related to such variability. The previous assumption that "rest" was equivalent to sleep cannot be made. These events alter biological rhythm cyclic properties and will influence other correlative measured periodic events such as body temperature and hormonal cycles. All previously reported studies have been of subjects maintained in time-free environments totally isolated from direct human contacts during the duration of their stay. We considered it important to study subjects in temporal isolation but with human social communication. This has the major advantage of allowing us to make certain biological measurements and psychological observations not possible with the previous constraints. Finally, all previous studies of hormonal cycles in temporal isolation studies have only used urinary measurements of derived metabolic products (1,2). In a series of recent studies (20,40,41,43-45), we had developed methods of obtaining frequent plasma samples, and demonstrated that important relationships exist between hormonal blood concentrations, sleep and sleep stages.

We have carried out detailed and prolonged measurements of sleep-waking function in human subjects for time periods ranging from 25 to 105 calendar days. We measured polygraphic sleep-stage characteristics, minute-by-minute body temperatures and frequent (approximately 20 minutes) blood sampling for cortisol and growth hormone in normal adult men living in an environment free of all time cues, under entrained, free-running and re-entrained conditions. The results described are part of a comprehensive multi-variable study of the chronophysiology of man living in a time-free environment with a non-schedule daily pattern of living (11,45).

A special environment was established where the individual subjects lived for many weeks. A three-room apartment (study, bedroom and bathroom) was arranged without windows, the walls sound attenuated and a double-door entrance (Temporal Isolation Facility, TIF). A closed-circuit TV system and voice intercom monitored the subject's activities.

Ten male subjects were individually studied. The first group (3 subjects, FRO1, 2 and 3) was studied for 15 calendar days, the second group (6 subjects, FRO4,5,6,7,9,10) for 25 calendar days, and a single subject (PRO1) for an extended stay of 105 calendar days. No subject had psychopathology, medical illnesss, nor received drugs. Each subject kept a written daily diary of sleep times for at least 2 weeks and maintained a regular scheduled sleep-wake schedule in accord with his usual habits. After entry in the TIF, an entrained condition of 3 or 4 scheduled 24-hour sleep-wake periods preceded the non-scheduled "free-running" portion of the study. The entrained clock times were determined by the subject's recorded habitual lights off-lights on time at home. The subject was told that his sleep time would be scheduled for certain portions of the study but was not advised of the clock times nor the duration. Following the entrained portion, each subject was told that he could choose to go to sleep and awaken at any time he wished. He was asked not to "nap" although he was permitted to go to sleep whenever he wished. A decision to go to sleep, therefore, represented bedrest for that biologic "day." Food was available to the subject on demand as breakfast, lunch, dinner and a "snack". The subject could request any meal type at any time. A set of buttons was available which when pushed were coded on a paper punch tape and indicated the behavior the subject was about to initiate and the elapsed time from the beginning of the study. These behaviors included bedrest onset, activity onset, urination, defecation, shower, blood sample times, exercise and mealtimes. The paper punch tape structured the entire time series of each study.

The subject was totally isolated from contact with all non-laboratory persons but communicated by intercom and direct discussion with selected laboratory staff. The supervising staff members were scheduled on a random basis as to time of day and duration of work-shift to prevent the subject from obtaining a time cue.

The following measurements were made for each subject.

(1) Polygraphic Sleep Recording - The interval between the subject's decision to sleep and lights-out with full electrode application was less than 15 minutes. All polygraphic records were scored by standard methods (31).

(2) Rectal Temperature - A rectal thermistor probe was maintained by each subject throughout the entire study except for brief daily periods of defecation. The temperature was automatically recorded every minute on the punch paper tape and a print-out.

(3) Plasma Cortisol and Growth Hormones - Blood samples were obtained at approximately 20-minute intervals using an indwelling

LABORATORY OF HUMAN CHRONOPHYSIOLOGY

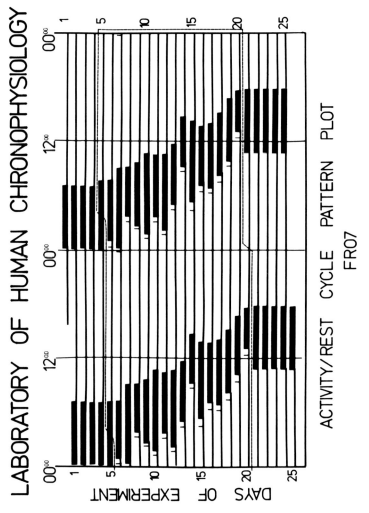

ACTIVITY/REST CYCLE PATTERN PLOT

FRO7

FIG. 1. Example of activity-rest cycle of subjects FR07 during entrained (days 1-5), free-running (days 6-20) and re-entrained (days 21-25). The abscissa is clock time. The short line just prior to the heavy black line represents the subject's request for sleep and the black line represents lights-out time. The free-running sleep-wake cycle period length is 24.72 (mid-sleep time).

specially designed catheter system(41,45). This venous catheter was changed at 2-5 day intervals using alternating arm veins without interrupting the sampling. Blood samples were not obtained from subject PRO1. Plasma control assays were performed using the competitive protein binding technique (26). The samples were assayed in duplicate using 25 ml aliquot for each assay. HGH was assayed in duplicate from each plasma sample by radioimmunoassay using 20 ml of plasma for each assay.

(4) Polygraphic Data Scoring - All scored data were transferred to a computer compatible format and analyzed for total sleep, lights out, and all sleep stages for each lights out-sleep period. The pattern of sleep stage sequences was visualized by a special display program. A quantitative determination was made for a set time period of the percent of each sleep stage and waking. The result of that analysis was also displayed utilizing a computer plotting technique.

(5) Special Mathematical Techniques and Computer Algorithms - In addition to the usual statistical method of analysis and computer plotting and display routines, several mathematical techniques were developed to assist in the analysis of the data. These include a) estimate of period length using a minimum variance fit, b) waveform eduction and c) averaged timed event relationship. This latter method consists of obtaining a mean value for the selected parameters of each individual time point before and after a specific defined event, e.g., 720-minute values of body temperature before and after sleep onset during the entire "free-running" experimental segment for a subject. (5) Special Mathematical Techniques and Computer Algorithms - In addition to the usual statistical method of analysis and computer plotting and display routines, several mathematical techniques were developed to assist in the analysis of the data. These include a) estimate of period length using a minimum variance fit, b) waveform eduction and c) averaged timed event relationship. This latter method consists of obtaining a mean value for the selected parameters of each individual time point before and after a specific defined event, e.g., 720-minute values of body temperature before and after sleep onset during the entire "free-running" experimental segment for a subject.

ACTIVITY-REST CYCLE AND SLEEP STAGE RESULTS

Each of the 10 subjects developed a free-running sleep-wake cycle following the entrained baseline condition (Figures 1 and 2). In each case the mean period length was longer than 24 hours. The subject population could be divided into two types [excluding the tenth subject (PRO1)]. In Type A (6 subjects - FRO1,2,5,6,7,9), the period lengths during FR averaged between 24.4 and 26.2 hours, whereas Type B (3 subjects - FRO3,4,10)

LABORATORY OF HUMAN CHRONOPHYSIOLOGY

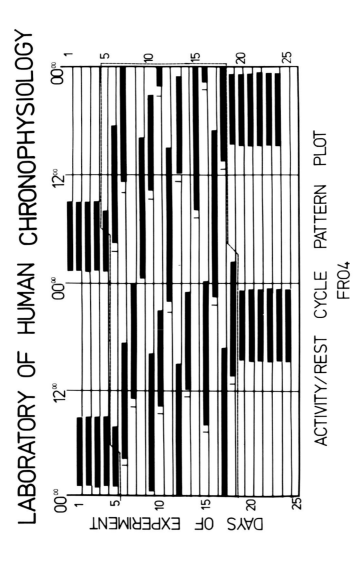

ACTIVITY/REST CYCLE PATTERN PLOT

FRO4

FIG. 2. Rest-activity cycle of subject FRO4. See legend Fig. 1. The free-running sleep-wake cycle period length is 37.60 (mid-sleep time).

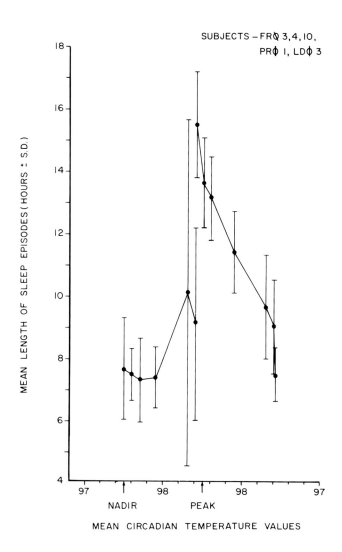

FIG. 3. Relationship between mean sleep duration and temperature for 5 subjects during free-running. The temperature values chosen were derived from an educed circadian temperature curve, each value representing sequential 30 degrees of circadian phase (12).

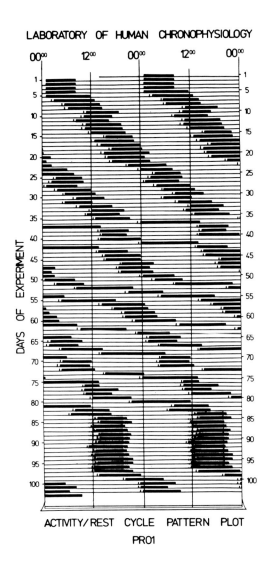

FIG. 4. Activity rest cycle of subject PRO1 during experimental period at 105 calendar days; entrained (days 1-5), free-running (days 6-85), 1, re-entrained by an imposed light-dark cycle (14) (days 86-97), free-running (days 98-103).

had consistently long periods greater than 37 hours (Fig. 2). The lights-out period for the Type B subjects ranged from 8 to 20 hours, with an average of 14 hours. Linear regression analysis through mid-sleep times demonstrated a very stable period length ($r^2 > 0.99$ for each). Short sleep periods recurred at a regular phase[1] of the circadian cycle with a period slightly longer than 24 hours. The long sleep episodes began at a phase angle approximately 180 degrees shifted from that of the short sleep episodes. That is, there was a phase difference of approximately 180^0 between the period line estimated from mid-sleep of the short sleep episodes and the estimated period line of the long sleep episodes (45). also used the circadian temperature curve as an indicator of the underlying circadian oscillator. When the mean core circadian body temperature (educed waveform) at sleep onset was related to the subsequent duration of sleep, it was found that long sleep episodes occurred when the average temperature was at the nadir (Fig. 3) (12). Thus variation in sleep length was related to the phase of the ongoing circadian oscillation at which the sleep episode occurs (11,45). When prior wakefulness lasted more than 1440 minutes, there was a clear increase of sleep length with sleep episodes lasting 600 to 1200 minutes (Figure 2).

Subject PRO1 lived under "free-running" conditions for 80 calendar days and demonstrated several important features (Fig 4). He maintained a regular free-running period length of approximately 25 hours for the first 30 activity-rest cycles. He then developed an activity-rest cycle pattern consisting of alternating long cycles (>36 hours) with a series of shorter cycles (approximately 25 hours). This alternating pattern persisted until it was interrupted by a special light-dark entrainment protocol on calendar day 84. The sleep time continued on an approximately 25-hour period length in spite of the interruption by very long non-circadian periods. These approximately 25-hour self-selected sleep-wake times were therefore entrained to an internal periodic process, which can be considered an "internal zeitgeber." The long sleep episodes (>600 minutes) occurred at a phase angle approximately 180 degrees shifted from the short sleep episodes but in parallel with the same period-length sleep regression line as was seen for subjects in Group B. Analysis of the relationship between length of sleep episode and length of prior wakefulness demonstrated that for only 7 out of 16 waking episodes lasting longer than 20 hours did the subsequent sleep episode exceed 12 hours in length. However, as was the case for the other subjects, no long sleep episode was preceded by a wake episode less than 20 hours in length.

---

[1]Phase is defined as a point in time on the circadian cycle (e.g., sleep-wake cycle, temperature cycle, cortisol cycle, etc.). A "phase shift" therefore is a change in timing of any specified event relative to another defined time point. This phase shift can be an "advance", *i.e.*, occurring earlier, or a "delay", *i.e.*, occurring later than the defined point.

There was a rapid phase delay of lights-out and sleep onset of at least 6 hours within 48 hours of the onset of the free-running condition for 8 of the 9 subjects. The ninth subject (FR07) delayed his sleep onset by 5 hours on the third biologic day. In addition, all 6 subjects in Group A had a characteristic "scalloped" appearance of the time of lights-out with a variable cycle of 3-4 days. This could not be explained as a "transient" process related to the onset of FR since it clearly continued throughout the FR condition in four subjects (FR02,5,7,9).

The lights-out episode in general corresponded with the sleep episode for each subject and for each night. However, it was found that at times there was a short delay from lights-out to sleep onset. The two older subjects (ages 50,51) (FR09,10) consistently interrupted their sleep episodes by awakening for short periods during the subjective night as well as remaining awake in the dark for durations up to one hour after awakening and prior to signaling "lights on." These waking interruptions were also present during the entrained segment as well. These findings emphasize the importance of defining sleep stages polygraphically when measurements of biological rhythm variables are made.

There was considerable variability in the mean total sleep time (TST) during FR. Two subjects averaged 13.8 and 12.8 hours (FR03, 4) as compared to other subjects who had average TST's of 7 to 8 hours. In spite of this variability in total sleep time per sleep period during FR, the ratio of sleep time to period length only varied between 0.24 and 0.35 across subjects with an average of 0.29. This compared with 0.30 during the entrained condition. When the entrained ratio was compared to the FR ratio for each subject, it was noted that two subjects with high entrained ratios (long sleepers) (0.31 and 0.32) increased the value to 0.35 and 0.34, respectively during free-running whereas four subjects with the lowest entrained ratios (0.27, 0.28, 0.29) and 0.29 (short sleepers) all decreased the ratio to 0.25, 0.25, 0.24 and 0.25 respectively, during FR. The 3 other subjects with intermediary entrained values had little change during FR.

The sleep stage characteristics for all subjects were compared as a function of sequential experimental nights during three experimental conditions (Entrained, Free-Running and Re-Entrained). The values of REM % of TST were remarkably constant throughout and did not differ significantly as a function of experimental conditions. The stages 3+4% of TST did increase to a small extent from the entrained (27.8%) to the FR (29.8%) condition, especially during the last 6 FR sleep episodes. A small average increment (32.2%) occurred during the five re-entrainment nights for 5 subjects.

An interesting result was obtained when comparisons were made for REM% of TST, by subject and by experimental condition. There was considerable variability in REM sleep across subjects (range 15 to 30%) during the entrained period. However, the intra-subject variability was very small as a function of experimental conditions. This was not the case for stages 3+4 since both the inter- and intra-subject variability was similar in all 3 experimental conditions. These results indicate that each subject maintained an individual control of REM% of sleep time which was independent of the entrained or free-running state. This does not appear to be the case for stages 3 and 4 sleep.

Three subjects (FRO3,4, 10) consistently had long sleep episodes associated with long sleep-wake cycle lengths. There were a total of 26 sleep episodes lasting 12 hours or longer. These long sleep episodes differed from the short sleep episodes. The timing of the onset of these long sleep periods occurred at a different phase of the subjects' circadian temperature rhythm (130 degrees to 270 degrees, 0 degrees = mid-trough) than the onset of the short sleep episodes (270 degrees to 120 degrees). In addition, during the long sleep episodes, sustained stages 3 and 4 sleep would characteristically occur between 12 and 18 hours after sleep onset. However, the first 4 hours of the long sleep episodes did not differ significantly in regard to the characteristic timing and amount of stages 3 and 4 sleep seen under entrained conditions. Thus despite normal amounts of 3-4 sleep present at the onset of these long sleep episodes, stage 3-4 would reappear after 12 to 16 hours of sustained sleep.

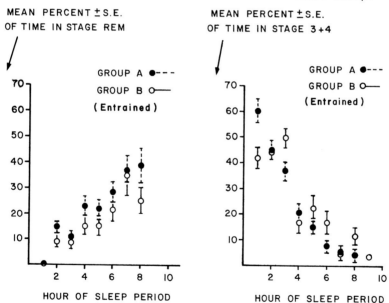

FIG. 5A. Graph of mean $\pm$ S.E. percent of sleep time in REM-sleep each sequential hour of lights out episode for subject Group A and B during entrainment (see text). The numbers adjacent to each point are the total number of sleep episodes averaged.

5B. Graph of mean $\pm$ percent of sleep time in stages 3 + 4 for each sequential hour of the lights out episode for subject Group A and B during entrainment.

Although occasional awake episodes interrupted these long sleep
times (especially for subject FR10), they were not sufficiently
long to explain the reoccurrence of stages 3 and 4.

The mean percent time of REM sleep per sequential hour of
sleep during the entrained (EN) condition demonstrated a
difference when comparing the two subject groups (A and B).
There was more REM during the 2nd through 8th hours of sleep for
Group B than for Group A (Fig. 5A). The sequential hourly pattern of the mean percent time of stages 3 + 4 sleep also differed
during the entrained conditions (Fig. 5B). There was more stage
3 + 4 initially for A (60% in first hour) than B (42% in first
hour). The decrease was then progressive for A (almost linear to
the 6th hour). In Group B, the percent stage 3+4 initially
increased, reaching a peak at the 3rd hour (51%) and then showing
a sharp fall at the 4th hour of the sleep period with a small
rise in percent again at the 5th and 6th hour after lights out.
The mean percent time awake in the entrained condition also
differed during the 1st hour after lights out, with Group A being
significantly less (15%) than B (28%). There was a tendency for
more wakefulness to occur later in the night (specifically hours
4,5 and 7) for Group B.

During the free-running (FR) as compared to the entrainment
condition for Group A, there were major changes in the sequential
hourly timing of REM sleep during the course of the subjects'
sleep episodes (Fig. 6). After the first hour of the night,

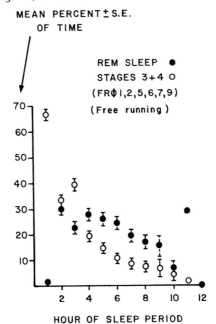

FIG. 6. Graph of mean + percent of sleep time in REM and stages
3 + 4 sleep for each sequential hour of the lights out episode
for subject Group A during free-running.

there was a major increment in the amount of stage REM during free-running, achieving a mean of 30% on the second hour and then gradually falling to reach a value of 15% on the 8th hour after lights out. (During entrainment, the 8th hour had a mean value of 40%.) Indeed, the curve of REM sleep % for the sleep period is a falling one during free-running compared to a rising one during entrainment. During free-running, a progressive decrement of the percent of stages 3+4 was clearly present, similar to that found in the entrained condition for Group A. The 2nd hour after lights out however did differ between the two conditions (Entrained 45%, Free-Running 33%). This difference at the 2nd hour was presumably related to the major increase of REM % during FR which occurred at this time. There was also a shift of the waking time to the latter third of the sleep period for Group A. The subjects had 9% of time awake in the first hour during FR compared to 16% when entrained. However at the 7th, 8th and 9th hours after lights out, the percent time awake dramatically increased (i.e., 7th hour, 4% EN, 26% FR; 8th hour, 11% EN, 37% FR; and 9th hour, 21% EN, 43% FR).

The shift of REM sleep to an earlier time within the circadian "daily" sleep episode during FR supports the concept that the timing of REM sleep is strongly influenced by an underlying endogenous biological rhythm oscillator and not by the sleep process itself. That is, the decision to go to sleep during the FR condition was followed by a very different temporally organized sleep pattern than when the subjects were told to go to sleep based on their habitual entrained daily sleep times. However, the timing of stages 3 and 4 sleep within the sleep episode did not differ substantially between entrained and free-running indicating that these sleep stages are primarily determined by the initiation and course of the sleep process itself when it occurs within a daily circadian sleep-wake cycle.

This clear shift of waking time within the daily sleep episode is also of considerable interest since this change is very similar to what is found in patients who complain of an "early morning arousal" insomnia. The characteristic pattern at night is to have a very short latency to sleep onset, but then awaken spontaneously "too early" followed by either fragmented sleep-wake pattern or frank full arousal with an inability to fall asleep again. The demonstration of a phase advance of waking time within the circadian sleep-wake cycle during free-running in normal subjects strongly suggests that there is a similar phase shift of the endogenous circadian "waking rhythm" in relation to the "sleeping rhythm."

Another characteristic difference between the long and short sleep episodes was the timing and amount of REM sleep within the first 3 hours after sleep onset. All of the sleep periods which had a very small REM latency (<20 minutes) were short sleep episodes during the FR condition. The mean REM latency (sleep onset to onset of first REM period) clearly decreased for 9 of the 10 subjects (FRO9 was the exception) comparing entrained to the

free-running condition. A partial recovery took place during the
re-entrainment conditions. In addition, the mean total minutes
of REM sleep in the first 3 hours of sleep increased for 8 of the
9 subjects (subject FRO9 excepted) between entrainment and free-
running. However, during re-entrainment these values did not return to baseline. The timing and amount of REM sleep during the
first 3 hours after sleep onset in PRO1 was determined during the
free-running condition when he had the alternating long and short
sleep-wake cycles. All the nights with a short REM latency (<10
minutes) and the nights with more than 30 minutes of REM sleep in
the first 3 hours, except one, occurred within 90 degrees of the
nadir (0 degrees) of the circadian temperature rhythm. In addition, for the 12 REM sleep onsets which took place within 10 minutes of sleep onset, 11 occurred within 60 degrees of a specific
phase (mid-trough) of the circadian temperature rhythm.

   In order to compare the sequential hourly pattern of sleep
stages and waking of short (<600 minutes) compared to long (>600
minutes) of sleep for all subjects (FRO1-10, PRO1) during free-
running, we determined the mean percent time in stages REM, 3+4
sleep and waking for each sequential hour.

   The graphs of mean percent of stages 3 + 4 were remarkably
similar for the long and short sleep episodes (Fig. 7). The 2nd,
3rd and 4th hours demonstrated a small decrease for the short
sleep episodes, reflecting the increase at those times of REM
sleep (see below) for the short sleep episodes. Of considerable
interest is the finding that there was an increase in stages 3+4
sleep after 14 hours of sleep, for the long nights, on the 14th,
17th, 18th and 20th hour. The values rose to 7, 6, 9 and 21% respectively for the few sleep episodes which lasted that long.

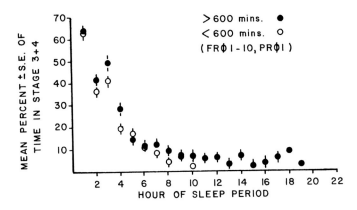

FIG. 7. Graph of mean + S.E. percent sleep time of stages 3+4
sleep for all subjects during free-running for each sequential
hour for all lights out episodes lasting less and greater than
600 minutes.

There was also a major difference in the REM sleep % comparing long and short sleep episodes (Fig. 8). REM sleep was higher by approximately twofold for hours 2, 3 and 4 for the short sleep episodes. However, for the 8th, 9th and 10th hours, the percent REM was higher for the long sleep episodes. Indeed, the usual polarity of REM sleep during the daily sleep episode was absent for the greater than 600 minute group, with the percent REM remaining at approximately 20% from the 2nd through the 18th hour.

The percent time awake was not different for the long sleep episodes for the first 6 hours, but of course remained small until the 11th hour. After that there was progressive increment of mean waking time, largely reflecting the variability in sleep length.

These additional results comparing long and short sleep episodes in a large group of subject-nights strongly supports the concept that the timing and amount of REM sleep during free-running is primarily determined by the phase angle of an underlying endogenous oscillator and that this process is closely associated

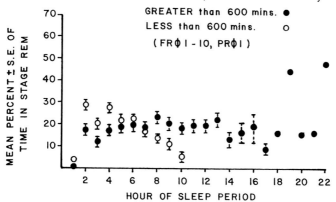

FIG. 8. Graph of mean ± S.E. percent sleep time of REM sleep for all subjects during free-running for each sequential hour for all lights out episodes lasting less and greater than 600 minutes.

with the duration of total sleep (11,45,46). Recent studies by our group have also indicated that cumulative REM minutes but not REM density (number of eye movements/min.) is phase-advanced relative to sleep onset (50). We therefore conclude that the initiation and subsequent sleep process itself is not a major determinant of the timing of the underlying brainstem REM mechanism. The timing and amount of stages 3+4 sleep, however, do appear to be directly linked to the initiation and course of the sleep process (for the subsequent 9 hours at least) and are not dependent on the total duration of sleep. It should be emphasized that a small amount of stages 3+4 sleep was present for each hour, even up to the 18th to 20th hour if the sleep episode persisted.

An analysis was made of REM-NREM sleep cycling during the different experimental conditions. The latency in minutes from sleep onset to first mid-REM period, first mid REM to second mid-REM, etc., was determined. It was found that except for a shortened latency from sleep onset to the mid-first REM period during free-running, there were no differences in cycle lengths as a function of experimental condition. There was a consistent decrease in cycle length for the fourth and fifth cycle for each condition. The sleep cycle length remained stable (x 85 minutes) for up to 11 cycles during the long sleep periods (>10 hours). Thus there is no evidence that sleep stage cycle length is altered by the increased sleep-wake period length during free-running conditions. In addition, previously reported results (18) of a stable but slightly reduced cycle length when sleep is extended are confirmed by these data for those long sleep episodes which extend from 10 to 20 hours.

BODY TEMPERATURE RHYTHM

The mean core (rectal) temperature for all subjects as a group was essentially the same in all three conditions. However, for each subject in Group A there was an increase in the mean temperature during FR whereas there was a decrease for each subject in Group B. During re-entrainment, the mean value of most subjects had returned to that of the entrained section.

During the entrained condition, the rectal temperature curve (values obtained every minute) demonstrated the well described sharp fall ($1-2°$ F) following sleep onset (5,9,36). A small decrease in temperature typically occurred at approximately 3 hours before sleep onset with a sharp elevation of temperature at the end of the sleep episode. During "free-running" for all subjects in Group A a change in both phase and shape of the curve occurred (47). The temperature began to decrease 6 to 8 hours prior to sleep onset. At the time of choosing sleep, the body temperature was close to the lowest value of the circadian rhythm. An additional small fall of temperature ($0.5°$ F) took place just after sleep onset during FR. During re-entrainment, the curve was similar to that found in the entrained condition, although it had not fully established the original shape. In two subjects with long sleep periods, a wave shape pattern was educed during the FR condition at the same period length as the sleep-wake cycle [39.1h (FRO3) and 37.6h(FRO4)]. The curves at the long period lengths resembled those in the entrained conditions (normalized to $360°$), both in shape and phase relationship to the average sleep time. These results suggest that the approximately 40-hour component in the temperature rhythm was a "response" to sleep onset in the sleep-wake cycle rather than an independent self-sustained rhythm. In each of these cases (FRO3, 04,10), as mentioned above, there was also an approximately 25-hour component in the circadian temperature rhythm. The amplitude was small (approximately 1° F) compared to the entrained condition (approximately 2.0 ° F) and compared to subjects FR with a sleep-wake cycle of approximately 25 hours (1.5 to 2.0° F).

Subject PRO1 had a small drop of overall mean temperature in the entrained compared with the free-running condition (98.42° to 98.17 ° F). He had a circadian temperature period length of 25.0 hours during the first 30 free-running days which shortened to 24.55 during the next 50 days.

PLASMA CORTISOL PATTERNS

We have been successful in obtaining plasma samples at 20-minute intervals for each of 9 subjects during the experimental conditions (FRO1-FR10; total samples obtained, 15,000).

During the entrained condition all subjects demonstrated the normal episodic pattern of secretion during each 24-hour period. The typical pattern was evident with very low values just prior to and during the first 3 hours of sleep, followed by a series of secretory episodes during the latter half of the night. An intermittent, eposidic secretory pattern was present during the waking day (40,41). The educed waveform for the entrained condition also demonstrated this circadian pattern of hormonal activity (45). During the free-running condition, a clear phase advance and change of wave shape of the circadian cortisol curve was evident for subjects FRO5,06,07. The nadir of the curve was now occurring 100 to 150° in advance of that during entrainment with respect to sleep onset. In addition, the average rate of rise of cortisol after the low point was much more gradual, nevertheless reaching the highest value at approximately the same time, namely, the end of the sleep period.

It is important to emphasize that the process of waveform eduction produces an overall mean curve at a defined period length and therefore will "smooth out" specific point-related events. Examination of the cortisol time series itself revealed that on many "free-running" days, especially with a progressive phase delay of sleep time, cortisol would be secreted just before sleep onset and then would stop being secreted for several hours just after sleep onset.

The duration of this inhibition was 1-3 hours at the beginning of sleep and did not continue throughout sleep. Sleep onset was therefore used as a "zero" point about which a time-locked response cortisol curve was obtained in several subjects. All demonstrated a clear pattern of cortisol inhibition following sleep onset. Therefore during the free-running condition, a phase advance of cortisol occurred in relation to the sleep episode, the overall wave shape was changed and a specific sleep-related inhibition of cortisol secretion was apparent. During the re-entrainment condition, a similar pattern was evident since the phase relationship between the cortisol rhythm and sleep had not yet returned to normal. Therefore the subject was going to sleep when the concentration of the hormone was high. Evidence that this sleep-related inhibition may well be operative even in subjects habitually living on a 24-hour routine may be deduced from the data obtained during the transition from the entrained to the free-running condition on those nights when a phase delay of sleep onset on a single night exceeded 2-3 hours. On those occasions, the hormone was released just before sleep and then immediately inhibited after sleep onset.

It thus appears that the episodic pattern of cortisol secretion is influenced both by an endogenous rhythmic component not directly related to the behavioral sleep-wake cycle and a specific sleep (or lights out in bed) related component. Whether other daily behavioral events such as sleep onset, lights on, out of bed, meal time, etc., are also determinants of the episodic pattern will require further detailed analysis of the extensive data we have obtained in these studies.

GROWTH HORMONE PATTERNS

HGH was found to be secreted in an episodic normal manner in all subjects with the typical pattern of brief episodes of secretion (1-2 hours) followed by long inter-episode intervals (6-12 hours) with no HGH detectable (42). The hormonal concentraion was less than the overall average (1 ng/ml) 80% of the time. A striking, highly consistent relationship between sleep onset and an episode of HGH secretion was found for all 3 experimental conditions for all subjects (32,35,43). A clear secretory episode followed sleep onset approximately 90% of the time. Thus far no independent rhythm of HGH could be detected but further analysis for an ultradian, or specific behavioral related event, will be searched for.

CORTISOL AND GROWTH HORMONE PATTERNS FOR "LONG" AND "SHORT" SLEEP EPISODES AT SLEEP ONSET AND WAKING

We compared the mean cortisol, GH, and temperature curves before and after sleep onset for the short (<600 minutes) and long (>600 minutes) sleep episodes during free-running for subjects O1- O9. All three parameters differed between the two sleep length conditions. The mean body temperature began to fall prior to (beginning at 6 hours before) sleep onset and reached a lower value by the time of sleep onset for the short compared to the long sleep episodes. The rate of temperature fall was also greater for short sleep episodes, reaching its nadir at 1 hour after sleep onset, but then rose rapidly during the next 9 hours, reaching a greater mean value (Δ.95 degree) at 10 hours after sleep onset for the short compared to the long sleep episodes. A difference in pattern was also found for the mean cortisol concentration curve (Fig. 9). The pre-sleep concentrations were lower for the short sleep episodes but did not fall as low after sleep onset as for the long sleep episodes. Within 2 hours the curve then began to rise and reached a peak value (11 mg/100ml) 6 hours after sleep onset for the short sleep group. By contrast during the long sleep periods the cortisol concentration achieved a much lower value during sleep, remained low longer and then only achieved a high mean value of 6.5 mg/100 ml at 10 to 12 hours after sleep onset. The growth hormone pattern was also different between the long and short sleep episodes (Fig. 10). Although the main daily secretory episode occurred just after sleep onset and reached a peak at the same time for the two conditions (70 minutes after sleep onset), the mean peak concentration was greater for the long (9.5 + 1.7 S.E. mcg/ml) in comparison to the short (6.6 + 0.6 S.E. mcg/ml) sleep episodes

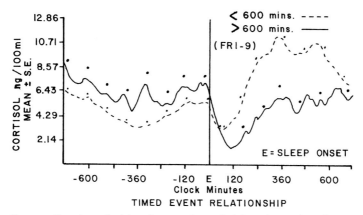

FIG. 9.   Graph of timed event relationship (TER) of mean + S.E.M. of plasma cortisol (μg/100ml) for subjects FR01-09 at sleep onset for all dark episodes lasting less than and greater than 600 minutes during free-running (FR).

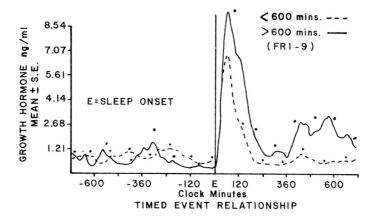

FIG. 10.   Graph of TER of mean + S.E.M. of plasma growth hormone (μg/ml) for subjects FR01-09 at sleep onset for all dark episodes lasting less than and greater than 600 minutes during FR.

(Fig. 10). In addition, during the long sleep episodes, this
sleep-related GH secretory episode lasted longer, not returning
to baseline until approximately 250 minutos after sleep onset.
These results again demonstrate the effect of sleep onset and
sleep duration on body temperature, cortisol and GH when they
occur at different phases within the free-running circadian
period. The increased amount of slow-wave sleep present during
the first 4 hours during the long sleep episodes is associated
with an increased amount of secreted GH and a greater degree and
length of cortisol inhibition.

We also compared the temporal pattern of cortisol secretion
during free-running for sleep episodes of less than 30 minutes to
those with greater than 30 minutes of REM sleep during the first
3 hours after sleep onset. There was a difference in both the
mean concentration just prior to and the nadir after sleep onset,
the values being lower for the sleep episodes with less REM
during the first 3 hours. We compared those sleep episodes with
a short REM latency (<30 minutes) to those with a long REM
latency (>30 minutes). A much higher value of cortisol was found
at the time of sleep onset for the short REM latency sleep
episodes (11 + 2.0 mcg/100 ml) compared to long REM latency
nights (6 + 0.5 mcg/100 ml). The decrease of concentration was
therefore much greater after sleep onset for the shorter REM
latency group, since the nadir value was essentially the same for
the two groups (3.0 mcg/100 ml). These results indicate that the
timing and amount of REM sleep within the circadian period is
correlated with the pattern of cortisol secretion in relation to
sleep onset.

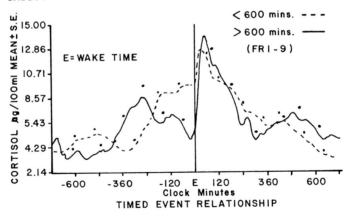

FIG. 11. Graph of TER mean + S.E.M. of plasma cortisol (μg/100
ml) for subjects FR01-09 at wake time for all dark episodes
lasting less than and greater than 600 minutes during FR.

We also analyzed the temporal pattern of the change of
cortisol and GH at the time of waking for both entrained and

free-running conditions for <u>long</u> and <u>short</u> sleep episodes. All subjects had a sharp but brief elevation of cortisol, assoicated with an inhibition of GH at awakening when entrained. This waking response was also present during free-running but differed for <u>long</u> and <u>short</u> sleep episodes. The mean cortisol was much higher just prior to awakening from short (9.0 + 0.5 mcg/100 ml) compared to long sleep (5.4 + 0.6 mcg/100 ml) (Fig. 11). The time of the peak value after awakening however was similar for the two sleep period groups (approximately 30-40 minutes) as was the mean peak concentration achieved (approximately 13 mcg/100 ml). The increment of cortisol secretion was therefore much greater for the long sleep episode (Δ = 8.0 mcg/100 ml) compared with the short sleep periods (Δ = 3.5 mcg/100 ml).

The opposite pattern was found for GH. Just prior to awakening for the long sleep periods, there was a progressive rise of GH to a mean concentration of approximately 3.5 mcg/ml. For the short sleep episodes, however, the mean GH concentration remained low at 0.5 mcg/ml for 3 hours prior to the final awakening. A dramatic fall of concentration to undetectable values then occurred, lasting for 4 hours after awakening for the long sleep episodes. The short sleep episodes remained at approximately 0.5 mcg/ml after awakening.

Thus there appear to be brain mechanisms which simultaneously control the yoked release of ACTH-cortisol and inhibition of GH as well as the simultaneous inhibition of cortisol and release of GH in relation to sleep onset and awakening. The former appears to be related to CNS activation (arousal?) and the latter to sleep (slow wave?). When the generator(s) is in an "activated" state (waking-REM sleep), there is secretion of ACTH-cortisol and inhibition of GH; when the CNS generator produces a non-REM (stages 3-4) sleep state, there is inhibition of ACTH-cortisol and secretion of GH.

SUMMARY AND CONCLUSIONS

We confirm previous studies that biological rhythms of human beings free-run at period lengths greater than 24 hours, typically at approximately 25 hours, but with individual variability. After a variable time of free-running, many normal humans will spontaneously develop "long" biologic days (>35 hours) and often these will alternate with "short" days (approximately 25 hours).

During free-running, although the sleep to total time ratio remains remarkably constant (approximately 0.30), short sleep episodes (<10 hours) occur at a specific phase angle of an internal circadian rhythm (e.g.,/body temperature) whereas long sleep periods (>10 hours) take place approximately 180 degrees out of phase with the short sleep episodes but maintain the same period length. Sleep stage organization changes during "free-running" such that REM sleep advances to an earlier time during sleep, with a shortened REM latency (occurring at times less than 10 minutes after sleep onset) and increased amounts during the first 3 hours of sleep. The total REM amount and percent for the entire sleep episode, however, remains constant. The timing and amount of REM sleep following sleep onset also occurred preferentially at a specific phase of the circadian temperature cycle, strongly supporting the concept that certain sleep processes in the brain are endogenous biological rhythms. The stage 3-4 sleep distribution remains essentially the same during the 3 experimental conditions. During the long sleep periods (>10 hours), stages 3 and 4 recur following 14 to 16 hours of sleep, indicating that these stages are not dependent on length of prior waking but may be related to length of prior elapsed time.

The core (rectal) temperature develops an approximate 25-hour rhythm in humans during free-running, but the wave-shape changes such that a phase advance (6-8 hours) of the falling phase develops in relation to the onset of sleep. The subject usually then selects sleep when the circadian temperature approaches its lowest value of the day. In addition, at the time of sleep onset (lights out and in bed) there is an additional drop of body temperature. This is espeically noted when sleep onset occurs when the immediately preceding core temperature is high (e.g., for the long sleep episodes).

Measurements of plasma cortisol throughout each study demonstrated two components of the circadian cortisol curve during free-running. One component has a phase advance (6-8 hours) relative to sleep onset whereas a second component clearly follows sleep onset. This second component appears to be a sharp inhibition of cortisol secretion during the first 2-3 hours of sleep interrupting a rising phase of the hormonal curve. Growth hormone secretion, on the other hand, was intimately related to the first 2 hours after sleep onset. A sharp episode of hormonal secretion occurs just after sleep onset for almost all sleep periods. No other independent circadian rhythm of GH has been detected thus far.

There appear to be brain mechanisms which simultaneously control the yoked release of ACTH-cortisol and inhibit GH, as well as the simultaneous inhibition of cortisol and release of GH. When there is "activation" (waking-REM sleep), there is inhibition of GH and secretion of cortisol, and when stages 3+4 sleep occur, there is inhibition of ACTH-cortisol and secretion of GH.

These and previously reported studies emphasize the lawfulness of biological rhythm functions in man and demonstrate the importance of the methodology using temporal isolation and the analysis of "free-running" rhythms to unravel these chronobiological processes.

ACKNOWLEDGEMENTS

This work is supported in part by USPHS grants AG00792-04 and MH28460-04.

REFERENCES

1. Aschoff, J. and Wever, R. (1962): Die Naturwissenschagten, 49:337-342.
2. Aschoff, J. (1967): Life Science and Space Research, V., p. 159-173.
3. Aschoff, J. Gerecke, U. and Wever, R. (1967): Jap. J. Physiol., 17:450-457.
4. Aschoff, J. (1969): Aerospace Medicine, 40:844-849.
5. Aschoff, J. (1970): In: Physiological and Behavior Temperature Regulation, edited by J.D. Hardy, A.P. Gagg and J.A. Stolwijk, pp. 905-919. Springfield, Ill., Charles C. Thomas.
6. Aschoff, J., Gerecke, U., Jureck, A., Pohl, H., Reiger, P., V. Saint Paul, U. and Wever, R. (1971): In: Biochronometry, edited by M. Menaker, p. 3-29. Washington, National Academy of Science.
7. Aschoff, J. (1973): In: Proceedings of the XXI International-al Congress of Aviation Space Medicine, p. 255.
8. Aschoff, J. and Wever, R. (1976): Fed. Proc., 35:2326-2332.
9. Aschoff, J., Gerecke, U. and Wever, R. (1976): Eur. J. Physiol., 295:173-183.
10. Chouvet, G., Mouret, J., Coindet, J., Siffre, M. and Jouvet, M. (1974): Electroencephalography and Clinical Neurophysiology, 37:367-380.
11. Czeisler, C.A., (1978): Ph.D. Dissertation, Stanford University.
12. Czeisler, C.A., Weitzman, E.D., Moore-Ede, M.C., Zimmerman, J. and Knauer, R. (1980) Submitted for publication.
13. Czeisler, C.A., Zimmerman, J.C., Ronda, J., Moore-Ede, M.C. and Weitzman, E.D. (1980): Sleep (in press).
14. Czeisler, C.A., Richardson, G.S., Zimmerman, J.C., MooreEde. M.C. and Weitzman, E.D. (1980): Photobiology Photochemistry (in press).
15. Daan, S. and Pittendrigh, C.S. (1976): J. Comp. Physiol., 106:253-266.
16. De Candolle, A.P. (1832): Physiologie Vegetale Vol. 2, pp. 854-862. Paris, Bechet Jeune.
17. DeMairan, J. (1729): Histoire de L'Academie Royal des Science, p. 35. Paris.
18. Feinberg, I. (1974): J. Psychiat. Res., 10:283-306.
19. Findley, J.D., Mialer, B.M. and Brady, J.V. (1963): In: Session of the 7th Interntl. Space Sci. Sym. Life Sci. and Space Res. V., edited by A.H. Brown and F.G. Favorite, pp. 159-163. Amsterdam, North Holland Publishing Co.

20. Hellman, L., Nakada, F., Curti, J., Weitzman, E.D., Kream, J., Roffwarg, H., Ellman, S., Fukushima, D.K. and Gallagher, T.F. (1970): J. Clin. Endocrin., 30:411-422.

21. Jouvet, M., Mouret, J., Chouvet, G., and Siffre, M. (1974): In: Experimental Bicircadian Rhythm in Man in the Neurosciences: Third Study Program, edited by L.E. Scheving, F. Halberg, and J.E. Pauly, pp. 560-563. Tokyo, Igaku Shoin Ltd.

22. Mills, J.N. (1966): Physiol. Rev., 46:128-171.

23. Mills, J.N. (1974): In: Chronobiology, edited by L.E. Scheving, F. Halberg, and J.E. Pauly, pp. 560-563. Tokyo, Igaku Shoin Ltd.

24. Mills, J.N., Minors, D.S., and Waterhouse, J.M. (1974): J. Physiol. (London), 240:567-594.

25. Mills, J.N., Minors, D.S., and Waterhouse, J.M. (1976): J. Physiol. (London), 257:54-55.

26. Murphy, B.P., Engelberg, W., and Pattee, C.J. (1963): J. Clin. Endocrin., 23:293-300.

27. Pittendrigh, C.S. (1961): Cold Spring Harbor Symposium on Quantitative Biology, 25: 159-184.

28. Pittendrigh, C.S., and Daan, S. (1974): Science, 186:548-550.

29. Pittendrigh, C.S., and Daan, S. (1976): J. Comp. Physiol., 106:223-252.

30. Pittendrigh, C.S., and Daan, S. (1976): J. Comp. Physiol., 106:291-331.

31. Rechtschaffen, A., and Kales, A. editors, (1968): A Manual of Standardized Terminology, Techniques and Scoring System for Sleep Stages of Human Subjects. Brain Information of Service/Brain Research Institute, University of California, Los Angeles.

32. Sassin, J.F., Parker,D.C., Mace, J.W., Gotlin, R.W., Johnson, L.C., and Rossman, L.G. (1969): Science, 165:513-515.

33. Siffre, M. (1965): Beyond Time. London, Chatto and Windus.

34. Siffre, M., Reinberg, A., Halberg, F., Chata, J., Perdriel, G., and Slind, R. (1966): La Presse Medicale, 74:915-920.

35. Takahashi, Y., Kipnis, D.M., and Daughaday, W.H. (1968): J. Clin. Invest., 47:2079-2090.

36. Timball, J., Colin, J., Boutelier, C., and Guieu, J.D. (1972): European J. Physiol., 335:97-108.

37. Webb, W.B., and Agnew, H.W. Jr. (1972): Phsychophysiology, 9:133.

38. Webb, W.B., and Agnew, H.W. Jr. (1974): Aerospace Med., 45: 617-622.

39. Webb, W.B., and Agnew, H.W. Jr. (1974): Aerospace Med., 45: 701-704.

40. Weitzman, E.D., Schaumburg, H. and Fishbein, W. (1966): J. Clin. Endocrin., 26: 121-127.

41. Weitzman, E.D., Fukushima, D., Nogeire, C. Roffwarg, H., Gallagher, T.F., and Hellman, L. (1971): J. Clin. Endocrinol., 33:14-22.

42. Weitzman, E.D., and Hellman, L. (1974): In: Biorhythms and Human Reproduction, edited by M. Ferin, T. Halberg, R. Richart and R. VandeWiele, pp. 371-395. Wiley, New York.
43. Wetizman, E.D., Boyar, R.M., Kapen, S. and Hellman, L. (1975): Recent Prog. Hor. Res., 31:399-446.
44. Weitzman, E.D. (1976): Annual Review of Medicine, 27: 225-243.
45. Weitzman, E.D, Czeisler, C.A., and Moore-Ede, M.D. (1979): In: Biological Rhythms and Their Central Mechanisms, edited by M. Suda, O. Hayaishi, and H. Nakagawa. Elsevier/North Holland Biomedical Press.
46. Weitzman, E.D., Czeisler, C.A., Zimmerman, J.C., and Ronda, J. (1980): Sleep (in press).
47. Wever, R. (1973): Internatl. J. Chronobiol., 1:371-390.
48. Wever, R. (1975): Internatl. J. Chronbiol., 3:19-55.
49. Wever, R. (1977): Proceedings XII Internatl. Conf. Inter- Natl. Soc. for Chronobiol., II Ponte, pp. 525-535. Milan, Italy.
50. Zimmerman, J.C., Czeisler, C.A. Laxminarayan, S., Knauer, R. and Weitzman, E.D. (1980) Sleep (in press).

Neurosecretion and Brain Peptides,
edited by J. B. Martin, S. Reichlin, and K. L. Bick.
Raven Press, New York © 1981.

Introduction to Section VIII:
Applicability of Studies of Pituitary Function in Neurological Disease

Seymour Reichlin

Endocrine Division, Department of Medicine, Tufts University School of Medicine, New England Medical Center, Boston, Massachusetts 02111

Pituitary hormone deficiency or excess frequently leads to disturbed brain function. This fact alone would justify efforts to evaluate pituitary secretion in patients with neurological and psychiatric disease. But recent neuroendocrine studies of such patients have not been prompted by this rationale. Rather, most of the current interest in pituitary function in patients with brain disease has been motivated either explicitly or implicitly by the assumption that the pituitary is a "window to the brain", and that the study of pituitary regulation may provide insight into the nature of abnormalities of central neurotransmitter regulation in specific neurological diseases.

Historically, this idea originated with the observation that dopamine concentration was low in the basal ganglia of patients with Parkinson's disease, that the administration of L-dopa, a precursor of brain dopamine, had therapeutic effects, and that the administration of L-dopa led to an increase in blood levels of growth hormone. These findings emphatically and forcefully drew to the attention of the scientific community that biogenic amines are important in the regulation of both motor function, and in the regulation of the pituitary. Soon thereafter detailed studies of central bioaminergic pathways revealed the wealth of amino-containing nerve endings in basal ganglia and in the hypothalamus and these findings reinforced the view that the detailed study of bioamine regulation of the pituitary might provide a clue to bioamine function of the brain generally. In the past decade extensive studies have confirmed that the tuberoinfundibular neuron system that constitutes the final common pathway of hypothalamic-pituitary control is, in turn, influenced by every known hypothalamic neurotransmitter, including all biogenic amines, acetylcholine and amino acids. Because pituitary secretions are relatively easy to measure by radioimmunoassay, and are readily perturbed by various pharmacological agents and physiological maneuvers, and because the neurotransmitter functions of the rest of the brain, by contrast, are relatively inaccessible, there has been continuing interest in the study of pituitary function as an indicator of central neurotransmitter regulation in brain disease.

Some observations provide support for this assumption, but other observations suggest that this approach to the elucidation of brain dysfunction is based on an excessively simplistic idea of hypothalamic-pituitary control. One of the purposes of the participants in this section of the conference was to evaluate critically the neuroendocrine approach to understanding neurological disease. Two hormonal systems are emphasized in the first two chapters, those regulating prolactin secretion and the other growth hormone secretion. These hormones were chosen for evaluation because of the large volume of investigation that has already been accomplished and because more is known about the neuropharmacological basis of regulation of these hormones than of the other pituitary hormones. The greatest emphasis in these reports was given to studies of Parkinson's disease and of Huntington's disease. Parkinson's disease is important as a model because the deficits have been so clearly localized to dopaminergic fibers. Huntington's disease is also important as a possible neuroendocrine model because the disease is accentuated by dopamine agonists and deficits in other neurontransmitters have been detected in autopsy specimens.

The third presentation in this section deals with newer insights into the chemistry and regulation of pituitary ACTH secretion of ACTH-related peptides. ACTH secretion by the pituitary has been well recognized for many years and is responsible for activation of the adrenal cortex in stress. More recently ACTH has been shown to be secreted by a certain population of brain cells and to have a direct effect on behavior in experimental animals. Further, ACTH is formed biosynthetically as part of a much larger precursor molecule that includes the potent opiate receptor binding peptide β-endorphin. Therefore, understanding of the nature of blood ACTH and ACTH-related peptides has bearing on several important aspects of brain function.

The last paper in this section deals with the role of neuropeptides in pituitary regulation. This topic was chosen because the wide variety and extensive distribution of neuropeptides in all parts of the central nervous system is now recognized to be an important component of brain regulation and because the neuropeptides have recently been shown to influence pituitary secretion. It was hoped that through a consideration of these issues, the pituitary "window" might be opened more widely to include a prophetic glimpse of central neuropeptide regulation.

Neurosecretion and Brain Peptides,
edited by J. B. Martin, S. Reichlin, and K. L. Bick.
Raven Press, New York © 1981.

Prolactin Secretion as an Index of Brain Dopaminergic Function

Michael O. Thorner and Ivan S. Login

*Departments of Internal Medicine and Neurology, University of Virginia School of Medicine,
Charlottesville, Virginia 22908*

Why should we consider prolactin to be a potential marker of brain dopaminergic function? The predominant hypothalamic influence on the release of prolactin from the anterior pituitary is mediated largely by dopaminergic inhibition (49,71). Dopamine is synthesized by tuberoinfundibular dopaminergic neurons (TIDA) with cell bodies in the arcuate and periventricular nuclei of the medio-basal hypothalamus and terminals ending on the hypophyseal portal system at the median eminence. These neurons secrete dopamine into the portal capillaries which transport it to cells of the anterior pituitary; dopamine binds to plasma membrane dopamine receptors on the lactotrope (14), activating as yet undefined mechanisms(s), to tonically inhibit prolactin release. Thus, serum prolactin levels inversely reflect both the tone of this dopamine pathway and the sensitivity of the dopamine receptor at the lactotrope. Prolactin increases as the effective dopamine concentration at the lactotrope is reduced: e.g., by dopamine receptor antagonists, such as butyrophenones or phenothiazines; catecholamine depletion (e.g.,after reserpine); secretion of false neurotransmitters (e.g., α methyl dopamine) or by physical disruption of the dopamine delivery system (e.g., hypothalamic disease, pituitary stalk section or pituitary tumors). Estrogens and thyrotropin releasing hormone each stimulate prolactin release (49,71). Very recently vasoactive intestinal peptide has also been shown to directly stimulate prolactin release (31,43). Prolactin secretion is suppressed in the presence of an enhanced dopaminergic tone, as seen during therapy with dopamine agonists, e.g., the treatment of hyperprolactinemia (72).

The TIDA is one of several distinct dopaminergic pathways in the central nervous system (29,59,74). Others include: (1) the nigrostriatal pathway with cell bodies in the pars compacta of the substantia nigra (A9) and axons which terminate in the caudate and putamen of the corpus striatum; (2) the mesolimbic pathway, which arises from the central tegmental area in the region of the interpeduncular nucleus (A10) and innervates the limbic forebrain structures including nucleus accumbens, paraolfactory, prefrontal and septal areas (29,37); (3) the mesocortical ascending tract which originates in the same mesencephalic region (A10) and innervates frontal, cingulate and entorhinal cortex.

Are disorders in any one of these four pathways reflected in the others since they all utilize dopamine as their neurotransmitter? Thus, if this were the case, serum prolactin should be a marker for disease located in diverse brain regions represented by striate, limbic, and cortical areas. The serum prolactin appears to be a simple and accurate measurement of TIDA tone and lactotrope DA receptor function. This contrasts with the great difficulty in biochemically characterizing the dopaminergic activity or the status of the post-synaptic dopamine receptor in vivo in the nigrostriatal, mesolimbic or mesocortical tracts. One method of assessing this is by measurement of catecholamine metabolites in the cerebrospinal fluid. Even these measurements, however, do not reflect overall CNS dopamine activity and do require invasive techniques.

These four tracts have in common their neurotransmitter but there are a multitude of differences among them. First, within the nigrostriatal pathway, for example, at least five distinct dopamine receptors exist as described by their pre- or post-synaptic localization and the presence or absence of a dopamine sensitive adenylate cyclase system (41). By comparison the lactotrope represents a post-synaptic receptor unlinked to adenylate cyclase (41,51). Second, the striatal pathway is approximately 100-fold and 10-fold as sensitive to activation of dopamine synthesis after acute administration of a dopamine receptor blocking drug, thioproperazine, than the mesocortical and mesolimbic dopaminergic tracts, respectively (39). Third, chronic neuroleptic treatment leads to tolerance in the nigrostriatal pathway with respect to the increased dopamine synthesis seen acutely. Tolerance, however, is not observed after chronic neuroleptic treatment in the mesolimbic and mesocortical (39) or hypothalamic TIDA tracts (9,75). Fourth, in contrast to classical neuronal feedback mechanisms in the striatal or limbic tracts, the dopamine turnover in the TIDA pathway is modulated by prolactin itself as a shortloop negative feedback system (3,50,65) (Fig. 1). And finally, we must consider the potential influence of other transmitters, e.g., serotonin, acetylcholine and glycine (15) and peptides (5) on regional dopamine function in each of these dopaminergic pathways.

Parkinson's disease, Huntington's disease, schizophrenia, tardive dyskinesia and manic depressive illness are among the most widely accepted diseases in which dysfunction of dopamineric transmission is considered to be significant, if not etiologic (59); however, a much larger tabulation of disorders in which dopamine agonist or antagonist drugs appear clinically effective can be constructed (8).

Despite the many dissimilarities in the four dopaminergic tracts, several investigators have indeed pursued the possibility that alterations in prolactin regulation may predict or reflect

CONTROL OF PROLACTIN SECRETION

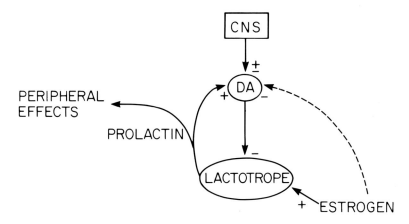

FIG. 1. Schematic representation of the inter-relationships between the lactotrope, dopamine (DA) and estrogen in the control of prolactin secretion.

the activity of several of the above mentioned diseases which are considered to be associated with alteration of dopaminergic receptors or transmission.

THE RELATIONSHIP OF DISORDERS OF PROLACTIN SECRETION TO NEUROLOGIC DISEASE

Parkinson's Disease

A fundamental abnormality in Parkinson's disease is a deficiency of dopamine in the nigrostriatal pathway (15,37). If this disorder extends to involve the TIDA tract, the basal serum prolactin level would be expected to be elevated. Pathologic changes have commonly been observed in hypothalamic tissue examined postmortem in patients with Parkinson's disease, although the tuberoinfundibular tract is usually relatively spared (45).

The majority of reports in the literature on the serum prolactin levels in patients with Parkinson's disease refer to patients who had already been treated with various drugs for their extrapyramidal symptoms. Many of these drugs may have directly affected prolactin secretion. The neuroendocrine investigation of previously untreated patients would offer the optimum study conditions from which to draw conclusions on TIDA activity in Parkinson's disease. Together with Drs. D.B. Calne and T. Eisler of the National Institutes of Health and Dr. R.M. MacLeod at the University of Virginia, we evaluated six men and six women with untreated Parkinson's disease and a normal control group

(24).   The basal serum prolactin was measured over three hours
after which the patients received thyrotropin releasing hormone
(TRH) 500 μg i.v. (a potent direct pituitary stimulus for prolactin release).   The basal prolactin levels in these patients and
their prolactin response to TRH were the same as those in the
age-matched control group.   When patients and normal controls received a single oral 2.5 mg dose of bromocriptine (a dopamine
agonist), the expected prolactin suppression was observed and in
both groups the subsequent response to TRH was markedly inhibited
(Fig. 2).

Prolactin secretion normally increases in episodic bursts
during and directly related to sleep (53).   In another study in
which prolactin levels were measured in untreated patients with
Parkinson's disease, Barreca et al. (7) found that during natural
sleep the total prolactin secretion was significantly reduced
compared to an appropriate control group despite the pattern of
prolactin secretion and several sleep parameters, as defined by
electroencephalography, being equal between the two groups.

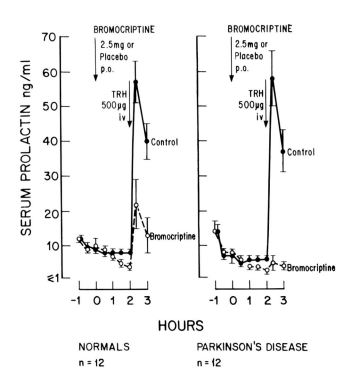

FIG. 2. Serum prolactin levels in 12 age and sex matched control
subjects and 12 patients with Parkinson's disease after 2.5 mg
oral bromocriptine or placebo with administration of TRH at 2
hours.(From ref. 24, with permission.)

From these studies it appears that the basal prolactin level is not elevated in untreated Parkinson's disease. Furthermore, the lactotrope is sensitive to inhibition by bromocriptine and stimulation by TRH. The administration of TRH allows the inhibitor effect of bromocriptine to be more easily documented, since the stimulation of prolactin release by TRH amplifies the effect of bromocriptine.

In contrast to the limited studies of nontreated patients, many reports of prolactin responsiveness are available in previously treated patients. In Parkinsonian patients being treated with various dopamine agonist drugs the basal serum prolactin level was found to be within normal limits by Ruggieri et al. (67) and Malarkey et al. (52). Among seven patients with Parkinson's disease chronically treated with dopamine agonists, basal serum prolactin was normal in four patients who had shown a good clinical response to treatment; however, it was elevated in three patients whose clinical symptoms had not responded to similar drug therapy (1). In another study, Parkinsonian patients demonstrated uniform suppression of prolactin after each of several daily doses of oral levo-dopa (52). If prolactin is being inhibited by dopamine released from the TIDA tract, the administration of a dopamine antagonist should stimulate prolactin release and thereby reflect the degree of endogenous dopaminergic inhibition. Fifteen chronically treated patients demonstrated both the expected rise in prolactin after the acute administration of the dopamine antagonist, haloperidol, and the expected inhibition of prolactin release after a single oral dose of bromocriptine 2.5 mg (2). The three patients with a poor clinical response of Parkinsonian symptoms to dopamine agonist therapy who had slightly elevated basal prolactin levels demonstrated only partial prolactin stimulation after acute haloperidol treatment and only partial suppression after apomorphine and bromocriptine (1). The serum prolactin responses to these three drugs were normal in the four patients whose clinical symptoms were well controlled with dopamine agonist therapy. Six other patients had a normal suppression of prolactin after chronic bromocriptine administration, but surprisingly had no response to a single 25 mg dose of this drug although prolactin levels were followed for four hours after administration of the drug (40).

Two recent preliminary reports are of theoretical interest. In one study domperidone (Janssen Pharmaceuticals), which acts as a dopamine antagonist, was shown not only to stimulate prolactin release but also to significantly reduce the clinical signs of tremor, rigidity, and bradykinesia in fourteen patients with chronic Parkinson's disease (2). The authors suggested that the primary effect of the drug was to induce hyperprolactinemia. They suggested that the prolactin then increased dopamine turnover, or release in the nigrostriatal pathway (64) which secondarily improved the clinical symptoms. As domperidone is considered not to cross the blood brain barrier, it is difficult to postulate any direct action of the drug in the striatum to ameliorate the clinical signs. However, the ability of prolactin

to influence the dopamine turnover in any tract other than the
TIDA is controversial, and a clear explanation of the observations made in these patients is lacking (34,65).

Secondly, Ruggieri et al. (67) studied the effects of Madopar
(L-dopa plus benserazide, a peripheral decarboxylase inhibitor,
Hoffman-LaRoche). They found that this fixed dose combination
drug effectively reduced Parkinsonian symptoms while stimulating
prolactin release. Benserazide given to normal subjects was
found to stimulate prolactin release itself. There are several
explanations for these apparently paradoxical results, but none
established with certainty.

Patients with neuroleptic induced Parkinsonism and amenorrhea,
who had elevated basal prolactin levels, demonstrated an adequate
and appropriate hormonal response to administration of three
dopamine agonists, including apomorphine, bromocriptine, and a
combination of L-dopa and a decarboxylase inhibitor (1).

From these observations it appears that in most patients with
Parkinson's disease the TIDA tract is functionally normal and the
dopamine receptors at the lactotrope are intact and respond normally. Thus the basal prolactin levels and the response of the
lactotrope to dopamine agonists, antagonists, and TRH are normal.
In the few patients with Parkinson's disease whose clinical symptoms respond poorly to dopamine agonist therapy, and in whom
prolactin levels are abnormal in the basal state and following
dynamic testing, there may indeed be a correlation between
altered prolactin regulation and nigrostriatal dopamine
deficiency. Possibly these patients may comprise a clinical
subgroup who demonstrate Parkinsonism (not Parkinson's disease)
as a manifestation of other neurodegenerative disorders such as
striatonigral degeneration or the Shy-Drager syndrome for which
diagnosis may require either prolonged observation or
pathological verification.

Huntington's Disease

Huntington's disease is an autosomal dominant neurodegenerative disorder whose clinical manifestations of chorea and
dementia typically appear in adult life with inexorable progression (63). Currently, an impression of functional dopaminergic
hyperactivity is held, based solely on the observation of clinical improvement of the chorea with drugs which block or deplete
dopaminergic activity, and exacerbation after administration of
dopamine agonist agents. In Huntington's disease neurochemical
abnormalities have been found for numerous compounds including
trace metals and peroxidases, amino acids, proteins and neurotransmitters (6,38). Within the latter group are deficiencies in
the striatum of the enzymes that synthesize GABA and acetylcholine, a reduction in receptor number for GABA, dopamine and
serotonin, and reduced Substance P and angiotension converting
enzyme (38). Dopamine levels in the basal ganglia have been
reported to be normal, increased or decreased and CSF homovanillic acid is probably normal (6,38).

This vast array of abnormal biochemical findings has led to speculation that the etiological basis of Huntington's disease is not primarily a defect of a single neurotransmitter, but rather a cellular membrane defect (6). The impression in Huntington's disease of "excessive dopaminergic tone" may theoretically reflect any or all of the following: (1) post-synaptic dopamine receptor hypersensitivity; (2) increased dopamine turnover and/or release; (3) a reduction in dopamine catabolism, (4) the influence of alteration in the complex interrelationships among several of these compounds within the striatum leading to functional secondary dopaminergic imbalance (37,38).

The characteristic neuropathologic changes of Huntington's disease including depletion of small diameter interneurons and gliosis are present in the hypothalamus (12). If TIDA activity were influenced in a manner similar to that in the basal ganglia, causing at least a functional dopaminergic excess, serum prolactin would be expected to be lowered in the basal state. However, in untreated patients with Huntington's disease, basal prolactin levels have been reported to be normal (19,61) or increased (16). Two reports have shown a significantly lower basal prolactin concentration in patients with Huntington's disease in whom dopamine antagonist medications were discontinued at least two weeks earlier (36,62). These two studies do not stand unchallenged, however, since basal prolactin levels have also been reported to be both normal (17,19,61) or elevated (13, 16,26) in patients who were similarly treated with dopamine antagonists and recently withdrawn.

Single dose trials of dopamine receptor agonist drugs (bromocriptine, L-dopa or combined L-dopa/decarboxylase inhibitor) have either failed to adequately lower prolactin (13,16,26) or have shown the expected inhibition (17,19,46,61). Kartzinel et al. (40) reported a poor response to acute, but a good response to chronic, dopamine agonist administration in the same patient group.

Pharmacologic manipulation to elevate prolactin in Huntington's disease patients by direct lactotrope stimulation (TRH) or dopamine receptor blockade (phenothiazines) has demonstrated both the expected prolactin increase (61) as well as a poor or blunted response (13,36).

Two neuroendocrine studies directed towards less well defined aspects of prolactin regulation in Huntington's disease have also been reported: prolactin levels in cerebrospinal fluid from such patients were normal (62); and dipropylacetic acid (valproic acid) which increases brain GABA levels by an uncertain mechanism failed to lower the serum prolactin (26).

In this genetically transmitted illness, a disease marker in asymptomatic persons at risk would be of great importance. Hayden et al. (36) studied 23 potentially affected first generation relatives of symptomatic patients. The basal prolactin was normal in all, as was the response of prolactin to TRH in 18 of the group. Five relatives showed blunted responses to TRH. Eleven had normal stimulation of prolactin after chloropromazine, but 12 had abnormal responses – 7 had blunted and 5 exaggerated responses compared to controls. No follow-up on these patients has been given and it is, therefore, unknown whether measurement of serum prolactin levels could be a useful predictive test (36).

No firm conclusion can be drawn about TIDA function in Huntington's disease. Only two studies support the concept of excessive TIDA dopaminergic activity in a few patients (36,62). The reports of elevated basal prolactin levels and hyporesponsiveness to DA agonists suggests hyposensitivity of lactotrope dopamine receptors (13,16,26), which is contrary to the clinically apparent situation in the basal ganglia. Friedhoff (27) has suggested that the sensitivity of receptors can be modified by the intensity or the duration of the presence of the agonist.

Theoretically, early in the clinical presentation of Huntington's disease dopaminergic excess could yield data comparable to that of Hayden et al. (36), while in the chronic state desensitization of the lactotrope dopamine receptor may occur as a result of prolonged excessive dopamine stimulation giving rise to the other conflicting data. The majority of studies show some dysfunction of prolactin secretion; their conflicting nature hampers interpretation. Up to the present time no consistent abnormality of prolactin secretion has been demonstrated in this disease.

Schizophrenia

The treatment of schizophrenia has been revolutionized with the introduction of neuroleptic agents, which have enabled many previously institutionalized patients to be rehabilitated into the general community. The discovery that these compounds are potent dopamine antagonists has led to the hypothesis that schizophrenia is associated with functional increased central dopaminergic activity (18,32,33). The mesolimbic and mesocortical dopaminergic tracts, at least if not the site of the etiologic neurochemical abnormality, are the possible targets for antipsychotic actions of neuroleptic drugs used in the treatment of the condition (18). Prolactin secretion has been studied in this condition, in the hope of detecting an abnormality in the TIDA tract as a marker of the disease. However, in untreated schizophrenia, basal prolactin levels have been found to be normal, although the possibility of prolactin levels being oversuppressed has not been excluded.

The mechanism of the antipsychotic effect of neuroleptic drugs is still poorly understood. It is unlikely that it is due directly or solely to dopamine antagonism since stimulation of prolactin secretion occurs in minutes to hours, while antipsychotic effects are often delayed for 6-8 weeks even though changes in dopamine turnover in the mesolimbic and mesocortical tracts occur within miniutes to hours after administration of neuroleptic drugs to rats (39). In one study, prolactin was elevated in all 27 schizophrenic patients within 72 hours of the initiation of neuroleptic therapy, while the psychiatric response was delayed for many weeks (56). Furthermore, following withdrawal of phenothiazines from 8 patients after 2-3 months of treatment, prolactin levels were back to normal within 2-3 days without deterioration in the psychiatric state in all but one patient (56).

Antipsychotic drugs generally possess anti-dopaminergic properties which are responsible for elevation of prolactin levels (18,33). The observation that clinically equipotent antipsychotic doses of different medications can have variable stimulatory effects on prolactin secretion emphasizes the differences between the dopaminergic pathways, e.g., clozapine is a potent antipsychotic drug, is devoid of a stimulatory effect on prolactin secretion and does not lead to tardive dyskinesia (33,68). Further, some compounds that are very potent in elevating prolactin have only poor antipsychotic or extrapyramidal effects, e.g., metoclopramide (58,68).

The development of tardive dyskinesia with chronic neuroleptic therapy is a serious complication reflecting nigrostriatal dopaminergic dysfunction, which is considered to be associated with dopamine receptor supersensitivity. However, the TIDA lactotrope axis in man does not demonstrate post-synaptic denervation hypersensitivity after chronic neuroleptic treatment (25,70).

The results of neuroendocrine studies in schizophrenic patients argue against the TIDA being involved in a more widespread disorder of dopaminergic transmission in the disease. In schizophrenic patients who were untreated prolactin levels were normal (56,57,68). Similarly, in those patients who were chronically treated but recently withdrawn, whether or not tardive dyskinesia was present or absent, the basal prolactin levels were normal (25,33,66,70). Schizophrenic patients treated with DA agonist drugs demonstrated a reduction in prolactin which was normal (33,66) or blunted compared to controls (70). In addition to confirming the presence of prolactin in human cerebrospinal fluid (48), Sedvall has further shown normal CSF prolactin in untreated schizophrenic patients (68).

Even if the theory of mesolimbic/mesocortical dopamine hyperactivity in schizophrenia is correct, this does not appear to be accompanied by dysfunction of the TIDA pathway.

Many neurological disorders are classified by the systems which are predominantly affected clinically, e.g., motor system disease, spinocerebellar degenerations, combined systems disease. Elucidation of a common cause for the involvement of these multiple tracts or systems in any particular disease would be a major advance in neurology. As we have shown, there are four anatomically distinct pathways in the brain that share a special property, each subserving dopaminergic neurotransmission. Having then identified this common biochemical function of the TIDA, nigrostriatal, mesocortical and mesolimbic systems, we speculated that if a clinical disorder was apparent in one tract, a comparable abnormality might be detected in another. The ease with which the TIDA and pituitary dopamine receptor sensitivity can be biochemically and functionally assayed and quantitated in vivo by measurement of the serum prolactin has led to great interest in whether this hormone could be a monitor or reflection of disease in the other three dopaminergic systems. As we have shown, with few exceptions, it appears this is not the case. However, prolactin is an excellent marker for neurologic disease within the hypothalamic/pituitary axis and the remainder of this paper will deal with this topic.

Prolactin Secretion and Hypothalamic/Pituitary Disorders

The clinical features of hyperprolactinemia have been extensively reviewed (28,72) and will only be briefly outlined. Women usually have gonadal dysfunction - either with infertility or menstrual disorders - amenorrhea, oligomenorrhea or, more rarely, polymenorrhea. Galactorrhea is present in 30-80% of hyperprolactinemic patients and may be the presenting complaint; however, in women this symptom is often seen with normal prolactin levels, limiting its specificity as a marker of disordered prolactin secretion. In contrast to women, men rarely complain of symptoms of gonadal dysfunction, but instead present with symptoms due to expansion of a large pituitary tumor, e.g., headaches and visual field defects. However, in retrospect, these men often have impotence and loss of libido, although only rarely have they been infertile.

In individual patients with hyperprolactinemia it may be difficult to determine the cause of the condition. Since prolactin secretion from the anterior pituitary is under tonic hypothalamic inhibition by dopamine from the TIDA neurons, any disease of the hypothalamus or any disruption of the hypothalamic hypophyseal portal circulation may lead to hyperprolactinemia. Thus granulomatous infiltraton (as in eosinophilic granuloma and sarcoidosis), primary CNS tumors (craniopharyngioma, glioma, pinealoma) and metastatic disease involving the hypothalamic region have been associated with hyperprolactinemia as has stalk section, both surgical and traumatic. In the pituitary, the majority of tumors are prolactin secreting and even those patients who have growth hormone or ACTH secreting tumors are not infrequently hyperprolactinemic as well. In acromegaly and Cushing's disease, abnormal secretion of prolactin may be due to one tumor also secreting prolactin, or to the remaining natural pituitary losing its inhibitory hypothalamic input secondary to intrapituitary vascular disruption.

As described earlier, drugs that deplete catecholamines, disrupt catecholamine synthesis or block dopamine receptors commonly cause hyperprolactinemia. Further, estrogens may increase prolactin secretion in a dose related manner, but the low dose in oral contraceptives has no effect on prolactin secretion in normal women. Hyperprolactinemia is also seen in some patients with primary hypothyroidism and with renal failure. Drug ingestion can be excluded by history and hypothyroidism and renal insufficiency should be excluded by appropriate investigations. The remaining differential diagnosis includes pituitary tumor or idiopathic hyperprolactinemia.

A patient with a large pituitary tumor with local symptoms from the expanding lesion of headaches and visual field defects, presents no diagnostic problem. However, many patients, particularly young women, present with symptoms of their endocrine disturbance rather than local neurological symptoms. It is these patients who present the greatest problem, since the neuroradiologic appearances of the pituitary fossa may be normal or only demonstrate equivocal abnormalities which, in euprolactinemic subjects, would be considered to be normal variants (4,11,23). The single most useful index of the presence of a tumor is the basal prolactin level (28,42,72); if this is greater than 200 ng/ml it is likely that the patient harbors a pituitary tumor, even if it is not radiologically evident; patients with stalk transection or hypothalamic disease usually have prolactin levels of less than 100 ng/ml.

Many patients with hyperprolactinemia, however, have no evidence of hypothalamic disease, have a normal radiologic appearance of the pituitary fossa, and a serum prolactin level of less than 200 ng/ml. Do these patients harbor a microadenoma, or do they have a functional hypothalamic/pituitary lesion? Dr. Tyson and colleagues (78) believe that psychogenic stress may cause hyperprolactinemia and its accompanying gonadal dysfunction; when the underlying conflict is resolved, prolactin levels fall to normal with resolution of gonadal abnormality. This hypothesis is yet to be proven. The infuence of environmental stress can be observed in the unique neuroendocrine status of the talipoin monkeys (10). These monkeys live in colonies. Of the females, one is dominant and she has normal prolactin levels and normal positive feedback of estrogen on luteinizing hormone secretion. The remaining females are subordinates and are hyperprolactinemic with impaired positive feedback. Both positive feedback and normal prolactin levels can be restored in subordinate female monkeys either by a change of their role to become dominant or by administration of bromocriptine which lowers the prolactin levels (10). Hyperprolactinemia in this situation may be an adaptive behavior to the environment! In the human, it is difficult to determine whether the apparent distress arises from the hyperprolactinemic state and its consequences or vice versa. If no underlying cause for hyperprolactinemia can be identified, we suggest such patients be considered idiopathic, although we recognize some may harbor a small pituitary tumor and others may have a functional illness which will resolve spontaneously.

The management of hyperprolactinemia is beyond the scope of this paper. The subject is controversial and has been extensively reviewed (28,72) and three modalities alone or in combination are available; radiotherapy, surgery, and medical therapy. We would like to highlight some recent advances with respect to medical therapy. Irrespective of the cause of hyperprolactinemia, prolactin levels can be lowered, usually to normal, by medical therapy with dopamine agonists of which the widest experience has been had with bromocriptine. Not only are prolactin levels

lowered, but normal gonadal function is restored in the vast
majority of these patients, even if a large pituitary tumor is
present. The accompanying hypogonadism appears to be a functional abnormality that occurs as a result of the hyperprolactinemia.

There is a growing impression among many endocrinologists
cited below that medical therapy of large prolactin secreting
tumors may not only suppress prolactin secretion and restore
gonadal function, but also lead to reduction of pituitary tumor
size in the majority of these patients. Although the specific
mechanism for tumor size reduction is unclear, several relevant
observations have been made. In man, as well as in animals,
estrogen leads to an increase in mitosis of pituitary cells
(probably lactotrope) in the in situ gland (22). This effect of
estrogen can be inhibited by bromocriptine (22,47). Bromocriptine also inhibits DNA synthesis in certain pituitary cells (22).

Clinical evidence of reduction in the size of large prolactin
secreting tumors by administration of dopamine agonists, particularly bromocriptine, has recently been presented. In 24 patients
changes in either the contour of the pituitary fossa on skull
x-rays or evidence of reduction in tumor size based on pneumoencephalography, metrizamide cisternography and CT scans have been

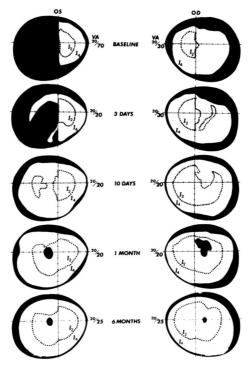

FIG. 3. Visual activity and visual field plots before and during
6 months' bromocriptine therapy in patient 1 with macroadenoma.
(From ref. 73, with permission.)

observed (20,21,30,44,54,55,60,69,76,77). Other patients have
shown bromocriptine induced improvement in visual field defects
due to chiasmatic compression. Documentation of tumor regression
has been observed over a period in excess of three months. In
our experience, tumor regression may occur as early as two weeks;
we have treated two young men (ages 24 and 25 years) with large
prolactinomas (3900 and 2350 ng/ml respectively) with bromocriptine, 7.5 mg per day the usual dose used in the treatment of
hyperprolactinemia (73). One patient had a large suprasellar extension causing bitemporal hemianopsia, while the second patient
had only a small suprasellar extension. The visual fields in the
first patient before and during six months bromocriptine treatment demonstrate the dramatic improvement commencing within days
of starting therapy (Fig. 3). The CT scans in this patient before and after two weeks of therapy demonstrate reduction in
tumor size at two weeks (Fig. 4).

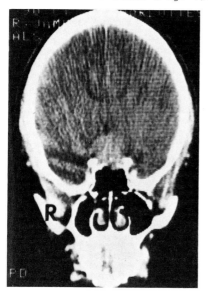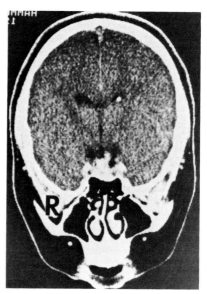

FIG. 4. Coronal CT head scan before (left) and after 2 weeks on
bromocriptine (right) showing regression of suprasellar extension
in patient 1. (From ref. 73, with permission.)

At six months metrizamide cisternography revealed a partially
empty fossa. In the second patient metrizamide cisternography
was performed before and after six weeks on bromocriptine. During
this short interval he had developed a partially empty fossa
showing reduction in volume of the pituitary tumor. In both
patients prolactin fell by 85% and their pubertal development,
which had been arrested, recommenced with improvement in sexual
function, rise in testosterone levels and increased sperm counts
(73).

The successful results of surgical intervention or medical therapy with bromocriptine are comparable in patients with well defined microadenomata; however, the results of surgery are very poor in patients with macroadenomas. Of patients with macroadenomas whose serum prolactin levels are greater than 500 ng/ml, only 19% are cured by surgery alone (35).

Thus, if medical therapy not only lowers prolactin levels but also reduces tumor size and preserves or restores other anterior pituitary functions, it may become the treatment of choice for large prolactin secreting macroadenomas as well as for microadenomas and idiopathic hyperprolactinemia. Demonstration of reduction in tumor size at two and six weeks in our two patients makes a short therapeutic trial of bromocriptine under close supervision in the hospital a viable therapy even in patients with visual field defects. This may possibly enable subsequent surgery, if indicated, to be more successful. At the present time it is not clear how often prolactin secreting macroadenomas will regress with bromocriptine therapy. However, in 16 patients treated in Newcastle upon Tyne (England), all 16 patients with large prolactinomas demonstrated reduction in tumor size with bromocriptine therapy at three months (Dr. A.M. Macgregor, personal communication).

Conclusions

The predominant inhibitory hypothalamic control of prolactin secretion is mediated by dopamine released from the TIDA neurons. The TIDA tract controlling prolactin release is unique since dopamine turnover in this pathway is controlled by the circulating prolactin levels rather than by feedback through dopamine autoreceptors.

Hyperprolactinemia is caused either by a pituitary tumor or disease of the hypothalamus. In some patients a definitive etiology cannot be identified and a diagnosis of idiopathic hyperprolactinemia is made, although some of these patients presumably harbor an occult microadenoma. In most patients medical therapy with dopamine agonists results in lowering of prolactin levels, restoration of impaired gonadal function and, in those patients with a large pituitary macroadenoma, reduction in tumor size. This latter observation may alter the future approach to the therapy of large prolactin secreting tumors, making medical therapy the initial treatment which it already is in the management of patients with either microadenomas or idiopathic hyperprolactinemia.

We believe that presently the bulk of evidence refutes the hypothesis that disorders of the nigrostriatal, mesolimbic, and mesocortical dopamine tracts are associated with a functional abnormality of the TIDA pathway or prolactin secretion.

REFERENCES

1. Agnoli, A., Ruggieri, S., Baldassare, M., Forchetti, C., Cerone, G., Falaschi, P., Fraiese, G., Rocco, A., and D'Urso, R. (1979): In: Neuroendocrinology: Biological and Clinical Aspects, edited by A. Polleri, and R.M. MacLeod, pp. 329-343. Academic Press, London.
2. Agnoli, A., Ruggieri, S., Falaschi, P., Baldassarre, M., Pompei, P., Del Roscio, M., Rocco, A., and Frajese, G. (1980): In: Central and Peripheral Regulation of Prolactin Function, edited by R.M. MacLeod and U. Scapagnini. Raven Press, New York. (In press).
3. Annunziato, L. (1979): Neuroendocrinology, 29:66-76.
4. Banna, M., Nicholas, W., and McLachlan, M. (1978): Neuroradiology, 16:440-442.
5. Barbeau, A. (1979): Adv. Exp. Med. Biol., 113:101-110.
6. Barbeau, A. (1979): In: Advances in Neurology., Vol. 23, edited by T.N. Chase, N.S. Wexler, and A. Barbeau, pp. 449-461. Raven Press, New York.
7. Barreca, T., Murri, L., Lamorgese, In: N. Murialdo, G., Masturzo, P., and Muratorio, A. (1979): In: Neuroendocrinology: Biological and Clinical Aspects, edited by A. Polleri and R.M. MacLeod, pp. 321-327. Academic Press, London.
8. Bianchine, J., Shaw, G., Greenwald, J., and Dandalides, S. (1978): Fed Proc., 37:2434-2439.
9. Bowers, M.B., and Rozitis, A. (1974): J. Pharm. Pharmacol., 26:743-745.
10. Bowman, L.A., Dilley, S.R., and Keverne, E.B. (1978): Nature, 275:56-58.
11. Bruneton, J.N., Drouillard, J.P., Sabatier, J.C., Elie, G.P., and Tavernier, J.F. (1979): Radiology, 131:99-104.
12. Bruyn, G.W. (1968): In: Handbook of Clinical Neurology, Vol. 6, edited by P.J. Vinken, and G.W. Bruyn, pp. 298-378. North-Holland, Amsterdam.
13. Caine, E., Kartzinel, R., Ebert, M., and Carter, A.C. (1978): Life Sci., 22:911-918.
14. Calabro, M.A., and MacLeod, R.M. (1978): Neuroendocrinology, 25:32-46.
15. Calne, D.B. (1977): Ann. Neurol., 1:111-119.
16. Caraceni, T., Panerai, A.E., Parati, E.A. Cocchi, D., and Muller, E.E. (1977): J. Clin. Endocrinol. Metab., 44:870-875.
17. Caraceni, T.A., Parati, E.A., and Cocchi, D., Mainini, P., and Muller, E.E. (1979): In: Neuroendocrine Correlates in Neurology and Psychiatry, Developments in Neurology, Vol. 2, edited by E.E. Muller, and A. Agnoli, pp. 167-168. Elsevier/ North-Holland, Amsterdam.
18. Carlsson, A. (1978): Am. J. Psychiatry, 135:164-173.
19. Chalmers, R.J., Johnson, R.H., Keogh, H.J., and Nanda, R.W. (1978): J. Neurol. Neurosurg. Psychiatry, 41:135-139.

20. Chiodini, P.G., Liuzzi, A., and Verde, G. (1978): International Symposium on Pituitary Microadenomas, Abstract 64, Milan.
21. Corenblum, B. (1978): Lancet, 2:786.
22. Davies, C., Jacobi, J., Lloyd, H.M., and Meares, J.D. (1974): J. Endocrinol. , 61:411-417.
23. Dubois, P.J., Orr, D.P., Hoy, R.J., Herbert, D.L., and Heinz, E.R. (1979): Radiology, 131:105-110.
24. Eisler, T., Calne, D.B., MacLeod, R.M., and Thorner, M.O.: (in preparation).
25. Ettigi, P., Nair, N.P.V., Lal, S., Cervantes, P., and Guyda, H. (1976): J. Neurol. Neurosurg. Psychiatry, 39:870-876.
26. Frattola, L., Albizzati, M.G., Bassi, S., and Trabucchi, M. (1979): In: Neuroendocrine Correlates in Neurology and Psychiatry, Developments in Neurology, Vol. 2, edited by E.E. Muller, and A. Agnoli, pp. 159-165. Elsevier/North-Holland, Amsterdam.
27. Friedhoff, A.J. (1977): Comp. Psychiatry, 18:309-317.
28. Friesen, H.G. (1978): Ann. R. Coll. Physicians Surg. Can., 11:275-281.
29. Fuxe, K., Andersson, K., Schwarcz, R., Agnati, L.F., Perez de la Mora, M., Hokfelt, T., Goldstein, M., Ferland, L., Possani, L., and Tapia, R. (1979): In: The Extrapyramidal System and its Disorders, Advances in Neurology, Vol. 24, edited by L.J. Poirier, T.L. Sourkes, and P.J. Bedard, pp. 199-215. Raven Press, New York.
30. George, S.R., Burrow, G.N., Zinman, B., and Ezrin, C. (1979): Am. J. Med., 66:697-702.
31. Gourdji, G., Bataille, D., Vauclin, N., Grouselle, D., Rosselin, G., and Tixier-Vidal, A. (1979): FEBS Lett., 104: 165-168.
32. Gruen, P.H. (1978): Med. Clin. North Am., 62:409-424.
33. Gruen, P.H., Sachar, E.J., Langer, G., Altman, N., Leifer, M., Frantz, A., and Halpern, F.S. (1978): Arch. Gen. Psych., 35:108-116.
34. Gudelsky, G.A., Simpkins, J., Mueller, G.P., Meites, J., and Morre, K. E. (1976): Neuroendocrinology, 22:206-215.
35. Hardy, J., Beauregard, H., and Robert, F. (1978): In: Progress in Prolactin Physiology and Pathology, edited by C. Robyn and M. Harter, pp. 361-370. Elsevier/North Holland Biomedical Press, New York.
36. Hayden, M.R., Paul, M., Vinik, A.I., and Beighton, P. (1977): Lancet, 2:423-426.
37. Hornykiewicz, O. (1977): Adv. Exp. Med. Biol., 90:1-20.
38. Hornykiewicz, O. (1979): In: Advances in Neurology, Vol. 23, edited by T.N. Chase, N.S. Wexler, and A. Barbeau, pp. 679-686. Raven Press, New York.
39. Julou, L., Scatton, B., and Glowinski, J. (1977): Adv. Biochem. Psychopharmacol., 16:617-624.
40. Kartzinel, R., Perlow, M.D., Carter, A.C., Chase, T.N., and Calne, D.B. (1976): Trans. Am. Neurol. Assoc., 101:53-56.
41. Kebabian, J.W. and Calne, D.B. (1979): Nature, 277:93-96.

42. Kleinberg, D.L., Noel, G.L., and Frantz, A.G. (1977): N. Engl. J. Med., 296:589-600.

43. Kordon, C., and Enjalbert, A. (1980): In: Central and Peripheral Regulation of Prolactin Function, edited by R. M. MacLeod, and U. Scapagnini, Raven Press, New York. (In Press).

44. Landolt, A.M., Wuthrich, R., and Fellmann, H. (1979): Lancet, 1:1082-1083.

45. Langston, J.W., and Forno, L.S. (1978): Ann. Neurol., 3: 129-133.

46. Leopold, N.A., and Podolsky, S. (1979): In: Advances in Neurology, Vol. 23, edited by T.N. Chase, N.S. Wexler, and A. Barbeau, pp. 299-403. Raven Press, New York.

47. Lloyd, H.M., Meares,J.D., and Jacobi, J. (1975): Nature, 255:497-498.

48. Login, I.S., and MacLeod, R.M. (1977): Brain Res., 132: 477-483.

49. MacLeod, R.M. (1976): In: Frontiers in Neuroendocrinology, Vol. 4, edited by L. Martini, and W.F. Ganong, pp. 169-195. Raven Press, New York.

50. MacLeod, R.M., and Login, I.S. (1977): Adv. Biochem. Psychopharmacol., 16:147-157.

51. MacLeod, R.M., Nagy, I., Login, I.S., Kimura, H., Valdenegro, C.A., and Thorner, M.O. (1980): In: Central and Peripheral Regulation of Prolactin Function, edited by R.M. MacLeod, and U. Scapagnini. Raven press, New York. (In Press).

52. Malarkey, W.B., Cyrus, J. and Paulson, G.W. (1974): J. Clin. Endocrinol. Metab., 39:229-235.

53. Martini, J.B., Reichlin, S., and Brown, G.M., editors (1977): Clinical Neuroendocrinology, pp. 129-147. F.A. Davis Co., Philadelphia.

54. McGregor, A.M., Scanlon, M.F., Hall, K., Cook, D.B., and Hall, R. (1979): N. Engl. J. Med., 300:291-293.

55. McGregor, A.M., Scanlon, M.F., Hall, R., and Hall, K. (1979): Br. Med. J., 2:700-703.

56. Meltzer, H.Y., and Fang, V.S. (1976): Arch. Gen. Psych. 33:279-286.

57. Meltzer, H.Y., Sachar, E.J., and Frantz, A.G. (1974): Arch. Gen. Psych. 31:564-569.

58. Mielke, D.H., Gallant, D.M., and Kessler, C. (1977): Am. J. Psych. 134:1371-1375.

59. Moskowitz, M.A., and Wurtman, R.J. (1975): N. Engl. J. Med., 293:274-280; 332-338.

60. Muhlenstedt, D., Osmers, F., and Schneider, H.P.G. (1978): Arch. Gyn. (Germ.), 226:341-346.

61. Muller, E.E., Parati, E.A., Cocchi, D., Zanardi, P., and Carcaceni, T. (1979): In: Advances in Neurology, Vol. 23, edited by T.N. Chase, N.S. Wexler, and A. Barbeau, pp. 319-334. Raven Press, New York.

62. Paulson, G.W. (1979): In: Advances in Neurology, Vol. 23, edited by T.N. Chase, N.S. Wexler, and A. Barbeau, pp. 177-184. Raven Press, New York.
63. Paulson, G.W., Malarkey, W.B., and Shaw, G. (1979): In: Advances in Neurology, Vol. 23, pp. 797-801. Raven Press, New York.
64. Perkins, N.A., and Westfall, T.C. (1978): Neuroscience, 3:59-63.
65. Perkins, N.A., Westfall, T.C., Paul, C.V., MacLeod, R.M., and Rogol, A.D. (1979): Brain Res., 160:431-444.
66. Rotrosen,J., Angrist, B., Clark, C., Gershon, S., Halpern, F.S., and Sachar, E.J. (1978): Am. J. Psych., 135:949-951.
67. Ruggieri, S., Falaschi, P., Baldassarre, M., D'Urso, R., Frajese, G., and Agnoli, A. (1979): In: Neuroendocrine Correlates in Neurology and Psychiatry, Developments in Neurology, Vol. 2, edited by E.E. Muller, and A. Agnoli, pp. 127-137. Elsevier/North-Holland, Amsterdam.
68. Sedvall, G. (1979): In: Neuroendocrine Correlates in Neurology and Psychiatry, Developments in Neurology, Vol. 2, edited by E.E. Muller, and A. Agnoli, pp. 195-209. Elsevier/North-Holland, Amsterdam.
69. Sobrinho, L.G., Nunes, M.C.P., Santos, M.A., and Mauricio, J.C. (1978): Lancet, 2:257-258.
70. Tamminga, C.A., Smith, R.C., Pandey, G., Frohman, L.A., and Davis, J.M. (1977): Arch. Gen.Psych. 34:1199-1203.
71. Thorner, M.O. (1977): Clin. Endocrinol. Metab., 6:201-223.
72. Thorner, M.O., Evans, W.S., MacLeod, R.M., Nunley, W.C., Rogol, A.D., Morris, J.L., and Besser, G.M. (1980): In: Ergot Compounds and Brain Function - Neuroendocrine and Neuropsychiatric Aspects, edited by G. Goldstein, D.B. Calne, A. Lieberman, and M.O. Thorner, Raven Press, New York. (In press.)
73. Thorner, M.O., Martin, W.H., Rogol, A.D., Morris, J.L., Perryman, R.L., Conway, B.P., Howards, S.S., Wolfman, MG., and MacLeod, R.M. J. Clin. Endocrinol. Metab., (in press).
74. Ungerstedt, U. (1971): Acta Physiol. Scand., 367S:1-48.
75. van Praag, H.M. (1977): Br. J. Psych., 130:463-474.
76. von Werder, K., Brendel, C., and Eversmann, T. (1980): In: Pituitary Microadenomas, edited by G. Faglia, M.A. Giovanelli, and R.M. MacLeod, Academic Press, London.(In press.)
77. Wass, J.A.H., Moult, P.J.A., Thorner, M.O., Dacie, J.E., Charlesworth, M., Jones, A.E., and Besser, G.M., (1979): Lancet, 2:66-69.
78. Zacur,H.A., Chapanis, N.O., Lake, C.R., Ziegler, M., and Tyson, J.E. (1976): Am. J. Obstet. Gynecol., 125:859-862.
79. Zervas, N.T., and Martin, J.B. (1980): N. Engl. J. Med., 302:210-214.

Neurosecretion and Brain Peptides,
edited by J. B. Martin, S. Reichlin, and K. L. Bick.
Raven Press, New York © 1981.

Growth Hormone Secretion in Neurological Disorders

A. Martinez-Campos, *P. Giovannini, D. Cocchi, *P. Zanardi,
*E. A. Parati, *T. Caraceni, and **E. E. Müller

*Department of Pharmacology, University of Milan, and *Istituto Neurologica C. Besta,
Milan, Italy; and **Institute of Pharmacology and Pharmacognosy, University of Cagliari,
Cagliari 09100, Italy*

There is now unanimous acceptance of the concept that the control exerted by the central nervous system (CNS) over pituitary
functions operates through a family of neuropeptides, the hypothalamic releasing and inhibiting or regulatory hormones, which
are synthesized in and released from hypothalamic neurons (67).
In turn, the secretion of the peptidergic regulatory hormones is
controlled by complex networks of neurons, mainly aminergic and
peptidergic in nature (43,54). A logical corollary of this notion is that evaluation of anterior pituitary (AP) function can
provide a sensitive measure of alterations arising primarily in
the hypothalamus or relayed to this area from higher brain centers (52).

As CNS control of growth hormones (GH) secretion has emerged
as the primary regulatory mechanism (47,51), studies of GH regulation appear particularly suited as a probe for hypothalamic
function in patients with neurological disorders. It is now
clear that GH secretion is a highly labile function, reacting to
an intrinsic neural cycle related to sleep, to altered intake of
food, to exercise, to psychological stimuli, and to many stressful, metabolic and endocrine factors (Table 1). In primates,
most of the stressful or metabolic and endocrine conditions which
elicit GH release act through the intervention of brain noradrenergic function, α- and β-noradrenergic receptors playing a fac-
ilitatory or inhibitory role, respectively (47,51). Aside from
norephinephrine (NE), dopamine (DA) has emerged in the last few
years as an important neurotransmitter for human GH (hGH) control. Not only the DA precursor amino acid levodopa, but a
series of directly and indirectly acting DA agonists, were shown
capable of inducing a rise in hGH levels; this effect is counter-
acted by antagonists to DA function (51).

The recognition of the important interrelationship between some neurological disorders, e.g., Parkinson's and Huntington's disease, and dopaminergic pathways (2) on one hand, and GH and DA on the other, is the concept which stimulated examination of hGH regulation in these disorders (52). In this respect, neuropharmacologic studies on GH secretion in neurological disorders should be more rewarding than investigations aimed at studying the secretion of other AP hormones, e.g., prolactin (PRL). In fact, while the control of PRL secretion is mainly exerted at the level of the medial-basal hypothalamus, a system with peculiar neurobiological features (50), higher brain centers play a pivotal role in hGH control (47,51).

TABLE 1. Factors influencing release of growth hormone

Stimulate	Inhibit
Glucoprivation	Glucose
Falling blood glucose	Elevated free fatty acids
Fasting	Elevated GH levels
Protein meal	Obesity
Exercise	Hypercortisolism
Stress:	
Physical	
Psychological	
Slow wave sleep	
α-Receptor agonists	α-Receptor antagonists
β-Receptor blockers	β-Receptor agonists
Dopamine agonists	Dopamine antagonists
5-OH tryptophan	
Cholinergic agonists	
GABA agonists	
Vasopressin	Dopamine agonists[++]
α-MSH	Antihistaminics
ACTH	Anticholinergics
Glucagon	Melatonin
TRH[+], LH-RH[+]	Methysergide
Opioid Peptides	Cyproheptadine
Neurotensin, Substance P	

+Only in some pathological conditions
++In acromegaly

Space does not permit exhaustive review of GH secretion in neurological disorders. Therefore, we report original data on the dopaminergic control of hGH secretion in Parkinson's disease (PD) and Huntington's disease (HD) and critically evaluate the relevant literature.

HYPOTHALAMIC INVOLVEMENT IN PARKINSON'S AND HUNTINGTON'S DISEASE

Parkinson's Disease

Though loss of DA substantia nigra neurons and ensuing striatal DA deficiency are crucial to the pathogenesis of PD (17), neuronal and biochemical lesions are not limited to the nigrostriatal system but involve also mesocortical and mesolimbic pathways (65) and the hypothalamus (36,57). Lewy bodies, whose presence in the pigmented brainstem nuclei is a constant feature of PD (3), are also found in the posterior and lateral portions of the hypothalamus (57). In this area, concentrations of DA and homovanillic acid (HVA), its main product of catabolism, are reduced and binding of ^{3}H-spiroperidol is significantly decreased (65). However, in PD an impairment in neuroendocrine function may not necessarily be linked to anatomical alterations of the hypothalamus. In rats, a lesion in the pars compacta of the substantia nigra results in decreased DA levels in the ventro-medial nucleus and median eminence (ME), implying the existence of a nigra-hypothalamic ME pathway (33).

Huntington's Disease

Neural damage in the paraventricular and ventromedial nucleus and the lateral hypothalamus has been described in HD (2,9). A host of vegetative symptoms of the disease, such as hyperhidro-sis, hyperphagia and progressive cachexia as well as disturbance of carbohydrate metabolism have been attributed to hypothalamic involvement (see ref. 4).

The presence of increased gonadoptropin-releasing hormone concentrations in the hypothalamus of female choreic patients (5) is also consistent with functional hypothalamic involvement and has been related to the increased fertility (64) and libido (4) reported in these patients.

PROPER EVALUATION OF DA-STIMULATED hGH SECRETION

Many studies have investigated GH responses to DA-mimetic drugs in neurological disorders, and especially to levodopa in PD, due to the concern that drug-induced GH rises may engender metabolic and endocrine alterations (70) or change the therapeu-tic potential of the drug (73). However, many of these studies were conducted with no attention paid to factors or conditions which may normally affect dopaminergically-medicated hGH release (Table 2).

Estrogens enhance hGH release following some provocative stim-uli (51) but instead hinder the hGH response to DA-mediated stimuli, so that in women the GH responses are lower than in men (19). Age is also a contributing factor, especially for studies in PD subjects. Lower spontaneous resting and sleep-entrained

TABLE 2. Factors influencing DA-stimulated hGH release

Patient-related	Physiologic	Sex Age Individual variability
	Pathologic	Hyperprolactinemia Acromegaly Hypercortisolism Diabetes mellitus
Drug-related		Routine administration Absorption Transport Degradation Penetration into BBB[*] Timing of administration Schedule of administration Antecedent treatment(s) Neuronal specificity Metabolic alterations

[*]BBB = Blood-brain barrier

secretory rates of GH are seen in older as compared with younger subjects (21) and the GH response to levodopa (66) or apomorphine (42) is consistenly lower in aged than in young individuals. The problem is further compounded by individual variability in the secretory response. Thus, GH unresponsiveness to levodopa has been reported in 13 to 40% of nonhypopituitary subjects (25). Coexistence of endocrine abnormalities, such as hyperprolactin-emia (48), acromegaly (51), or hypercortisolism (35), likely due to the associated brain DA defect (53), limits the effectiveness of DA agonist drugs.

Whether a given DA agonist drug administered orally is suffi-cient to induce a reliable hGH stimulation in a normal individual will depend on many factors. Absorption and transport across the gastrointestinal wall, rapidity of degradation (e.g., levodopa) and degree of penetration into the brain parenchyma are but a few variables. Timing and schedule of drug administration are worth considering. Thus, presence of spontaneously elevated resting hGH levels when the drug is administered may, by a short-loop feedback effect (26), blunt the drug-induced GH rise. Repeated oral doses of the drug may also lead to decreased responsiveness, a finding that has been observed after levodopa administration and ascribed to a central rather than a peripehral mechanism (58).

Antecedent treatments with neuroactive drugs may markedly affect receptor sensitivity, thus leading to misinterpretations of the neuroendocrine response. In PD subjects, loading doses of levodopa alone or with inhibitors of L-aromatic amino acid decarboxylase (L-AAAD), when administered for long periods, are expected to blunt the susceptibility of the DA synapse. In HD subjects who are normally under chronic treatment with DA antagonist neuroleptic drugs, impairment of the DA-mediated GH response, if proper drug withdrawal is not carried out, may be present. An opposite condition, i.e., supersensitivity of DA receptors, may be generated, on theoretic grounds, in HD patients "washed out" from neuroleptics (24). However, from results obtained in schizophrenic subjects "washed out" from neuroleptics (20), the likelihood of such an occurrence is small.

Neuropharmacologic profiles of the compounds administered also deserve consideration. While apomorphine, piribedil and some of the therapeutically useful ergot derivatives, e.g., bromocriptine, lisuride, pergolide, are principally direct stimulants of DA receptors (51), this is not true for levodopa which can also be converted in the body to NE, thus causing NE-mediated neuroendocrine effects.

Finally, the drug-induced hGH response may be influenced by the occurrence of secondary metabolic alterations. Ingestion of levodopa increases blood glucose and free fatty acid levels (70), which may blunt GH response to provocative stimuli (Table 1).

BASAL AND DA-STIMULATED hGH LEVELS IN PARKINSON'S DISEASE

Normal basal GH levels have been reported in untreated PD subjects (11,41); however, in one report, the 24-hour records from 8 untreated patients were flat except for a very few short-lived rises above the basal levels occurring mostly at night. This pattern contrasted that present in PD subjects treated with levodopa alone or associated with an L-AAAD inhibitor, in whom a pulsatory type of release was manifested (49). However, neither age nor sex of the patients is reported, and polygraphic recordings of the sleep phases were not performed. Moreover, the study on levodopa effect was confined to a few hours of the day. In similar studies, Passouant et al. (60) noticed that in PD subjects sleep-entrained GH rises were present in the male but not in the female. They could not detect any change in hGH secretion following institution of bromocriptine therapy.

In the study of Hyppa et al. (28), chronic treatment with bromocriptine (30 mg daily X 20 weeks) induced a very slight, though significant, decrease in hGH levels in 15 patients. In contrast, Shaw et al. (69) found no significant difference in basal hGH levels when compared to pretreatment values in 9 patients treated chronically with 40 mg daily of the same drug.

Contradictory findings have also been reported in PD subjects on chronic treatment, challenged by acute dopaminergic stimulation. In 24 out of 26 patients treated with levodopa for 1 to 5 years there were consistent hGH rises 30 to 120 minutes after a single dose of the drug (59), and similar results were obtained after acute ingestion of the drug in 4 patients who received levodopa for 6 to 11 months (37). In contrast to these findings, Malarkey et al. (44) found only 1 of 10 patients treated for 2.5 to 4 years to be responsive to levodopa. Further evidence of disturbed hGH regulation was provided by an unexpected suppression of GH following levodopa in 3 patients, and a paradoxical GH rise following glucose in 2 patients. None of these subjects had a defective GH response to insulin-induced hypoglycemia.

Levodopa appears to be a more consistent stimulus for hGH release in PD subjects than more specific stimulants of DA receptors which elicit comparable therapeutic effects. In the study of Parkes et al. (59), only 4 of 8, and in that of Galea Debono et al. (22), only 2 of 7 patients showed a GH secretory response after acute administration of bromocriptine. Similarly, apomorphine increased GH levels in only 1 of 4 male patients investigated by Brown et al. (1973), though lack of proper age-matched controls limits the validity of this study.

In sum, scrutiny of some of the data appearing in the literature leads to the conclusion that both resting and stimulated hGH levels are not altered in subjects with PD. Presence of occasionally low basal plasma GH concentrations must be evaluated in relation to the age of the patients and the state of akinesia induced by the disease. In spite of long-term treatment and different treatment schedules, in most of the cases, levodopa appears competent to induce clear-cut GH rises whereas direct DA agonists appear less effective in this context. Thus, a possible noradrenergic component in the hGH releasing effect of levodopa cannot be excluded.

A problem hampering proper evaluation of hGH response to DA-mimetic drugs in PD is the lack of "true" control subjects, i.e., neurologically intact individuals submitted to chronic and loading doses of DA agonists before acute testing is performed. The possibility should be kept in mind that in the "true" controls, blunted receptor sensitivity, with ensuing lowering of the neuroendocrine response, may be generated by repeat drug dosing. If so, the hGH response to the different DA-mimetic drugs seen in PD subjects, at least for levodopa, may be actually greater than that evoked in controls. Proof of the existence of hypersensitive DA receptors for GH release in PD subjects may be provided by acute stimulation tests performed in unmedicated subjects.

There does not seem to be any practical consequence of the periodic changes in hGH secretion induced by levodopa in PD subjects and the danger of marked endocrine-metabolic alternations (70) has no sound justification. It is also unlikely that the increase in plasma hGH concentrations induced by levodopa is important in the antiparkinsonian effect of this drug (73) since patients who did not have any endocrine response did have a therapeutic response (23).

EFFECT OF ACUTE ADMINISTRATION OF LEVODOPA PLUS BENSERAZIDE
ON hGH LEVELS IN PATIENTS WITH PARKINSON'S DISEASE

Due to intense peripheral catabolism only a small proportion of oral levodopa is metabolized to DA in the brain. The percentage is significantly increased by the concurrent use of inhibitors of the L-AAAD, the enzyme that converts the aromatic amino acids into the corresponding amine neurotransmitters (1).

Treatment of PD subjects has progressed considerably with the combined use of levodopa and L-AAAD inhibitors, that has led to a marked reduction in levodopa doses while maintaining cerebral levels of DA unchanged (71). The most commonly used L-AAAD inhibitors are carbidopa (MK-486, Merck Sharp & Dohme) and benserazide (Ro-4-4602, Hoffmann LaRoche). Reportedly, they do not cross the blood-brain barrier (BBB) when administered in a certain dose range; hence, both are considered mainly peripheral inhibitors of L-AAAD (1). As a result, higher amounts of the unmetabolized amino acid are shunted to the CNS. In the human, the addition of carbidopa (100 mg) with an oral dose of levodopa (250 mg) resulted in a fivefold augmentation of plasma levodopa and a twofold increase in hGH levels (45). The GH stimulation induced by levodopa-carbidopa (Sinement, Merck Sharpe & Dohme) is so consistent that this pharmacological "cocktail" has been proposed and used as a reliable screening method for hGH deficiency (68).

We report here a study conducted on PD subjects given levodopa (200 mg) acutely, in association with the other well-known L-AAAD inhibitor, benserazide (50 mg) (Madopar, Hoffmann LaRoche). Seventeen PD subjects (15 women and 2 men), aged 37 to 74 years, with a mean age of 56.3 years, were studied. Of these patients, 6 women, aged 42 to 74 years (mean age, 54.2 years), were under chronic Madopar therapy for 6 months to 8 years, while 4 women, aged 37 to 67 years (mean age, 41.2 years), had never been treated. Seven PD subjects (5 women and 2 men), aged 62 to 70 years (mean age, 66.5 years), were withdrawn from specific therapy for at least 15 days. Ten nonobese hospitalized subjects (8 women and 2 men) aged 42 to 71 years (mean age, 48.2 years), who were recovered from minor disease and were without endocrine and metabolic disorders, were selected as controls.

All experiments were performed in the morning after an overnight fast and about 12 hours after the last drug administration. Basal hGH levels in untreated PD patients were not different from those present in controls (0.4 $\pm$ 0.1 vs 1.6 $\pm$ 1.0 ng/ml). Chronic treatment with Madopar did not significantly affect basal hGH levels (0.8 $\pm$ 0.2 ng/ml in patients under Madopar therapy and 0.4 $\pm$ 0.2 ng/ml in patients withdrawn from therapy).

Fig. 1 shows that Madopar (250 mg, orally) induced an inconsistent rise in plasma hGH in control subjects (peak levels, 3.4 $\pm$ 1.1 and 3.5 $\pm$ 1.6 ng/ml at 60 and 90 minutes; range from 0.4 to 15.5 ng/ml). The GH-releasing effect of Madopar was instead clear-cut in PD subjects receiving chronic Madopar therapy; plas-

ma hGH peaked at 90 minutes (9.4 $\pm$ 0.7 ng/ml) and hGH values significantly different from control values were present at 60 and
90 minutes (p <0.005). Unlike subjects under chronic Madopar
therapy, the effect of Madopar on hGH release in previously unmedicated PD subjects was not significantly different from that
present in normal controls. A statistically significant difference was found at 60 minutes vs Madopar-treated subjects
(p <0.05). In patients "washed out" from drug therapy, Madopar
induced a rise in plasma hGH levels more delayed and sustained
than that present in control subjects. However, the overall hGH
response was not statistically different (data not shown).

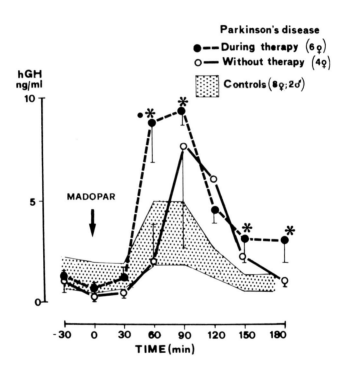

FIG. 1.   Plasma GH response to oral administration of Madopa
(levodopa 200 mg + benserazide 50 mg) in control and Parkinsonian
subjects. Each point is the mean $\pm$ SEM of 4 to 10 determinations. Number and sex of subjects in brackets.   *P <0.05 vs
control subjects; P <0.05 vs unmedicated PD patients.

The flat hGH response of control subjects given acute administration of Madopar is difficult to interpret, especially when

viewed in relation to the consistent hGH rises reported in normal individuals taking Sinemet (45,68). Although levodopa alone was not administered to these subjects, the hGH response elicited by Madopar was consistently lower than tha which can be expected following administration of levodopa to normal old adults (66).

The median eminence and the AP are two areas unprotected by the BBB (74); thus our results might be attributable to a decreased availability of DA operated by benserazide at the level of these structures. However, this hypothesis is difficult to reconcile with: 1) the different patterning of hGH response induced by Sinemet; 2) the evidence, though circumstantial, favoring the nonresponse of ME-DA neurons to GH regulation (7,16) and disproving the presence of DA receptors on pituitary somatotrophs (56). It may be proposed, as an alternative hypothesis, that benserazide at the dose used in the present investigation penetrates inside the BBB and hinders, at least partially, the conversion of levodopa to DA also in areas located above the ME. Supporting this proposition are data obtained by us (unpublished results) and others (63) on the PRL-releasing effect of Madopar in normal individuals.

If this mechanism is truly operative for benserazide, the more consistent GH response to Madopar of PD patients under chronic Madopar treatment may be related either to a lower penetration of benserazide into the CNS of PD patients or to DA, formed from levodopa, acting on supersensitive (hypothalamic?) DA receptors. That in previously unmedicated patients or in patients "washed out" from therapy for at least 15 days the GH response to Madopar was not significantly different from that present in control subjects would indicate that the patterning of hGH response to acute Madopar administration is related in PD subjects to chronic exposure to the drug, and not to an inherent supersensitivity of DA receptors.

PLASMA GH LEVELS AND GH RESPONSE TO DIFFERENT STIMULI IN PATIENTS WITH HUNTINGTON'S DISEASE

Evidence has been presented for a variety of hypothalamic abnormalities in patients with HD. Abnormal regulation of neuroendocrine control of GH secretion was first reported by Podolsky and Leopold (62) who noted failure of suppression of GH release after oral glucose and, in another study (38), an exaggerated GH response to arginine. In addition, Keogh et al. (32) reported an enhanced GH response following insulin-induced hypoglycemia (see Table 3).

Even though HD clinically represents in many ways the mirror image of PD and the existence in this syndrome of a functional predominance of monoaminergic, notably dopaminergic, mechanisms within the extrapyramidal centers has been postulated (2), an increased sensitivity to intracerebral DA is present, as manifested by the exacerbation of the choreiform movements after levodopa at doses that have hardly any significant effect in normal subjects (34).   Data obtained by our group would suggest that behavioral hypersensitivity to dopaminergic stimulation is accompanied in choreic subjects by hyper-responsiveness of the neuroendocrine system that controls GH secretion (Table 3).

TABLE 3.   Growth hormone responses to different stimuli in HD vs normal subjects

Study	Base line	Glucose load	Insulin	Arginine	CB-154	Apo	Levodopa
Normal response		↓	↑	↑	↑	↑	↑
Podolsky et al (62)	normal	paradox. rise					↑↑
Leopold et al (38)	normal			↑↑			
Phillipson et al (61)	increased normal	→	↑↑				
Keogh et al (32)	normal		↑↑				
Caraceni et al (10)	normal				↑↑		
Chalmers et al (12)	normal				↓↓		
Levy et al (39)						↓↓	↑
Miller et al (55)	normal				↑↑	↑↑	↑↑

↑↑ = Greater than controls; ↓↓ = Less than controls; → = The same as controls
CB-154 = bromocriptine   Apo = apomorphine

Studies were conducted on 25 randomly selected hospitalized HD patients.   Diagnosis depended upon the presence of choreic movements, progressive dementia of varying degree and a positive family history for the disease.   Nonobese hospitalized subjects without endocrine or metabolic disease were selected as controls (for details see refs. 10,55).

Administration of the DA agonist bromocriptine (CB-154 or Parlodel, Sandoz) (2.5 mg, orally) induced in patients with HD a greater and earlier hGH rise than in controls. A factorial analysis of variance showed that there was a significant difference between the increase in hGH levels induced by bromocriptine in patients and controls (10,55). Similarly to bromocriptine, oral administration of levodopa (500 mg) or subcutaneous administration of apomorphine (1.0 mg) resulted in a significantly greater and, in the case of levodopa, more prompt increase in plasma GH in patients than in controls. No consistent changes in plasma hGH levels were present in 5 HD patients who received a placebo when compared to the pattern present in 5 normal controls (10, 55).

The more consistent and prompt GH rise in response to dopaminergic stimuuli present in HD patients suggests the existence of supersensitive DA receptors, probably due to the deterioration of a dopaminergic pathway impinging on neurosecretory neurons for the control of GH release. In some of our patients, preceding treatment with neuroleptic drugs had been discontinued for at least one month, and some patients had never been treated, a fact which argues against the possibility that supersensitive DA receptors had been generated by drug-induced chemical denervation (24). In contrast to our findings, a blunted GH rise after bromocriptine administration in HD patients untreated or who had stopped medication with phenothiazines 72 hours earlier had been found by Chalmers et al. (12). Lewy et al. (39) have reported no GH response to apomorphine in HD patients who had discontinued therapy for two weeks.

Although no valid explanation can be offered for these contrasting data, it is noteworthy that also in our studies 3 out of 14 subjects exhibited either a normal (2 cases) or no (1 case) GH response to bromocriptine. Conversely, of the 12 patients of Chalmers et al. (12), 2 untreated and 3 phenothiazine-pretreated subjects had prompt and appropriate GH rises after the dopaminergic stimulus.

Recently, a further study was performed by us in HD patients with the use of another potent DA agonist, lisuride (27). Six women and one man participated in this study. The age range of the patients was 21 to 58 years, with a mean age of 40 years. The average length of documented clinical signs was 4.2 years (range 3 to 7 years). One patient suffered depression. Criteria followed for the diagnosis of the disease were those previously reported (10,55). Six nonobese hospitalized female subjects without endocrine or metabolic diseases were selected as controls. Their age was between 30 and 54 years with a mean age of 40 years. Fig. 2 shows that lisuride (Lisenil, Spofa, 200 μg orally) induced in control subjects a modest rise in hGH levels with peak levels (5.5 ng/ml) occurring 90 and 120 minutes post-drug administration. In HD patients lisuride evoked no rise in hGH levels (p <0.005 and p <0.05 at 90 and 120 minutes, respectively, vs control subjects).

Growth hormone unresponsiveness to lisuride present in HD patients is reminiscent of the findings obtained by Chalmers et al. and Lewy et al. with bromocriptine and apomorphine, respectively. This makes it unlikely that GH non-response is due to lisuride itself acting, for example, at the level of a distinct population of brain DA receptors (30). Rather, it may suggest that, although grouped together on clinical grounds, patients with HD may represent a large spectrum of individual entities whose variability we are disclosing.

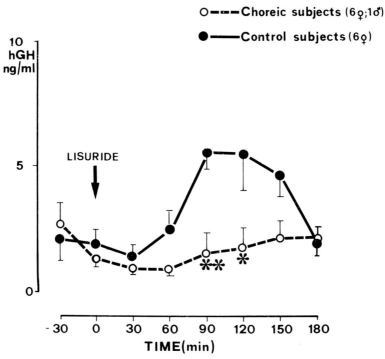

FIG. 2.   Plasma GH response to oral administration of lisuride (200 μg) in control and choreic subjects.   Each point is the mean $\pm$ SEM of 6 and 7 determinations, respectively.   Number and sex of subjects in brackets.   *P <0.05,   **P <0.005 vs control subjects.

STUDIES ON GH REGULATION IN AN ANIMAL MODEL
MIMICKING HUNTINGTON'S DISEASE

The altered GH responsivenss to dopaminergic stimulation present in HD subjects prompted studies to ascertain quantitative or qualitative alterations in GH release in an animal model mimicking the cardinal biochemical and behavioral features of the disease.   It is now established that direct injection of kainic acid, a rigid analogue of glutamate, into the rat caudate nucleus produces histological and biochemical changes mimicking those of HD (13).   Although minor differences do exist, in nearly all respects the similarities are striking (46).

Male adult (300 g) Sprague-Dawey rats were anesthetized with sodium pentobarbital (50 mg/kg) and received 3 nmoles (0.63 µg) of kainic acid in a volume of 1 µl into each striatum.   Coordinates according to DeGroot steroetaxic atlas (15) were: anterior, 8.8 mm; lateral, 2.6 mm; and ventral, 4.5 mm.   The kainic acid was dissolved in isotonic saline and injected at a rate of 1 µl/5 min.   After the injection, the needle was left in the striatum for 5 minutes before removal.

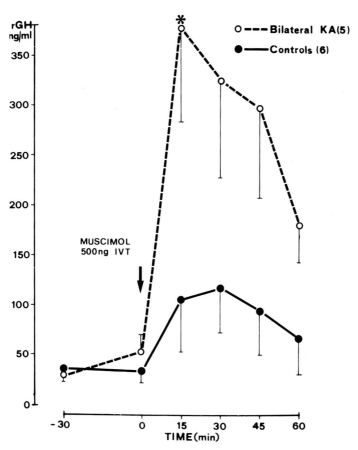

FIG. 3.   Plasma GH reponse to intraventricular administration of muscimol (500 ng/rat) in sham-operated rats or rats lesioned bilaterally in the head of the caudate nucleus with kainic acid. Each point is the mean ± SEM of 6 and 5 determinations, respectively.   *P <0.05 vs sham-operated rats.

Control rats underwent similar procedures, but instead were given microinjections of isotonic saline. Postoperatively, the kainic acid-lesioned animals became aphagic and adipsic, and had to be fed intragastrically for four to five days, after which they recovered and maintained themselves on ad libitum food and water. Kainic acid-lesioned animals lost weight for a few days following the operation but regained slowly the original weight. Fifteen days after lesioning, rats were prepared with chronic indwelling jugular cannulae which were inserted into the right atrium, and underwent experiments 2 days later.

In the first experiment, apomorphine (0.1 mg/kg) was administered acutely by intravenous route and blood was sampled at 15-minute intervals for 75 minutes for GH determinations in plasma. Resting GH levels in experimental animals were not significantly different from sham-operated animals. Apomorphine induced a clear-cut GH rise in sham-operated rats, starting at 45 minutes with peak levels occurring at 60 minutes. This pattern was not present in the kainic-acid treated rats ($p < 0.05$ at 75 minutes) (data not shown).

In another experiment, rats injected bilaterally with kainic acid and sham-operated controls received an intraventricular injection of muscimol, a potent agonist for GABA receptors (14). Administration of muscimol (500 ng/20 μl) induced in sham-operated controls a rise in plasma GH, with peak levels at 30 minutes. Muscimol induced in the lesioned rats a more prompt and striking rise in plasma GH levels, with peak levels occurring at 15 minutes ($p < 0.05$ vs corresponding values of control subjects) (Fig. 3).

Studies in an animal model that mimics HD reveal the existence of alterations in the secretion of GH following application of DA- and GABA-mimetic drugs. Administration of apomorphine elicited a GH secretory response in sham-operated controls, an effect already observed in freely moving unanesthetized rats (40), but did not do so in rats with lesions placed in the head of the caudate nucleus. Absence of GH responsiveness to apomorphine in kainic acid-lesioned rats is reminscent of the blunted GH response to bromocriptine (12), apomorphine (39) or lisuride (this study) present in some HD patients. On the contrary, the GH-releasing effect induced by functional activation of the GABA-ergic pathways was greatly magnified in kainic acid-lesioned rats. Muscimol also elicits a GH response when administered to subjects with HD (72).

These data are preliminary and need corroboration and extension before any firm conclusion can be drawn as to the mechanism(s) underlying the altered GH responsiveness of kainic acid-lesioned rats. Considering the discreteness of the lesion which appears histologically and biochemically confined to the head of the caudate nucleus (unpublished results), it seems unlikely that the altered GH responsiveness was the result of mere diffusion of the toxic amino acid from the locus of injection to the adjacent (hypothalamic?) structures. Rather, a possible explanation of our findings is that kainic acid destroyed a pathway relaying impulses from the striatum to the medio-basal hypothalamus. Whatever their interpretation, the potential import of these findings for understanding the mechanism(s) underlying the deranged GH regulation of HD patients should not escape attention.

TABLE 4. Paradoxical hGH secretory responses

	High basal levels	Paradoxical rise after glucose	TRH-induced release	Suppression by DA drugs
I. Immediate post-natal period	+	+	−	NI
II. Neurological disorders				
Diencephalic syndrome	+	+	NI	NI
Parkinson's disease	−	+*	−	+ *
Huntington's disease	−	+	−	−
Freidreich's ataxia	+*	+*	NI	−
III. Psychiatric disorders				
Depression	−	−	+	−
Anorexia nervosa	+	−	+	−
Schizophrenia	−	−	+	−
IV. Endocrine-metabolic disorders				
Acromegaly	+	+	+	+
Liver cirrhosis	+	+	+	−
Acute intermittent porphyria	−	+	NI	NI
Renal failure	+	+	+	−
Protein-calorie malnutrition	+	+	+	NI
Primary hypothyroidism	−	−	+	−
Diabetes melltius	−	−	+	−
Endometrial carcinoma	−	+	NI	NI

*Present in some patients
**Young schizo-affective subjects
  NI = Not investigated

PARADOXICAL hGH SECRETORY RESPONSES IN NEUROLOGICAL DISORDERS

Evaluation of GH responsiveness to dopaminergic stimulation in PD and HD does not reveal consistent alteration in the former disorder and discloses instead the existence of GH hyper- or hyporesponsiveness in individual patients with the latter disorder, who, on clinical grounds, cannot be discriminated. Also in PD, however, a few patients have been described in whom GH regulation is disturbed as evidenced by failure of levodopa to provoke GH release, paradoxical GH suppression by the amino acid and paradoxical increase following glucose (Table 4). A paradoxical rise in plasma GH after a glucose load is also present in many subjects with HD, though this feature is shared also by other neurological and endocrine-metabolic disorders (Table 4). It is noteworthy that another abnormal hGH response, that following administration of TRH (5), is lacking in patients with extrapyramidal disorders (authors' unpublished results), but it is a hallmark of some psychiatric disorders (see ref. 6). The latter do not exhibit instead paradoxical GH rise after glucose. The abnormal TRH-induced GH rise may be due either to a direct action of the peptide at the level of pituitary sommatotrophs lacking normal neurohormonal influences for GH regulation (51) or to a CNS-mediated mechanism (31). Whatever the physiopathology of this abornmal GH response is, altered brain monoamine function would be the common denominator (6). Supporting this suggestion is the presence of a TRH-induced GH rise in metabolic-endocrine disorders, e.g., renal failure, liver cirrhosis, in which an altered brain amine metabolism has been demonstrated (29). Extension of this neuroendocrine maneuver to other disorders of the brain function seems a promising approach to a better understanding of the underlying neurotransmitter disturbance.

ACKNOWLEDGEMENTS

We gratefully acknowledge Mrs. Lucia Scimone for technical assistance, Miss Isabella Zago for preparation of the manuscript, and Dr. E. Rainer, Schering Spa, Italy, for supplying lisuride. The work was supported by a grant of the Hereditary Disease Foundation and CNR grant 790236465.

REFERENCES

1. Bartholini, G., and Pletscher, A. (1975): Pharmacol. Ther. [B], 3:407-421.
2. Bernheimer, H., Birkmayer, W., Hornykiewica, O., Jellinger, K., and Seitelberger, F. (1973): J. Neurol. Sci., 20:415-455.
3. Berthlem, J., and den Hartog Jager, W.Z. (1960): J. Neurol. Neurosurg. Psychiatr., 23:74-90.
4. Bird, E.D. (1979): In: Advances in Neurology, Vol. 23, edited by T.N. Chase, N.S. Wexler, and A. Barbeau, pp. 291-297. Raven Press, New York.
5. Bird, E.D., Chiappa, S.A., and Fink, G. (1976): Nature, 260: 536-538.

6. Brambilla, F., Smeraliddi, E., Sacchetti, E., Negri, F., Cocchi, D., and Muller, E.E. (1978): Arch. Gen. Psychiatr., 35:1231-1238.
7. Brown, G.M., Garfinkel, P.E., Warsh, J.J., and Stancer, H. (1976): J. Clin. Endocrinol. Metab., 43:236-239.
8. Brown, W.A., Van Woert, M.H., and Ambani, L.M. (1973): J. Clin. Endocrinol. Metab., 37:463-465.
9. Bruyn, G.W. (1968): In: Handbook of Clinical Neurology, Vol. 6, edited by P.J. Vinken, and G.W. Bruyn, pp. 379-396. North-Holland, Amsterdam.
10. Caraceni, T., Panerai, A.E., Parati, E.A., Cocchi, D., and Muller, E.E. (1977): J. Clin. Endocrinol. Metab., 44:870-875.
11. Cavagnini, F., Peracchi, M., Scotti, G., Raggi, U., Pantir-oli, A.E., and Bana, R. (1972): J. Endocr., 54:425-433.
12. Chalmemrs, R.J., Johnson, R.H., Keogh, H.J., and Nanda, R.N. (1978): J. Neurol. Neurosurg. Psychiatr., 41:135-139.
13. Coyle, J.T., and Schwartz, R. (1976): Nature, 263:244-246.
14. Curtis, D.R., Duggan, A.W., Felix, D., and Johnston, A.R. (1971): Brain Res., 32:69-96.
15. DeGroot, J. (1959): J. Comp. Neurol., 113:389-400.
16. Delitala, G., and Masala, A. (1977): Biomedicine, 27:219-222.
17. Ehringer, H., and Hornykiewicz, O. (1960): Klin. Wschr., 38:1236-1239.
18. Enna, S.J., Stern, L.Z., Wastek, G.J., and Yamamura, H.I. (1977): Life Sci., 20:205-212.
19. Ettigi, P., Lal, S., Martin, J.B., and Friesen, H.G. (1975): J. Clin. Endocrinol. Metab., 40:1094-1098.
20. Ettigi, P., Nair, N.P.V., Lal, S., Cervantes, P., and Guyda, H. (1976): J. Neurol. Neurosurg. Psychiatr., 39:870-876.
21. Finkelstein, J.W., Roffwarg, H.P., Boyar, R.M., Kream, J., and Hellman, L. (1972): J. Clin. Endocrinol. Metab., 35:667-670.
22. Galea-Debono, A., Donaldson, I., Marsden, C.D., and Parkes, J.D. (1975): Lancet, 2:987-988.
23. Galea-Debono, A., Jenner, P., Marsden, C.D., Parkes, J.D., Tarsy, D., and Walters, J. (1977): J. Nerol. Neurosurg. Psychiatra., 40:162-167.
24. Gianutsos, G., Hynes, M.D., and Lal, H. (1974): Biochem. Pharmacol., 24:581-582.
25. Gommez-Sanchez, C., and Kaplan, N.M. (1972): J. Clin. Endo-crinol. Metab., 34:1105-1107.
26. Hagen, T.C., Lawrence, A.M., and Kirsteins, L. (1972): Met-abolism, 21:603-610.
27. Horowski, R., and Wachtel, H. (1976): Eur. J. Pharmacol., 36: 373-383.
28. Hyyppa, M.T., Langvik, V., and Rinne, U.K. (1977): J. Neu-ral. Trans., 42:151-157.
29. Jellinger, K., and Riederer, P. (1977): J. Neural. Trans., 41:275-286.

30. Kebabian, J.W., and Calne, D.B. (1979): <u>Nature (London)</u>, 277: 93-96.

31. Keller, H.H., Bartholini, G., and Pletscher, A. (1974): <u>Nature</u>, 248:528-529.

32. Keogh, H.J., Johnson, R.H., Nanda, R.N., and Sulaiman, W.R. (1976): <u>J. Neurol. Neurosurg. Psychiatra.</u>, 39:244-248.

33. Kizer, J.S., Palkovits, M., and Brownstein, M. (1976): <u>Brain Res.</u>, 108:363-370.

34. Klawans, H.L., and Weiner, W.J. (1976): <u>Prog. Neurobiol.</u>, (N.Y.) 6:49-80.

35. Krieger, D.T. (1973): <u>J. Clin. Endocrinol. Metab.</u>, 36:277-284.

36. Langston, J.W., and Forno, L.S. (1978): <u>Ann. Neurol.</u>, 3: 129-133.

37. Lebovitz, H.E., Skyler, J.S., and Boyd, A.F. (1974): In: <u>Advances in Neurology</u>, Vol. 5., edited by F. McDowell, and A. Barbeau, pp. 461-469. Raven Press, New York.

38. Leopold, N., and Podolsky, S. (1975): <u>J. Clin. Endocrinol. Metab.</u>, 41:160-163.

39. Lewy, C.L., Carlson, H.E., Soiwers, J.R., Goodllett, R.E., Tourtellotte, W.W., and Hershman, J.M. (1979): <u>Life Sci.</u>, 24: 743-750.

40. Locatelli, V., Cocchi, D., Gil-Ad, I., Mantegazza, P., and Muller, E.E. (1977): <u>59th Ann. Meet. Endocr. Soc., Abs.</u>, 246.

41. Lundberg, P.G. (1972): <u>Acta Neurol. Scandinav.</u>, 48:427-432.

42. Maany, I., Frazier, A., and Mendels, J. (1975): <u>J. Clin. Endocrinol. Metab.</u>, 40:162-163.

43. McCann, S.M., Krulich, L., Ojeda, S.R., Negro-Vilar, A., and Vijayan, E. (1979): In: <u>Central Regulation of Endocrine System</u>, edited by K. Fuxe, T. Hokfelt, and R. Luft, pp. 329-347. Plenum Press, New York.

44. Malarkey, W.B., Cyrus, J., and Paulson, G.W. (1974): <u>J. Clin. Endocrinol. Metab.</u>, 39:229-235.

45. Mars, H., and Genuth, S.M. (1973): <u>Clin. Pharmacol. Ther.</u>, 14:390-395.

46. Marsden, C.D. (1979): In: <u>Advances in Neurology</u>, Vol. 23, edited by T.N. Chase, N.S. Wexler, and A. Barbeau, pp. 567-576. Raven Press, New York.

47. Martin, J.B. (1976): In: <u>Frontiers in Neuroendocrinology</u>, Vol. 4, edited by L. Martini, and W.F. Ganong, pp. 129-168. Raven Press, New York.

48. Martin, J.B., Lal, S., Tolis, G., and Friesen, H.G. (1974): <u>J. Clin. Endocrinol. Metab.</u>, 39:180-182.

49. Mena, I., Cotzias, G.G., Brown, F.C., Papavasiliou, P.S., and Muller, S.T. (1973): <u>N. Engl. J. Med.</u>, 288:320-321.

50. Moore, K.E., Annunziato, L. and Gudelski, G.A. (1978): In: <u>Advances in Biochemical Psychopharmacology</u>, Vol. 19, edited by E. Roberts, G. Woodruff, and L.L. Iverson, pp. 193-204. Raven Press, New York.

51. Muller, E.E. (1979): In: <u>Hormonal Proteins and Peptides</u>, Vol. 7, edited by C.H. Li, pp. 123-204.

52. Muller, E.E. and Agnoli, A, editors (1979): <u>Neuroendocrine Correlates in Neurology and Psychiatry</u>, Devel. Neur., Vol. 2. Elsevier/North-Holland, Amsterdam.

53. Muller, E.E., Camanni, F., Cocchi, D., Genazzani, A.R., Massara, F., Casanueva, F., DeLeo, V., and LaRosa, R. (1980): In: <u>Neuroactive Drugs in Endocrinology: Physiologic, Diagnostic and Therapeutic Applications</u>, edited by E.E. Muller. Elsevier/North-Holland, Amsterdam. (In Press.)

54. Muller, E.E., Nistico, G., and Scapagnini, U. (1978): <u>Neurotransmitters and Anterior Pituitary Function</u>. Academic Press, New York.

55. Muller, E.E., Parati, E.A., Panerai, A.E., Cocchi, D., and araceni, T. (1979): <u>Neuroendocrinology</u>, 28:313-319.

56. Muller, E.E., Salerno, F., Cocchi, D., Locatelli, V., and Panerai, A.E. (1979): <u>Clin. Endocrinol.</u>, 11:645-656.

57. Ohama, E., and Ikuta, F. (1976): <u>Acta Neuropath.</u>, 34:311-319.

58. Parker, K.M., and Eddy, R.L. (1976): <u>J. Clin. Endocrinol. Metab.</u>, 42:1188-1191.

59. Parkes, J.D., Debono, A.G., and Marsden, C.D. (1976): <u>Lancet</u>, 1:483-484.

60. Passouant, P., Besset, A., Descomps, B., Bonardet, A., Billiard, M., and Negre, Ch. (1979): <u>La Nouv. Presse Med.</u>, 8: 3237-3242.

61. Phillipson, O.T., and Bird, E.D. (1976): <u>Clin. Sci. Mol. Med.</u>, 50:551-554.

62. Podolsky, S., and Lepold, N. (1974): <u>J. Clin. Endocrinol. Metab.</u>, 39:36-39..

63. Polleri, A., Masturzo, P. Murialdo, G., and Carolei, A. (1980): <u>Acta Endocrinol.</u>, (Kbh.) 93:7-12.

64. Reed, T.E., and Neel, J.V. (1959): <u>Am. J. Human Genet.</u>, 11: 107-112.

65. Rinne, U.K. (1979): In: <u>Neuroendocrine Correlates in Neurology and Psychiatry</u>, edited by E.E. Muller, and A. Angoli, pp. 119-125. Elsevier/North-Holland, Amsterdam.

66. Sachar, E.J., Mushrush, G., Perlow, M, and Weitzman, E.D. (1972): <u>Science</u>, 178:1304-1305.

67. Schally, A.V., Arimura, A., Coy, D.H., Kastin, A.J., Meyers, C.A., Redding, T.W., Chihara, K., Huang, W.Y., Chang, R.O.O., Pedroza, E., and Vilchez-Martinez, J. (1979): In: <u>Central Regulation of the Endocrine System</u>, edited by K. Fuxe, T. Hokfelt, and R. Luft, pp. 9-29. Plenum Press, New York.

68. Schonberger, W., Grimm, W., and Ziegler, R. (1977): <u>Europ. J. Pediatr.</u>, 127:15-19.

69. Shaw, K.M., Lees, A.J., Hayes, S., Ross, E.J., Stern, G.M., and Thompson, B.D. (1976): <u>Lancet</u>, 1:194.

70. Sirtori, C.R., Bolme, P., and Azarnoff, D.L. (1970: <u>N. Engl. J. Med.</u>, 287:729-733.

71. Smith, R.C., Tamminga, C.A., Haraszti, J., Pandey, G., and Davis, J.M. (1977): <u>Am. J. Psychiatr.</u>, 134:763-768.

72. Tamminga, C.A., Neophytides, A., Chase, T.N., and Frohman, L.A. (1978): J. Clin. Endocrinol. Metab., 47:1348-1351.
73. Tang, L.C., and Coptzias, G.C. (1976): Arch. Neurol., 33: 131-134.
74. Weindl, A., and Joynt, R.J. (1972): In: Brain-Endocrine Interaction. Median Eminence: Structure and Function, edited by K.M. Knigge, D.E. Scott, and A. Weindl, pp. 280-297. Karger, Basel.

Neurosecretion and Brain Peptides,
edited by J. B. Martin, S. Reichlin, and K. L. Bick.
Raven Press, New York © 1981.

Human Plasma ACTH, Lipotropin, and Endorphin

Dorothy T. Krieger, Hajime Yamaguchi, and Anthony S. Liotta

Department of Medicine, Division of Endocrinology, Mt. Sinai School of Medicine,
New York, New York 10029

In the three decades that have followed the isolation of adrenocortocotropic hormone (ACTH), there have been numerous studies characterizing its regulation and biological action. In recent years, it has become evident that ACTH is synthesized as part of a precursor molecule (10,13) (Fig. 1) which also contains within it the structure of β -lipotropin (a pituitary peptide originally isolated by Li) (7), as well as other sequences such as γ -MSH, a substance of thus far uncharacterized biological activity. Both ACTH and β -lipotropin can serve as precursors for other peptides, some of which may possess additional biological activity (Fig. 1). In view of the coexistence of all these peptides in the precursor molecule, it is important to establish whether these peptides are regulated and secreted in a manner similar to that previously described for ACTH.

METHODOLOGICAL CONSIDERATIONS

In order to understand the secretion and ultimately the factors regulating secretion of ACTH and ACTH-related peptides, it is necessary to compare all the molecular species and the concentrations thereof, within the pituitary, to those released (in vivo and in vitro), and to characterize forms present in normal versus pathological states. Additionally, in evaluating the kinetics of hormone secretion in vivo, problems of metabolic clearance must be considered. Characterization of the pituitary forms should take into consideration the species studied (some species have both anterior and intermediate lobes, others lack an intermediate lobe). The precursor molecule is processed mainly to ACTH and β-LPH in the anterior lobe, and to α -MSH and β -endorphin in the intermediate lobe. Another consideration in elucidating the form of ACTH related peptides is the method of extraction. Some methods can generate forms which are not normally present within the pituitary or can result in poor extraction efficiency of some forms and the possible conversion

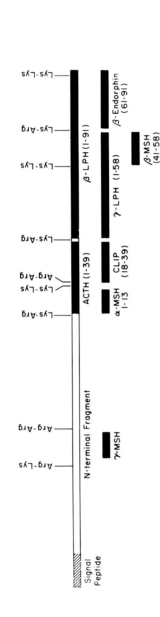

FIG. 1. Schematic representation of the bovine precursor molecule as proposed by Nakanishi et al. Following the signal peptide, there is an N-terminal fragment followed by ACTH (1-39) which is followed by the β-lipotropin (1-91) sequence. γ-MSH refers to the sequence of amino acid residues -55 to -44 (amino acid residues N-terminal to the ACTH sequences are given minus designation) in the N-terminal fragment as determined from the cloned complementary DNA sequence of a mRNA molecule. ACTH-(1-39) contains within it the sequence of α-MSH and CLIP (corticotropin-like intermediate lobe peptide). γ-lipotropins contains within the sequences of β-lipotropin, α-MSH and β-endorphin. Note that all the indicated peptide fragments are flanked on both sides by a pair of basic amino acid residues which can serve as a potential site of proteolytic cleavage to yield component peptides.

in vitro to forms not normally present in vivo. Such changes could conceivably occur by alteration of the activity of processing enzymes secondary to loss of tonic stimulatory or inhibitory factors present in situ but not under in vitro culture conditions. Changes in molecular form may also occur due to post-mortem autolysis of pituitary tissue. For a complete delineation of the naturally occurring pituitary forms, it is necessary to characterize the precursor-product pathway so that one may observe the sequences in which they appear (i.e., during the course of pulse-labeling studies). The present paper considers only those forms present in human plasma. Reference to pituitary forms will be made only when appropriate.

Since the adult human pituitary has no intermediate lobe (although some question has been raised as to whether there is invasion of the posterior lobe by cells with characteristics of intermediate-lobe cells), consideration only of the nature of forms present in anterior lobe is appropriate. In the anterior lobe, the precursor molecule is sequentially cleaved (presumably by trypsin and carboxypeptidase-like activities) to first yield two fragments, one being β-lipotropin and the other ACTH joined to a thus far not completely characterized amino terminal segment. This latter fragment is then cleaved to yield ACTH-(1-39) and the N-terminal segment. Unlike other species, in the human, ACTH-(1-39) exists only in an unglycosylated form. The N-terminal segment contains within it a sequence termed γ-MSH (because of its homology to previously described α- and β-MSH); the remainder of this portion of the molecule has not yet been characterized. Imura et al. (personal communication) have presented evidence of secretion of γ-MSH in the human with evidence that it is regulated in response to stimulation and feedback in a manner similar to that described for ACTH. Although, as indicated in Fig. 1, ACTH can potentially be cleaved to α-MSH and CLIP (ACTH 18-39), corticotropin-like intermediate lobe peptide) as occurs in the intermediate lobe, there is no evidence that this occurs to a significant degree in the anterior pituitary. Analogously, as indicated in Fig. 1, β-lipotropin can potentially be cleaved to γ-LPH and β-endorphin (as occurs in intermediate lobe), and γ-LPH further cleaved to β-MSH. This does not occur in anterior lobe. β-lipotropin is the predominant lipotropin present in anterior pituitary, constituting more than 90 percent of lipotropin immunoreactivity on a molar basis. β-endorphin immunoreactive-like material is present in human anterior pituitary at less than 1/10 the molar concentration of β-lipotropin.

These observations have to be tempered by the recent observations of Zakarian and Smythe (19) that rat pituitary material co-eluting with β-endorphin on cation exchange chromatography filtration represents a heterogeneous group of compounds comprising 61-87 β-endorphin and acetylated forms of this compound, as well as of native β-endorphin (Fig. 2). We have confirmed the finding of Zakarian and Smythe utilizing non-equilibrium iso-

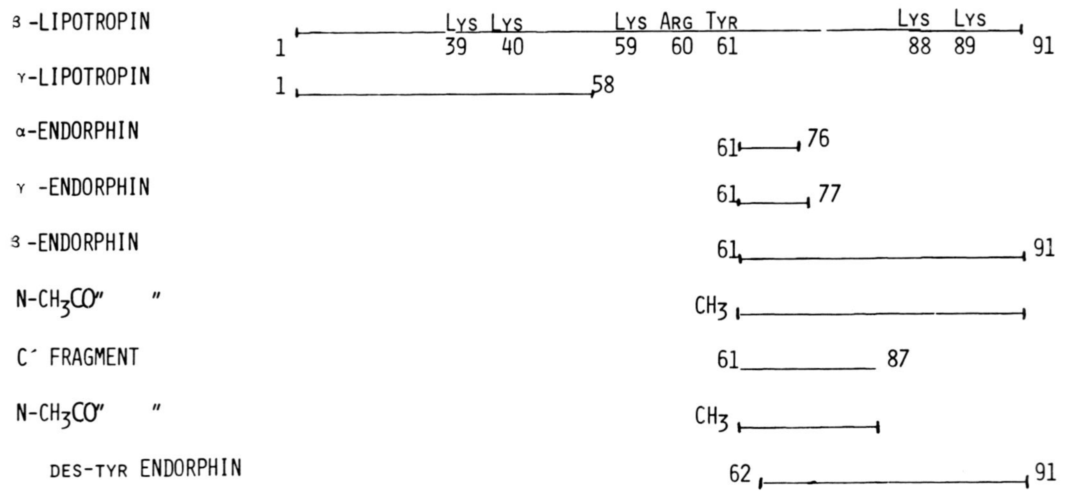

FIG. 2. Schematic representation of the β-lipotropin molecule and various peptide fragments that may potentially be contained within.

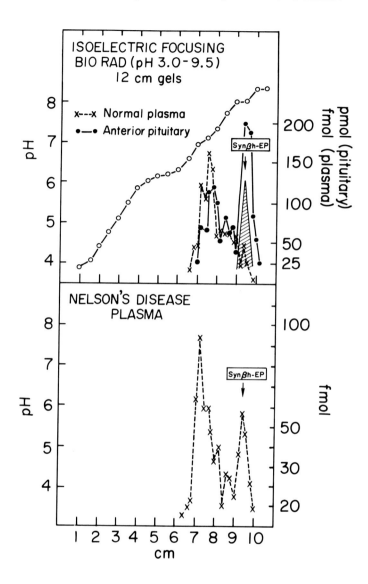

FIG.3. Charge heterogeneity of human plasma and pituitary immunoreactive "β-endorphin" as demonstrated by non-equilibrium isoelectric focusing (∿ 1000 V hrs). *Hatched area indicates position of synthetic β$_h$-endorphin. In both pituitary and plasma from normal subjects and a patient with Nelson's syndrome, the major portion of such immunoreactive material consists of components which are more acidic than β-endorphin (61-91).
*(Plasma and pituitary extracts were submitted to Sephadex G-50 filtration, and immunoreactive β-endorphin eluting with a K$_{av}$ similar to synthetic β-endorphin was pooled, concentrated, and applied to the focusing column).

electric focusing. Rat plasma reflects this heterogeneity and β-endorphin (61-91) appears to be a minor species with regard to total β-endorphin immunoreactive material (unpublished observations). Preliminary studies suggest similar findings for human plasma (9) (Fig. 3). Both N-terminal and midportion directed β-endorphin antisera would be expected to cross-react with these other endorphin molecules (though not necessarily on an equimolar basis). Since only the unacetylated β-endorphin (61-91) molecule exhibits significant opiate-like activity, two points should be emphasized: (1) dissociation of immunoreactive and opiate-like activity (bioassay, radioreceptor assay) should be expected, and (2) pituitary "β-endorphin" may have a physiological role unrelated to opiate activity.

Use of acid extraction of plasma as commonly used to concentrate peptides and remove substances which contribute nonspecific interference was designed to optimize the extraction of basic compounds. Using such extraction, the more acidic endorphin fragments may not be extracted with the same efficiency as β-endorphin (personal observation). Such fragments may also not react with equal affinity with β-endorphin antisera. It can be expected, therefore, that measurements of "β-endorphin" concentrations in plasma using β-endorphin as standard may underestimate or overestimate the concentrations of these other peptides.

It should also be noted that most currently available β-lipotropin antisera exhibit full or partial cross-reactivity with γ-lipotropin, so that "β-lipotropin" values reported represent the presence of these two peptides. We currently quantify immunoreactive β-LPH, γ-LPH and "β-endorphin" by an immune-affinity chromatography technique (manuscript submitted for publication) (Fig. 4). Since many β-endorphin antisera also exhibit cross-reactivity with β-LPH, this technique is useful even if only "β-endorphin" is to be quantified (8). The ACTH antibody employed in the present studies was one which reacted with the midportion of the ACTH molecule with significant cross-reactivity with either extreme N-terminal, C-terminal fragments or α-MSH. Values obtained correlated well with values obtained by bioassay.

STUDIES IN NORMAL SUBJECTS (BASAL LEVELS)

Morning ACTH and lipotropin concentrations are shown in Fig. 5 (4). These are in agreement with other published studies on concentrations of both of these peptides. There is a significant correlation (R = 0.65) between ACTH and "β-lipotropin" concentrations of unchromatographed plasma extracts. (As noted above, the antibody measures both β- and γ-lipotropin.) With the use of immune-affinity chromatography (Fig. 6), we have been able to demonstrate (Yamaguchi et al., submitted for publication) that considerable variability exists in the relative concentrations of β- and γ-LPH in normal subjects under basal conditions. Tanaka et

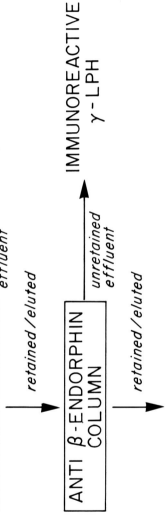

FIG. 4. Detection of plasma β-LPH, γ-LPH, and β-endorphin. To achieve separation of these peptides utilizing available antibodies (i.e., β-endorphin antibody which cross-reacts with both β-lipotropin and β-endorphin, and LPH antibody which cross-reacts equally with β- and γ-LPH), the above scheme utilzing

FIG. 4. Legend con't.
resin immobilized antibodies was employed. The extracts were
percolated through an anti-β-LPH column (which recognized β- and
γ-LPH with equal affinity, but does not react at all with
β-endorphin). The effluent from this column could be assayed
with the β-endorphin antibody, since any β-endorphin-like
material present in the plasma would not be retained in this
column. The anti-β-LPH column was then eluted and this eluate
percolated through the anti-β-endorphin column. The eluate
containing γ-LPH could be assayed in the β-lipotropin immunoassay. The material retained on the column (β-lipotropin) was
eluted and assayed for β-lipotropin.

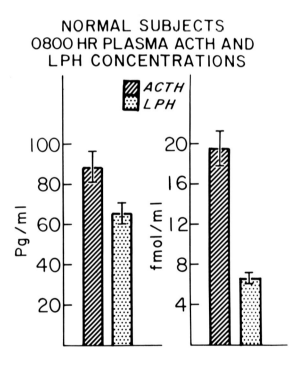

NORMAL SUBJECTS
0800 HR PLASMA ACTH AND
LPH CONCENTRATIONS

FIG. 5. Concentrations of plasma ACTH and lipotropin at 0800 in
normal subjects. Note: Lipotropin was measured with an antibody
that reacts with β-LPH and γ-LPH. (From ref. 6, with permission).

al. (16) have presented gel filtration data on two normal subjects also indicating the presence of significant quantities of
γ as well as of β-lipotropin. In view of this finding of significant amounts of γ-LPH in the plasma of normals subjects, one
would expect β-endorphin to be present as well. Recently (5),
using an affinity chromatographic concentration technique, we
have been able to detect low levels (2.2-15 pg/ml, mean 8.2
pg/ml) in 17 of 26 basal normal subjects. Similar concentrations
have been reported by Nakao et al. (12) and Wiedemann et al.
(18), while Wardlaw and Frantz (27) have reported concentrations
of 21.0 + 7.3 pg/ml. It is possible that such discrepancies
noted by different investigators may also reflect, as noted
above, different cross-reactivity of their antibodies with the
"β-endorphin" present in plasma (see Fig. 2).

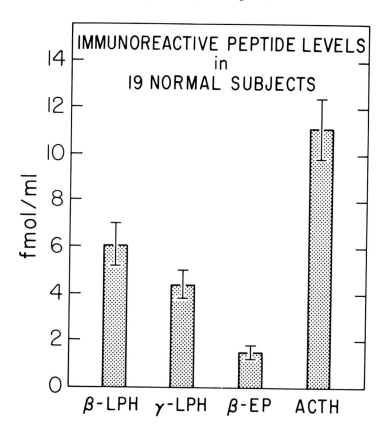

FIG. 6.   Basal levels of peptides in plasma as measured after
immune-affinity chromatography (see Fig. 3).

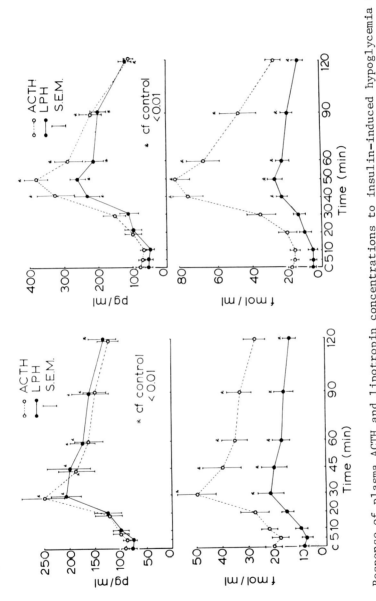

FIG. 7. Response of plasma ACTH and lipotropin concentrations to insulin-induced hypoglycemia (left-hand figure) and to Pitressin[R] administration (right-hand panel) in 7 normal subjects. Parallelism of plasma ACTH and lipotropin concentrations is evident. Note apparent longer half-life of lipotropin. Results are depicted on a gravimetric and molar basis. (From ref. 6, with permission).

RESPONSE TO STIMULATION AND SUPPRESSION IN NORMAL SUBJECTS

We have previously reported (Fig. 7) that plasma ACTH and lipotropin concentrations rise concomitantly following insulin-induced hypoglycemia, Pitressin[R] (4), and metyrapone administration (3). In preliminary studies, we have recently observed lesser increments of γ-LPH concentrations following such stimuli (Fig. 8). From this figure, it is also apparent that increments of β-endorphin are likewise less than those of β-LPH following such stimuli. Metyrapone administration is associated with greater γ-LPH/ β-LPH and β-endorphin/β-lipotropin ratios than seen following either hypoglycemia or Pitressin[R] administration (3). This may be related to the more prolonged stimulus that this agent provides or to the possible effect of decreased corticoid concentrations on peptide processing. With such decreased corticoid concentrations, there may be less stabilization of lysosomal membranes or altered activity of membrane-associated enzymes so that increased rates of proteolytic processing may take place which would not be seen under basal conditions. There are less data with regard to the effect of corticosteroid administration on levels of these peptides.

Suppression of both plasma ACTH and lipotropin concentrations is seen following dexamethasone administration (4). There appears to be a similar response of plasma β-endorphin concentrations, although this is more difficult to ascertain because of the lower basal levels present and even their undetectability in some normal subjects. Studies are not yet available with regard to γ-LPH concentrations.

There have been no detailed studies on the circadian periodicity of these peptides, or their episodic secretion in normal subjects. We have, however, noted the resumption of episodic secretion of ACTH and of LPH in an Addisonian subject following suppression of such concentrations by cortisol administration, and have also noted parallel episodic secretion in a patient with Nelson's syndrome.

PEPTIDE CONCENTRATIONS IN DISEASE

ACTH and plasma lipotropin concentrations are elevated in patients with Cushing's disease, Nelson's syndrome, ectopic ACTH secretion, and Addison's disease, with lower molar ratios of ACTH/lipotropin than seen in normal subjects (Fig. 9). An extremely high correlation coefficient ($R > 0.9$) for these two peptides exists. We have recently found greater concentrations of γ-lipotropin than of β-lipotropin in two pools of plasma from a patient with Nelson's syndrome and there are two other studies which report similar data (2,16). Patients with Addison's disease also have greater relative plasma concentrations of γ-lipotropin compared to those of β-lipotropin. Lesser increases in γ-LPH concentrations relative to those of β-LPH have been noted in patients with ectopic ACTH secretion and in patients with Cushing's disease. Plasma "β-endorphin" concentrations are

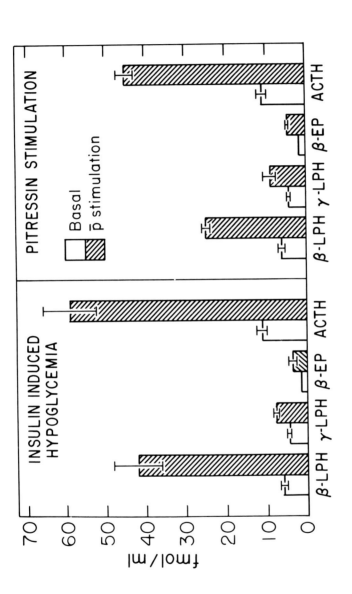

FIG. 8. Response of peptides deprived from pro-opiocortin to insulin-induced hypoglycemia and Pitressin[R] as measured by immune-affinity chromatography.

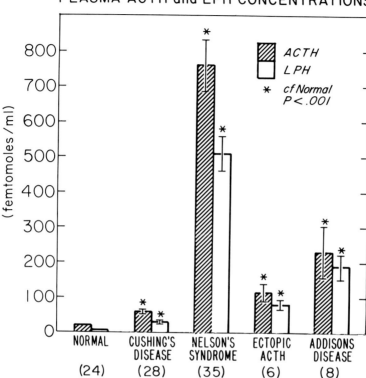

FIG. 9. Molar concentrations of plasma, ACTH and lipotropin in patients with Cushing's disease, Nelson's syndrome, ectopic ACTH-secreting tumors, and Addison's disease. In all cases, molar concentrations of ACTH were greater than those of LPH. (From ref. 6, with permission.)

also markedly elevated in all of these disease categories (Fig. 10) (15,18). The extent of elevation of "β-endorphin" concentrations in these conditions is similar to that noted for γ-LPH.

Suda <u>et al.</u> (14) have demonstrated that adenomatous tissues obtained from patients with Nelson's syndrome and Cushing's syndrome contain significant concentrations of β-endorphin in addition to LPH and ACTH. This is in contrast to the findings in normal human pituitary. Of great interest is their observation that ACTH and [β-LPH + β-endorphin] concentrations were suppressed in tissues surrounding the adenomas, but even in the suppressed tissues the proportion of β-endrophin present relative to that of β-LPH was greater than seen in normal pituitary tissue. This could indicate some residual abnormalities in the surrounding tissues as well as in the adenoma, which might suggest an extra-pituitary (<u>i.e.</u>, CNS) etiology.

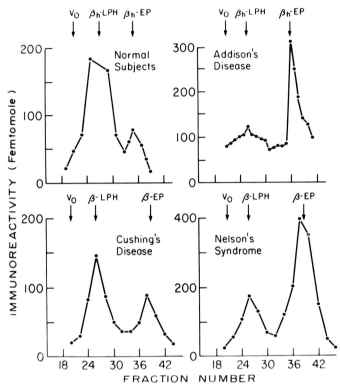

FIG. 10. Sephadex G-50 gel filtration of human plasma extracts from (A) normal subjects under basal conditions; (B) a patient with Addison's disease; (C) a patient with Cushing's disease associated with a macroscopic pituitary tumor; and (D) a patient with Nelson's syndrome. It is apparent that, in the normal subject, only a small amount of β-endorphin is present, whereas marked increases are seen in patients with these endocrine diseases. (From ref. 6, with permission.)

Patients with Cushing's disease and Nelson's syndrome show equivalent and concomitant percentage increases in plasma lipotropin and ACTH concentration in response to Pitressin[R] administration (4). The previously reported anomalous response of plasma ACTH concentrations following TRH administration to such patients (5) is also seen with regard to lipotropin concentrations.

In view of numerous reports of elevated concentrations and altered clearance of peptide hormones (such as growth hormone, parathormone, and prolactin) in patients with chronic renal disease, and the previously reported presence of "β-MSH" concentrations in such patients, it was of interest to study levels of the pro-opiocortin family of peptides in such patients. We have

reported elevation of " β -lipotropin" and "β -endorphin" concentrations in patients with chronic renal disease (1), with a
correlation coefficient of R = 0.93. Plasma ACTH concentrations
are much less strikingly elevated, and there is a poor correlation (R = 0.21) between plasma ACTH and "β-lipotropin" concentrations. The reason for such discrepancies is unclear, and may
possibly reflect differences in feedback mechanisms, intrapituitary processing, peripheral degradation and/or clearance.

CONCLUSIONS

It is apparent that in both normal subjects and patients with
hypersecretion of ACTH due to a variety of conditions, there is
concomitant secretion of all of the known peptides comprising the
pro-opiocortin molecule. There is also suggestive evidence that
the relative proportions of the peptides secreted may differ in
normal subjects under conditions of stimulation when compared to
basal levels, and that in states of pituitary hypersecretion relative concentrations of these peptides are altered from those
under basal conditions. It is further noteworthy that the proportions of these peptides in pituitary disease seem to differ
depending on whether there is hypersecretion based in response to
a lowering of plasma corticosteroid concentration or whether such
hypersecretion is secondary to tumor, either within the pituitary
or from an ectopic location. Thus far, of all peptides comprising the pro-opiocortin molecule, ACTH is the only one for which a
definite physiological role has been described. Elucidation of
the physiological role of the other peptides and their contribution to the pathophysiology of endocrine disease and perhaps in
other as yet uncharacterized states may provide exciting new
insights into one of the body's major homeostatic systems.

ACKNOWLEDGMENTS

These studies were supported in part by NIH grant 02893-13 and
the Lita Annenberg Hazen Charitable Trust.

REFERENCES

1. Aronin, N., Shickmanter, B., Liotta, A.S., and Krieger, D.T.
    (1979): Endocrine Soc. 61st Ann. Mtg., 225:129.
2. Gilkes, J.J.H., Rees, L.H., and Besser, G.M. (1977): Brit.
    Med. J., 1:996-998.
3. Krieger, D.T., Liotta, A.S., and Suda, T. (1978): In:
    Endorphins '78, edited by L. Graf, M. Palkovits, and A.Z.
    Ronai, pp. 275-294. Akademiai Kiado, Budapest.
4. Krieger, D.T., Liotta, A.S., Suda, T., Goodgold, A., and
    Condon, E. (1979): J. Clin. Endocrinol. Metab., 48:566-571.
5. Krieger, D.T., Liotta, A.S., Suda, T., and Yamaguchi, H.
    (1980): In: Brain and Pituitary Peptides, edited by W.
    Uttke, A. Weindl, K.H. Ziott, R.R. Dries, pp. 34-45. Karger
    Basel.

6. Krieger, D.T., Liotta, A.S., Brownstein, M.J., and Zimmerman, E.A. (1980): Rec. Prog. Horm. Res. Academic Press, New York. (In Press).
7. Li, C.H., Barnafi, L., Chretien, M., and Chung, D. (1965): Nature (Lond.), 208:1093-1094.
8. Liotta, A.S., Gildersleeve, D., Brownstein, M.J., and Krieger, D.T. (1979): Proc. Natl. Acad. Sci.USA, 76:1448-1452.
9. Liotta, A.S., and Houghten, R. (1980): Clin. Res., 28:270A.
10. Mains, R.E., Eipper, B.A., and Ling, M. (1977): Proc. Natl. Acad. Sci. USA, 74:3014-3018.
11. Nakanishi, S., Inoue, A., Kita, T., Nakamura, M., Chang, A.C.Y., Cohen, S.N., and Numa, S. (1979): Nature, 278: 423-427.
12. Nakao, K., Nakai, Y., Oki, S., Horii, K., and Imura, H. (1978): J. Clin. Invest., 62:1395-1398.
13. Roberts, J.L., and Herbert, E. (1977): Proc. Natl. Acad. Sci.USA, 74:5300-5304.
14. Suda, T., Demura, H., Demura, R., Jibiki, K., and Shizume, K. (1980): Sixth Int. Cong. Endocrinol., 762:590.
15. Suda, T., Liotta, A.S., and Krieger, D.T., (1978): Science, 202:221-223.
16. Tanaka, K., Nicholson, W.D., and Orth, D.N. (1978): J. Clin. Invest., 62:94-104.
17. Wardlaw, S.L., and Frantz, A.G. (1979): J. Clin. Endocrinol. Metab., 48:176-180.
18. Wiedemann, E., Saito, T., Linfoot, J.A., and Li, C.H. (1979): J. Clin. Endocrinol. Metab., 49:478-480.
19. Zakarian, S., and Smythe, D. (1979): Proc. Natl. Acad. Sci. USA, 76:5972-5976.

Neurosecretion and Brain Peptides,
edited by J. B. Martin, S. Reichlin, and K. L. Bick.
Raven Press, New York © 1981.

Effect of CNS Peptides on Hypothalamic Regulation of Pituitary Secretion

H. Imura, Y. Kato, H. Katakami, and N. Matsushita

*Department of Medicine, Kyoto University Faculty of Medicine,
Sakyo-ku, Kyoto 606, Japan*

In the past decade, a variety of peptides has been identified from the nervous tissue of various species. Immunohistochemical and radioimmunoassay studies in recent years have shown that these peptides in general are widely distributed in the central and periphral nervous systems and even in the endocrine cells of the gastrointestinal tract and pancreas. However, most of the neuropeptides are highly concentrated in the hypothalamus, suggestng their important role in regulating endocrine function.

This article will briefly review the effects of neuropeptides on the release of pituitary hormones, especially growth hormone (GH) and prolactin (PRL), and then focus on the effect of opioid peptides and their interaction with brain monoamines in regulating pituitary hormone secretion.

EFFECT OF NEUROPEPTIDES OTHER THAN OPIOID PEPTIDES ON GH AND PRL SECRETION

Although numerous reports on the effect of neuropeptides on GH and PRL secretion have appeared in recent years, the results have been sometimes contradictory. This is probably due to a variety of factors. For example, experimental conditions, such as species, sex, age, hormonal pretreatment and castration, may considerably affect the results. Anesthesia usually has profound effect on the hypothalamic pituitary axis and may reverse the response to certain drugs. In addition, many neuropeptides are not freely permeable through the blood-brain barrier and, therefore, the route of administration becomes an important factor. For these reasons, unanimous results have not always been obtained.

Thyrotropin Releasing Hormone (TRH)

TRH is known to enhance TSH and PRL secretion and, under certain conditions, GH secretion too. Our previous study showed that the intraventricular injection of TRH inhibited the chlor-promazine-induced increase of plasma GH in urethane-anesthetized rats (6).   GH and PRL secretion induced by pentobarbital, β-endorphin, morphine, and suckling was likewise inhibited by the intraventricular or systemic administration of TRH in anesthe-tized and unanesthetized rats (2,7). We have noted that the in-travenous infusion of TRH lowers secretion of GH induced by a variety of stimuli in normal human subjects (5,22). These results suggest that TRH has dual effects, one acting directly on the pituitary to enhance GH and PRL secretion and the other acting on the central nervous system to inhibit GH and PRL secretion.

Somatostatin

Somatostatin has a potent inhibitory effect on the secretion of GH and TSH from the anterior pituitary. We observed that the intraventricular injection of somatostatin increased plasma GH levels in urethane-anesthetized rats (1).   This stimulatory effect of somatostatin on GH secretion was less remarkable in rats bearing hypothalamic lesions, so we considered that it acted through the central nervous system.

Vasoactive Intestinal Polypeptide (VIP)

We observed that the intravenous or intraventricular injection of VIP caused a rise of plasma PRL in urethane-anesthetized rats (17).   This effect of VIP was suppressed significantly by the injection of naloxone, an opiate receptor antagonist (17).   Later studies confirmed the PRL stimulatory action of VIP in conscious rats and further demonstrated the enhancement of GH release by intraventricular injection of VIP (33). There are two possible sites where VIP might act: the pituitary and the hypothalamus. VIP is known to stimulate PRL release or to antagonize the inhib-itory action of dopamine on PRL release from the monolayer cul-tures of pituitary cells (17).   On the other hand, VIP was re-ported to inhibit somatostatin release from the hypothalamus.

Substance P

Intravenous injection of Substance P enhances release of GH and PRL in urethane-anesthetized rats (15,27).   The site of action of Substance P in this stimulatory action on GH and PRL release might be the pituitary, because this action has been observed in rats with large hypothalamic lesions (15) and because Substance P stimulates PRL release from the pituitary *in vitro* (30).
On the other hand, Chihara *et al.* (4) reported that the intraventricular injection of Substance P significantly inhibited GH release but not PRL release in urethane-anesthetized rats. Since this effect was abolished by the injection of anti-somato-statin sera and not by anti-Substance P sera, Chihara *et al.* (4)

speculated that Substance P injected into the ventricle enhances release of endogenous somatostatin, thus inhibiting GH secretion.

Neurotensin

Neurotensin, which shares several biologic actions with Substance P, also stimulates GH and PRL in urethane-anesthetized rats, when administered intravenously (27). The effect was blunted by diphenylhydramine, a histamine antagonist. However, a different route of administration gave an opposite effect; intraventricular injection of neurotensin was reported to lower plasma GH and PRL in anesthetized rats (21). In conscious rats, only a smaller dose of neurotensin is reported to lower plasma PRL when injected into the ventricle (30). Maeda and Frohman (21) compared the neuroendocrine action of neurotensin with that of histamine, somatostatin and Substance P, and suggested a peptidergic neurotransmitter role for neurotensin in the central nervous system. It is possible that intraventricularly administered neurotensin increases the release of somatostatin and PRL inhibiting factor, thus inhibiting GH and PRL secretion. On the other hand, the mechanism by which intravenously administered neurotensin stimulates PRL and GH release is not clear, although a direct effect on pituitary lactotropes has been suggested (30).

Cholecystokinin (CCK) and Gastrin

Intraventricular injection of CCK has been reported to raise plasma levels of GH at all doses used and of PRL at a certain dose in conscious, ovariectomized rats (32). On the other hand, intravenously administered CCK raised only plasma PRL in a dose-dependent fashion but did not raise plasma GH. Since CCK had no effect on the hemipituitaries in vitro, Vijayan et al. (32) suggested that CCK might act on the hypothalamus.

Vijayan et al. (31) reported that intraventricular injection of gastrin lowered plasma PRL and raised plasma GH especially at higher doses in conscious, ovariectomized rats. On the other hand, intravenously administered gastrin failed to modify plasma GH and PRL levels (31). These results indicate the hypothalamic site of action of gastrin in altering pituitary hormone secretion.

Summary

Effects of various neuropeptides on GH and PRL secretion are summarized in Table 1. Although contradictory results have been obtained in some details, it is evident that a variety of peptides affect pituitary hormone secretion. The observations that these peptides exist in high concentrations in the hypothalamus seem to suggest the physiological significance of the peptides in regulating pituitary function. However, it is difficult at the moment to determine whether the observed effects of peptides on pituitary hormone secretion are really physiological or rather

TABLE 1.   Effects of neuropeptides on plasma GH and PRL levels in rats

	GH		PRL	
	IV	ICV	IV	ICV
TRH	↑	↓	↑	↓
Somatostatin	↓	↑	→	→
VIP	↑	↑	↑	↑
Substance P	↑	↓	↑	→
Neurotensin	↑	↓	↑	↓
CCK	→	↑	↑	↑
Gastrin	→	↑	→	↓
Opioid Peptides	↑	↑	↑	↑

IV: intravenous injection.   ICV: intracerebroventricular injection.
↑ :enhancement.   ↓ :inhibition.   → :no change.

pharmacological, because relatively large doses of peptides might pharmacologically affect neurotransmitter release or turnover. Further studies are required to clarify the physiological role of these neuropeptides.

EFFECT OF OPIOID PEPTIDES ON GH AND PRL SECRETION

It is well-known that morphine has a variety of actions on pituitary hormone secretion. Therefore, the recent discovery of endogenous opioid peptides in the brain has prompted investigators to study the effect of opioid peptides on neuroendocrine function.

Effect of Opioid Peptides and Their Antagonist on GH and PRL Secretion in Anesthetized and Unanesthetized Rats

We have compared the effect on GH and PRL secretion of several opioid peptides administered into the ventricle of urethane-anesthetized rats. All opioid peptides used, β -endorphin, α -endorphin, Met-enkephalin and Leu-enkephalin, dose-dependently raised plasma PRL and GH (16). Fig. 1 shows the dose-related rises of plasma GH and PRL following the intraventricular injection of β -

endorphin. When compared on both a weight and molar basis, β - endorphin was most effective, followed by α-endorphin and en- kephalins in that order. When the increment of GH and PRL above the basal levels was compared, PRL was found to have increased more than GH. Smaller doses of opioid peptides raised only PRL but not GH. This stimulatory effect of opioid peptides was com- pletely inhibited by the administration of naloxone, an opiate receptor antagonist.

Many studies so far reported agree that opioid peptides and morphine injected intraventricularly have a stimulatory effect on GH and PRL secretion in anesthetized and conscious rats (8) and that β-endorphin is more potent than Met-enkephalin. Systemic administration of Met-enkephalin or its analogues such as FK 33-824 (D-Ala ^{2}, MePhe ^{4}, Met(o) ^{5}-ol)-enkephalin, also elicited a rise of plasma GH and PRL in urethane-anesthetized or in unanes- thetized, decapitated rats (3,18,19,25). It is clear, therefore, that all opioid peptides and morphine, regardless of the route of administration, enhance secretion of GH and PRL and that this effect is antagonized by naloxone.

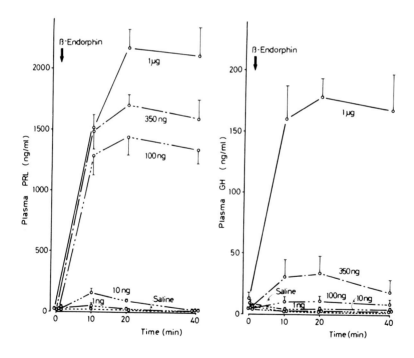

FIG. 1. Effect of intraventricular injection of varying doses of β-endorphin on plasma PRL and GH levels in urethane-anesthetized rats. (From ref. 16, with permission).

Effect of Opioid Peptides and Their Antagonists in Conscius Rats
Studied by Repeated Blood Samplings

Recent studies by Martin and his co-workers (24) have demonstrated that GH secretion is of pulsatile nature in unanesthetized, freely behaving rats bearing the intrajugular catheter.
The pulsatile or episodic release of GH occurs with an interval
of 3-4 hours. Plasma PRL also shows episodicity, although less
remarkable than plasma GH. If such pulsatile secretion is not
taken into consideration, experiments on unanesthetized rats may
give erroneous results. We have also measured plasma GH in unanesthetized rats by repeated blood samplings and confirmed the
presence of pulsatile bursts of GH secretion occurring at nearly

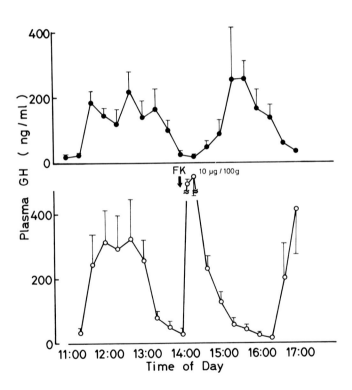

FIG. 2. Changes of plasma GH in conscious, freely moving rats
bearing the intrajugular catheter, as determined by serial blood
samplings (upper frame). Effect of intravenous injection of FK
33-824, 10 µg/100 g body weight, on plasma GH levels is shown in
lower frame. Mean ± SEM of 8 (upper) and 6 (lower) rats are
shown.

regular intervals (Fig. 2). Therefore, a potent enkephalin analogue, FK 33-824, was injected at the interval between two
pulses. As shown in Fig. 2, intravenous injection of FK 33-824,
10 µg/100 g body weight, raised plasma GH abruptly, followed by a
rapid decline. The following GH pulse seemed to appear later than
expected. Similar results were obtained when β-endorphin was
given into the ventricle. Plasma PRL also rose significantly
following the injection of FK 33-824 or β-endorphin (Fig. 3). We
then studied the effect of naloxone on FK 33-824-induced GH and
PRL secretion. Naloxone, given intravenously, significantly
suppressed the rise of plasma GH and PRL.

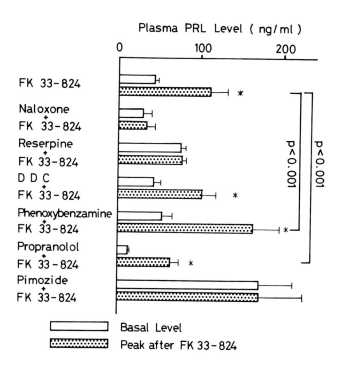

FIG. 3. Effect of FK 33-824 with or without various monoaminergic
agents and naloxone on basal and peak plasma PRL levels in
conscious rats. An asterisk represents a significant increase
over basal levels after the injection of FK 33-824. Mean ± SEM
of 4-8 rats are shown.

These experiments indicate that a rise of plasma GH after
injection of opioid peptides is not due to a fortuitous pulse of
GH secretion. It is also clear that opioid peptides raise plasma
PRL in conscious rats.

Effect of Opioid Peptides on GH and PRL Secretion in Species
Other Than Rats

Graffenried et al. (13) reported that a single intramuscular
injection of FK 33-824, a potent analogue of Met-enkephalin,
raised plasma GH and PRL dose-dependently in normal human subjects.   We also confirmed this observation and further observed
that rises of plasma GH and PRL induced by FK 33-824 are blunted
by the concomitant administration of naloxone (18,19). These
results suggest that opioid peptides enhance GH and PRL secretion
even in humans.

INTERACTIONS OF BRAIN MONOAMINES WITH OPIOID PEPTIDES
IN MODULATING SECRETION OF GH AND PRL

Brain monoamines, such as norepinephrine, dopamine and serotonin, have been demonstrated to exert an important influence on
GH and PRL secretion. Dopamine is a potent inhibiting factor
on PRL release from the pituitary, whereas it enhances GH release
probably acting on the hypothalamus. Norepinephrine seems to be
involved in modulating GH and PRL release, since adrenergic
blocking agents alter the hormone release.   Serotonin has a
stimulating efffect on GH and PRL secretion.   Based on these
observations, it can be assumed that the effect of opioid
peptides is mediated or modified by brain monoamines.

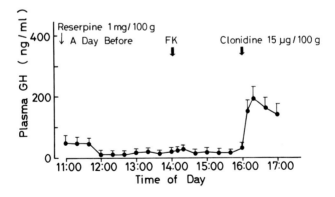

FIG. 4. Effect of pretreatment with reserpine, 1 mg/100 g body
weight given 24 h previously, on plasma GH levels and the
response to FK 33-824, in conscious rats. Means ± SEM of 6 rats
are shown.

Interaction of catecholamines with opioid peptides in modulating secretion of GH and PRL

In conscious, freely moving rats bearing the intrajugular catheter, we observed the effect of various monoaminergic agents on pulsatile GH bursts and on FK 33-824-induced GH and PRL release. Pretreatment wiht reserpine, 1 mg/100 g body weight given 24 hr previously, abolished GH bursts and the plasma GH response to intravenous administration of FK 33-824 (Fig. 4). Intravenous administration of clonidine re-established the pulse

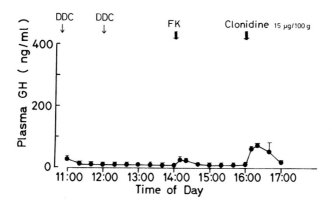

FIG. 5. Effect of diethyldithiocarbamate (DDC), 100 mg/100 g body weight in two divided doses, on plasma GH levels and the response to FK 33-824 in conscious rats. Means ± SEM of 8 rats are shown.

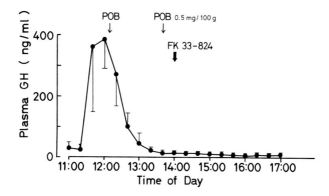

FIG. 6. Effect of phenoxybenzamine (POB), 1 mg/100 g body weight in two divided doses, on plasma GH levels and the response to FK 33-824 in conscious rats. Means ± SEM of 4 rats are shown.

of GH secretion. These results indicate that either catecholamines or serotonin or both are involved in pulsatile and FK 33-
824-induced GH secretion. PRL secretion induced by FK 33-824 was
also abolished by reserpine (Fig. 3).

We then studied the effect of two dopamine- β-hydroxylase
inhibitors, diethyldithiocarbamate (DDC) and fusarate. DDC, 100
mg/100 g body weight in two divided doses, given at 11:00 h and
12:00 h completely abolished pulsatile GH secretion as well as
the plasma GH rise induced by FK 33-824. Intravenous injection
of clonidine significantly raised plasma GH in DDC-treated rats
(Fig. 5). The plasma PRL response to FK 33-824 was not significantly affected by pretreatment with DDC (Fig. 3).   Similar
results were obtained with fusarate, another dopamine-β-hydroxylase inhibitor.   Pretreatment with two doses of fusarate, 5
mg/100 g body weight given at 12:00 h and 13:40 h suppressed
spontaneous GH bursts and significantly blunted the plasma GH
response to FK 33-824.

These results seem to suggest the importance of norepinephrine
but not of dopamine in regulating GH secretion. This is further
confirmed by the observation that intravenous injection of two
doses of a dopamine antagonist, pimozide, 0.05 mg/100 g body
weight, did not alter pulsatile GH bursts nor plasma GH response
to FK 33-824. On the other hand, basal plasma PRL levels were
significantly elevated by pimozide treatment and did not further
increase with FK 33-824 (Fig. 3).

In order to elucidate whether an α or β-adrenergic mechanism
is involved in FK 33-824-induced GH release, phenoxybenzamine and
propranolol were given to rats. As shown in Fig. 6, intravenous
injection of 2 doses of an α-adrenergic blocking agent, phenoxybenzamine, 0.5 mg/100 g body weight, suppressed spontaneous
pulsatile GH bursts and abolished the response to FK 33-824.
Plasma GH response to β-endorphin was also blunted by phenoxybenzamine. On the other hand, intravenous injection of propranolol, a β-adrenergic blocking agent, 0.5 mg/100 g body weight for
2 doses, did not significantly affect pulsatile GH release nor
plasma GH response to FK 33-824.   On the contrary, plasma PRL
response to FK 33-824 was blunted by propranolol and rather
augmented by phenoxybenzamine (Fig. 3). These results indicate
that norepinephrine, but not dopamine, is involved in both pulsatile GH release and GH release induced by opioid peptides, and
that norepinephrine exerts its effect through α-adrenergic
receptors. However, the adrenergic receptor mechanisms for FK
33-824-induced PRL release seems to be the opposite of that for
GH release, although further studies are required to obtain
conclusive evidence.

In recent years, evidence has been obtained for an α-adrenergic mechanism for rat GH secretion. Durand et al. (10) observed
that clonidine, an alpha agonist, was able to restore pulsatile
GH secretion in rats pretreated with α-methyl-p-tyrosine which
blocked episodic GH secretion. Willoughby et al. (34) reported
that (+) butaclamol, a dopamine receptor antagonist, did not

alter the basic configuration of pulsatile GH secretion, although the rise was partially reduced. Our results essentially agree with these earlier observations.

Our current studies have shown further that the α-adrenergic mechanism is involved in GH secretion induced by opioid peptides, whereas the dopaminergic mechanism has little if any effect. This is in contrast with the observations that PRL secretion induced by opium or opioid peptides is altered by dopamine agonists or antagonists as shown in our experiment mentioned above and by others (26). A decrease in turnover of dopamine in the hypothalamus (9) and a reduction in dopamine concentrations in the pituitary portal vessels induced by the intraventricular injection of opioid peptides (14) also suggest the involvement of dopamine in opium or opioid peptide-induced PRL secretion. On the other hand, Fuxe et al. (11) reported that ß-endorphin acted principally by increasing norepinephrine turnover in many hypothalamic nuclei. It is possible, therefore, that opium or opioid peptides stimulate release of norepinephrine from noradrenergic neurons, by exerting an inhibitory effect on inhibitory interneurons which are assumed to exist in the synapses of noradrenergic neurons. A similar stimulatory mechanism of opioid peptides has been proposed in hippocampal pyramidal neurons (35). The increased release of norepinephrine in the hypothalamus may act through growth hormone-releasing factor, since GH release induced by opioid peptides is not affected by anti-somatostatin antisera (4,8).

Interaction of Serotonin with Opioid Peptides in Modulating Secretion of GH and PRL

Recent studies have suggested the close interactions of brain serotonin with opioid peptides or morphine in their various central nervous system effects. Involvement of the serotonergic mechanism in pulsatile GH secretion has been demonstrated by using parachlorophenylalanine (PCPA) which is an inhibitor of serotonin biosynthesis or methysergide which is a serotonin antagonist (23). However, Tache et al. (28) reported that treatment with PCPA did not alter the plasma GH response to opioid peptide in rats.

On the other hand, morphine-induced PRL release in rats was reported to be antagonized by metergoline and cyproheptadine, which are antiserotonergic agents, as well as PCPA, and the destruction of neurons by 5, 7-dihydroxytryptamine (20). Studying the interrelationship between morphine and serotonergic agents on PRL release in rats, Meites et al. (26) reached the conclusion that opioid peptides and morphine act to increase PRL release by stimulating serotonin activity in the hypothalamus. It was reported that morphine or opioid peptides increase the turnover of brain serotonin (29). However, Tache et al. (28) failed to demonstrate the reduction of PRL response to opium and opioid peptides by PCPA treatment. Studies are now in progress in our laboratory to determine the interaction of serotonin in GH and PRL release induced by opioid peptides.

SUMMARY

A variety of neuropeptides, such as TRH, somatostatin, VIP, Substance P, neurotensin, CCK, gastrin, and opioid peptides, alter secretion of GH and PRL from the pituitary. These actions differ according to the route of administration or with experimental conditions, especially anesthesia. Among these peptides, the most consistent results have been obtained with opioid peptides, which stimulate GH and PRL release.

Both β-endorphin and enkephalins are capable of stimulating GH and PRL release in anesthetized and unanesthetized, freely moving rats. The effect is blocked by naloxone, an opiate receptor antagonist. GH secretion induced by opioid peptides seems to be mediated by an α-adrenergic mechanism, since treatment with DDC and fusaric acid, which are dopamine-β-hydroxylase inhibitors, reserpine, and phenoxybenzamine which is an α-adrenergic blocking agent, blunted GH secretion. However, pimozide, a dopamine receptor antagonist, and propranolol, a β-adrenergic blocking agent, were without effect. On the other hand, basal PRL secretion was augmented by pimozide, suggesting the possible involvement of dopamine. It is also possible that serotonin is involved in the GH and PRL release induced by opioid peptides.

The physiological significance of opioid peptides in regulating GH and PRL secretion is still unclear. Contradictory results (12,25) have been obtained concerning the effect of naloxone on basal or stimulated GH and PRL secretion in rats, monkeys and humans when tested by the continuous blood sampling method, which rules out the erroneous evaluation of results caused by episodicity of plasma hormone levels. Further studies should clarify the physiological role of opioid peptides in regulating pituitary function.

ACKNOWLEDGEMENTS

These studies were supported in part by grants from the Ministry of Education, the Ministry of Health and Welfare, and the Foundation for Growth Science, Japan. We are indebted to the National Institute of Arthritis, Metabolism and Digestive Diseases for supplying materials for rat GH and PRL radioimmunoassay. We thank Sandoz Ltd. (Basel), Endo Labs. (New York), and Daiichi Pharmaceutical Co. Ltd. (Tokyo) for the gift of FK 33-824, naloxone and β-endorphin, respectively. Thanks are also due to Misses Y. Mitsuda and K. Horii for secretarial assistance.

REFERENCES

1. Abe, H., Kato, Y., Iwasaki, Y., Chihara, K., and Imura, H. (1978): Proc. Soc. Exp. Biol. Med., 259:346-349.
2. Brown, M., and Vale, W. (1975): Endocrinology, 97:1151-1156.
3. Bruni, J.F., Van Vugut, D., Marshall, S., and Meites, J. (1977): Life Sci., 21:461-466.
4. Chihara, K., Arimura, A., Coy, D.H., and Schally, A.V. (1978): Endocrinology, 102:281-290.
5. Chihara, K., Kato, Y., Maeda, K., Abe., H., Furumoto, M., and Imura, H. (1977): J. Clin. Endocrinol. Metab., 44:78-84.
6. Chihara, K., Kato, Y., Ohgo, S., Iwasaki, Y., Abe, H., Maeda, K., and Imura, H. (1976): Endocrinology, 98:1047-1053.
7. Collu, R., and Tache, Y. (1979): In: Central Nervous System Effects of Hypothalamic Hormones and Other Peptides, edited by R. Collu, A. Barbeau, J.R. Ducharme, and J.G. Rochefort, pp. 97-121. Raven Press, New York.
8. Cusan, L., Dupon, A., Kledzik, G.S., Labrie, F., Coy, D.H., and Schally, A.V.(1977): Nature, 268:544-547.
9. Deyo, S.N., Swift, R.M., and Miller, R.J. (1979): Proc. Natl. Acad. Sci. USA, 76:3006-3009.
10. Durand, D., Martin, J.B., and Brazeau, P. (1977): Endocrinology, 100:722-728.
11. Fuxe, K., Andersson, K., Hokfelt, T., Mutt, V., Ferland, L., Agnati, L.F., Ganten, D., Said, S., Eneroth, P., and Gustafsson, J.-A. (1979): Fed. Proc., 38:2333-2340.
12. Gold, M.S., Redmond, D.E., Jr., and Donabedian, R.K. (1979): Endocrinology, 105:284-289.
13. Graffenried, B., Del Pozo, E., Roubicek, J., Krebs, E., Poldinger, W., Burmeister, P., and Kerp, L. (1978): Nature, 272:729-730.
14. Gudelsky, G.A., and Porter, J.C. (1979): Life Sci., 25:1697-1702.
15. Kato, Y., Chihara, K., Ohgo, S., Iwasaki, Y., Abe, H. and Imura, H. (1976): Life Sci., 19:441-446.
16. Kato, Y., Iwasaki, Y., Abe, H., Ohgo, S., and Imura, H. (1978): Proc. Soc. Exp. Biol. Med., 158:431-436.
17. Kato, Y., Iwasaki, Y., Iwasaki, J., Abe, H., Yanaihara, N., and Imura, H. (1978): Endocrinology, 103:554-558.
18. Kato, Y., Katakami, H., and Imura, H. (1980): In: Growth and Growth Factor, edited by K. Shizume, pp. 159-169. Tokyo Univ. Press, Tokyo.
19. Kato, Y., Matsushita, N., Katakami, H., and Imura, H. (1980): In: Advances in Prolactin, edited by M.L 'Hermite and S.L. Judd. Karger, Basel. (In Press).
20. Koenig, J.I., Mayfield, M.A., McCann, S.M., and Krulich, L. (1979): Life Sci., 25:853-864.
21. Maeda, K., and Frohman, L.A. (1978): Endocrinology, 103:1903-1909.

22. Maeda, K., Kato, Y., Chihara, K., Ohgo, S., Iwasaki, Y., and Imura, H. (1975): J. Clin. Endocrinol. Metab., 41:408-411.

23. Martin, J.B., Durand, D., Gurd, W., Faille, G., Audet, J., and Brazeau, P. (1978): Endocrinology, 102:106-113.

24. Martin, J.B., Renaud, L.P., and Brazeau, P. (1974): Science, 186:538-540.

25. Martin, J.B., Tolis, G., Woods, I., and Guyda, H. (1979): Brain Res., 168:210-215.

26. Meites, J., Bruni, J.F., and Van Vugut, D.A. (1979): In: Central Nervous System Effects of Hypothalamic Hormones and Other Peptides., edited by R. Collu, A. Barbeau, J.R. Ducharme, and J.-G. Rochefort, pp. 261-271. Raven Press, New York.

27. Rivier, C., Brown, M., and Vale, W. (1977): Endocrinology, 100:751-754.

28. Tache, Y., Charpenet, G., Chretien, M., and Collu, R. (1979): In: Central Nervous System Effects of Hypothalamic Hormones and Other Peptides, edited by R. Collu, A. Barbeau, J.R. Ducharme, and J.-G. Rochefort, pp. 301-313. Raven Press, New York.

29. Van Loon, G.R., and DeSouza, E.B.(1978): Life Sci., 23:971-978.

30. Vijayan, E., and McCann, S.M. (1979): Endocrinology, 105:64-68.

31. Vijayan, E., Samson W.K., and McCann, S.M. (1978): Life Sci., 23:2225-2232.

32. Vijayan, E., Samson, W.K., and McCann, S.M. (1979): Brain Res., 172:295-302.

33. Vijayan, E., Samson, W.K., Said, S.I., and McCann, S.M. (1979): Endocrinology, 104:53-57.

34. Willoughby, J.O., Brazeau, P., and Martin, J.B. (1977): Endocrinology, 101:1298-1303.

35. Zieglgansberger, W., French, E.D., Siggins, G.R., and Bloom, F.E. (1979): Science, 205:415-417.

Neurosecretion and Brain Peptides,
edited by J. B. Martin, S. Reichlin, and K. L. Bick.
Raven Press, New York © 1981.

Introduction to Section IX:
New Frontiers in Peptides

Seymour Reichlin

*Endocrine Division, Department of Medicine, Tufts University School of Medicine,
New England Medical Center, Boston, Massachusetts 02111*

This section includes a wide range of papers, some relating to concrete problems in evaluation of peptide secretory abnormalities in man, and some to speculative articles about potential implications of peptides in neurological disease, potential usefulness of peptide agonists and antagonists, and non-neural systems for study of neuropeptide regulation.

In his chapter, Dr. Reichlin asks whether the mode of control of secretion of neuropeptides in neural tissue resembles that in other tissues in which these substances are found. As a corollary, he has tried to assess the extent to which study of non-neural tissues is relevant to the study of brain. He focuses on TRH and somatostatin, but the questions are equally relevant to the numerous other neuropeptides, most of which have extensive distribution outside of the brain.

Dr. Nathanson summarizes current knowledge at the cellular level of the mode of actions of biogenic amines and peptides and the extent to which they are mediated through "second messengers".

Crucial tools for elucidating the role of specific neuropeptides in physiological (and pathological) states are agents that selectively stimulate and antagonize specific receptors. Work with biogenic amine agonists and antagonists illustrates the progress that can be made with this approach. Some examples of similar compounds in the field of peptide research are naloxone (an opiate receptor antagonist), DDAVP (a vasopressin analogue that resists degradation and has a favorable ratio of antidiuretic to pressor action), and saralasin (an angiotensin II receptor antagonist). Dr. Vale describes work with analogues of LHRH and somatostatin, and general principles of peptide modification for pharmacologic effects.

Dr. Reichelt and coworkers present some of their provocative studies of peptide-containing fractions obtained from the urine of patients with schizophrenia and childhood autism. The basis of this work is the hypothesis that "the genetic inheritance of several psychiatric disorders may be due to key peptidase insufficiency. Increased loading or strain on the peptidergic system regulated by such a key peptidase could reach a point where peptide release would exceed breakdown, with peptide excretion, in addition to pathology, resulting." Using an ethanol-benzoic acid precipitation method, he and his collaborators have extracted a mixture of peptides, glycoproteins, uric acid and unidentified material from urine, separated them by column chromatography, and empirically defined patterns of excretion peaks whose chemical nature is as yet unknown. The separate peaks have been concentrated and tested for biological activity in a wide variety of in vitro and in vivo neuropharmacological test systems. In the evaluation of his work, it is well to point out that the specificity of test responses given by relatively crude biological materials, the site of origin of the biologically active material, i.e., the brain, gut or elsewhere, and the general thesis that brain peptides would accumulate due to loss of specific peptidases all remain to be validated. Nevertheless, the identification of new biological activities in urine, elucidation of their chemical nature, and possible correlations with states of psychiatric abnormality are of continuing interest.

Dr. Bird deals with practical issues of obtaining, standardizing, and validating collection of human brain tissue for studies of neuropeptides in neurological and psychiatric disease, and summarizes current knowledge.

Issues of ontogeny, and differentiation of brain are considered by Dr. Caviness. He speculates that a system of brain peptides, analogous to nerve growth factor, selectively distributed in brain, might act as local, highly restricted tissue regulators. Several diseases of man and genetically determined mutant neurological disorders of the mouse are used as examples.

Finally, Dr. Martin speculates to provide an outline of potential implications of peptides in neurological disease. Melding chemical insights with newer neurobiological information and techniques, has provided a valuable blueprint for future work in this area.

Neurosecretion and Brain Peptides,
edited by J. B. Martin, S. Reichlin, and K. L. Bick.
Raven Press, New York © 1981.

Systems for the Study of Regulation of Neuropeptide Secretion

Seymour Reichlin

Endocrine Division, Department of Medicine, Tufts University School of Medicine,
New England Medical Center Hospital, Boston, Massachusetts 02111

Among the most dramatic insights of the "peptide revolution" considered in this volume has been the recognition that the same peptide substance may play different roles in different sites. Noradrenaline has long been recognized as both a neurotransmitter and a hormone, but other neurotransmitters and hormones have enjoyed a certain exclusivity of function defined either by the anatomical distribution of nerve endings or by the distribution of specific receptors on cells by which circulating trophic hormones might exert specific effects. Not so with many of the newly discovered peptides that may serve in some locations as hormones, in others as neurotransmitters, in still others as paracrine regulators, and as "autocrine" controllers of function. The wide distribution and varied functions of somatostatin best exemplify this assertion (cf., 96, for recent review). The first recognized hormonal function of somatostatin was as a regulator of pituitary function, an action mediated by release of the peptide as a neurosecretion into the hypophysial-portal circulation. Although this vascular system is small and circumscribed, its interposition between effector cell (the hypophysiotropic peptidergic neuron) and the regulated cell (the pituitary cell) allows us to speak of it as a true hormone, in a strict sense; that is, a secretion formed by one organ that exerts an effect on a remote site after circulation in the blood. In this situation (as with all hormones), specificity of action is further ensured by the presence of specific membrane receptors for somatostatin on only certain types of pituitary cells. Somatostatin also acts as a hormone in the rare circumstance in which tumors secrete excess peptide into the blood to produce inhibition of growth hormone, insulin and glucagon secretion. In brain sites outside the hypothalamus, somatostatin may act as a neurotransmitter or as a neuroregulator. It is present in a population of brain cells diffusely distributed in cerebral cortex, midbrain and spinal cord, is found in the dorsal ganglia of sensory neurons and in the vagal and sympathetic nerves. Its presence in cerebrospinal fluid is evidence that the cellular secretory product leaves the neuron and enters the perineural space. Within the pancreas and GI tract, somatostatin acts as a paracrine regulator, but the finding that the hormone enters both the blood draining the gut and the lumen of the gut suggests that it may also exert effects by one or both of these other routes. The idea that gut somatostatin may act as a true hormone in regulating gut function is supported by the recent studies of Schusdziarra et al. (110). The peptide may also act to regulate gut secretion after release into the lumen as proposed by Uvnas-Wallensten and collaborators (124). Peptides that act after entry into the lumen were termed by Vinick "Lumones", i.e., lumenal hormones (126).

Recently noted are the possible <u>autocrine</u> effects of somatostatin. In the parafollicular (C cell) of the thyroid are two populations of calcitonin secreting cells, one of which contains both somatostatin and calcitonin and another that contains only calcitonin (125). Since infusions of somatostatin inhibit calcitonin secretion (42), this may be by a paracrine mechanism (release of somatostatin into tissue space, and regulation of neighboring C cells) or by an autocrine mechanism (feedback control within the cell itself).

Multifunctional actions of the other neuropeptides have also been recognized. For example, thyrotropin releasing hormone (TRH) in hypophysial-portal blood is a hormone, in neurons of the rest of the brain, a neurotransmitter or neuromodulator, and in the gut and pancreas it is presumed to act as a paracrine secretion (57,74). Luteinizing releasing hormone (LHRH) which acts as a hormone in hypophysial-portal blood has recently been identified as a neurotransmitter in frog sympathetic ganglia (58). The list of peptides with multiple sites of action is as long as the list of recognized neuropeptides. This chapter deals with biological significance of the fact that individual peptides may mediate different actions in different sites. We would like to know the extent to which the secretion of a given neuropeptide is regulated by the same stimuli in all sites and whether the response to the peptide in all tissues is fundamentally similar. Can the mode of regulation of secretion of neuropeptides in brain be inferred from a study of regulation of non-neural systems? An even more restricted question of importance is whether the control of neuropeptide secretion in one part of the brain (and the neuronal response to that peptide) is the same in other parts of the nervous system. Is the chemical nature of the receptors for the neuropeptide the same in all responsive tissues? Remarkably little information is available to help answer these questions. I will consider in this paper regulation of TRH secretion in mammalian hypothalamus as compared with frog skin, somatostatin synthesis and regulation in brain, pancreas and gut, and the interaction of vasopressin and somatostatin in toad bladder. My objective is to define what is known about these topics, and to call attention to promising possibilities for future investigation.

HYPOTHALAMIC SECRETION-SKIN SECRETION

One approach to determination of the factors that regulate the secretion of TRH from the hypothalamus has been to analyze effects of neuropharmacologic agents on control of TSH secretion. This approach is subject to the important limitations that TSH secretion is determined by two hypophysiotropic hormones in addition to TRH (somatostatin and dopamine) and that neuropharmacologic agents modify the secretion of these substances as well as of TRH (17,95,103).

Reviews of neurotransmitter control of TSH release have been published by Müller and collaborators (77), by Tuomisto et al. (118), Jackson et al. (57), Krulich (67) and Kordon (60,63). The most convincing data (from whole animal studies) is that norepinephrine is stimulatory to TSH release. Intraventricular injection of norepinephrine or of clonidine (an α-adrenergic agonist) stimulates TSH secretion (2,67); blockade of synthesis of NE with disulfiram or blockade of α-adrenergic receptors bring about a decrease in serum TSH (64,118). Cold exposure-induced release of TSH is blocked by central NE deficiency brought about by disulfiram, DDC, or by α receptor blockage (65,118). Less certain from whole animal studies is the effect of dopamine because dopamine releases somatostatin, a TSH inhibitory factor (see below) and acts directly on the pituitary (in some studies) to inhibit TSH secretion. Most, but not all, workers have found that serotonin is inhibitory to TSH secretion, and has no direct effect on the pituitary. Administration of serotonin precursors lowered TSH levels in hypothyroid humans, inhibited TSH release in response to cold in rats (118), and lowered basal levels of TSH in rats (75). Intraventricular injection of serotonin has given contradictory findings, Jordan et al. (59) reporting stimulation and Krulich et al. (73) inhibition. Procedures that reduce central serotonin "tone" were found by Ruzsas et al. (100) to increase plasma TSH levels, thyroid hormone levels, and hypothalamic TRH content. Procedures that modify central acetylcholine actions appear to be without effect on TSH release.

The first efforts to define neurotransmitter control of TRH secretion in isolated hypothalamic fragments were those of Grimm and Reichlin (38). We incubated mouse hypothalamic fragments in radiolabeled histidine to allow [^{3}H] histidine-TRH to form, and then determined the effect of several neurotransmitters on the release of labeled material separated chromatographically. In this work, the identity of the released labeled TRH was established by chromatography, and isolation to constant specific activity of product with synthetic ^{3}H-TRH. However, subsequent advances in TRH isolation methods by McKelvy and Grimm (now Grimm-Jorgensen) cast doubt on the validity of proof of identity of the labeled material as being TRH exclusively (78,83). Nevertheless, in the light of later findings, the results of the mouse incubation work merits consideration. We found that the release of "IRH" was stimulated by norepinephrine and by dopamine. The stimulatory effect of dopamine was inhibited by exposure to disulfiram, an agent that blocks conversion of dopamine to norepinephrine. In this system, serotonin was inhibitory to "TRH" release, while acetylcholine had no effect. More recently, three studies on regulation of TRH from hypothalamic fragments in vitro have been published utilizing specific radioimmunoassay methods. Hirooka et al. (44) reported that norepinephrine releases radioimmunoassayable TRH, while Maeda and Frohman (71) found no effect from this treatment, although they did find that dibutyryl cAMP did stimulate TRH release. Joseph-Bravo et al. (60) did not observe any effects of norepinephrine either. Hirooka et al. reported that somatostatin inhibited TRH release, while Maeda and Frohman could demonstrate no effect of this peptide, nor an effect of serotonin or acetylcholine. The latter workers found that dopamine stimulated TRH release and believe this to be through a direct dopamine receptor agonist effect, since it could be duplicated by bromocriptine and blocked by haloperidol. Of a variety of putative transmitters assayed, Joseph-Bravo et al. (60) found that only histamine released TRH.

Thus, the limited in vitro work available fails to clarify the nature of neurotransmitter control of hypothalamic TRH secretion, but virtually all in vivo studies and some of the in vitro work suggests a stimulatory role for norepinephrine.

Knowledge about the neurotransmitter regulation of extrahypothalamic brain TRH is limited. Catecholamine depletion by reserpine, or α-methyl-p-tyrosine had no effect on either hypothalamic or cerebral cortical TRH concentration (61), but high doses of 6-OH dopamine caused a striking increase in TRH concentration of forebrain and posterior cerebral cortex (128). Although these effects may be attributable to its action in inhibiting norepinephrine synthesis, Winokur et al. (128) point out that 6-OH dopamine in the high doses used has effects on brain serotonin and acetylcholine. Parachlorophenylalanine, an inhibitor of serotonin synthesis, also was without effect on cortical or hypothalamic TRH content (61). Amphetamine administration lowers striatal TRH concentration by an action on nigrostriatal dopaminergic pathways (116). Unfortunately none of these studies differentiate effects on release from those on synthesis, or take into account that the same cell may contain both a biogenic amine and the neuropeptide as has been shown for some sympathetic ganglia cells (norepinephrine and somatostatin; norepinephrine and enkephalin) and cells in the lower medulla oblongata (serotonin and Substance P) (36).

The recent demonstration from this laboratory by Jackson and collaborators that frog skin was rich in TRH (12,56) (Fig. 1), and the widely held view that neural tissue and skin are derived from a common ectodermal origin (90), led our group to study the neuropharmacological control of TRH from frog skin in tissue incubates (76). Norepinephrine stimulated release of TRH in a dose-related fashion; the effect was blocked by prior treatment with phenoxybenzamine, an α-receptor blocker, leading to the

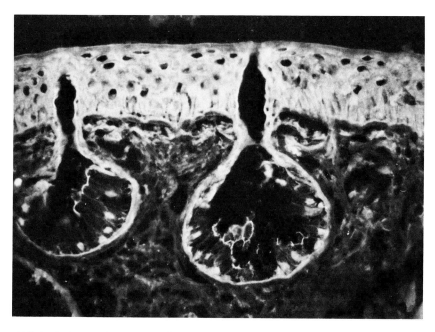

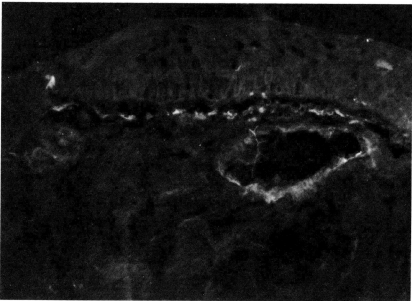

FIG. 1. Immunocytochemical localization of TRH in frog skin. Upper panel: anti-TRH. Lower panel: anti-sera absorbed with excess TRH (from ref. 11, with permission).

conclusion that the response was due to α-adrenergic stimulation
(Fig. 2).   Acetylcholine and epinephrine were without effect in
this system.   Subsequent studies by Bolaffi and Jackson (13)
confirmed the effects of norepinephrine and further showed that
release was unaffected by histamine or somatostatin.

DOSE RESPONSE EFFECTS OF NOREPINEPHRINE ON
THE RELEASE OF TRH AND 5 HT BY FROG SKIN

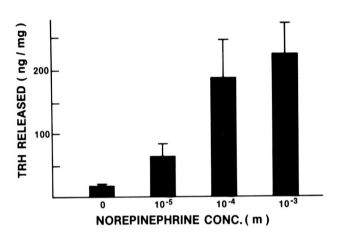

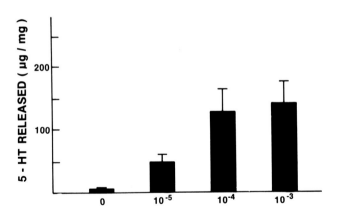

FIG. 2.   Release of TRH and of serotonin (5HT) from frog skin
incubates in the presence of various concentrations of
norepinephrine.

When we take all the available data into account, we are left with the one comforting thought that norepinephrine may be a releaser of TRH in two systems as diverse as rat hypothalamus and frog skin and that acetylcholine is without effect in both. It would be premature to generalize from these observations that norepinephrine is the universal TRH releaser, but, at the very least, findings of this kind point out the kind of studies that are needed. These include the systematic analysis of transmitter effects, of characterization of the receptors and of the mechanisms of neurotransmitter action in each of the known TRH secreting systems. Since TRH secretion is crucial to the feedback regulation of the pituitary-thyroid axis, determination of effects of thyroid hormone on TRH secretion in hypothalamus and all other sites in which it is localized are of importance.

SYSTEMS FOR THE STUDY OF SOMATOSTATIN SYNTHESIS, TRANSPORT AND SECRETION

Somatostatin Biosynthesis

The widespread distribution of somatostatin, and its importance in modulating function in many tissues, including the pancreas, has encouraged great interest in mechanisms of somatostatin biosynthesis. There is now overwhelming evidence that somatostatin is formed as part of a precursor molecule that is subsequently processed to its secreted form. In this respect, it resembles all other peptide secretions that have been well characterized (see chapter by Habener, this volume). The first indication that somatostatin was formed as part of a prohormone was the finding that tissue extracts contain immunoreactive somatostatin molecules of larger size than the tetradecapeptide (3). This has now been shown for extracts of rat median eminence, anterior hypothalamic-preoptic area, amygdala and parietal cortex, pancreas and gastrointestinal tract in several species (68,80,99,130). In general, there now appear to be at least four immunoreactive forms, one corresponding to cyclic somatostatin (1.6K) and others corresponding to approximately 6K,10K, and 15K (68). The precise size of the larger somatostatin molecules has not been fully settled, different authors assigning somewhat different apparent MW to the products. One of the larger somatostatins isolated from porcine intestine has been sequenced and found to be an N-terminally extended molecule with 28 amino acid residues (92). This substance has a calculated molecular weight of about 3.6K. More recently, Schally and collaborators (104) report that a molecule with a sequence identical to the pig intestine somatostatin has been isolated from pig hypothalamus. Mention should be made of the report that somatostatin molecules extracted from angler fish pancreas and pigeon pancreas (114) have the same sequence as typical mammalian tetradecapeptide, but that the prosomatostatin sequence corresponding to the 28 residue somatostatin, exclusive of the tetradecapeptide, is different

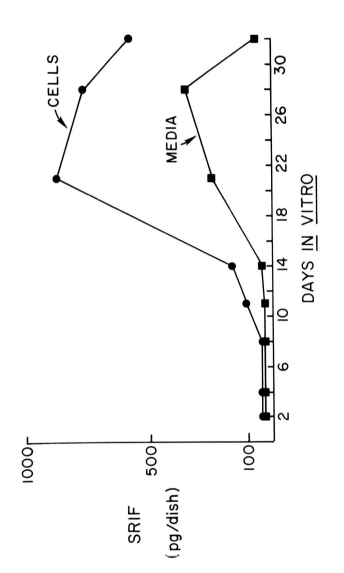

FIG. 3. Synthesis of immunoreactive somatostatin by dispersed cerebral cell culture from fetal rat (27).

from the porcine material (Goodman, personal communication). Catfish pancreas contains two distinct somatostatin molecules with different sequences, neither of which is identical to mammalian somatostatin tetradecapeptide (80). The findings indicate that at least four different genes must be involved in controlling somatostatin synthesis. Two of the larger forms can be processed to smaller forms by incubation in the presence of trypsin (68), or tissue extracts (129), or secretory granules (33), further supporting the presumption that post-translational processing must occur.

Incorporation studies of amino acids into somatostatin-like substances have now been reported for pancreas (31,80) and for hypothalamic fragments (31), and these also show initial formation of a larger molecular form compound of immunoreactive somatostatin.

Working initially in collaboration with Dichter and Delfs (see this volume), our group demonstrated that dispersed cell cultures of fetal rat cerebral cortex synthesize somatostatin actively after a lag period of approximately 6-7 days (Fig. 3) (27). Using this preparation we then studied the incorporation of tritiated phenylalanine into immunoreactive somatostatin and determined that within an hour the principal radioactivity was associated with one of the "large" somatostatin peaks (Fig. 4).

Axoplasmic Transport of Somatostatin

As outlined in this volume by Gainer, neuropeptides are generally synthesized on endoplasmic reticulum, assembled in the Golgi apparatus, and transported by axoplasmic flow to secretion sites in axon terminals. At least in the case of the neurohypophysial peptides, post-translational processing continues even after the nascent peptide has been packaged in granule form, but it is not known with certainty whether other peptides are similarly transformed. Owing largely to the inaccessibility of central somatostatinergic pathways to study, not much is known about axoplasmic transport of this peptide, nor of the factors that control its transport. Surgical isolation of the anterior-preoptic area (a principal site of somatostatinergic cell bodies) leads to a depletion of the peptide in the median eminence (16), a finding that is interpreted quite reasonably to mean that the peptide is transported there from perikarya situated outside the cut.

The discovery of somatostatin in peripheral nerves by Hokfelt and collaborators (46,47) suggested to us that standard methods for evaluation of axon transport might be applicable to the study of this compound. Hokfelt showed that section of the dorsal root lowered the concentration of somatostatin in the spinal cord corresponding to the segments of section (46). Our group has studied somatostatin transport in sciatic nerve by the ligation method (93). Following ligation, somatostatin accumulates proximally, and the rate of accumulation can be used to estimate synthesis and transport of the peptide, provided that appropriate corrections are made for the "fixed" vs the "labile" phase of the

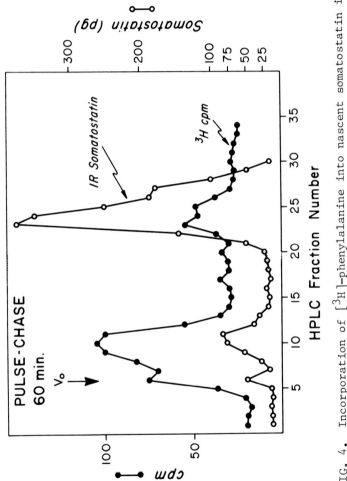

FIG. 4. Incorporation of [³H]-phenylalanine into nascent somatostatin in cerebral cell culture. Dispersed cells, incubated for 13 days, exposed to phenylalanine-free media for 16 hours, followed by addition of 2250 μCi[³H]-phenylalanine for 120 minutes and a 60 minute "chase" with phenylalanine containing media.

peptide, and consideration of retrograde transport rate (Table 1). We found that hypothyroidism produced in rats reduced the rate of accumulation of somatostatin proximal to the ligature, a finding analogous to that seen by us in acrylamide-induced neuropathy. If transport and/or synthesis are reduced in hypothyroidism, it is reasonable to speculate (but without supporting data) that secretion rate from nerve endings might also be reduced. I emphasize this point here because of the recent findings of Berelowitz and collaborators (9) that hypothyroidism reduces the rate of release of somatostatin from hypothalamic fragments in vitro, and that treatment with T_3 in vivo, or in vitro, stimulated somatostatin release. It may be that control of somatostatin secretion by thyroid hormone in both hypothalamus and peripheral nerve are subject to the same regulatory influences. In the hypothalamus, thyroid deficiency-induced low somatostatin secretion is compatible with the known hypersensitivity of the pituitary of hypothyroid animals to injection of TRH.

TABLE 1.   Axoplasmic transport of somatostatin in rat sciatic nerve

Apparent Transport Rate	79.8mm/24 hr
Mobile Fraction	19.2%
Calculated Absolute Transport Rate	415.6mm/24 hr
Amount Transported Distally per nerve (assuming no degradation in situ)	27ng/24 hr
Sciatic Dorsal Ganglia content	0.3ng per nerve
Calculated turnover in ganglia (assuming that proximal support is equal to distal transport)	18X per 24 hr

Recently Gilbert and collaborators (36) and Lundberg et al. (70) have demonstrated axoplasmic transport of somatostatin (as well as several other peptides) in the vagus nerve using the crush technique. It remains to be seen whether the factors regulating transport in this system are the same as in brain and other peripheral nerves.

Chromatographic analysis of the immunoreactive somatostatin in the sciatic nerve indicates that it is present in a form corresponding to the tetradecapeptide, whereas in dorsal ganglia approximately 16% is in a larger form (Fig. 5). We conclude from this work that processing of peptide is essentially complete by the time it reaches the axon process and that the tetradecapeptide is the secreted form. However, there is good evidence that median eminence contains both large and small forms of immunoreactive somatostatin (99), as is also true for the immunoreac-

tive material released into the hypophysial portal blood (122).
It is thus apparent that different neurons may process somatostatin differently.   All forms, both large and small, of
somatostatin are biologically active (99,104,115).   Different
forms may have different half lives in the synapse, may be
metabolised differently, and perhaps have different potencies.

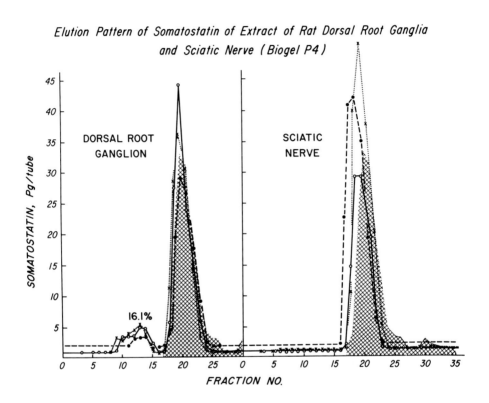

FIG. 5.   Chromatography on Biogel P10 of somatostatin
immunoreactivity extracted from dorsal root ganglia and sciatic
nerve (93).

    The molecular size of somatostatin secreted by the gut is not
certain, but probably is mainly the 1600 dalton size, much of
which becomes complexed with a carrier protein of about 150,000
daltons.   This conclusion is based on the results of chromatography of peripheral plasma (26), the principal site of origin which
is the GI tract.

Secretion of Somatostatin

A relatively large amount of information is now available about the factors that regulate the release of somatostatin from nerve and glandular cells. From the standpoint of finding generalizing principles of release regulation, however, much less information has been gained in a systemic way. As would be expected, scientists have generally been interested in learning how somatostatin secretion is regulated in relation to specific functions, such as control of growth hormone secretion, of glucagon secretion, and of gastric acid secretion. The emphasis in the literature reflects this emphasis.

I have reviewed a large number of papers dealing with regulation of somatostatin release, summarizing the findings in Table 2 and in two Ven diagrams (Fig. 6). I was able to find only two factors, glucagon and secretin, that had been tested for effects on hypothalamus, pancreas and gut, and found to be stimulatory in all three sites. Interaction of glucagon and somatostatin is obviously important for understanding pancreatic islet control because glucagon secretion is inhibited by somatostatin. Hence the two factors may be linked in a negative feedback loop. Likewise, release of secretin, a secretory product of the duodenum (and historically the first hormone to be described), is also inhibited by somatostatin (11) and, like glucagon, is conceivably part of a negative feedback loop mechanism. Secretin has not, as yet, been found in brain (90), but secretion is believed (on the basis of amino acid sequence homologies) to be related in an evolutionary sense to vasoactive intestinal peptide (VIP) and glucagon (28), and these peptides are widely distributed in both nervous system, gut and pancreas (101). Glucagon has been found in the brain, in highest concentration in the hypothalamus (25), and hence may be acting as a modulator of pituitary function.

The exclusivity of glucagon and secretin as pan-stimulators may be as much a reflection of the limited data that has accumulated thus far as it is of selectivity of function.

Several factors have been shown to stimulate somatostatin release from both hypothalamus and pancreas. They are depolarization (either by high K^+ in the presence of Ca^{++}, or by veratridine), norepinephrine (or other α-adrenergic agonists), by Substance P, or by cyclic AMP (or its more active analogues). Depolarization is a general stimulator of secretion by all cells, and cyclic AMP the widely distributed second messenger of many hormone actions. Substance P is present in both the hypothalamus and the pancreas and, conceivably, is part of the intrinsic regulatory system in both sites.

Only one factor has thus far been shown to stimulate somatostatin secretion by hypothalamus and gut, and that is dopamine. This neurotransmitter stimulates the release of somatostatin into hypophysial portal blood after intraventricular injection, release from hypothalamic blocks in vitro, release into the antral lumen of cats, and, as we have found, release from perfused duodenum of rats (Table 3).

TABLE 2.   Summary of factors regulating somatostatin secretion

Hormone State	Hypothalamus	Pancreas	Gut
GH deficiency	↓ 4,45,85		
GH	↑ 4,8,21,45,85, 91,11		
Thyroidectomy	↓ 9 ↔37		
TH	↑ 9 ↔37		
Electrical stimulation	↑ 24,119		↑ 39,122
Depolarization (Ca^{++}) dependent	↑ 7,29,55,71,82, 83,116		
Neurotransmitters			
Catechol depletion	↓ 119		
Dopamine	↑ 23,71,78	↓ 6	↑ 123
Norepinephrine	↑ 23,78,119 ↔71		
Serotonin	↔23,71,119		
ACH	↑ 23 ↓ 98	↓ 102	
GABA	↑ 119		
Cyclic AMP	↔71	↑ 5,6,86,105	↑ 6
Peptides			
Glucagon	↑ 1 ↔32	↑ 6,43,86,88, 89,127	↑ 20
VIP	↓ 32	↑ 43,51	
Secretin	↓ 32	↑ 50	↑ 19
Substance P	↔32 ↑113	↑ 43	↓ 19
Opiates	↔113	↓ 54	↓ 19
Gastrin		↑ 50	↑ 6,19
GIP		↑ 50	
CCK		↑ 43,50	
Bombesin		↔43,54	↑ 19
Insulin		↑ 54,↔89	↓ 20
Somatostatin		↓ 53	
Pancreatic			↓ 20
Neurotensin	↑ 71,90,113		↔19
Nutrients			
Amino acids		↑ 30,41,49,86, 87,89,106	↔20
Glucose		↑ 30,49,86,89,105, 107,108,127	
HCL			↔108,124
Feeding			↑ 108

Code for table: ↑ = increase,   ↓ = decrease, ↔ = no change

Data on hypothalamic secretion summarizes studies of portal vessel secretion in vivo, or incubates of fragments in vitro. No synaptsome release studies are included. Depolarization releases

Table 2.   Legend con't.
somatostatin from spinal cord (112).   Studies on GH regulation
include several papers where only content, not secretion was
measured.   Pancreatic studies are mainly of perfused pancreas,
isolated islet preparations, and in a few instances of blood
draining the pancreas.   Gastric studies are mainly of blood
release in whole animals or from isolated stomach or stomach
mucosa incubates in vitro.

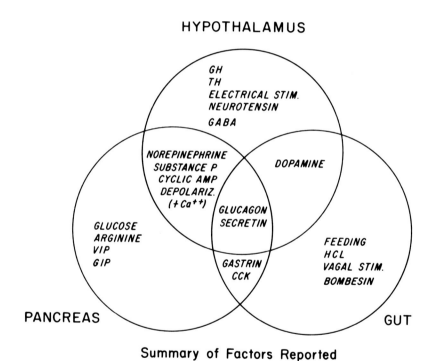

HYPOTHALAMUS

GH
TH
ELECTRICAL STIM.
NEUROTENSIN
GABA

NOREPINEPHRINE
SUBSTANCE P
CYCLIC AMP
DEPOLARIZ.
$(+ Ca^{++})$

DOPAMINE

GLUCAGON
SECRETIN

GLUCOSE
ARGININE
VIP
GIP

FEEDING
HCL
VAGAL STIM.
BOMBESIN

GASTRIN
CCK

PANCREAS

GUT

Summary of Factors Reported
to Stimulate Somatostatin Release.

FIG. 6.   Diagrams to illustrate overlap of factors that modify
release of somatostatin from hypothalamus, gut and pancreas (see
Table 2).   A.   Factors that stimulate somatostatin release.
(con't. next page).

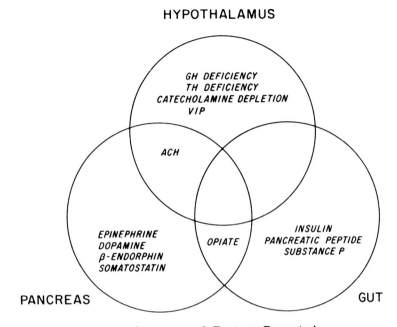

Summary of Factors Reported
to Inhibit Somatostatin Release.

B. Factors that <u>inhibit</u> somatostatin release. Not all stimuli
have been tested in all tissues.

TABLE 3.   Release of immunoreactive somatostatin into lumen of
ileum of the rat following infusion of dopamine*

Control period	After infusion
6.6	15.5
8.2	18.1
0	25.4
15.4	15.9

*Data expressed as pg/ml perfusate.

Gastrin and cholecystokinin are two stimuli that bring about somatostatin release in both pancreas and gut. In turn, the secretion of both peptides is inhibited by somatostatin, and hence (as in the cases of VIP, secretin, and glucagon noted above) is possibly indicative of a negative feedback loop as well.

Even without doing further systematic studies of all known stimuli in all known tissues, enough information has accumulated about the commonality of certain somatostatin regulatory stimuli to warrant detailed comparison of control mechanisms at each site.

INTERACTION OF SOMATOSTATIN AND VASOPRESSIN IN THE CONTROL OF WATER TRANSPORT BY TOAD BLADDER AND KIDNEY

An important aspect of neuropeptide effects on brain cells is that particular populations of nerve cells may be exposed to a multitude of interacting chemical influences, some stimulatory, some inhibitory, some synergistic, that may operate over short or long time scales. Bloom (110) has dealt with these questions recently, pointing out that nerve cells cannot be viewed simply as "a highly complex logic gate with properties of a living transistor." Future research on neuropeptide effects has to deal with these interactions. As in the case of regulation of neuropeptide synthesis and secretion, the neurobiologist may gain insight from study of model systems outside the brain. One such system that has recently come to light is the interaction of somatostatin and vasopressin in the regulation of water transport in toad bladder, a tissue embryologically derived (like brain) from primitive ectoderm. Vasopressinergic neurons (as pointed out in the chapter by Zimmerman, this volume) have widely distributed trajectories beyond the traditionally recognized neurohypophysis. Undoubtedly (in view of the ubiquitous distribution of the somatostatinergic systems), there are many sites of potential interaction in the brain.

In studies of water transport by toad bladder in vitro, Forrest observed that the stimulating effect was blocked by treatment with somatostatin (34). To determine whether the effects were "pharmacological", or a manifestation of a normal regulatory system, he suggested to me that we attempt to identify somatostatin as an endogenous constituent of toad bladder epithelium.

Indeed, somatostatin was found to be present in this tissue by both radioimmunoassay and immunohistochemical techniques (34,14) (Table 4) (Fig. 7). We do not know as yet whether the material is found in situ (as in the case of other tissues), or is taken up selectively from blood or urine. If the latter possibility is true, somatostatin would be acting as a regulatory hormone similar to vasopressin, which is also formed ex situ. The relevance of this effect to water balance in the mammalian kidney is being studied. In the rat, Reid and Rose (97) found that intra-aortic injection of somatostatin caused an increase in urine flow and free water clearance (suggesting block of endogenous vasopres-

TABLE 4.   Tissue somatostatin in toad (Bufo marinus)

Hypothalamus	51.7	± 12.6 ng SRIF/mg protein
Forebrain	4.2	± 0.6
Intestine	6.3	± 2.4
Bladder	2.1	± 0.3
Kidney	0.015	± 0.002
Liver	0.009	± 0.001

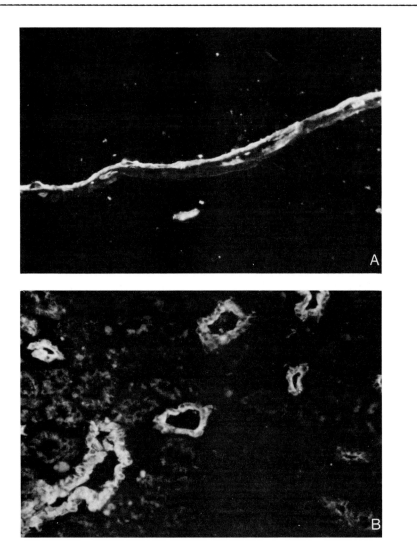

FIG. 7.   A. Localization of immunoreactive somatostatin in toad bladder epithelia.   B.   Toad kidney

sin), while Brautbar et al. (15) made similar observations in dogs. However, Gerich et al. (35) point out that patients given somatostatin show no alterations in renal function or electrolyte metabolism. We were not able to detect somatostatin in extracts of whole kidney (84), but have more recently identified immunore-active material in collecting ducts of human kidney (unpublished). The mechanism of interaction of the two peptides has not been elucidated.

Though of interest to renal physiologists, these observations of interaction between vasopressin and somatostatin in toad bladder may give some direction to future study of interaction of these two peptides in control of neurons. In the endeavor to unravel neuropeptide actions in the brain, neurobiologists will need every bit of biological help they can get.

ACKNOWLEDGEMENTS

Studies from the author's laboratory were supported by USPHS Grants AM 16684 and AM 07039.

REFERENCES

1. Abe, H., Kato, Y., Chiba, T., Taminato, T., and Fujita, T. (1978): Life Sci., 23:1647-1654.
2. Annunziato, L., DeRenson, G., Lombardi, G., Scopacasa, F., Schettini, G., Preziosi, P., and Scapagnini, U. (1977): Endocrinology, 100:738-744.
3. Arimura, A., Sato, H., Dupont, A., Nishi, N., and Schally, A.V. (1975): Science, 189:1007-1009.
4. Baker, B.L., and Yen, Y-Y (1976): Proc. Soc. Exp. Biol. Med., 151:599-602.
5. Barden, N., Alvarado-Urbina, G., Cote, J., and Dupont, A. (1976): Biochem. Biophys. Res. Commun., 71:840-844.
6. Barden, N., Cote, J-P., Lavoie, M., and Dupont, A. (1978): Metabolism (Suppl. 1), 27:1212-1218.
7. Berelowitz, M., Kronheim, M.S., Pimstone, B., and Sheppard, M. (1978): J. Neurochem., 31:1537-1540.
8. Berelowitz, M., Harris, S.J., and Frohman, L.A. (1980): Prog. 62nd Ann. Meeting, The Endocrine Soc., Washington, D.C., Abstr. 748.
9. Berelowitz, M., Maeda, K. Harris, S., and Frohman L. (1980): Endocrinology, 107:24-29.
10. Bloom. F.E. (1978): In: Central Regulation of the Endocrine System, edited by K. Fuxe, T. Hokfelt, and R. Luft, pp. 173-187. Plenum Press, New York.
11. Boden, G., Sivitz, M.D., Owen, O.Ee., Essa-Koumar, N., and Landor, J.H. (1975): Science, 190:163-165.
12. Bolaffi, J.L., and Jackson, I.M.D. (1979): Cell Tis. Res., 202:505-508.
13. Bolaffi, J.L., and Jackson, I.M.D. (1979): Amer. Zool., 19: 939.

14. Bolaffi, J.L., Reichlin, S., Goodman, D.P.B., and Forrest, J. Jr. (1980): Science, (in press).

15. Brautbar, N., Levine, B.S., Coburn, J.W., and Kleeman, C.R. (1979): Am. J. Physiol., 237:E428-436.

16. Brownstein, M.J., Arimura, A., Fernandez-Durango, R., Schally, A.V., Palkovits, M., and Kizer, J.S. (1977): Endocrinology, 100:246-249.

17. Burger, H., and Patel, Y.C. (1977): In: Clinical Neuroendocrinology, edited by L. Martinia, and G..M. Besser, pp. 67-131. Academic Press, New York.

18. Charli, J.L., Joseph-Bravo, P., Palacios, J.M., and Kordon, C. (1978): Eur. J. Pharmacol., 52:401-403.

19. Chiba, T., Taminato, T., Kadowaki, Y., Inoue, Y., Mori, K., Seino, H., Abe, K., Chira, S. Matsukura, S., Fujita, T., and Goto, Y. (1980): Endocrinology, 106:145-149.

20. Chiba, T., Taminato, T., Kadowski, S., Abe, H., Chihara, K., Matsudur, S., Goto, Y., Seino, Y., and Fujita, T. (1980): Diabetes, 29:292-295.

21. Chihara, K. (1979): Prog. 61st Ann. Meeting, The Endocrine Soc., Anaheim, Calif., Abstr. 290.

22. Chihara, K., Arimura, A., and Schally, A.V. (1979): Endocrinology, 104:1434-1441.

23. Chihara, K., Arimura, A., and Schally, A.V. (1979): Endocrinology, 104:1656-1662.

24. Chihara, K., Arimura, A., Kubli-Garfias, C., and Schally, A.V. (1979): Endocrinology, 105:1416-1418.

25. Conlon, J.M., Samson, W.K., Dobbs, R.E., Orci, L., and Unger, R.H. (1979): Diabetes, 28:700-702.

26. Conlon, J., Srikant, C., Ipp, E., Schusdziarra, V., Vale, W., and Unger R. (1978): J. Clin. Invest., 62:1187-1193.

27. Delfs, J., Robbins, R., Connolly, J.L., Dichter, M., and Reichlin, S. (1980): Nature, 283:676-677.

28. Dockray, G.J. (1978): In: Gut Hormones, edited by S.R. Bloom, pp. 64-67. Churchill Livingston, Edinburgh.

29. Drouva, S.V., Epelbaum, J., Hery, M., Laplante, E., and Kordon, C. (1980): Prog. 62nd Ann. Meeting, Endocrine Soc., Washington, D.C., Abstr. 38.

30. Efendic, S., Nylen, A., Roovete, A., and Uvnas-Wallensten, K. (1978): FEBS Lett., 92:33-35.

31. Ensinck, J.W., Laschansky, E.C., Kanter, R.A., Fujimoto, W.J., Koerker, D.J., and Goodner, C.J. (1978): Metabolism (Suppl. 1), 27:1207-1210.

32. Epelbaum, J., Tapia-Arancibia, L., Besson, J., Rotsztejn, W.H., and Kordon, C. (1979): Eur. J. Pharmacol., 58:493-505.

33. Fletcher, D.J., Noe, B.D., Bauer, G.E., and Quigley, J.P. (1980): Diabetes, 29:593-590.

34. Forrest, J.N., Jr., Reichlin, S. and Goodman, D.B.P. (1980): Proc. Natl. Acad. Sci. USA, (in press).

35. Gerich, J., Schultz, T., Tsalikian, E., Lorenzi, M., Lewis, S., and Karam, J. (1976): Diabetologia, 13:537-544.

36. Gilbert, R.F.T., Emson, P.C., Fahrenkrug, J., Lee, C.M., Penman, E., and Wass, J. (1980): J. Neurochem., 34:108-113.
37. Gillioz, P., Geraud, P., Conte-Devolox, B., Jaquet, P., Codaccioni, J.L., and Oliver, C. (1979): Endocrinology, 104:1407-1410.
38. Grimm-Jorgensen, J., and Reichlin, S. (1973): Endocrinology, 93:626-631.
39. Guzman, S., Chayvialle, J.A., Banks, W., Rayford, P.L., and Thompson, J.C. (1979): Surgery, 86:329-336.
40. Guzman, S., Lonovics, J., Chayvialle, J.A., Hejtmancik, K.E., Rayford, P.L., and Thompson, J.C. (1980): Endocrinology, 107:231-236.
41. Hara, M., Patton, G., and Gerich, J. (1979): Life Sci., 24:625-628.
42. Hargis, C.K., Williams, G.A., Reynolds, W.A., Chertow, B.S., Kukreja, S.C., Bowser, E.N., and Henderson, W.J. (1978): Endocrinology, 102:745-750.
43. Hermansen, K. (1980): Endocrinology, 107:256-261.
44. Hirooka, Y., Hollander, C.S., Suzuki, S., Ferdinand, P., and Juan, S.I. (1978): Proc. Natl. Acad. Sci. USA, 75:4509-4513.
45. Hoffman, D.L., and Baker, B.L. (1977): Proc. Soc. Exp. Biol. Med., 156:265-271.
46. Hokfelt, T., Elde, R., Johansson, O., Luft, R., and Arimura, A. (1975): Neurosci. Lett., 1:231-235.
47. Hokfelt, T., Elde, R., Johansson, O., Luft, R., Nilsson, G., and Arimura, A. (1976): Neuroscience, 1:131-136.
48. Hokfelt, T., Johansson, O., Ljungdahl, A., Lundberg, J., Schultzberg, M., Fuxe, K., Goldstein, M., Steinbusch, H., Verhofstad, A., and Elde, R. (1978): In: Central Regulation of the Endocrine System, edited by K. Fuxe, T. Hokfelt, and R. Luft, pp. 31-48. Plenum Press, New York.
49. Ipp, E., Dobbs, R., Arimura, A., Vale, W., Harris, V., and Unger, R. (1977): J. Clin. Invest., 60:760-765.
50. Ipp, E., Dobbs, R.E., Harris, V., Arimura, A., Vale, W., and Unger, R.H. (1977): J. Clin. Invest., 60:1216-1219.
51. Ipp,E., Dobbs, R., and Unger, R. (1978): FEBS Lett., 90:7678.
52. Ipp, E., Dobbs, R., and Unger, R. (1978): Nature, 276:190-191.
53. Ipp, E., Rivier, J., Dobbs, R., Brown, M., Vale, W., and Unger, R.H. (1979): Endocrinology, 104:1270-1273.
54. Ipp, E., and Unger, R.H. (1979): Endocr. Res. Commun., 6:37-42.
55. Iversen, L.L, Iversen, S.D., Bloom, F., Douglas, C., Brown, M., and Vale, W. (1978): Nature, 273:161-163.
56. Jackson, I.M.D., and Reichlin, S. (1977): Science, 198:414-415.
57. Jackson, I.M.D., and Reichlin, S. (1979): In: Central Nervous Systems of Hypothalamic Hormones and Other Peptides, edited by R. Collu, A. Barbeau, J.R. Ducharme, and J-G. Rochefort, pp. 3-54. Raven Press, New York.

58. Jan, U.N., Jan, L.Y., and Kuffler, S. (1978): Proc. Natl. Acad. Sci. USA, 76:1501-1505.
59. Jordan, D., Ponsin, G., and Mornex, R. (1976): Fifth Int. Cong. Endocrinol., Hamburg, Abstr. 151.
60. Joseph-Bravo, P., Charli, J.L., Palacios, J.M., and Kordon, C. (1978): Endocrinology, 104:801-806.
61. Kardon, F., Marcus, R.J., Winokur, A., and Utiger, R.D. (1977): Endocrinology, 100:1604-1609.
62. Kordon, C. (1979): In: Central Regulation of the Endocrine System, edited by K. Fuxe, T. Hokfelt, and R. Luft, pp. 473485. Plenum Press, New York.
63. Kordon, C., Enjalbert, A., Epelbaum, J., and Rotsztejn, W. (1979): In: Brain Peptides: A New Endocrinology, edited by A.M. Gotto, Jr., E.J. Peck, Jr., and A.E. Boyd III, pp. 277293.   Elsevier/North Holland Biomedical Press, Amsterdam.
64. Kronheim, S., Berelowtiz, M., and Pimstone, B.L. (1976): Clin. Endocrinol., 5:619-630.
65. Krulich, L., Giachetta, A., Marchlewska-Koj, A., Hefco, E., and Jameson, H.E. (1977): Endocrinology, 100:496-505.
66. Krulich, L., Vijayan, E., Coppings, R.J., Giachetta, A., McCann, S.M., and Mayfield, M.A. (1979): Endocrinology, 105: 276-283.
67. Krulich, L. (1979): Ann. Rev. Physiol., 41:603-615.
68. Lauber, M., Camier, M., and Cohen, P. (1979): Proc. Natl. Acad. Sci. USA, 76:6004-6008.
69. Lee, S.L., Havlicek, V., Panerai, A.E., and Friesen, H.G. (1979): Experientia, 35:351-352.
70. Lundberg, J.M., Hokfelt, T., Nilsson, G., Terenius, L., Rehfeld, J., Elde, R., and Said, S. (1978): Acta Physiol. (Scand.), 104:499-501.
71. Maeda, K., and Frohman, L.A. (1980): Endocrinology, 106: 1837-1842.
72. McKelvy, J.F. (1974): Brain Res., 65:489-502.
73. McKelvy, J.F. (1977): In: Hypothalamic Peptide Hormones and Pituitary Regulation, edited by J.C. Porter, pp. 77-98. Plenum Press, New York.
74. Morley, J.E. (1979): Life Sci., 25:1539-1550.
75. Mueller, G.P., Twohy, C.P., Chen, H.T., Advis, J.P., and Meites, J. (1976): Life Sci., 8:715-724.
76. Mueller, G.P., Alpert, L., Reichlin, S., and Jackson, I.M.D. (1980): Endocrinology, 106:1-4.
77. Muller, E.E., Nistico, G., and Scapagnini, U. (1977): Neurotransmitters and Anterior Pituitary Function,   pp. 291-295. Academic Press, New York.
78. Negro-Vilar, A., Ojeda, S.R., Arimura, A., and McCann, S.M. (1978): Life Sci., 23:1493-1497.
79. Noe, B.D., Speiss, J., Rivier, J.E., and Vale, W. (1979): Endocrinology, 105:1410-1415.

80. Noe, B.D., Fletcher, D.J., and Speiss, J. (1979): Diabetes, 28:724-730.
81. Oyama, H., Bradshaw, R.A., Bates, O.J., and Permutt, A. (1980): J. Biol. Chem., 255:2251-2254.
82. Patel, Y.C. (1977): Nature, 267:852-853.
83. Patel, Y.C., Zingg, H.H., and Dreifuss, J.J. (1977): Metabolism, 24:1589-1594.
84. Patel, Y.C., and Reichlin, S. (1978): Endocrinology, 102: 523-530.
85. Patel, Y.C. (1979): Life Sci., 24:1589-1574.
86. Patel, Y.C., Amherdt, M., and Orci, L. (1979): Endocrinology, 104:676-679.
87. Patton, G., Ipp, E., Dobbs, R., Orci, L., Vale, W., and Unger, R. (1976): Life Sci., 19:1957-1960.
88. Patton, G., Dobbs, R., Orci, L., Vale, W., and Unger, R. (1976): Metabolism (Suppl. 1), 25:1499-1500.
89. Patton, G., Ipp, E., Dobbs, R., Orci, L, Vale, W., and Unger, R. (1977): Proc. Natl. Acad. Sci. USA, 74:2140-2143.
90. Pearse, A.G.E. (1979): In: Brain Peptides: A New Endocrinology, edited by A.M. Gotto, Jr., E.J. Peck, Jr., and A.E. Boyd III, pp. 89-101. Elsevier/North Holland Biomedical Press, Amsterdam.
91. Pimstone, B.L., Sheppard, M., Shapiro, B., Kronheim, S. Hudson, A., Hendricks, S., and Waligora, K. (1979): Fed. Proc., 38:2330-2333.
92. Pradayrol, L., Jornvall, H., Mutt, V., and Ribet, A. (1980): FEBS Lett., 109:55-58.
93. Rasool, C.G., Schwartz, A.L., Bollinger, J.A., Reichlin, S., and Bradley, W.G. (1980): Endocrinology, (in press).
94. Ravazzola, M., Brown, D., Lippaluoto, J., and Orci, L. (1979): Life Sci., 25:1331-1334.
95. Reichlin, S., Martin, J.B., and Jackson, I.M.D. (1978): In: The Endocrine Hypothalamus, edited by S.L. Jeffcoate, and J.S.M. Hutchinson, pp. 230-269. Academic Press, London.
96. Reichlin, S. (1980): In: Peptides: Integrators of Cell and Tissue Functions, edited by F.E. Bloom. Raven Press, New York.
97. Reid, I., and Rose, J. (1977): Endocrinology, 100:782-785.
98. Richardson, S.B., Hollander, C.S., D'Eletto, R., Greenlead, P.W., and Thaw, C. (1980): Endocrinology, 107:122-129.
99. Rorstad, O.P., Epelbaum, J., Brazeau, P., and Martin, J.B. (1979): Endocrinology, 105:1083-1092.
100. Ruzsas, C.S., Jozsa, R., and Mess, B. (1979): Endocrinologia Experimentalis, 13:9-18.
101. Said, S.I. (1980): In: Frontiers in Neuroendocrinology, Vol. 6, edited by L. Martini and W.F. Ganong, pp. 293-331. Raven Press, New York.
102. Samols, E., Weir, G.C., Ramseur, R., Day, J.A., and Patel, Y.C. (1978): Metabolism (Suppl. 1), 27:1219-1222.

103. Scanlon, M.F., Lewis, M., Weightman, D.R., Chan, V., and Hall, R. (1980): In: Frontiers in Neuroendocrinology, Vol. 6, edited by L. Martini, and W.F. Ganong, pp. 333-380. Raven Press, New York.

104. Schally, A.V., Huang, W-Y., Chang, R.C.C., Arimura, A., Redding, T.W., Miller, R.P., Hunkapiller, M.W., and Hood, L.E. (1980): Proc. Natl. Acad. Sci. USA, (in press).

105. Schauder, P., McIntosh, C., Arends, J., Arnold, R., Frerichs, H., and Creutzfeldt, W. (1976): FEBS Lett., 68:225-227.

106. Schauder, P., McIntosh, C., Arends, J., Arnold, R., Frerichs, H., and Creutzfeldt, W. (1977): Biochem., Biophys. Res. Commun., 75:630-635.

107. Schauder, P., McIntosh, C., Panten, V., Arneds, J., Frerichs, H., and Creutzfeldt, W. (1977): FEBS Lett., 81:355-358.

108. Schusdziarra, V., Harris, V., Conlon, J.M., and Arimura, A. (1978): J. Clin. Invest., 62:509-518.

109. Schusdziarra, V., Rouiller, D., Pietri, A., Harris, V., Zyznar, E., Conlon, J.M., and Unger, R.H. (1979): Am. J. Physiol., 237:E555-560.

110. Schusdziarra, V., Zyznar, E., Rouiller, D., Boden, G., Brown, J.C., Arimura, A., and Unger, R.H. (1980): Science, 207:530-532.

111. Sheppard, M.C., Kronheim, S., and Pimstone, B.L. (1978): Clin. Endocrinol., 9:583-586.

112. Sheppard, M., Kronheim, S., Adams, C., and Pimstone, B. (1979): Neurosci. Lett., 15:65-70.

113. Sheppard, M.C., Kronheim, S., and Pimstone, B.L. (1979): J. Neurochem., 32:647-649.

114. Speiss, J., Rivier, J.E., Rodkey, J.A., Bennett, C.D., and Vale, W. (1979): Proc. Natl. Acad. Sci USA, 76:2974-2978.

115. Speiss, J., Villarreal, J., and Vale, W. (1979): Prog. 61st Ann. Meeting Endocrine Soc., Abstr. 146.

116. Spindel, E. (1980): Ph.D. Thesis, Massachusetts Institute of Technology, (unpublished).

117. Terry, L.C., Rorstad, O.P., and Martin, J.B. (1978): Prog. 60th Ann. Meeting Endocrine Soc., Miami, Abstr. 86.

118. Tuomisto, J., Ranta, T., Mamisto, P., Saarinine, A., and Lippaluota, J. (1975): Eur. J. Pharmacol., 30:221-229.

119. Turkelson, C.M., Chihara, K., Kubli-Garfias, C., and Arimura, A. (1979): Prog. 61st Ann. Meeting Endocrine Soc., Anaheim, Abstr. 285.

120. Unvas-Wallensten, K., Efendic, S., and Luft, R. (1977): Acta Physiol. (Scand.), 99:126.

121. Uvnas-Wallensten, K., Efendic, S., and Luft, R. (1978): Horm. Metab. Res., 10:173.

122. Uvnas-Wallensten, K., Efendic, S., and Luft, R. (1978): Metabolism, 27:1233-1234.

123. Unvas-Wallensten, K., Lundberg, J., and Efendic, S. (1978): Acta Physiol. (Scand.), 103:343-345.

124. Uvnas-Wallensten, K., Larsson, I., Johansson, C., and Efendic, S. (1979): Scand. J. Gastroenterol., 14:797-800.
125. VanNoorden, S., Polak, J., and Pearse, A. (1977): Histochem., 53:243-247.
126. Vinik, A.I. (1978): In: Gut Hormones, edited by S.R. Bloom, pp. 156-157. Churchill Livingston, Edinburgh.
127. Weir, G., Samols, E., Day, J., and Patel, Y. (1978): Metabolism (Suppl. 1), 27:1223-1226.
128. Winokur, A., Kreider, M.S., Dugan, J., and Utiger, R.D. (1978): Brain Res., 152:203-208.
129. Zingg, H.H., and Patel, Y.C., (1979): Biochem. Biophys. Res. Commun., 90:466-472.
130. Zyznar, E.S., Conlong, J.M., Schusdziarra, V., and Unger, R.H. (1979): Endocrinology, 105:1426-1431.

Neurosecretion and Brain Peptides,
edited by J. B. Martin, S. Reichlin, and K. L. Bick.
Raven Press, New York © 1981.

Cellular Interactions of Biogenic Amines, Peptides, and Cyclic Nucleotides

James A. Nathanson

Departments of Neurology and Pharmacology, Harvard Medical School and Massachusetts General Hospital, Boston, Massachusetts 02114

There is now considerable evidence that biogenic amines, peptides and cyclic nucleotides may serve, individually, as regulators of neuronal communication. According to current thinking, amines and peptides most likely act extracellularly as neurotransmitters or neuroregulators, while cyclic nucleotides, particularly cyclic AMP, appear to function at the level of the plasma membrane, as second messengers of neurotransmitter-receptor binding, or possibly intracellularly, as regulators of cellular metabolism (17). The present paper describes how these three classes of cellular regulators - biogenic amines, peptides and cyclic AMP - may interact with one another to create sophisticated, and often complex, regulation of interneuronal communication.

For didactic purposes, Fig. 1 attempts to classify amine-peptide-cyclic AMP interactions into four major groups. No doubt, there may exist still additional types of functional relationships, but the groupings shown are applicable to much of what is currently known about amine-peptide-cyclic AMP interaction and provide a framework for predicting future types of interaction.

TYPE I INTERACTION: PEPTIDERGIC RECEPTORS, ASSOCIATED OR UNASSOCIATED WITH ADENYLATE CYCLASE, PRESENT ON AMINERGIC NEURONS

For a number of years there has been considerable evidence that certain peripheral polypeptide hormones act at the receptor level to stimulate specific hormone-sensitive adenylate cyclases, i.e., enzymes which synthesize cyclic AMP at an increased rate in the presence of a particular polypeptide (27). These hormones include (among others) glucagon, ACTH, and vasopressin (ADH), for each of which there exists a well-described hormone-sensitive adenylate cyclase: ACTH-activated adenylate cyclase in adrenal

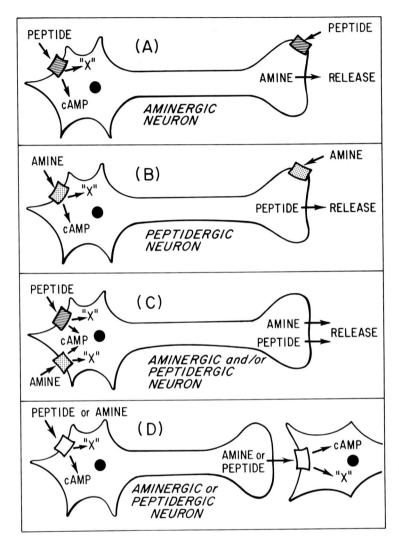

FIG. 1. Proposed classification of amine-peptide-cyclic AMP interactions into four groups. A) Type I represents those interactions in which peptidergic receptors, present on aminergic neurons, have their actions mediated through adenylate cyclase or some noncyclic AMP mechanism ("X"); B) Type II, the converse of Type I, represents those interactions which take place between aminergic receptors (± cyclic AMP) present on peptidergic neurons; C) Type III interactions are between peptidergic and aminergic receptors both present on the same neuron or between peptides and amines released from the same neuron; D) in Type IV, interaction between peptides, amines and cyclic AMP results from transsynaptic changes between neurons. See text for details.

cortex and adipose tissue, glucagon-sensitive adenylate cyclase in liver, and ADH-sensitive adenylate cyclase in kidney. There is additional physiological evidence that, for each of these polypeptides, cyclic AMP is the second messenger which translates hormone receptor binding into intracellular changes in metabolism.

With such precedent for the mediation of certain peripheral hormone actions through cyclic AMP, it is not surprising that a number of polypeptide-activated adenylate cyclases have recently been isolated from brain.

One of the first to be described was Substance P-activated adenylate cyclase, present in human and rat brain (7,8). Stimulation of enzyme activity by the peptide occurs at concentrations as low as 10^{-8} M, and half-maximal activation at 10^{-7}M, similar to the concentration of Substance P normally found in brain.

Because of the long-lasting membrane effects of both Substance P and cyclic AMP (17,21), a possible association between cyclic AMP metabolism and Substance P's role as a neurotransmitter is an attractive possibility. However, further questions need to be answered. For example, can the effects of Substance P on spinal cord and brain neurons be mimicked by the application of cyclic AMP or its derivatives? Also, can these electrophysiological effects be augmented by phosphodiesterase inhibitors, which inhibit the breakdown of cyclic AMP? Finally, if specific Substance P agonists and antagonists can be developed, will the pharmacological characteristics of Substance P-sensitive adenylate cyclase mimic those of the Substance P receptor described by physiological means?

Despite such questions, Substance P remains a good candidate with which to illustrate a Type I amine-peptide-cyclic AMP interaction. Specifically, an example of a prototype Type I synapse would be that formed by the Substance P-containing striatonigral axons which appear to make contact with dopaminergic neurons in the substantia nigra. Since Substance P-sensitive adenylate cyclase is present in the substantia nigra (8) and the peptide has been shown to excite nigral neurons (19), it is possible that the actions of a postsynaptic Substance P receptor present on dopaminergic neurons might be mediated through stimulation of cyclic AMP formation in these nigral neurons. In other words, this would be an interaction between a peptidergic receptor, associated with adenylate cyclase, located on an aminergic neuron (Fig. 1A).

Several investigators have reported that the direct application of opiates to brain homogenates can inhibit (5, 29-31) and, at least in one case, stimulate (23) neurotransmitter-sensitive adenylate cyclase in vitro. Inhibitory effects have been observed, also, in glioma x neuroblastoma hybrid cell lines (28). Because electrophysiological studies have shown that opiates can inhibit the firing of noradrenergic neurons located in the nucleus locus coeruleus (which contains opiate receptors)

(1,3,13,20), this opiate-amine synapse is another example of a Type I amine-peptide-cyclic AMP interaction. That is, it is possible that the endogenous peptides which interact with the opiate receptors on the norepinephrine-containing locus coeruleus neurons have their effects mediated through inhibition or activation of adenylate cyclase.

Stimulation of brain adenylate cyclase activity has been reported for vasoactive intestinal peptide (VIP) (24). Such stimulation is additive to that due to catecholamines, adenosine, and prostaglandin E_1; is greatest in the cortex and hippocampus; and is not inhibited by α-adrenergic, β-adrenergic, opiate or dopamine blockers. VIP-activated adenylate cyclase has also been found in the anterior pituitary (26). As with Substance P, these biochemical studies as yet lack the additional physiological evidence which would support an actual mediation by cyclic AMP of the events following peptide-receptor interaction.

In a survey of the effects of various peptides on adenylate cyclase activity in cat forebrain homogenates (Table 1), we have recently found that lys-ADH causes modest (approximately 40% above control) but consistent increases in enzyme activity. In

TABLE 1. Effect of various peptides on adenylate cyclase activity in cat brain homogenates.[*]

| Peptide (5μM) | Enzyme stimulation (% increase above control)[*] | | | | | |
| | +EGTA (0.5mM) | | −EGTA | | +CaCl₂ (10μM) | |
	−GTP	+GTP[@]	−GTP	+GTP	−GTP	+GTP
VIP	12	NS[#]	10	NS	NS	10
LHRH	20	NS	23	7	NS	13
PROL	8	NS	34	4	22	18
ADH	43	26	51	26	28	14
OXYTOCIN	NS	NS	NS	NS	NS	NS
ACTH	NS	6	15	9	NS	NS
ISOPROTERENOL	33	13	44	18	14	16

[*]The values shown represent the percent increase over basal activity caused by various peptides under different incubation conditions: i.e., in the presence or absence of GTP (30μM), EGTA (0.5mM), or exogenous calcium (10μM). Basal enzyme activities (in pmol/mg protein/min) were: +EGTA,−GTP = 40; +EGTA,+GTP = 190; −EGTA,−GTP = 49; −EGTA,+GTP = 231; +CaCl₂,−GTP = 79; +CaCl,+GTP = 262. Values shown are the mean for replicate samples, each assayed for cyclic AMP content in triplicate. [@]GTP = 30μM.
NS = not significantly different from control.

these same experiments, under certain conditions, VIP, prolactin and luteinizing hormone releasing hormone (LHRH) caused small stimulations, while oxytocin and ACTH had little effect. In studies with rat brain, others have reported both stimulatory and inhibitory effects of ACTH on adenylate cyclase (32). In neuroblastoma x glioma hybrids, both secretin and glucagon cause activation of adenylate cyclase (22). For none of the above peptides has sufficient biochemical information yet been obtained to ascribe a receptor-associated role for adenylate cyclase. However, because (as mentioned) well described ACTH-, glucagon- and ADH-sensitive adenylate cycalses exist in the periphery, it would not be surprising if similar enzymes were found in the brain.

Of course, it should be pointed out that peptidergic modulation of aminergic neurons may occur in the absence of any involvement of cyclic AMP. Not all peptide receptors need be coupled to adenylate cyclase, and even among those that are, there may exist subclasses of peptide receptors, some associated with adenylate cyclase and some not (similar to that which has recently been found for the D-1 and D-2 dopamine receptors (12).) Based on what is known about other non-cyclic AMP-associated aminergic receptors, such peptidergic receptors might be associated with presynaptic or autoregulation at aminergic nerve terminals (see Fig. 1A). Indeed, there is recent evidence that peptides can, in fact, affect amine neurotransmitter release (2,16).

TYPE II INTERACTION: AMINERGIC RECEPTORS, ASSOCIATED OR UNASSOCIATED WITH ADENYLATE CYCLASE, PRESENT ON PEPTIDERGIC NEURONS

The supposition that this type of interaction exists (Fig. 1B) is based on the well established presence of amine-sensitive adenylate cyclases in various areas of the brain (17) and on the probable existence of aminergic receptors on peptide-containing neurons. For example, it is possible that some of the dopamine-sensitive adenylate cyclase activity present in the striatum might be located on Substance P-containing neurons, which are known to be present in this brain area. If so, then dopaminergic neurotransmission to such peptidergic neurons could be mediated through stimulation of cyclic AMP synthesis.

Aminergic modulation of peptidergic neurons appears better established in the area of hypothalamic control of neuroendocrine releasing factors. For example, in the median eminence, noradrenergic and dopaminergic terminals have been shown to be in close proximity to somatostatin-containing nerve terminals (9). Furthermore, application of dopamine and norepinephrine to median eminence fragments stimulates somatostatin release, an effect which is blocked by dopamine and norepinephrine blockers, respectively (18). Presumably, catecholamine receptors present on somatostatin terminals are involved in this modulation. Similar anatomical relationships and effects on release have been observed in the hypothalamus for catecholamines and LHRH (9).

Whether cyclic AMP is involved in such releasing effects is not known. As mentioned above, certain data suggest that presynaptic catecholamine receptors present in other brain areas lack an association with adenylate cyclase. On the other hand, although normal pituitary cells probably contain a dopamine receptor unlinked to adenylate cyclase (12), it has been reported, in pituitary adenomas, that dopaminergic inhibition of prolactin release is proportional to the degree of inhibition of adenylate cyclase activity by dopamine in this same tissue (6). Also, there is some but not complete evidence that the releasing factors themselves might stimulate pituitary trophic hormone release through alteration of cyclic AMP levels (14).

TYPE III INTERACTION:

A. Peptidergic and Aminergic Receptors Present on the Same Neuron

Experimental evidence indicates that many neurons contain multiple neurotransmitter receptors, including those for peptides and biogenic amines. Multiple receptors on the same neuron would presumably interact through mutual or opposing regulation of dendritic or somatic postsynaptic potentials (Fig. 1C). A good example of this type of interaction is the simultaneous presence of opiate and α_2-adrenergic receptors on locus coerulus neurons. Agonists of either receptor result in locus coerulus hypoactivity which can be blocked by appropriate opiate or α_2-adrenergic receptor antagonists (1). Although stimulation by either opiates or α_2-adrenergic agonists has the same effect on neuronal activity, the receptors for the two transmitters appear to function independently. Thus, inhibition of opiate action by naloxone has no effect on the electrophysiological effects of the α_2-adrenergic antagonist, clonidine, and, conversely, inhibition of clonidine's action by the α_2-adrenergic antagonist, piperoxan, has no effect on the electrophysiological effects of opiates. Such independence can lead to interesting interactions. For example, during chronic opiate administration, the locus coeruleus neurons are initially inhibited but then their activity gradually returns toward normal due both to the development of opiate receptor subsensitivity as well as to non-opiate receptor compensatory mechanisms. That such other compensatory mechanisms occur is shown by the fact that neuronal hyperactivity occurs if an opiate antagonist is given after a period of morphine addiction. Such hyperactivity is still inhibited by low doses of clonidine, indicating that chronic morphine treatment has had little effect on the α_2-adrenergic receptor. Whether, beyond the receptor level, opiates and adrenergic agonists act through a common intracellular mechanism (such as a cyclic nucleotide) is not yet known.

B. Peptides and Amines in the Same Neuron

Recent reports have suggested that certain peptides may coexist with other neurotransmitters in nerve terminals (4,10,11). Such a common localization suggests another level of potential peptide-amine interaction. For example, two such co-transmitters could affect each other's release or, more likely, once released, act to exert different effects on the same post-synaptic neuron. Such dual regulation might be beneficial in the case in which neurotransmission had the purpose of creating both short term and long term postsynaptic changes.

TYPE IV INTERACTION: PEPTIDE, AMINE AND CYCLIC AMP INTERACTIONS RESULTING FROM TRANS-SYNAPTIC CHANGES BETWEEN NEURONS

Changes in the activity of neuronal pathways can sometimes lead to reciprocal changes in the sensitivity of receptors located on postsynaptic neurons. Thus, for example, lesions of nigrostriatal neurons lead to the development of supersensitivity of dopamine receptors in the caudate nucleus, as well as to the alteration of striatal dopamine-activated adenylate cyclase (15). Such transsynaptic changes in sensitivity may also be applicable to interactions between peptide receptors and aminergic neurons.

For example, in collaborative studies done with Dr. Eugene Redmond of Yale University, we have recently found that chronic stimulation of opiate receptors in the locus coeruleus can lead to the development of postsynaptic supersensitivity of norepine-phrine receptors, as manifested by an increase in the activity of norepinephrine-sensitive adenylate cyclase in the cerebellum. As described earlier, opiates, such as morphine, inhibit the firing of locus coeruleus neurons (1,3,13). During opiate withdrawal, these neurons become hyperactive, a finding consistent both with the symptoms (such as anxiety and hypertension) observed during withdrawal as well as with increases in norepinephrine turnover seen in areas receiving projections from the locus coeruleus (25). In our studies of addiction in sub-human primates we were interested in determining whether opiate administration and opiate withdrawal by causing, respectively, locus coeruleus hypoactivity and hyperactivity, would result in reciprocal changes in receptor sensitivity and adenylate cyclase activity in areas receiving projections from the locus coeruleus. We chose to look at the noradrenergic pathway from locus coeruleus to cerebellum because of the considerable evidence that norepine-phrine released from locus coeruleus nerve terminals in the Purkinje cell layer alters the activity of Purkinje neurons through activation of a β-adrenergic receptor associated with an adenylate cyclase (17).

For these experiments, vervet monkeys were given morphine sul-phate, 3 mg/kg, subcutaneously three times a day for ten days and then sacrificed two hours after their last injection. Control animals received saline injections. Cerebellums were dissected,

frozen, and then later assayed for   -adrenergic-sensitive adenylate cyclase activity. As shown in Table 2, adenylate cyclase
activation by isoproterenol (a β-adrenergic agonist) was greater
among morphine-treated animals than among controls at all concentrations of isoproterenol tested. Basal enzyme activity, i.e.,
that in the absence of agonist, was similar in the two groups, as
were the activation constants (K_a) for isoproterenol stimulation.
    These findings are consistent with the pattern which has been
observed in other types of lesion or drug-induced receptor supersensitivity (33). Before concluding, however, that our results
were due to the chronic effects of morphine, it was necessary to
rule out the possibility that morphine, still present in the
cerebellum at the time of sacrifice, might have exerted a direct
stimulatory effect on enzyme activity. That this did not occur
was shown by other experiments in which morphine, added directly
to homogenates of control cerebellum, had no stimulatory effect
on either basal or   β-adrenergic-stimulated adenylate cyclase
activity.
    In other experiments we looked at the effect of morphine with-
drawal on   β-adrenergic-sensitive adenylate cyclase activity.
Since locus coeruleus neuronal activity is increased during morphine withdrawal, one might predict that enzyme activity would be
decreased relative to that seen in chronically addicted monkeys.
This was exactly what was found (Table 2) in a group of animals
which was first given morphine as described above and then sacrificed 50 hours after their last dose (at which time they had all
demonstrated morphine withdrawal). In fact, the average maximum

TABLE 2. Effect of chronic morphine administration and morphine
    withdrawal on monkey cerebellum adenylate cyclase activity
      stimulated by varying concentrations of isoproterenol

Isoproterenol (M)	Chronic Morphine	Saline Control	Morphine Withdrawal
10^{-7}	13.2 + 4.7[*]	8.2 + 3.5	5.7 + 3.7
10^{-6}	52.8 + 3.9	42.5 + 6.2	27.3 + 9.8
10^{-5}	91.8 + 4.7	73.9 + 9.0	66.0 + 11.4
10^{-4}	111.9 + 8.8	75.2 + 7.5	57.9 + 16.4

[*]Shown are the mean increases in pmol/mg protein/min (+ SEM)
above basal activity for each group. Basal activities were:
Chronic Morphine = 89.9 + 7.1; Saline Control = 85.2 + 9;
Morphine Withdrawal = 95.5 + 7.

enzyme velocity for the withdrawal group was less than that for the saline-treated controls, suggesting that locus coeruleus hyperactivity led to a subsensitivity of noradrenergic receptors.

The above results are consistent with the hypothesis that alteration of neuronal activity through activation of an opiate (and therefore, presumably, a peptidergic) receptor on one neuron can lead to reciprocal changes in the sensitivity of a β-adrenergic, i.e., an aminergic, receptor linked to adenylate cyclase and present postsynaptically on a second neuron (Fig. 1D). This type of interaction among amines, peptides, and cyclic AMP serves to illustrate the potential complexity of relationships which can occur between neurons which utilize different transmitters, which contain multiple receptors, and which may or may not utilize an intracellular second messenger.

ACKNOWLEDGEMENTS

Supported by NIH grant NS16356, by a PMAF starter grant, and by a grant from the McKnight Foundation.

REFERENCES

1. Aghajanian, G.K. (1978): Nature, 276: 186-188.
2. Arbilla, S., and Langer, S.Z. (1978): Nature, 271:559-560.
3. Bird, S.J., and Kuhar, M.J. (1977): Brain Res., 129:366-370.
4. Chan-Palay, V., Jonsson, G., and Palay, S.L. (1978): Proc. Natl. Acad. Sci. USA, 75:1582-1586.
5. Collier, H.O.J., and Roy, A.C. (1974): Nature, 248:24-27.
6. DeCamilli, P., Macconi, D., and Spade, A. (1979): Nature, 278:252-254.
7. Duffy, M.J., and Powell, D. (1975): Biochem. Biophys. Acta, 385:275-280.
8. Duffy, M.J., Wong, J., and Powell, D. (1975): Neuropharmacology, 14:615-618.
9. Hokfelt, T., Elde, R., Fuxe, K., Johansson, O., Ljungdahl, A., Goldstein, M., Luft, R., Efendic, S., Nilsson, G., Terenius, Ganten, D., Jeffcoate, S.L., Rehfeld, J., Said, S., Perez de la Mora, M., Possani, R., Tapia, R., Teran, L. and Palacios, R. (1978): In: The Hypothalamus, edited by S. Reichlin, R.J. Baldessarini, and J.B. Martin, pp. 69-135. Raven Press, New York.
10. Hokfelt, T., Elfvin, L.G., Elde, R., Schultzberg, M., Goldstein, M., and Luft, R. (1977): Proc. Natl. Acad. Sci. USA, 74:3587-3591.
11. Hokfelt, T., Johansson, O., Ljungdahl, A., Lundberg, J.M., and Schultzberg, M. (1980): Nature, 284: 515-521.

12. Kebabian, J.W., and Calne, D.B. (1979): Nature, 277:93-96.
13. Korf, J., Bunney, B.S., and Aghajanian, G.K. (1974): Eur. J. Pharmacol., 25:165-169.
14. Labrie, F., DeLean, A., Borgeat, P., Borden, N., Poirier, G., and Drouin, J. (1975): In: Anatomical Neuroendocrinology, edited by W.E. Stumpf, and L.D. Grant, pp. 62-68. Karger, Basel.
15. Mishra, R.K., Gardner, E.L., Katzman, R., and Makman, M.H. (1974): Proc. Natl. Acad. Sci USA, 71:3883-3887.
16. Montel, H., Starke, K. and Taube, H.D. (1975): N.S. Arch. Pharmacol., 288: 427-433.
17. Nathanson, J.A. (1977): Physiol. Rev., 57:157-256.
18. Negro-Vilar, A., Ojeda, S.R., Arimura, A. and McCann, S.M. (1978): Life Sci., 23:1493-1498.
19. Nicoll, R.A., Schenker, C., and Leeman, S.E. (1980): Ann. Rev. Neurosci., 3:227-268.
20. Pert, C.B., Kuhar, M.J., and Snyder, S.H. (1975): Life Sci., 16:1849-1854.
21. Phillis, J.W., and Limacher, J.J. (1974): Brain Res., 69: 158-163.
22. Propst, F., Moroder, L., Wunsch, E., and Hamprecht, B. (1979): J. Neurochem., 32:1495-1500.
23. Puri, S.K., Cochin, J., and Volicer, L. (1975): Life Sci., 16:759-768.
24. Quik, M., Iversen, L.L., and Bloom, S.R. (1978): Biochem. Pharmacol., 27:2209-2213.
25. Redmond, D.E. Jr., Roth, R.H., Hattox, S.E., Stogin, J.M., and Baulu, J. (1979): Abst. Soc. Neurosci., 5:348.
26. Robberecht, P., Deschodt-Lanckman, M., Camus, J.C., DeNeef, P., Lambert, M., and Christophe, J. (1979): FEBS Lett., 103:229-233.
27. Robison, G.A., Butcher, R.W., and Sutherland, E.W. (1971): Cyclic AMP. Academic Press, New York.
28. Sharma, S.K., Nirenberg, M., and Klee, W.A. (1975): Proc. Natl. Acad. Sci. USA, 72:590-594.
29. Traber, J., Fischer, K., Latzin, S., and Hamprecht, B. (1975): Nature, 253:120-122.
30. Tsang, D., Tan, A.T., Henry, J.L., and Lal, S. (1978): Brain Res., 152:521-527.
31. Walczak, S.A., Wilkening, D., and Makman, M.H. (1979): Brain Res., 160:105-116.
32. Wiegant, V.M., Dunn, A.J., Schotman, P., and Gispen, W.H. (1979): Brain Res., 168:565-584.
33. Wolfe, B.B., Harden, T.K., and Molinoff, P.B. (1977): Ann. Rev. Pharmacol. Toxicol., 17:575-604.

Neurosecretion and Brain Peptides,
edited by J. B. Martin, S. Reichlin, and K. L. Bick.
Raven Press, New York © 1981.

Pharmacology of Gonadotropin Releasing Hormone: A Model Regulatory Peptide

Wylie W. Vale, Catherine Rivier, Marilyn Perrin, Mark Smith, and Jean Rivier

Peptide Biology Laboratory, The Salk Institute, San Diego, California 92138

Much of our understanding of the central nervous system roles of biogenic monoamines and of a few selected peptides such as the opioid peptides is attributable to the availability of a vast armamentarium of pharmacologic agents. It is essential to the appreciation of the significance of other peptides like gonadotropin releasing hormone (GnRH), somatostatin, neurotensin and Substance P that such tools be established for all of the regulatory peptides.

The development of this pharmacology of GnRH and other peptides is directed toward basic as well as practical goals. Peptide analogues have been used to probe the structural requirements for receptor recognition and intrinsic activity (or ability of ligand-receptor complex to activate membrane effector systems). Conformationally restricted yet active analogues are particularly useful as subjects of physiochemical studies directed toward the elucidation of "preferred" conformations and the topography of the receptor. Insight concerning possible sites of catabolic cleavage of the peptides is gained by finding stable long-acting peptide analogues. Such long-acting analogues can also be used to study the consequences of over-production of the peptide which, by virtue of desensitization, can lead to an inhibition rather than a stimulation of the processes dependent upon the endogenous peptide. Analogues which, by virtue of altered distribution or receptor affinities, exhibit selectivity for a discrete population of receptors provide important tools for "sorting out" what is often a complex interaction of the multiple target sites of an endogenous or administered peptide. The physiologic roles of endogenous peptides are perhaps best investigated with specific antagonists of sufficient potency and duration of action for meaningful <u>in vivo</u> experiments. Finally, analogues are useful in the characterization of bioassays and immunoassays, and can be tailor-made for a variety of purposes including their use as immunogens, radioligands, or affinity labels.

The distribution and the important physiologic and perhaps patho-physiologic roles of receptors for regulatory peptides provide a strong rationale for the development of agents which can modulate the functions of these receptors. Analogues which are highly potent, long-acting, orally active, safe, selective in their affinities for or distributions to particular cell types, or which behave as antagonists are being developed which may have diagnostic or therapeutic significance.

The basic and practical utilities of peptide analogues can be illustrated by a discussion of the development of a pharmacology of the decapeptide (4,32), gonadotropin releasing hormone, GnRH (also referred to as the LRF or LHRH). GnRH plays a key role in the neuroregulation of the secretion of the gonadotropins (Gn), luteinizing hormone (LH) and follicle stimulating hormone (FSH). The essential function of GnRH in the regulation of reproductive processes has been supported by the marked reductions in plasma gonadotropins and sex steroid levels, gonadal atrophy and inhibition of gametogenesis observed following active or passive immunoneutralization of GnRH in experimental animals (1,29).

GnRH SUPERAGONISTS

Numerous analogues involving modification to each residue in GnRH have been synthesized and tested in vitro and in vivo for abilities to modulate Gn secretion (53,55). The structures of several selected superagonists are shown in Fig. 1. The abilities of these peptides to stimulate the secretion of LH by primary anterior pituitary cell cultures is the basis of the in vitro bioassay. Modifications to three residues have resulted in peptides with enhanced agonist potencies: Gly^6, Leu^7 and $Gly^{10}-NH_2$. The substitution of D-Ala for Gly^6 increases potency fourfold while L-Ala in that position decreases potency considerably (33). The C-terminal modification of Fujino (17) increases potencies fourfold and can be combined with a D-amino acid in position 6 to yield analogues with potencies which are approximately the product of the singly modified peptides (56). The most potent GnRH agonists have aromatic D-amino acids such as D-Trp or imidazole benzyl-D-His in position 6 and Pro^9-NEt at the C-terminus. In vitro [D-Trp^6,Pro^9-NEt]-GnRH (56) and [(imBzl)-D-His6,Pro^9-NEt]-GnRH (60) are ca. 140 and more than 200 times as potent as GnRH. In some analogues, N^αmethylation of Leu^7 slightly increases potency (30), e.g., [D-Ala^6,N^αMeLeu^7,Pro^9-NEt]-GnRH with a potency of ca. 30 is twice as potent as [D-Ala6,Pro^9-NEt]-GnRH. These examples illustrate the remarkable capabilities of the biochemist to modify structure-function relationships.

Several groups have reported that GnRH superagonists are degraded more slowly by tissue extracts than is GnRH (29,31,56,58). Relative stability to inactivation may in fact contribute to high

POTENCIES OF LRF AGONISTS

	IN VITRO BIOASSAY	RADIOLIGAND MEMBRANE BINDING ASSAY
pGlu^1-His^2-Trp^3-Ser^4-Tyr^5-Gly^6-Leu^7-Arg^8-Pro^9-Gly^{10}-NH$_2$	1	1
pGlu-His-Trp-Ser-Tyr-Gly-Leu-Arg-Pro^9-NEt	4	–
pGlu-His-Trp-Ser-Tyr-D-Ala^6-Leu-Arg-Pro-Gly-NH$_2$	4	4
pGlu-His-Trp-Ser-Tyr-D-Ala^6-Leu-Arg-Pro^9-NEt	14	10
pGlu-His-Trp-Ser-Tyr-D-Ala^6-MeLeu^7-Arg-Pro^9-NEt	31	19
pGlu-His-Trp-Ser-Tyr-D-Trp^6-Leu-Arg-Pro-Gly-NH$_2$	36	30
pGlu-His-Trp-Ser-Tyr-D-Trp^6-Leu-Arg-Pro^9-NEt	144	50
pGlu-His-Trp-Ser-Tyr-D-His$^6_{(Bzl)}$-Leu-Arg-Pro^9-NEt	210	55

FIG. 1. Potencies determined in vitro (55,53) and by competitive radioligand membrane binding assay (35,38,37) of selected GnRH agonists.

in vivo and in vitro potencies of GnRH analogues.   Of the peptides tested, we have shown only the triply modified analogue
[D-Ala ^{6},N$^\alpha$MeLeu ^{7},Pro ^{9}-NEt]-GnRH as being completely stable during
incubations with tissue extracts (56,58).

The most important determinant of the potency of the GnRH
superagonists appears to be their high affinities for GnRH receptors.   We have reported that GnRH superagonists are more potent
than GnRH in displacing ^{3}H-GnRH from pituitary membrane preparations (35).   Recently, Clayton et al. (9) and our group (37,38)
have obtained similar results when an iodinated superagonist is
employed as the radioligand.   We chose to iodinate [D-Ala ,N$^\alpha$Me-
Leu ^{7},Pro ^{9}-NEt]-GnRH because of its resistance to degradation.
The radioligand was prepared by mild chloramine T iodination and
purified by ion exchange chromatography on carboxymethyl cellulose and high pressure liquid chromatography (HPLC).   The radioligand thus obtained was shown by analytical HPLC to be identical
to [mI-Tyr ^{5},D-Ala ^{6},N$^\alpha$MeLeu ^{7},Pro ^{9}-NEt] -GnRH which was synthesized
de novo and, furthermore, to be stable under conditions of the
binding assays.   The radioligand binds to anterior pituitary
membranes with a K_D of 0.3 nM.   The high affinity sites have been
shown to be saturable and specific.   As illustrated in Fig. 1,
there is a correlation between the potencies of these superagonists in the in vitro bioassay and in the pituitary membrane
binding assay.

Similar rank orders in the binding assay and in the secretion
assay are observed with the more biologically active agonists exhibiting the higher potencies in the binding assay, suggestive of
those peptides having higher affinities for the GnRH receptor.
However, the actual potencies relative to the GnRH are higher in
the bioassay than in the radioligand binding assay.   Perhaps the
increased hydrophobicity of these analogues gives them an advantage in the intact cell but not in the membrane preparations.   It
is also possible that relative resistance to degradation contributes to the higher potencies seen with intact cells during a 4
hour incubation at 37°C.

Consistent with their in vitro activities and reported resistance to degradation, the more potent GnRH agonists exhibit high
potency and prolonged duration of action in vivo in experimental
animals and human beings (5,12,13,48).   Based upon integrated
plasma gonadotropin levels, these agonists, often referred to as
superagonists, such as [D-Leu ^{6},Pro ^{9}-NEt] GnRH, [D-Trp ^{6},Pro ^{9}-NEt]-
GnRH and [imBzl-D-His ^{6},Pro ^{9}-NEt]-GnRH, are hundreds of times more
effective to stimulate LH and FSH in vivo than is GnRH in the rat
(20,58).   The superagonists are effective when given by various
routes, including intranasally and orally (20,58).

While the acute administration of superagonists by various
routes is associated with increases in gonadotropins and sex
steroid levels as well as induction of ovulation (57), the long-
term administration of superagonists, in contrast, decreases
reproductive functions in males and females.   In the female,
repeated GnRH administration is associated with a decrease in
ovarian and uterine weights, and in sex steroid levels.   GnRH

agonists can prevent ovulation, delay puberty and terminate pregnancy (10,28,42,58). The daily administration of GnRH agonists to male rats results in dose-related decreases in testes, seminal vesicle and prostrate weights, and the levels of testosterone and prolactin (2,28,43,47,58). The analogue [D-Trp[6],Pro[9]-NEt]-GnRH is more than 1000 times more potent than GnRH (43) for these inhibitory effects can be observed within 3 days of the initiation of a daily injection schedule.

The decreases in the weights of the accessory sex organs, the seminal vesicles and prostrate are primarily related to the inhibition of androgen production by Leydig cells. Consistently, the administration of high doses of exogenous testosterone to super-agonist treated rats can overcome the decrease in accessory sex organ weights (43). Recently, however, Bardin and colleagues (52) have reported that [D-Trp[6],Pro[9]-NEt]-GnRH can reduce accessory sex organ weights in hypophysectomized castrated male rats maintained on moderate doses of testosterone. These data were interpreted to suggest the possibility that GnRH agonists could, through an as yet undefined mechanism, antagonize the actions of sex steroids.

Morphologically there is evidence for considerable inhibition of spermatogenesis (43,2) which can also be observed as soon as 3

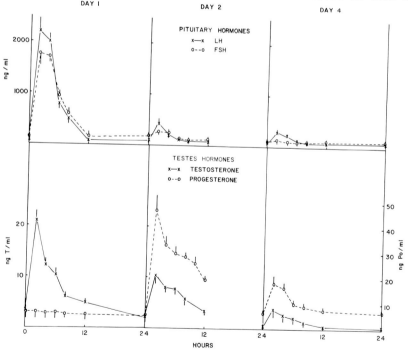

FIG. 2. Plasma levels of pituitary (LH and FSH) and gonadal (testosterone, progesterone) hormones in male rats administered 1 mg [D-Trp[6],Pro[9]-NEt]-GnRH subcutaneously. Responses on days 1, 2 and 4 are illustrated.

days after beginning of treatment, at which time, in one study
(43), tubular diameter was already reduced and only 20% of tubules showed evidence of mitotic activity.   After 5 days of treatment, most tubules were empty; membranes were disrupted and
polynucleated cells were present.   Thus both Leydig cells which
produce testosterone and sperm producing tubular elements are
suppressed by exposure to superagonists.

The mechanisms whereby superagonists inhibit steroidogenesis
in the rat probably involve desensitization at both the pituitary
and gonadal levels.   The initial injection of a superagonist
induces a prolonged (up to 5 hours) elevation of gonadotropin
levels.   However, the Gn secretory responses to subsequent
injections of LRF or a superagonist are blunted considerably for
at least 72 hours following the first dose of superagonist (43).
The plasma levels of pituitary and gonadal hormones on days 1, 2
and 4 in animals receiving daily injections of $[D-Trp^6,Pro^9-NEt]-$
GnRH are shown in Fig. 2.   It is noteworthy that, while the LH
secretory response is blunted considerably, the FSH secretory
response disappears entirely, thereby leading to a shift in the
LH/FSH ratio produced by the pituitary.

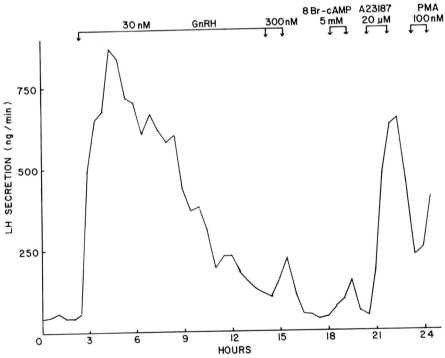

FIG. 3.   Secretion of LH by superfused rat anterior pituitary
cells attached to Cytodex[TM] beads (49), continuous delivery of
GnRH followed by administration of 5 mM ^{8}Br-cyclic AMP, 20 μM
A-23187 and then 100 nM PMA (phorbol myristate acetate).   Data
from ref. 50.

We have developed method (49) suitable for the investigation of pituitary desensitization phenomenon <u>in vitro</u>. Dissociated anterior_{TM} pituitary cells have been cultured <u>in the</u> presence of Cytodex^{TM} beads to which they attach. The bead-attached cells have been placed in small columns immersed in a 37°C water bath and superfused with a low bicarbonate complete medium. As shown in Fig. 3, the continuous superfusion of such cells with GnRH causes a rapid increase in LH secretory rates. After more than 3 hours, the rate of LH secretion gradually decreases until by 9 hours, the cells are completely refractory to further GnRH even though <u>ca.</u> 50% of the cells LH content remained. The finding that <u>cells</u> desensitized to GnRH are still capable of responding to other secretagogues such as the co-carcinogen, phorbol myristate acetate, and the Ca^{++} ionophore, A-23187, suggests that depletion of readily releasable LH pools does not account for the desensitization. Studies are underway to determine whether pituitary GnRH desensitization involves changes in receptor numbers or affinity, in membrane effector systems, or in the cellular response to intracellular mediators.

Knobil and associates (3) have shown that the pulsatile delivery of LRF (6 minutes per hour) to castrated rhesus monkeys with

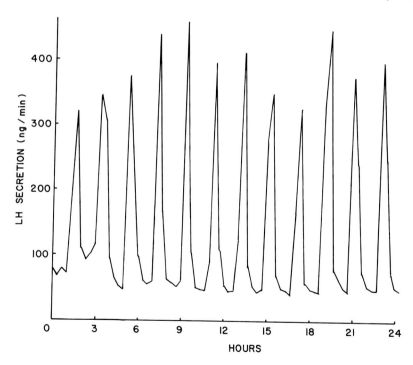

FIG. 4. Secretion of LH by superfused rat anterior pituitary cells attached to Cytodex^{TM} beads (49). 30 nM GnRH administered in 15-minute pulses every 2 hours. Data from ref. 50.

hypothalamic lesions results in a physiological-type pulsatile
pattern of gonadotropin secretion which could be maintained
chronically. However, when LRF was delivered continuously, the
LH secretory rates as well as integrated amounts produced over
given time periods were drastically reduced. This phenomenon has
also been observed in vitro (Fig. 4) (50); the superfused bead-
attached cells release LH and FSH (not shown) for 48 hours in response to the pulsatile delivery of GnRH.

Recent clinical studies by Crowley and Kelch (15,61) and their
associates have shown improvements in GN and testosterone production in hypogonadal males following the delivery of LRF in a pulsatile manner to hypogonadotropic hypogonadal males. Crowley and
colleagues (14) similarly found that Gn and testosterone levels
and testicular size in such subjects could be elevated by the administration of [D-Trp^6,Pro^9-NEt]-GnRH every other day. When the
peptide was given every day, the improvements were reversed.

As shown in Fig. 2, the nature and amounts of steroids produced in response to each administration of GnRH superagonist
changes throughout the regimen. While the secretion of testosterone becomes progressively lower, progesterone, which is not
stimulated by the first injection of the superagonist, is markedly elevated by the injection on the second day.

Several laboratories have demonstrated that the administration
of high doses of gonadotropins (LH or hCG, human chronic gonadotropin) or of LRF agonists result in prolonged loss of gonadal LH
receptors (2,6,20,22). Furthermore, since the Leydig cells'
steroidogenic responses to hCG, as well as other stimuli, such as
dibutyryl 3'5' cyclic AMP, are also blunted, post-receptor mechanisms are probably also affected (7).

The administration of high levels of exogenous Gn can, however, reverse the effects of superagonists on testosterone production as demonstrated by the high weights of the testosterone-
dependent organs (such as seminal vesicles) of rats receiving
[D-Trp^6,Pro^{9}NEt]-GnRH and the gonadotropins, hCG and the pregnant
mare's serum gonadotropin (PMSG). Thus, the high amounts of Gn
secreted acutely in response to GnRH agonists are not the exclusive reason for the inhibition of steroidogenesis induced by
superagonists. We have proposed that a combination of pituitary
and Leydig cell desensitization conspires to produce the fall in
testosterone production (43,59). The administration of a superagonists results in a 4-6 hour elevation in LH followed by an
ensuing period with nondetectable LH levels. The Leydig cells,
desensitized as a consequence of the initial burst of LH, would
be subjected to long periods of low LH levels during a time when
sensitivity to LH is minimal.

Catt and associates (6) have shown that testes from hCG-
treated male rats exhibit specific steroidogenic defects in vitro
at the 17,20 desmolase and earlier steps. We have demonstrated
the presence of similar enzymatic blocks in incubated Leydig
cells of rats treated with the superagonist [D-Trp^6,Pro^9 -NEt]
-GnRH (4). Following the intense gonadotropin stimulation in-

duced by the initial injections of a superagonist, intermediates such as progesterone (Po), continue to be synthesized and secreted, even though the development of blocks of specific enzymes progressively suppresses testosterone secretion (Fig. 2). Eventually, through mechanisms that are not defined, production of Po and other precursors are also inhibited.

Another component of the paradoxical action of the LRF agonists was first suggested by the work of Rippel and Johnson (41), who showed that superagonists decreased uterine weight in hypophysectomized female animals. Hseuh and Erickson (24) showed that superagonists could inhibit steroid production by FSH-treated cultured ovarian cells, thus confirming the possibilities of direct effects of superagonists on the gonads. These workers also demonstrated an inhibition of Gn-stimulated testosterone production in hypophysectomized male rats (23).

Consistent with these direct gonadal effects of GnRH, our group and Catt's have demonstrated the presence of high affinity,

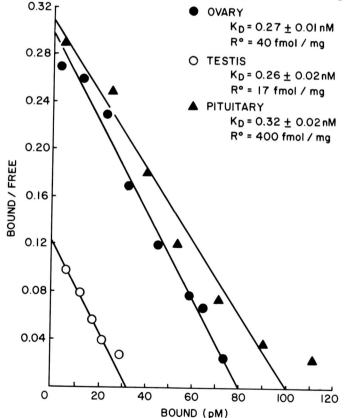

OVARY
$K_D = 0.27 \pm 0.01$ nM
$R^o = 40$ fmol / mg

TESTIS
$K_D = 0.26 \pm 0.02$ nM
$R^o = 17$ fmol / mg

PITUITARY
$K_D = 0.32 \pm 0.02$ nM
$R^o = 400$ fmol / mg

FIG. 5. Scatchard analysis of binding of [^{125}-Tyr^5,D-Ala^6, N$^\alpha$ MeLeu^7,Pro^9-NEt]-GnRH to rat membranes derived from ovary (●), testis [interstitial cell preparation (o)], or anterior pituitary (▲).

saturable GnRH binding sites on not only pituitary but on ovarian
(8,36,38) and testicular interstitial cell membranes (Fig. 5)
(37).   Pituitary, ovary and testis membranes contain approximately 400, 50 and 20 fmoles binding sites per mg membrane protein, respectively.   The K_D's for binding the stable radioligand
are very close for the membrane from the three different tissues.
All of the agonists have shown close agreement between their affinities for pituitary than for ovarian binding sites (Table 1)
(8).   It is possible that differential rates of analogue degradation by the two membrane preparations could explain this discrep-
ancy.   Alternatively, these results might reflect differences
between the GnRH receptors in the two tissues.   If the latter

TABLE 1.   Dissociation constants (K_D) for binding of GnRH
           to pituitary or ovarian membrane preparations

PEPTIDE	K_D(nM) PITUITARY	KD(nM) OVARY
GnRH	6	4
[D-Ala[6],NαMeLeu[7],Pro[9]-NEt]-GnRH	0.32	0.27
[m-I-Tyr[5],D-Ala[6],NαMeLeu[7],Pro[9]-NEt]-GnRH	0.29	0.32
[D-Trp[6]]-GnRH	0.20	0.18
[D-Trp[6],Pro[9]-NEt]-GnRH	0.12	0.17
[imBzl,D-His[6],Pro[9]-NEt]-GnRH	0.11	0.10
[D-Phe[2],D-Trp[3,6]]-GnRH	0.80	1.20
[D-pGlu[1],D-Phe[2],Trp[3,6]]-GnRH	0.54	0.69
[Ac-dehydro-Pro[1],D-Phe[2],D-Trp[3,6]]-GnRH	0.17	0.37
[Ac-dehydro-Pro[1],pCl-D-Phe[2],D-Trp[3,6]]-GnRH	0.29	1.00
[Ac-dehydro-Pro[1],pCl-D-Phe[2],D-Trp[3,6],NαMeLeu[7]]-GnRH	0.20	1.20
Des-Arg[8]-GnRH	>100	>100
Somatostatin	>100	>100
TRF	>100	>100

interpretation were correct, then it may be feasible to develop a series of agonists and antagonists of greater selectivity. These selective analogues would be useful in the characterization of the various receptors. Furthermore, such selective analogues will permit us to determine at what level LRF acts in intact animals to decrease gonadal activities. This disruption in tubular structure and spermatogenesis might in part be due to the fall in testosterone production. However, it is noteworthy that while Gn restores testosterone production in the superagonist-treated male, Gn does not reverse or prevent damage to seminifer-ous tubules. In fact, Gn itself can inhibit spermatogenesis whether or not superagonists are co-administered (43). Consist-ent with a possible role of Gn in permitting or mediating tubular damage in superagonist-treated rats is the observation that superagonists do not alter the tubular morphology of testoster-one-treated hypophysectomized rats but cause tubular disruption in testosterone-treated rats with intact pituitary glands (60).

Since progesterone has been reported both to have deleterious effects on seminiferous tubules of male rats (17) and yet to maintain qualitatively normal spermatogenesis in hypophysectomiz-ed animals (21,51), the role played by the abnormal elevation of Po (Fig. 2) is not defined. However, the intratesticular levels of Po in these studies are probably astronomical and may well, in part, mediate the deleterious effects of Gn and GnRH superagon-ists on seminiferous tubules.

The failure of exogenous testosterone to prevent the effects of superagonists on tubular morphology might be partially ex-plained by an inhibitory effect of high intratesticular Po. It may be relevant that superagonist-treated rats exhibit increased LH to FSH ratios in view of the established role of FSH in the initiation of spermatogenesis. The possibility of direct effects of GnRH on tubular elements should also be considered.

Studies of the antigonadal actions of GnRH superagonists have not only led to an expanded appreciation of the multiple poten-tial roles of GnRH, but may be of practical significance. For example, superagonists have already been shown to prevent ovula-tion and induce luteolysis in normal women. It is plausible that a contraceptive regimen of superagonist plus testosterone might be used to maintain secondary male sex characteristics in the presence of an inhibition of spermatogenesis.

In summary, the inhibition of Leydig cell steroidogenesis by superagonists may be caused by desensitization at both gonadal and pituitary levels. The superagonist mediated inhibition of testosterone production can be overcome by exogenous gonado-tropins. The disruption of seminiferous tubular structure and spermatogenesis is pituitary-dependent, and may be related to supernormal production of progesterone. Furthermore, GnRH may exert direct effects at gonadal and accessory sex organ levels. Though additional studies are needed to assess the precise roles played by the nonphysiological secretions of LH, FSH, testos-terone and progesterone and the gonadal effects of GnRH in the anti-reproductive effects of GnRH superagonists, these peptides are providing important investigational and perhaps therapeutic tools. The application of these agents to the regulation of fertility is promising.

GnRH ANTAGONISTS

Some analogues with modifications to the N-terminal three amino acids in GnRH have behaved as partial agonists (exhibiting lowered response maxima) or antagonists. The first reported (59) competitive antagonist of GnRH was des-His^2-GnRH which reduced the amounts of LH released by GnRH by 50% at a molar ratio ([antagonist] : [GnRH] = ICR$_{50}$) 2500/1. The reciprocal ICR$_{50}$ values, which provide a basis for comparing the apparent potencies of antagonists both with each other as well as with GnRH, are shown for several antagonists in Fig. 6. This figure includes the potencies of the peptides in the competitive radioligand membrane binding assay and the dose of antagonist required when given on noon of proestrus to block ovulation in freely cycling rats by more than 90%.

Improvements in antagonist potencies have come primarily by virtue of the incorporation of six position substitutions, such as D-Trp^6, which enhance the potencies of both agonists and antagonists along with substantial modifications to the N-terminal region of the molecule. The more recent N-terminal alterations such as found in [Acetyl-dehydro,Pro^1,pCl-D-Phe^2,D-Trp^{3,6} N$^\alpha$Me-Leu^7]-GnRH (Antag) reduce intrinsic activity of the molecule while returning high potency. The development of these powerful antagonists was the result of the combined efforts and interactions of several groups in this field (7,11,25,40,60).

A representative of another series of antagonists is shown here: D-Cys-D-Phe-D-Trp-Ser- Tyr
                       | |
         H$_2$N-Cys- Pro - Arg -Leu-D-Trp

We have found that this cyclic antagonist has higher affinity for pituitary GnRH binding sites than does GnRH which suggests that the conformation preferred for peptide binding to the GnRH receptor is one aligning the N- and C-terminal as we speculated earlier (19). Such less flexible compounds may be very useful for mapping the topography of the GnRH receptor and might represent the first step in the development of an "alkaloid" with affinity for the GnRH receptor. It has proven much easier to develop cyclic antagonists than cyclic agonists. The receptor fit may need to be more precise for agonists. If, as the "zipper" theory requires, the ligand undergoes a conformational change at the time of receptor-mediated transduction events, then a rigid or cyclic peptide would be less likely to make the structural transition required for the biologic response.

The peptide [Acdehydro,Pro^1,pCl,D-Phe^2,D-Trp^{3,6}N$^\alpha$MeLeu] -GnRH (Antag) exhibits ca. 25 times higher affinity for pituitary GnRH binding sites than does GnRH itself in agreement with the high 1/ICR$_{50}$ values for this antagonist (Fig. 6). Furthermore, Antag is a powerful and long-acting inhibitor of ovulation in the rat. A single subcutaneous injection of 20 mg on noon of proestrus blocks ovulation by 100%; when administered 24 hours earlier on diestrust-II, 250 mg blocked ovulation in 67% of rats (43). The administration of Antag to pregnant rats was found to terminate pregnancy (43).

Recently we have given Antag daily for 2 weeks to normal male rats and observed dramatic effects on the weights of androgen-dependent organs (reduced by > 90%) and on testicular weight (decreased by > 55%) (46). The Leydig cells of Antag-treated males were atrophied and tubular diameters were uniformly reduced by ca. 50% (46). The demonstration of such profound effects of GnRH antagonists in the male and female is strong support for the essential role of endogenous GnRH in normal reproductive functions in both sexes. Furthermore, the potential application of GnRH antagonists as contraceptives in females (as antiovulatory agents or menstrual regulators) and males (with androgen replacement) is being explored.

The GnRH antagonists are important tools for the study of the neurobiologic and other actions of endogenous GnRH. GnRH has been reported to have a variety of behavioral, biochemical and biophysical effects on the central nervous system. GnRH-like immunologic activities have been found by radioimmunoassay and immunohistochemical techniques to be in hypothalamic and extra-hypothalamic areas (amygdala, midbrain, spinal ganglia) and to be in the vicinity of GnRH responsive cells.

Recently Jan et al. have built a strong case for a role of GnRH as a neurotransmitter in the frog sympathetic ganglia. They found that the application of GnRH could duplicate the late slow excitatory postsynaptic potentials seen following preganglionic electrical stimulation. The antagonist [D-pGlu^1 D-Phe^2,D-Trp^{3,6}]-LRF was shown not only to block the response to exogenous GnRH but, more significantly, to inhibit the response to preganglionic stimulation. These studies, when considered with the demonstration of LRF immunoreactivity in the frog ganglia (27), are consistent with the hypothesis that GnRH may function as a transmitter in the automatic nervous system of the frog.

Studies of the groups of Moss (34) and Pfaff (39) showing that GnRH could enhance mating behavior in some animal preparations have recently been supported by the observation by Dudley et al. (16) that GnRH antagonists are capable of suppressing mating behavior in castrated estrogen progesterone-treated female rats. Such experiments are critical to our appreciation of the role of endogenous GnRH in mating behavior and in other central nervous system functions.

BIOLOGICAL ACTIVITIES OF LRF ANTAGONISTS

	IN VITRO $1/IDR_{50}$	MEMBRANE BINDING ASSAY	ANTIOVULATORY ED > 90
pGlu[1] - His[2] - Trp[3] - Ser[4] - Tyr[5] - Gly[6] - Leu[7] - Arg[8] - Pro[9] - Gly[10] - NH_2 (LRF)	-	1	-
pGlu ----[2] - Trp - Ser - Tyr - Gly - Leu - Arg - Pro - Gly - NH_2	.0004	-	-
pGlu - D-Phe[2] - Trp - Ser - Tyr - Gly - Leu - Arg - Pro - Gly - NH_2	.001	.15	-
pGlu - D-Phe[2] - Trp - Ser - Tyr - D-Trp[6] - Leu - Arg - Pro - Gly - NH_2	.07	5.4	> 1000 μg
pGlu - D-Phe[2] - D-Trp[3] - Ser - Tyr - D-Trp[6] - Leu - Arg - Pro - Gly - NH_2	.06	7.5	> 1000 μg
D-pGlu[1] - D-Phe[2] - D-Trp[3] - Ser - Tyr - D-Trp[6] - Leu - Arg - Pro - Gly - NH_2	.3	11	250 μg
Ac-dehydro-Pro[1] - pCL-D-Phe[2] - D-Trp[3] - Ser - Tyr - D-Trp[6] - Leu - Arg - Pro - Gly - NH_2	67	21	10 μg
Ac-dehydro-Pro[1] - pCL-D-Phe[2] - D-Trp[3] - Ser - Tyr - D-Trp[6] - MeLeu - Arg - Pro - Gly - NH_2	25	32	10 μg

FIG. 6. Biological activities of GnRH antagonists in: In vitro bioassay, ICR_{50} = molar ratio [antagonist] : [GnRH] required to reduce the GnRH mediated release of LH by 50% (53); competitive radioligand membrane binding assay (37, 48); in vivo antiovulatory assay.

ACKNOWLEDGEMENTS

The excellent technical assistance of Helene Laurent, Nancy Keating, Gail Laughlin, Karen von Dessonneck, Joan Vaughan, Ron Kaiser, Bill Schaber, John Porter and Sue Orloff and manuscript preparation by Susan McCall are gratefully acknowledged. We also are indebted to the NIAMDD Rat Pituitary Hormone Programs for supplying rat LH and FSH kits, and to Drs. W. Crowley and S. Yen for the generous gifts of steroid antisera. Research supported by NIH grant and contract, HD09690 and RFP CD78-5, The Rockefeller Foundation and The Salk Institute Texas Foundation. Research was conducted in part by The Clayton Foundation for Research, California Division. Drs. Vale, C. Rivier, Perrin and J. Rivier are Clayton Foundation investigators.

REFERENCES

1. Arimura, A., Sato, H. Kumasaka, T., Worobec, R., Debeljuk, L., Dunn, J., and Schally, A.V. (1973): Endocrinology, 93: 1092-1103.
2. Auclair, C., Kelly, P.A., Coy, D.H., Schally, A.V., and Labrie, F. (1977): Endocrinology, 101:1890.
3. Belchetz, P.E., Plant, T., Nakai, Y., Keogh, E., and Knobil, E. (1978): Science, 202:631-633.
4. Burgus, R., Butcher, M., Amoss, M., Ling, N., Monahan, M., Rivier, J., Fellows, R., Blackwell, R., Vale, W., and Guillemin, R. (1972): Proc. Natl. Acad. Sci. USA, 69:278-282.
5. Casper, R., and Yen, S.S.C. (1979): Science, 205:408-410.
6. Catt, K.J., Baukal, A.J., Davies, T.F., and Dufau, M.L. (1979): Endocrinology, 104:17-25.
7. Cigorraga, S.B., Dufau, M.L., and Catt, K.J. (1978): J. Biol. Chem., 253:4297-4304.
8. Clayton, R.N., Harwood, J.P., and Catt, K.J. (1979): Nature, 282:90-92.
9. Clayton, R.M., Shakespear, R.A., Duncan, J.A., and Marshall, J.C. (1979): Endocrinology, 104:1484-1494.
10. Corbin, A., Beattie, C.W., Reese, R., and Yardley, J. (1977): Fertil. Steril., 28:471-475.
11. Coy, D.H., Mezo, I., Pedroza, E., Nekola, M.V., Vilchez, J., Piyachaturawat, P., Schally, A.V., Seprodi, J., and Teplan, I. (1979): In: Peptides: Structure and Biological Functions, edited by E. Gross, and J. Meienhofer, pp. 775-781. Academic Perss, New York.
12. Crowley, W.F., Beitens, I., Vale, W., Kliman, B., Rivier, J., Rivier, C., and McArthur, J. (1979): N. Engl. J. Med. (submitted).
13. Crowley, W.F., Beitens, I.A., Kliman, B., McArthur, J.W., Rivier, J., Rivier, C., and Vale, W. (1979): 61st Ann. Endocrine Soc. Mtg., abstract.

14. Crowley, W., Vale, W., Beitens, I., Rivier, J., Rivier, C., and McArthur, J. (1979): 61st Ann. Endocrine Soc. Mtg., abstract 16, p. 76.

15. Crowley, W. (1980): (submitted).

16. Dudley, C.A., Vale, W., Rivier, J., and Moss, R.L. (1980): Proc. Soc. Neurosci., (abstract submitted).

17. Flickinger, C.J. (1977): Anat. Rec., 187:405-430.

18. Fujino, M., Yamazaki, I., Kobayashi, S., Fukuda, T. Shinawaga, S., and Nakayme, R. (1974): Biochem. Biophys. Res. Commun., 57:1248-56.

19. Grant, G., and Vale, W. (1971): Nature, 237:182-183.

20. Haour, F., and Saez, J.M. (1977): Mol. Cell. Endocrinol., 7:17-24.

21. Harris, M.E., and Bartke, A. (1975): Endocrinology, 96:1319-1323.

22. Hsueh, A., Dufau, M., and Catt, M. (1977): Proc. Natl. Acad. Sci USA, 74:592-595.

23. Hsueh, A., and Erickson, G. (1979): Nature, 281:66-67.

24. Hsueh, A., and Erickson, G. (1979): Science 204:854.

25. Humphries, J., Wan, Y.P., and Folkers, K. (1976): Biochem. Biophys. Res. Commun., 72:939-944.

26. Jan, L.Y., Jan, Y.N., and Kuffler, S. (1980): Proc. Natl. Acad. Sci. USA, 76:1501-1505.

27. Jan, L.Y., and Jan, Y.N. (1980): Fed. Proc., (in press).

28. Johnson, E.D., Gendrich, R.L., and White, W.F. (1976): Fertil. Steril., 27:853-860.

29. Koch, Y., Baram, T., Chobsieng, P., and Fridkin, M. (1974): Biochem. Biophys. Res. Commun., 61:95-103.

30. Ling, N., and Vale, W. (1975): Biochem. Biophsy. Res. Commun., 63:801-806.

31. Marks, N., and Stern, F. (1974): Biochem. Biophys. Res. Commun., 61:1458-1463.

32. Matsuo, H., Baba, Y., Nair, R.M.G., Arimura, A., and Schally, A.V. (1971): Biochem. Biophys. Res. Commun., 43:1334-1339.

33. Monahan, M., Amoss, M., Anderson, H., and Vale, W. (1973): Biochemistry, 12:4616-4620.

34. Moss, R.L., and McCann, S.M. (1973): Science, 178:417-419.

35. Perrin, M., Rivier, J., and Vale, W. (1980): Endocrinology, 106:1289-1296.

36. Perrin, M.H., Rivier, J.E., and Vale, W.W. (1980): Fed. Proc., 39:487.

37. Perrin, M.H., Vaughan, J.M., Rivier, J.E., and Vale, W. (1980): Life Sci., 26:2251-2255.

38. Perrin, M., Rivier, J., and Vale, W. (1980): J. Biol. Chem., (submitted).

39. Plotnikoff, N.P., Prange, A.J., Breese, G.R., Anderson, M.S., and Wilson, I.C.(1972): Science, 178:417-418.

40. Rees, R.W., Foell, T., Chai, S., and Grant, N. (1974): J. Med. Chem., 17:1016-1019.

41. Rippel, R., and Johnson, E. (1977): Proc. Soc. Exp. Biol. Med., 152:432-436.

42. Rivier, C., Rivier, J., and Vale, W. (1978): Endocrinology, 103:2299-2305.
43. Rivier, C., Rivier, J., and Vale, W. (1979): Endocrinology, 105:1191-1201.
44. Rivier, C., Rivier, J., and Vale, W. (1980): Endocrinology, (submitted).
45. Rivier, C., Rivier, J., and Vale, W. (1980): Int. J. Fertil., (in press).
46. Rivier, C., Rivier, J., and Vale, W. (1980): Science, (in press).
47. Sandow, J. (1976): In: Basic Aplications and Clinical Uses of Hypothalamic Hormones, edited by A.L. Chorro Salaado, R. Fernandez Durango, and L.G. Lopez del Campo, pp. 113-123. Excerpta Medica, Amsterdam.
48. Schally, A.V., and Coy, D. (1977): In: Hypothalamic Peptide Hormones and Pituitary Regulation, edited by J.C. Porter, pp. 99-122. Plenum Press, New York.
49. Smith, M.A., and Vale, W. (1980): Endocrinology, (in press).
50. Smith, M.A., and Vale, W. (1980): Endocrinology, (submitted).
51. Steinberger, E., Chowdhury, A.K., Tcholakian, R.K., and Roll, H. (1975): Endocrinology, 96:1319-1323.
52. Sundaram, K., Cao, Y.Q., Bardin, C.W., Rivier, J., and Vale, W. (1980): Nature, (submitted).
53. Vale, W., Grant, G., Amoss, M., Blackwell, M., and Guillemin, R. (1972): Endocrinology, 91:562-572.
54. Vale, W., Grant, G., Rivier, J., Monahan, M., Amoss, M., Blackwell, M., Burgus, R., and Guillemin, R. (1972): Science, 176:933-934.
55. Vale, W., Rivier, C., Brown, M., Chan, L., Ling, N., and Rivier, J. (1976): In: Hypothalamus and Endocrine Functions, edited by F. Labrie, J. Meites, and G. Pelletier, pp. 397-429. Plenum Press, New York.
56. Vale, W., Rivier, C., Brown, M., Leppaluoto, J., Ling, N., Monahan, M., and Rivier, J. (1976): Clin. Endocrinol., 5:261S.
57. Vale, W., Rivier, C., and Brown, M. (1977): Am. Rev. Physiol., 39:473-527.
58. Vale, W, Rivier, C., Brown, M., and Rivier, J. (1977): In: Hypothalamic Peptide Hormones and Pituitary Regulation, edited by J.C. Porter, pp. 123-156. Plenum Press, New York.
59. Vale, W., Brown, M., Rivier, C., Perrin, M., and Rivier, J. (1979): In: Brain Peptides: A New Endocrinology, edited by A.M. Gotto, Jr., E.J. Peck, Jr., and A.E. Boyd III, pp. 71-88. Elsevier/North Holland Biomedical Press, Amsterdam.
60. Vale, W., Rivier, C., Perrin, M., and Rivier, J. (1979): In: Peptides: Structure and Biological Function, edited by E. Gross, and J. Meienhofer, pp. 781-793. Academic Press, New York.
61. Valk, T.W., Corley, K.P., Kelch, R.P., and Marshall, J.C. (1979): 61st Ann. Endocrine Soc. Mtg., abstract 741, p. 258.

Neurosecretion and Brain Peptides,
edited by J. B. Martin, S. Reichlin, and K. L. Bick.
Raven Press, New York © 1981.

Biologically Active Peptide-Containing Fractions in Schizophrenia and Childhood Autism

K. L. Reichelt, *K. Hole, **A. Hamberger, G. Saelid,
P. D. Edminson, †C. B. Bræstrup, ‡O. Lingjærde, P. Ledaal,
and H. Orbeck

*Pediatric Research Institute, Rikshospitalet, Oslo 1, Norway; *Department of Physiology,
University of Bergen, N-5000 Bergen, Norway; **Department of Neurobiology, University of
Gothenberg, S-40053 Gothenberg, Sweden; † Department of Psychopharmacology,
Sct. Hans Hospital, DK-4000 Roskilde, Denmark; and ‡ The University Psychiatric
Hospital, Asgard Hospital, N-9010 Tromso, Norway*

The chromatographic patterns of peptides and peptide-associated proteins from the urines of schizophrenics and childhood autistics gave two reproducible patterns for each type of disorder. Various fractions from further purification steps showed a number of different biological activities both in <u>in vivo</u> and <u>in vitro</u> test systems. Several of the factors have been purified and confirmation of the peptides' structures is being carried out with synthetic compounds at present.

BACKGROUND

General Role of Peptides

Peptides as signal molecules are phylogenetically old, and can be found from bacteria to humans. At the intracellular level, peptides have an important role in the regulation of key enzyme activities, such as gramicidin inhibition of RNA polymerase in bacteria, and peptide inhibition of N-methyl transferase in rabbit brain (44). Activation of enzymes is shown by the control of glycogen phosphorylase in liver by angiotensin II. Recently, a very important concept has emerged. This is the existence of peptides binding to enzymes, protecting them from degradation, and also ensuring correct conformation. Similar peptides seem to be responsible for the stabilization of rat liver phosphofructokinase and ATP-citrate lyase (17,50). At a slightly higher level, chemotactic peptides for eosinophils and neutrophils have been described (23,6). The more complex and

sequential process of phagocytosis is initiated by a tetrapeptide, tuftsin. Complex behavioral processes in lower animals are
also regulated by peptides; eating behavior in hydra is started
by glutathione (39), and peptides control metamorphosis in
insects (22). Activity in primitive nervous systems is also
regulated by peptides, including bursting pacemaker activity in
ganglion cells (30) and the discharge of neurosecretory cells in
snails and aplysia (2).

The widespread existence of a phylogenetically old signal
system ought to predict the following: a) The initiation of
complex processes in higher organisms may be expected to be
peptide-dependent or mediated, b) peptides ought to have
extensive roles as regulatory agents, and c) if peptides are of
vital importance to the integrity of an organism, the peptides
would be expected to contain mainly non-essential amino acids.

Peptides in the CNS

Endogenous peptides have a major role in the preservation of
the integrity of higher animals, the most important being the
hypophysiotropic release and release-inhibitory peptides from the
hypothalamus (58). The hypothalamic peptidergic neurons function
as transducers for multisignal inputs leading to the control of
pituitary function. There are many examples in the literature of
peptides modifying the uptake, release and metabolism of neurotransmitters, thus exerting control over CNS function. The role
of peptides as neurotransmitters has recently been reviewed by
Emson (18). TRH potentiates acetylcholine sensitive neuronal
responses (71) and facilitates dopamine release in n. accumbens
(35). The turnover of acetylcholine is regulated by a number of
peptides including β-endorphin (46) α-MSH, ACTH and somatostatin
(70). β -endorphin exerts a number of effects on catecholamines,
inhibiting noradrenalin release from hippocampus (1), and also
inhibiting striatal dopamine-sensitive adenylate cyclase (48).
Homocarnosine, anserine and carnosine regulate the uptake and
metabolism of GABA in rat brain (65). Prolyl-leucyl-glycinamide
inhibits α-methyl-p-tyrosine induced catecholamine disappearance
in rat brain (68). Barker has suggested a novel mode of neuronal
communication based on the opiate peptide modulation of amino
acid responses in neurons, particularly in changing the threshold
value for the formation of excitatory potentials (3).

Several complex and sequential behavioral processes seem to be
initiated by peptides, as illustrated by the induction of head to
tail rotation in rats following intracranioventricular injections
of TRH (13), and barrel rotation by somatostatin, LHRH and vasopressin (13,37). Excessive grooming behavior is modulated by
neuropeptides (24), while mothering behavior in virgin rats can
be induced by oxytocin (52). Other basic functions modulated by
peptides include sexual behavior by LHRH (47), and sleep by several peptides (59,67). Learning and forgetting also seem to be
regulated by peptides, with vasopressin facilitating acquisition

and preventing extinction (11), while oxytocin inhibits conditioning and accelerates extinction. The effects of peptides on arousal are of great interest to psychiatry, particularly the response to ACTH and $ACTH_{4-10}$ (34), and the effect of TRH on emotion on the reduction of barbiturate depression (41). Endorphins have been postulated to have a role in psychiatric disorders following the demonstration of the induction of "catatonia" by β-endorphin (10). An improvement in schizophrenics following treatment with (des-tyr) γ-endorphin has recently been described (16). The fact that peptides are widely distributed in the CNS and may possibly function as neurotransmitters (18) suggests the possibility that peptides could be implicated in various CNS disorders.

OUR WORKING HYPOTHESIS

We have previously presented our working hypothesis in some detail (53,66), but it may be outlined as follows: peptides are formed in and/or are released from peptidergic neurons acting as transducers for multiple inputs (58). It is quite usual in biological systems that key, or rate-limiting, steps are subject to feedback regulation. It has been shown that, for typical control peptides such as the hypothalamic releasing factors, degradation is under extensive feedback control. Thus TRH breakdown is regulated by both thyroxine (4) and dopamine (43), while LHRH breakdown is controlled by gonadal hormones (25). On this basis we consider it reasonable to propose that the genetic inheritance of several psychiatric disorders may be due to key peptidase insufficiency (66). Increased loading or strain on the peptidergic system regulated by such a key peptidase could reach a point where peptide release would exceed breakdown, with peptide excretion, in addition to pathology, resulting. We have previously shown that genetic selection in rats for activity levels, and excessive physical exhaustion in normal healthy military cadets can lead to abnormal urinary peptide and associated protein patterns (66). The application of such a model to mental disorders could explain: a) the complex etiology of mental disorders, b) the lack of complete concordance in homozygous twins, c) the apparently relentless course of many psychiatric states, and d) could accommodate more psychodynamic data, where neuronal patterns corresponding to behavioral patterns impinge on the same transducing peptidergic neurons.

SCHIZOPHRENIA

Before embarking upon the presentation of our findings in connection with schizophrenia, we consider it necessary to present a number of discipline-defined domains, which are relevant to any discussion on the biological basis of mental disorders. The domains are shown in Table 1.

TABLE 1. <u>Evaluation of psychiatric data must be based on</u>:

Anatomy	Psychology
Genetics	Learning
Physiology	Hyper and hypofunction
Pharmacology	

In connection with schizophrenia we have chosen to concentrate on the first four only.

Anatomy

The symptomatology of schizophrenia points to an involvement of the basal ganglia (40), and a clearcut anatomical model relating the disorder to the mesolimbic and limbic brain has been proposed (61,62). Many years ago, Heath recorded pathological EEGs from the n. accumbens, septum, head of the caudate and amygdala (26) of schizophrenics. These nuclei are all terminal regions for the mesolimbic dopaminergic system originating in the ventral tegmental area (A 10). Fibers also reach the limbic forebrain from the A 10 region. A number of aminergic changes have also been described for these same regions: a) increased dopamine concentrations in the limbic areas (8), b) 40-50% reduction in serotonin binding in the frontal cortex (7), c) an increased concentration of noradrenaline in the frontal cortex of chronic paranoid schizophrenics (28), and d) postsynaptic supersensitivity of the dopaminergic system as indicated by the effect of the dopamine agonist apomorphine on growth hormone release (51).

Pharmacology

Many findings indicate that a major factor in schizophrenia is hyperfunction of dopaminergic systems. DOPA treatment of predisposed individuals leads to a stark schizophrenic psychosis (61), as may also be found following amphetamine treatment (5). It is also very indicative that the antipsychotic drugs available act mainly on dopaminergic transmission, by either inhibiting release (60), blocking postsynaptic receptors (14) or dopamine-dependent adenylate cyclase (12). However, other transmitters may also be critical, as it has been shown that compounds which block or inhibit serotonin function and increase dopamine activity, may lead to hallucinations, particularly of auditory type (31), probably through a confusion of reality with cortically stored information (31). Atropine intoxication leads to a condition which mimics schizophrenia (9) while GABA may be implicated by the observation that GABA agonists were detrimental for rating scales of schizophrenics (64).

Physiology

In addition to EEG changes (26), some hebephrenic and schizo-phrenia disposed persons show autonomic and/or emotional hyper-arousal as measured by evoked skin conductance tests (45). In these patients, amplitudes are larger than normal, and normaliza-tion occurs more rapidly (45). However, habituation to the stim-ulus does not occur. These findings have been confirmed by others, and, in addition, paranoid cases have been observed to show habituation, although dishabituation is lacking (29). In some 40% of seriously ill patients, no skin conductance orienting response is found, and this has been interpreted as an extreme case of Yerkes-Dodson's law of protective inhibition of sensory inputs (63,19). Another physiological finding related to motor disturbances is that 'pursuit eye movements' in psychotic schizo-phrenics are irregular and not continuous as is normal (28).

Genetics

A clear genetic disposition for schizophrenia has been indi-cated by monozygous twin studies (32,36), but the concordance varies from 0.33 to 0.92. This variation may be due to an environmental or birth-induced traumatic factor (45). Children, adopted shortly after birth and later developing schizophrenia, have been found to show greater correlation with their biological parents than with their adoptive parents (57).

Experimental

If our hypothesis that pathological conditions are reflected by increased excretion of peptides from petidergic neurons is correct, we would expect to find peptides in the urine of schizo-phrenics with extreme symptoms. There is some confusion in the diagnosis of schizophrenia but the subdivision into core schizo-phrenia and schizophreniform psychoses is useful (38). In order to limit the study, only hard core cases with clearcut symptoma-tology were investigated. The total number of patients was 13, and of these 6 had been without medication for a minimum of 5 weeks, and up to one year. The patients on drugs all showed definite symptoms (66). The diagnostic criteria presented in Table 2 were found applicable for all the untreated patients (21) with the exceptions indicated for motor disorders.

Chemical Methods

Urine samples (24 hours) were collected over 1 g thymol and stored frozen when possible. The details of the benzoic acid precipitation procedures have been published elsewhere (66). The precipitates were dissolved in 0.1 M ammonium bicarbonate, pH 9.4 and analyzed by gel filtration on a Sephadex G-25 column of di-mensions 3 x 143 cm and eluted with the same buffer at a rate of 36 ml/h. Fractions of 12 ml were collected and the UV absorption

TABLE 2. Diagnostic criteria for schizophrenics

A. PERCEPTIVE DISORDER: Visual, tactile, olfactory and sensory intrusion and lack of habituation. Attentional confusion of reality with memory.

B. COGNITIVE: Delusions and hallucinations with a clear consciousness. Ideational intrusion.

C. AFFECTIVE RESPONSES: Aplanation of emotions. Lack of drive or emotional investment. Emotional withdrawal.

D. ASSOCIATIVE FUNCTIONS: Autism and 'concrete' thinking. Disturbance of associational flow.

E. MOTOR DISORDERS: Catatonia at times (1/6). Mannerisms and rituals (4/6).

F. ABSENCE OF DEMONSTRABLE NEUROLOGICAL DEFECTS

---

at 280 nm monitored continuously. This absorption was due to glycoproteins, uric acid and unidentified UV material (unpublished data).

Patterns of urinary gel filtration for normals and schizophrenics were compared (Fig. 1). In the severe cases studied by us the total UV absorption of the precipitate was more than twice and sometimes up to ten times the average of normal urines. We defined two types of excretory patterns in schizophrenics based on visual examination of elution curves. Type 1 excretion pattern has a large increase in the late peak from 900-1500 ml at least three times above the average area for a control group made up of 17 normal adolescents and adults. Type 2 excretion pattern shows a minimal increase in peaks from 600-900 ml of two times the control area. We do not yet know if there is a relationship between peak area and severity of disorder.

We have not yet been able to distinguish between the two types clinically, but the two patterns are not related to any medication. However, most of the paranoid schizophrenics have a pattern of type II, while cataleptic hebephrenics have mainly a type I pattern, although two hebephrenics of type II have also been found.

The material corresponding to the various peaks was dried by rotary evaporation, treated with acetic acid and subjected to chromatography on Biogel P2. Biologically active fractions were then purified using a number of different techniques as described previously (54). The purification is carried out by ice cold acetic acid (0.5M) treatment of lyophilized material from each peak and fractionation on P2 gels, LH20 Sephadex, anion and cation exchangers. The ninhydrin reactive material is followed in each step. This is only possible by a) hydrolysis of an aliquot of each fraction to remove N-terminal groups and

b) increase the sensitivity by the number of amino groups liberated by hydrolysis. We finally end up with a factor which yields amino acids (3-7) only on hydrolysis and with either only 1

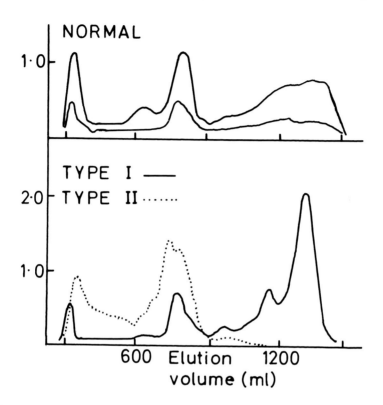

FIG. 1. The urinary gel filtration patterns for normals (n=17) and schizophrenics (n=13, of which 6 were without medication). The normal range is indicated by the two upper traces. Curves from pathological cases are constructed by averaging the UV absorption for each point corresponding to a certain elution volume. Column conditions: Sephadex G-25, dimensions 2.5x160 cm, eluted with 0.1 M ammonium bicarbonate, pH 8.4 at a flow rate of 0.7 ml/min. UV absorption monitored continuously at 280 nm.

N-terminal amino acid or an N substituted amino acid. Final proof awaits synthesis (in preparation). Table 3 illustrates increase in specific activity for GABA release. The purification steps were followed by both biological testing, and analyzing column eluates using ninhydrin coloring, after alkaline hydrolysis, to detect peptidic material. Table 4 lists the different activities found to date, together with collaborators. The details have either been published (27), are in preparation for publication, or are summarized below.

TABLE 3.   Increased release of labelled amino acids from hippocampal tissue following treatment with the same doses of nano-
molar amino acid equivalents (basal level - 100%) (Hamberger, in
preparation)

Purification step	GABA	K^+	Glu	K^+
Step 1 (Biogel P 2)	256	462	361	462
Step 7 (pure peptide)	1746	450	180	450

As the effect on GABA release could be dissociated from the
effect on Glu release, there are probably two factors present at
step 1 of the purification scheme.

TABLE   4.   In vitro and in vivo activities isolated from the
urine of schizophrenics

	Activity	G25 elution volume (ml)	Test System	Tested by
a)	Dopamine uptake inhibition	600-900	A	Hole (27)
b)	Dopamine hyper-activity	600-900	B	Hole (27)
c)	Analgesia	600-900	B	Hole (27)
d)	Catatonia	1100-1500	B	Hole[*]
e)	GABA release	600-900	A	Hamberger[*]
f)	glu release	600-900	A	Hamberger[*]
g)	QNB replacement	600-900	A	Ledaal[*]
h)	Loss of righting reflex	600-900	B	Hole (27)
i)	Hypothermia	600-900	B	Hole (27)

A - in vitro system, B - in vivo system, [*] - to be published.

The most recently discovered activity is a factor which
competes with QNB for binding to a muscarinic receptor at a
concentration of 10^{-9}M amino acid equivalents.

Brief Description of Unpublished Bioassays

GLU/Gaba release were studied by D. Hamberger in slices of hippocampus using a small perfusion chamber. Labelled Glu or Gaba were measured in the perfusate before and after addition of peptide. To make sure of the identity of released amino-acids precolumn derivatization with OPT/mercaptoethanol, HPLC and flourescence detection was carried out. When the effect of peptide had worn off a K^+ pulse was used to ensure intactness of synaptosomes (Hamberger et al., in preparation.) Peptide concentration $10^{-6}-10^{-9}$ M.

QNB replacement in binding assay. This was carried out with synaptosomal membrane preparation in hepes buffer. Atropine replacement was carried out as control as was the effect on guinea pig ileum. These two tests should ensure that the binding to the muscarinic receptor was physiological. Active range $10^{-7}-10^{-10}$ M peptide (Ledaal et al., in preparation). Muscarinic receptors were chosen because of a possible relationship to schizophrenia.

5 HT (serotonin) uptake stimulation. Uptake was measured in blood platelets with 5HT in 2 times Km concentrations using labelled compound. Both plasma and a balanced buffer were used and the uptake was measured after 2 min incubation at 37°C. Peptide fractions were found active in the range $10^{-6}-10^{-8}$ M (Lingdaerde et al., in preparation).

Spiroperidon binding replacement by peptide was carried out on synaptosomal membranes by Braestrup using ^{3}H-spiroperidon. Percent replacement induced by peptide was measured. Active peptide concentration $10^{-6}-10^{-10}$M (Braestrup et al., in preparation).

5HT binding replacement was also determined with labelled LSD, again using synaptosomal membranes. Active peptide concentration was $10^{-6}-10^{-8}$M (Braestrup et al., in preparation). Physiological activity has not been tested, but one fraction increased 5HT uptake in platelets.

Comments.
The following points should be noted:
After 4 purification steps, material isolated from the G 25 peak (600-900 ml) gave a number of changes in rats injected intracranioventricularly with 5 nanomol amino acid equivalents. The factor induced violent or explosive motor behavior, loss of righting reflex, loss of grasping reflex, catalepsy, and body and tail rigidity. Haloperidol potentiated the loss of righting reflex, rigidity and catalepsy. Long lasting effects were observed in the rats for up to 3 weeks (27). Specific points related to Table 4 are:

a) Dopamine uptake inhibition was found for striatal synaptosomes but not hippocampal.

a) and b) are the same factor, the hyperfunction being demonstrated in vivo by the induction of ipsilateral turning following unilateral nigrostriatal lesions, similar to the effect of amphetamine (27).

    c)   Rigidity and catatonia-like symptoms were induced by a tripeptide which has been characterized. A synthetic peptide is being tested at present for biolgoical activity, for the final confirmation of the structure (Hole, to be published).

    d)   The most recently discovered activity is a factor which competes with QNB for binding to a muscarinic receptor at a concentration of 10^{-9}M amino acid equivalents.

Discussion

    We cannot, as yet, draw any conclusions as to the significance of these findings in relation to schizophrenia. However, the anatomical and pharmacological data indicate that several neurotransmitters may be directly and indirectly (possibly as secondary compensatory effects) of major importance. Fig. 2. illustrates the known transmitters involved in the innervation of the n. accumbens (69).

    One important point is that the A 10 region does not receive any direct GABA fibers from the n. accumbens, particularly in relation to the effect of muscimol injected directly into n. accumbens potentiating the arousal effect of dopamine (64). We are in the process of isolating a group of factors (most probably peptides) which have the potential of inducing dopamine hyperfunction, GABA and Glu hyperfunction and acetylcholine modulation at the receptor level in normal animals. As can be seen from Fig. 2, these transmitters are of crucial importance in the function of the n. accumbens and also other limbic structures.

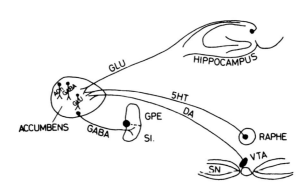

FIG. 2.   Possible innervations of the n. accumbens according to Walaas and Fonnum (69). (Reproduced by kind permission of Munksgard, Copenhagen.)

CHILDHOOD AUTISM

As there is a paucity of physiological and pharmacological data on childhood autism as compared to the information available for schizophrenia, we have chosen to group the symptoms function- ally as a means of indicating the possible anatomical site of the lesions of the CNS involved (15).

The disorder is manifest by the age of three and is character- ized by developmental retardation and social isolation (33).

a) <u>Emotional</u>: Emotional interaction with other persons is lacking in autistics, and, as it is accompanied by withdrawal from visual and auditory contact, parents often suspect deafness. Expression of affection is actively avoided and ignored. Neo- natal hippocampal anoxic damage in animal models may induce hyperarousal with emotional withdrawal, lack of habituation and social isolation (45). These findings point to limbic or meso- limbic damage or malfunction.

b) <u>Perception</u>: Perception is severely disturbed as shown by the excessive importance attached to sensory and internal sensa- tions. Reactive inhibition does not occur in autistics as they are able to focus on trivial things for hours. They also show repetitive activity and a fascination with rotary movements, often to the exclusion of other sensory inputs and resulting in an apparent deafness. Basal ganglia are extensively involved in perception and cognition (40) and the perceptual disturbances in autistics would point to damage or functional anomalies in this disorder (15).

c) <u>Linguistics</u>: These patients often have a poor, if any, spoken language, and peculiarities in syntax indicate aberrations in cognitive processes. Damage to the medial frontal lobe or the temporal lobes may often produce similar language deficits.

d) <u>Motor disturbances</u>: Posture and locomotion anomalies occur frequently, in addition to muscle tone aberrations (15). Very few autistics can perform two simple locomotor sequences simul- taneously, and epileptic fits are about 5-6 times more frequent than in the general population. Asymmetry of emotional facial expression is common (reversed facial paralysis). Animal models with damage to the mesolimbic and striatal systems show similar defects, and also ritualistic and stereotyped behavior (15).

e) <u>Social isolation</u>: This is probably secondary to the emotional withdrawal and lack of language.

Biological Findings

A few findings indicate a possible involvement of CNS trans- mitters. Urinary excretion of 3-methoxy-4-hydroxy-phenethy- lene-glycol (MHPG) is reduced in autistics (72). As 60% of urinary MHPG originates from CNS noradrenaline metabolism, the reduction could imply an abnormality in noradrenalin function. Increased levels of serotonin in blood have also been reported (49). Increased urinary levels of a number of aromatic compounds have been demonstrated, including hippuric acid derivatives, N-methylpyridin-5-carboxamide and purine derivatives (uric acid, dimethyluric acid) (42). At the anatomical level, asymmetric dilation of the third ventricle has also been found (Sagedahl, personal communication).

<u>Experimental</u>

We have investigated the urine of 25 autistics. The patients were diagnosed according to Rimland's criteria (55) as described elsewhere (66). All the patients had been without medication for at least 4 weeks prior to urine collection. The benzoic acid precipitates were analyzed on G 25 as described above, and gave the types of patterns shown in Fig. 3.

In autism type A, a sequence of two small peaks and one big final peak between 900 and 1500 ml was typical (Fig. 3). Also in childhood autism the minimal increase in the peak from 600-900 in type B is from 3-10 times the average of normals. Type B has little or no late peak found in type A. For the late peaks in type A (900-1500 mls) we find 3-15 times the normal average (Fig. 3). Our material is as yet small and a larger series is being run blind. Two more steps in peak purification are, however, probably necessary for a definite diagnosis of unknown samples (27).

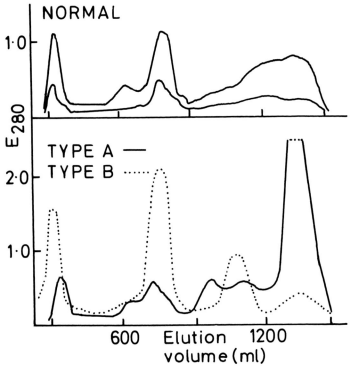

FIG. 3. G-25 patterns of urines from autistic subjects.

Of the 25 patients investigated, 2 had the normal type patterns (one of the patients was found to lack temporal lobes on CAT), and 2 had typical type II schizophrenia patterns. Autistic pattern type A was found in 13 of the patients, and type B in 9. The retarded peak (900-1300 ml) contains relatively large amounts of uric acid derivatives in addition to some peptidic material (Saelid, to be published). The material eluting between 600-900 ml, when subjected to further purification, gave peptide patterns which were very similar for both type A and type B. However, a corresponding analysis of the retarded material revealed considerable differences between the two types.

Biological testing of fractions from both types led to the discovery of a number of activities which were very similar to those described above for schizophrenia. This is not too surprising as schizophrenia is probably the most important consideration in differential diagnosis, and shares a number of symptoms with childhood autism. Table 5 lists the biological activities found in material from the urines of autistics.

TABLE 5. <u>In vitro activities isolated from the urine of autistic patients</u>

	Activity	Pattern Type	G25 elution volume (ml)	Tested by
a)	Dopamine uptake inhibition	A	900-1300	Hole[*]
b)	Noradrenalin uptake stimulation	A&B	900-1300	Hole[*]
c)	Spiroperidol replacement	A&B	600-900	Braestrup[*]
d)	5HT replacement	A&B	600-900	Braestrup[*]
e)	Stimulation of 5HT uptake in platelets	A&B	600-900	Lingjaerde[*]
f)	GABA release	A&B	600-900	Hamberger[*]
g)	Glu release	A&B	600-900	Hamberger[*]
h)	QNB replacement	A&B	600-900	Ledaal[*]

[*] - to be published.

Comments.

a) and b). The dopamine uptake inhibition and noradrenaline uptake and stimulation were carried out using a synaptosomal preparation from striatum and hippocampus, respectively.

c) and d). The replacement of spiroperidol and serotonin in a receptor binding system were found for peptidic material which was active in doses of about $1C_{50}=1\mu M$.

e) Approximately 50% increase in serotonin uptake was found for peptidic material which had been purified through 8 purification steps.

Discussion

The complexity of the symptoms in autism are suggestive of a comprehensive derangement of CNS function. At present there is no coherent neurochemical hypothesis for this disorder, which may be centered in the mesolimbic and striatal systems. However, it is clear that anything modifying the function of the major transmitters in the CNS will have extremely serious consequences for behavior. Several of the peptides isolated have been synthesized, and are being checked for biological activity.

SUMMARY

It is well documented that peptides have a major role in the effective functioning of higher animals at all levels from enzyme stabilization to homeostatic mechanisms governing essential functions such as eating, sexual behavior, and temperature regulation. The effects of exogenously administered peptides on neurotransmitter release, uptake, metabolism and behavioral consequences are also well established. We have attempted to extend these findings by postulating peptidergic neurons as transducers of multisignal inputs, and that development of pathological states may be due to genetically-determined reduced levels of activity of key peptidases, leading to excretion of regulatory peptides into the circulation. We have been able to demonstrate that, in schizophrenia and autism (in well defined clinical cases), the patterns of peptides and associated proteins from urinary samples differ considerably from each other and from normal controls. In addition to this, further purification of the material obtained has led to the discovery of a number of factors capable of modulating the function of major neurotransmitters. Some of these are in the final stages of characterization as peptides, while the remainder are also probably peptides, as purification has been followed by both biological testing and chemical analysis for peptidic material. We have outlined a number of parameters which we consider relevant in any attempt to put psychiatric disorders on a biological foundation. Any new advances in the neurochemical understanding of such disorders must take into consideration the observations of several different disciplines including genetics and psychology. However, at this stage of our research it is far too early to speculate on the relevance of the various biological activities to the etiology and symptomatology of schizophrenia and childhood autism.

REFERENCES

1. Arbilla, S., and Langer, S.Z. (1978): Nature, 271:559-560.
2. Barker, J.L., and Gainer, H. (1974): Science, 184:559-560.
3. Barker, J.L., Neal, J.H., Smith, T.G., Jr., and MacDonald, R.L. (1978): Science, 199:1451-1453.
4. Bauer, K. (1976): Nature, 259:591-593.
5. Beamish, P., and Kiloh, L.H. (1960): J. Ment. Sci., 106:337-343.
6. Becker, E.L. (1976): Amer. J. Pathol., 85:385-394.
7. Bennett, J.P., Enna, S.J., Bylund, D.B., Gillin, J.C., Wyatt, R.J., and Snyder, S.H. (1979): Arch. Gen. Psychiat., 36:927-934.
8. Bird, E.D., Spokes, E.G.S., and Iversen, L.L. (1979): Brain, 102:347-360.
9. Bleuler, M. (1979): Am. J. Psych. 136:1403-1409.
10. Bloom, F., Segal, D., Ling, N., and Guillemin, R. (1976): Science, 194:630-632.
11. Bohus, B., Urban, I., Van Wimmersma Greidanus, T.B., and DeWied, D. (1978): Neuropharmacology, 17:239-247.
12. Clement-Cormier, Y.C., Kebabian, J.W., Petzold, G.I., and Greengard, P. (1974): Proc. Natl. Acad. Sci. USA, 71:1113-1117.
13. Cohn, M.L., and Cohn, M. (1977): Psychoneuroendocrinology, 2:197-202.
14. Creese, I., Burt, D.R., and Snyder, S.H. (1976): Science, 192:481-483.
15. Damasio, A.R., and Maurer, R.G. (1978): Arch. Neurol., 35:777-786.
16. DeWied, D., Kovacs, G.L., Bohus, B., van Ree, J.M., and Greven, H.M. (1978): Eur. J. Pharmac., 49: 427-436.
17. Dunaway, G.A., and Segal, H.L. (1976): J. Biol. Chem., 251:2323-2329.
18. Emson, P.C. (1979): Progress in Neurobiology, 13:61-116.
19. Eysenck, H.J. (1977): The Biological Basis of Personality. C.C Thomas, Springfield, Ill.
20. Farley, I.R., Price, K.S., McCullough, E., Deck, J.H., Hordynski, W., and Hornykiewicz, O. (1978): Science, 200: 456-458.
21. Feighner, J.P., Robins, E., Guze, S.B., Woodruff, R.A., Winokur, G., and Munoz, R. (1972): Arch. Gen. Psychiat., 26:57-63.
22. Gersch, M. (1977): Naturwissenchaften, 64:417-426.
23. Getzl, E.J., and Austen, K.F. (1975): Proc. Natl. Acad. Sci. USA, 72:4123-4127.
24. Gispen, W.H., Wiegant, V.M., Greven, J.M., and De Wied, D. (1975): Life Sci., 17:645-652.

25. Griffiths, E.C., Hooper, K.C., Jeffcoate, S.L., and Holland, D.T. (1975): Brain Res., 88:384-388.
26. Heath, R. (1954): Studies on Schizophrenia. Harvard University Press, Cambridge, MA.
27. Hole, K., Bergslien, H., Jorgensen, H.A., Berge, O.G., Reichelt, K.L., and Trygstad, O.E. (1979): Neuroscience, 4:1883-1893.
28. Holzman, P.S., Kringlen, E., Levy, D.L., Proctor, L.R., and Haberman, S. (1978): J. Psychiat. Res., 14:111-120.
29. Horvath, T., and Meares, R. (1979): Brit. J. Psychiat., 134:39-45.
30. Ifshin, M.S., Gainer, H., and Barker, J.C. (1975): Nature, 254:72-73.
31. Jacobs, B.L., and Trulson, M.E. (1979): Trends Neurosci., 2:267-280.
32. Killman, H., (1946): Am. J. Psychiat., 203:309-322.
33. Kanner, L. (1943): Nerv. Child., 2:217-250.
34. Kastin, A.J., Coy, D.H., Schally, A.V., and Meyers, C.A. (1978): Ann. Rev. Biochem., 47:89-128.
35. Kerwin, R.W., and Bycock, C.J. (1979): Brit. J. Pharmacol., 67:323-325.
36. Kringlen, E.A. (1964): Acta Psychiat. Scand., Suppl. 40: 178:1-76.
37. Kruse, H., Van Wimmersma Greidanus, T.B., and DeWied, D. (1977): Pharmacol. Biochem. Behav., 5:665-669.
38. Langfeldt, G. (1939): The Schizophreniform States. Oxford University Press, London.
39. Lehnhoff, H.M. (1968): Science, 161:434-442.
40. Lidsky, T.I., Weinhold, P.M., and Levine, F.M. (1979): Biological Psychiat., 14:3-12.
41. Lipton, M.A., Breese, G.R., Prange, A.J., Wilson, I.C., and Cooper, B.R. (1978): In: Hormones Behaviour and Psychopharmacology, edited by E.J. Sachar, pp. 15-29. Raven Press, New York.
42. Lis, A.W., McLaughlin, D.I., McLaughlin, R.K., Lis, E.W., and Stubbs, E.G. (1978): Clin. Chem., 22:1528-1532.
43. MarcanodeCotte, D., DeMenezes, C.E.L., Bennett, G.W., and Edwardson, J.A. (1980): Nature, 283:487-489.
44. Marzullo, G., Rosengarten, H., and Friedhoff, A.J. (1977): Life Sci., 20:775-784.
45. Mednick, S.A., Schulsinger, F., Higgins, J., and Bell, B. (1974): Genetics, Environment and Psychopathology. North Holland Elsevier, Amsterdam.
46. Moroni, F., Cheney, D.L., and Costa, E. (1977): Naunyn-Schmicdeberg Arch. Pharmacol., 299:149-153.
47. Moss, R.I., and Foreman, M.M. (1976): Neuroendocrinology, 20:176-181.
48. Motomatsu, T., Lis, M., Seidah, N., and Chretien, M. (1977): Biochem. Biophys. Res. Comm., 77:442-447.
49. Ornitz, E.M., and Ritvo, E.R. (1975): Am. J. Psychiat., 113:609-621.

50. Osterlund, B., and Bridger, W.A. (1977): Biochem. Biophys. Res. Commun., 76:1-8.
51. Pandey, G.N., Garver, D.L., Tamminga, C., Ericksen, S., Ali, S.I., and Davis, J.M. (1977): Am. J. Psychiat., 134:518-522.
52. Pedersen, C.A., and Prange, A.J., Jr., (1979): Proc. Natl. Acad. Sci. USA, 76:6661-6665.
53. Reichelt, K.L., and Edminson, P.D. (1977): In: Peptides in Neurobiology, edited by H. Gainer, pp. 171-181. Plenum Press, New York.
54. Reichelt, K.L., Trygstad, O.E., Foss, I., and Johansen, J.H. (1979): In: Psychopharmacology of Aggression, edited by M. Sandler, pp. 159-172. Raven Press, New York.
55. Rimland, B. (1971): J. of Autism and Child Schizophrenia, 1:161-174.
56. Ritvo, E.R., Ornitz, E.M., and Eviatar, A. (1969): Neurology, 19:653-658.
57. Rosenthal, D., Wender, P., Kety, S., Weher, J., and Schulsinger, F. (1971): Am. J. Psychiat., 128:307-311. ·
58. Schally, A.V., Arimura, A., and Kastin, A.J. (1973): Science, 179:341-350.
59. Schoenenberger, G.H., and Monnier, M. (1977): Proc. Natl. Acad. Sci. USA, 74:1282-1286.
60. Seeman, P., and Lee, T. (1975): Science, 188:1217-1219.
61. Stevens, J.R. (1973): Arch. Gen. Psychiat., 29:177-189.
62. Stevens, J.R. (1979): Trends Neurosci., 2:102-105.
63. Straube, E.R. (1979): J. Nerv. Mental Dis., 167:601-611.
64. Tamminga, C.L., Crayton, J.W., and Chase, T.N. (1978): Am. J. Psychiat., 135:746-747.
65. Tardy, M., Rolland, B., Bardakdjian, J., and Gonnard, P. (1978): Experientia, 34:823-824.
66. Trygstad, O.E., Reichelt, K.L., Foss, I., Edminson, P.D., Saelid, G., Bremer, J., Hole, K., Orbeck, H., Johansen, J.H., Boler, J.B., Titlestad, K., and Opstad, P.K. (1980): Brit. J. Psychiat., 136:59-72.
67. Urban, I.V., and DeWied, D. (1978): Pharmacol. Biochem. Behav., 8:51-59.
68. Versteeg, D.H.G., Tanaka, M., DeKloet, E.R., Van Ree, J.M., and DeWied, D. (1978): Brain Res., 143:561-566.
69. Walaas, I., and Fonnum, F.(1979): GABA-Neurotransmitters, pp. 60-73. Munksgaard, Copenhagen.
70. Wood, P.L., Cheney, D.L., and Costa, E. (1979): J. Pharmacol. Exp. Ther., 209:97-103.
71. Yarbrough, G.C. (1976): Nature, 263:523-524.
72. Young, J.G., Cohen, D., Caparulo, B.K., Brown, S.L., and Mass, J.W. (1979): Am. J. Psychiat., 136:1055-1057.

Neurosecretion and Brain Peptides,
edited by J. B. Martin, S. Reichlin, and K. L. Bick.
Raven Press, New York © 1981.

Potential Role of Neural Peptides in CNS Genetic Disorders

Verne S. Caviness, Jr.

*Neurology Service, Massachusetts General Hospital, Boston, Massachusetts 02114; and
The Eunice Kennedy Shriver Center for Mental Retardation, Inc.,
Waltham, Massachusetts 02154*

Development of the mammalian central nervous system (CNS) progresses through a succession of complex cytologic events (17). Cytogenesis, or cell division, and histogenesis, or the assembly of laminate and non-laminate structures through cell migration, are the initial events. Dividing cell populations, which appear homogenous, give rise to progeny that differentiate into a host of different neuronal and neuroglial classes. As development continues through an extended period of cell growth and differentiation, the cells in each class evolve characteristic dendritic-somatic configurations and form class-characteristic patterns of connections with other cells.

It is a central tenet of biology that this sequence of cellular events is set in motion by instructions encoded within the genome of each cell. These instructions are probably implemented by a variety of molecular mechanisms. The molecular agents may include diffusible substances such as peptides or non-peptide transmitters (19). They may include substances integral to the plasma membranes of cells such as glycoproteins (38).

The present review will consider the hypothesis that diffusible peptides, among the many candidates, serve as regulators of developmental events. More specifically, it will consider the possibility that genetic mutations give rise to developmental disorders of the CNS by interfering with the regulatory functions of peptides. Three lines of evidence, all indirect, are consistent with this possibility. First, the function of peptides may be abnormal consequent to single gene mutations. Second, certain peptides are known to be present in the CNS during the developmental epoch, and some of these appear to be biologically active. Third, certain mutation-induced disorders of the developing CNS appear to be expressions of disturbed regulatory functions known to be exercised by peptides in other developing tissues.

PEPTIDE FUNCTION - A TARGET OF SINGLE GENE MUTATION

A variety of single gene mutations are recognized that give rise to malfunction in the regulatory action of one or more peptide hormones. The disturbance may range from a "general endocrine disorder" to abnormalities in the function of specified peptides. Examples of the former include panhypopituitarism in the Snell (dw) and Ames (df) dwarf mice (5,28), the human and mouse (pg) pigmy mutations (20,51), the diabetes (db) mutant human growth disorders (52). The little mouse (lit) (3) and the Brattleboro rat (54) illustrate disturbed function of specified peptide hormones consequent to autosomal recessive mutation. The first group of disorders is related to abnormal function of growth hormone while the second involves the vasopressin system and is associated with diabetes insipidus.

Disturbance in peptide function caused by each of these mutations is associated in a general way with restricted tissue growth and differentiation; none are associated, as far as is known, with abnormalities of the earlier events of cytogenesis or histogenesis in the CNS. Even among those where the mutation clearly involves functions of specific peptides, the mechanism of action of the mutation at the molecular level remains obscure. Thus, it is uncertain whether the abnormality of peptide function derives from an abnormality in the structure of the peptide itself, control of its synthesis and distribution or in the response of target tissues.

BIOLOGICALLY ACTIVE PEPTIDES IN THE DEVELOPING CENTRAL NERVOUS SYSTEM

Approximately 20 biologically active peptides are known to be widely distributed in the adult mammalian CNS including that of man (6,22,33). A partial list includes the pituitary hormones - thyrotropin (TSH) growth hormone (GH) ACTH, α-MSH; the hypothalamic releasing factors - luteinizing hormone releasing hormone (LHRH), thyroid releasing hormone (TRH); the posterior pituitary peptides derived from the hypothalamus- vasopressin and oxytocin; and a spectrum of peptides first recognized in the gastrointestinal tract - gastrin, cholecystokinin, vasoactive intestinal peptide. The list also includes angiotensin, neurotensin, the endorphins and enkephalins, β-LPH, Substance P and carnosine. In general, these peptide substances are most densely concentrated in the hypothalamus and brainstem but are widely distributed in other areas including the cerebral neocortex and cerebellum. The concentration of each is low, by and large in picograms, which is orders of magnitude less than concentrations achieved by some peptides in the pituitary (22).

Cell bodies containing peptides and, presumably, representing sites of synthesis, have been recognized only in the axial nervous system and hypothalamus (22). Although the capacity to synthesize biologically active peptide may be much more widely distributed among neurons of the CNS, the pattern of distribution may be only in part a reflection of local synthesis or axon transport (22). Another important and more general mechanism of distribution is diffusion from the cerebrospinal fluid, and possibly, from the blood stream as well.

Electrophoretic analysis suggests that the developing brain contains essentially the same spectrum of peptides and peptide-like substances that is found in the adult brain (66). However, only a few are actually known to be present in the brain during the developmental period. Among these are peptides associated with pituitary function (58). These peptides appear to be essential to normal body growth. In rare cases of decapitation, occurring during the first trimester of gestation (30) or in cases of an encephaly in which both hypothalamus and pituitary are destroyed early in gestation, body growth at term is less than 30% of normal (16,58). Furthermore, the peptides of hypothalamic-pituitary origin appear to be essential to the normal conduct of parturition and thus play an essential, though indirect, role in protecting the organism from perinatal injury associated with delayed or interrupted labor (58). Only endorphin, the two enkephalins and Substance P have been specifically identified, thus far, in the early developing CNS (2,43). With regard to endorphin, the concentration is substantially higher than in adult neural tissue; enkephalin by contrast, increases in concentration in a fashion which roughly parallels the rate of increase of tissue protein throughout the developmental epoch (2). There is a parallel development during this period of opioid binding sites (10). Substance P and enkephalin have been demonstrated by immunohistochemistry to be concentrated in the growth cones of growing axons in the axial regions of the nervous system (43). The role played by these peptides during this early developmental period is obscure.

DEVELOPMENTAL PROCESSES AND EVENTS REGULATED BY PEPTIDES AND DISRUPTED BY MUTATION AFFECTING CNS DEVELOPMENT

Developmental Processes and Events Regulated by Peptides

Diffusible peptide substances appear to be involved in the regulation of many of the biological processes critical to tissue development. The evidence is drawn from studies in non-neuronal tissue in some instances and from studies in the peripheral nervous system in others. Such studies have not as yet been extended to the central nervous system, however.

Cell division, the most elementary developmental event, has been demonstrated to be controlled in fibroblasts by at least two diffusible peptide substances. When the cell is quiescent, it requires a platelet-derived peptide growth factor in order to become competent to replicate DNA (45). Progression beyond this initial phase of competence requires the mediation of somatomedin C, a member of a family of insulin-like peptides whose plasma concentrations are governed by the pituitary (57).

Other observations suggest that peptides play a role in the regulation of synthesis of specific macromolecules, that they provoke and direct migration of cells and that they dictate the geometry of cell deployment in the assembly of complex tissues. Such functions appear to be exercised by a wide spectrum of hydroxyproline-containing peptides, smaller than the alpha chain of collagen, in the developing extremities of vertebrates (15). Abundant in early stages of extremity development, these peptides decrease in concentration with maturation of the tissue. They appear to regulate, in specific fashion, the synthesis of collagen (29). They exercise chemotactic regulation of the direction of movement of freely motile fibroblasts (46). Further, the alignment and polarity of certain cell populations appears to be determined by concentration gradients of such peptides in developing mesenchymal tissues (14).

Other families of peptides and polypeptides have been identified that specifically regulate the growth, differentiation and maintenance of a variety of tissues (4). Pre-eminent among these "trophic factors" is nerve growth factor (NGF) which is indispensable to the development of neurons of the peripheral sympathetic nervous system as well as to some somatosensory neurons that develop in the dorsal root ganglia (27). NGF is probably synthesized and released within the tissues targeted by the axons of these specific cell classes (4). As the peptide diffuses passively from its sites of synthesis, it appears to provoke and direct the growth of responsive axon terminals at specific receptor sites and is transported to the nucleus of the cell where it encounters additional high affinity receptors (4). It is thought that at the nucleus, by participation in the events of transcription, the peptide induces the synthesis of structural proteins essential to neurite growth (34). It appears, also, to induce synthesis of tyrosine hydroxylase and dopamine β-hydroxylase (62?r), enzymes which are specific catalysts in the chain of synthetic events for noradrenalin. This transmitter is employed by the sympathetic neurons which are very specifically dependent upon NGF. Although they have been less extensively characterized, there are also other classes of peptide substances, which regulate in specific fashion similar to NGF, critical events in the development of cholinergic systems (1).

NGF injected directly into the CNS provokes the entry along anomalous trajectories of axons of the periperhal autonomic nervous system (36). Whether this peptide exercises specific trophic regulatory control over the CNS stands, is the subject of vigorous controversy (42). Whatever the resolution of this question with regard to NGF, certain observations raise the possibility that substances structurally analogous to NGF play a similar regulatory role in the CNS. Thus, high affinity binding sites for NGF have been identified within the CNS (60,61). Further, NGF appears to cause the modification of cell surface "adhesive specificity" in the tectum of developing chicks (37).

Single Gene Mutations and CNS Development

Single gene mutations are associated with a wide spectrum of developmental anomalies in the mammalian CNS. Although the developmental history is, in general, not known in detail, the morphological evidence in certain examples is consistent with the hypothesis that malformation reflects disruption of one, or in some instances, more than one, of the following developmental processes and events: 1) the formation, maintenance, growth and differentiation of specific cell classes; 2) guidance of cell movement during migration and cell arrangement in the assembly of neural structures; and 3) guidance of specific classes of axons in the construction of neural systems. As reviewed in the prior section these events and processes are the same as those which appear to be regulated by peptides in developing peripheral tissues.

Cell generation.
Kallman's syndrome, a human disorder transmitted by autosomal recessive inheritance, is characterized by anosmia and infertility (31). The brain is abnormal in that there is complete absence of the olfactory bulbs. Apparently, the population of neurons which constitutes this structure, involving multiple classes, never forms. It is of interest in this regard that the polypeptide, carnosine, is a unique constituent of the developing olfactory bulbs (32). This peptide might therefore be a trophic factor regulating olfactory bulb development or may also be the target of this autosomal recessive mutation.

A second example is a restricted class of microcephaly (or small brain) in man which has been observed in some instances to be transmitted by autosomal recessive inheritance (12). The neocortex may be deficient in the neuronal classes which normally populate layers II-IV (12,63). These are the small and medium pyramidal cell classes as well as the granular interneuronal class. They are the last neuronal classes to be formed by the generative epithelium during development (53,48). Those neurons that are present in the neocortex are the earliest to be formed; that is, the large pyramidal cells and polymorphic cells of layers V and VI, respectively are expanded throughout the cellular zone of the cortex, immediately subjacent to a well-developed molecular layer. There may be little or no evidence of intercurrent destructive pathologic process. The appearance suggests, instead, that the germinal epithelium fails to produce those neuronal classes normally formed in the latter half of the epoch of cell generation.

Cell maintenance.
A variety of single gene mutations are associated with morbidity and death of restricted neuronal classes. Neurons at widely scattered points in the nervous system are vulnerable to such mutations. Particularly common are mutations affecting the

granular and Purkinje cells of the cerebellar cortex, on the one
hand, and the retinal photoreceptors, on the other. Different
mutations may affect the same cell class or classes, but they do
so in different fashion (9). Thus, the distribution in the brain
of affected cells of a class may vary significantly from mutation
to mutation. Further, the rate of cell death, and the appearance
of cells in the course of degeneration may be highly characteristic of a specific mutation.

Mutations affecting neurons of the cerebellar cortex.
     Three autosomal mutations have been identified in mice that
are associated with degeneration of the cerebellar Purkinje
cells; an additional two mutations in the same species have the
cerebellar cortical granular cells as their principal targets of
action.
     Those affecting the Purkinje cells include lurcher (Lc) a
mutation associated with a general loss of Purkinje cells evident
as early as the third or final week of gestation in the mouse
(59). The nervous mutation (nr) similarly causes a general (up to
90%) attrition of Purkinje cells (24). In contrast to Lc, however, cell death in nr occurs over a protracted interval extending from the third to the sixth postnatal week. It is the distinctive feature of the Purkinje cell affected by nr that its
mitochondria are swollen and spherical in configuration. Purkinje cell degeneration (pcd), yet another mutation, is associated
with death of Purkinje cells over an even more extended period,
continuing from the 15th postnatal day through the third month of
age (25).
     Both the nr and pcd mutations are associated also with degen-
eration of some of the granular cells of the cerebellar cortex as
well as with degeneration of photoreceptor cells of the retina
(40). Whereas the granule cell death seen in these two mutants
may be an indirect, or transneuronal effect consequence to loss
of Purkinje cells, the photoreceptor generation is probably a
manifestation of pleiotropic gene action.
     Leaner (tgla) (9,35) and weaver (wv) (49,50,55) are the two
mutations associated with degeneration of massive numbers of
granule cells in the developing cerebellar cortex. Degeneration
occurs, predominantly before or during cell migration. Degeneration in leaner occurs, for the most part, in a territory restricted to the anterior and nodular lobes of the cerebellum. By
contrast, the process is essentially generalized throughout the
cerebellar cortex in weaver. Both of these mutations are associated with loss of a few Purkinje cells, but again, this is
probably an indirect, or transneuronal consequence of the loss of
granule cells.

Retinal photoreceptor cells.
     It has been mentioned above that the retinal photoreceptor
cells as well as cerebellar cortical neurons degenerate in the
nervous and Purkinje cell degeneration mutant. The photoreceptor
cells degenerate in a more severe and restricted fashion in

association with the retinal degeneration (rd) mutants of both mice and rats. In the rat mutation, the degeneration proceeds in a protracted fashion from the 20th through the 60th postnatal days (39). It is associated with an inability of pigment epithelial cells to phagocytize degenerating material in normal fashion, a phenomenon which may play a role in receptor cell death (41). In the murine mutation, photoreceptor degeneration proceeds more rapidly and is largely completed by the end of the fourth postnatal week (26). In this species, there is no apparent abnormality in the phagocytic ability of the pigment epithelial cell.

Differentiation of specific organelles of specific cell classes.
A dramatic example of this phenomenon arising as a consequence of autosomal recessive mutation is illustrated in the staggerer (sg) mutation in mice (23,56). The mutation affects, primarily, the cerebellar Purkinje cell. This cell class is abundant in the mutant and lies in normal relationship to their cellular elements of the cerebral cortex. Further, parallel fiber axons are deployed in apparently normal fashion through the dendritic arbors of these cells. However, neither spines nor postsynaptic densities differentiate on the Purkinje cell dendrites. Further, the Purkinje cell is unable to accept the parallel fibers as presynaptic partners. Ultimately large numbers of granule cells die, apparently as a secondary consequence of their inability to form synaptic junctions with the Purkinje cells.

Tay-Sach's disease and Batten's disease, both autosomal recessive disorders in man, present bizarre disorders of neuronal differentiation which are similar to each other in some respects but not in others. As a consequence of both mutations, there appears to be increased insertion of membranous material at the inferior pole of the cell body of pyramidal cells of the neocortex. In Tay-Sach's disease, this leads to a massive deformity, the "meganeurite" (47). This bizarre structure has dendritic membranous properties in that it forms synapses with axon terminals. The pyramidal cell abnormality in Batten's disease is more modest - only a fusiform enlargement of the proximal axon segment (64). The abnormal membrane in this second disorder has not been observed to be postsynaptic to axon terminals.

Both of these mutations are associated with enzymatic defects which lead to accumulation of lipid materials in the lysosomes of all cells of the body. In Tay-Sach's disease there is decreased production of the enzyme hexosaminidase A, which leads to the accumulation of GM_1 gangliosides. In Batten's disease, there is an abnormality in the enzyme system responsible for degradation of retinoyl complexes (65). The relationship of the enzymatic defects and the stored metabolic products to the disorder of membranous differentiation of pyramidal cells is obscure.

Mutation-induced anomalies of cortical histogenesis.
Single gene mutations are recognized that interfere with the migration of neurons from the central generative zones of the

developing brain to their target cortical or subcortical structures.   Others disrupt mechanisms controlling the guidance of
specific axon systems.

Neuronal migration. The Zellweger malformation arises consequent to autosomal recessive mutation in man and is associated
with a dramatic though highly restricted disorder of neuronal
migration to the developing cerebral cortex (11). The anomaly is
limited to cortex of the centro-Sylvian region and is present in
mirror-symmetric fashion in both cerebral hemispheres. Although
the five principal neuronal laminae of the neocortex are present
in the abnormal regions, each layer is profoundly depleted of its
nprincipal neuronal class. This is increasingly the case with
progression from the deep to superficial layers. Neurons normally destined for each of these layers are apparently generated
during development; however, many fail to complete their migrations. Thus, neurons representing each of the classes characteristic of the five neocortical laminae, are scattered in hetero-
topic fashion within and below the cortex. Evidently, the malformation reflects the effect of a disturbance of neuronal migration which has acted in continuous but partial fashion throughout
the entire migratory epoch (11).
    Reeler (rl) is an autosomal recessive mutation occurring in
mice which disrupts migration and the deployment of neurons in
virtually all cortical structures of the forebrain, in the
cerebellum and in the dorsal cochlear nucleus of the brainstem as
well (7,9). For the purposes of the present discussion, attention is directed to the anomaly as it affects the neocortex of
the forebrain. As reviewed in a prior section, the earlier cells
to be generated and to undergo their migrations in the normal
brain are those destined for the deepest neocortical layers.
Progressively later migrating cells come to lie at successively
more superficial cortical layers so that the last cells to be
generated will come to occupy layer II of the neocortex, lying at
the interface of the cellular zone and the molecular layer.
    Although the mechanisms underlying this pattern of distribu-
tion of neurons is not well understood, two events appear to be
essential (8). First of all, the migrating neuron must be able
to bypass neurons which migrated before it. This maneuver
carries the cell fully to the external plexiform zone where its
migration is arrested. Secondly, once in position at the end of
migration, the cell must become fixed with respect to the tangentially and radially adjacent cellular elements that surround
it.
    In reeler, there is inversion of the normal radial order of
neuron clases (7). That is, the earliest cells to be formed come
to lie at the most superficial level of the cortex while subsequently generating and migrating cells occupy successively deeper
levels within the cortex. The migrating neuron in the mutant,
unlike its normal counterpart, appears unable to bypass post-

migratory cells. Rather as it encounters early arrivals within the cortex, its migration comes to a halt (8).

Axon trajectories. The reeler mutation in mice is associated with anomalies in the trajectories of major axonal systems (9) as well as with the abnormal pattern of cell position discussed in the previous section. From the earliest phases of cortical histogenesis, for example, extrinsic afferents, some identified as monoaminergic axons, extend across the full width of the cortex in reeler (8,44). These afferents form a plexus at the most superficial level of the mutant cortex as well as one at an intermediate level within the cortical plate. The pattern of early afferent deployment in the cortex of the mutant contrasts with that in the normal mouse where tangentially coursing fiber tracts are most conspicuous in the base of the cortex and in the molecular layer. The anomaly probably stems from the earliest encounters between afferent axons and psotmigratory neurons in the developing cortical zone at the outer margin of the hemisphere. However, the cellular events through which this abnormal pattern of axonal deployment arises have not yet been delineated.

In mice (21), autosomal recessive mutations are known which cause a failure of formation of the callosal commissural connections that normally bridge the two cerebral hemispheres. Similar malformations occur in man but the role of genetic transmission is uncertain (12). In this disorder in both mouse and man, neocortical cytoarchitecture appears to be normal. A large, rostrocaudally coursing fiber bundle lies medially in each hemisphere in close relation to the cingulate bundle. Presumably, the axons which follow this anomalous trajectory are re-routed counterparts of the callosal system of normal animals.

The Siamese mutation in cat, an example of multiple "pigmentary" mutations occurring in a variety of mammalian species, is associated with an abnormal routing of the axons of the retinal projection at the optic chiasm (13). An increased complement of rthose axons normally deflected to the ipsilateral lateral geniculate nucleus (LGN) is directed instead to the contralateral n?rucleus. Despite this, the re-routed axons terminate in the laminae of LGN appropriate to the eye of origin and create a topologically ordered projection. There are no recognized anomalies elsewhere in the brain of these mutants other than this restricted abnormality in the visual system.

FUTURE DIRECTIONS

The preceding review illustrates the principle that single gene mutations may be antecedent to the development of highly characteristic aberrations of the phenotype of the CNS. These aberrations in form are expressed as development progresses through a succession of regular but abnormal cellular interactions. The molecular machinery, disrupted by mutation, could in theory deflect development from its normal course via many different mechanisms. Some of these, for example those related to the maintenance or differentiation of individual neuronal classes, may follow close upon transcriptional events. Thus, as a consequence of a mutant gene, enzymatic systems or structural molecules might not be competent to maintain cell metabolism or to implement cell differentiation. Other developmental disorders reflecting the consequences of impaired cell migration or axonal guidance, may stem from abnormalities in the ways cells interact with each other, through their plasma membranes or through diffusible substances.

For the present, the view that peptides play a "trophic" or regulatory role in the course of normal CNS development or that they are related to abnormal CNS phenotype resulting from genetic mutation are hypotheses only. They are among multiple hypotheses that, for the present, might be given equal weight. Such hypotheses do offer points of entry into lines of research which will seek to identify and characterize the molecular mechanisms which assure a normal sequence of developmental events under circumstances favorable to the organism. Animals bearing single gene mutations will continue to be a valuable resource for such research. Increasingly, technological advances are bringing to the field, lectins, antibodies and radioactively labeled tracer substances which may serve as direct probes of the molecular events regulating the interaction between cells in the course of normal and abnormal development.

REFERENCES

1. Adler, R., Landa, K.B., Manthorpe, M., and Varon, S. (1979): rScience, 204:1434-1436.
2. Bayon, A., Shoemaker, W.J., Bloom, F.E., Mauss, A., and Guillemin, R. (1979): Brain Res., 179:93-101.
3. Beamer, W.H., and Eichler, E.M. (1976): J. Endocrinol., 71: 37-45.
4. Bradshaw, R.A. (1978): In: Ann. Rev. Biochem., 47:191-216.
5. Brasel, J.A., and Blizzard, R.M. (1974): In: Textbook of Endocrinology, edited by R.H. Williams, pp. 1038-1039. W.B. Saunders, Philadelphia.
6. Brownstein, M.J. (1977): In: Peptides in Neurobiology, edited by H. Gainer, pp. 145-170. Plenum Press, New York.
7. Caviness, V.S., Jr. (1977): In: Society for Neuroscience Symposia, Vol. 2, edited by W.M. Cowan and J. A. Ferrendelli, pp. 27-46. Society for Neuroscience, Bethesda.
8. Caviness, V.S., Jr. (1980): In: Morphogenesis and Pattern Formation, edited by L.L. Brinkley, B.M. Carlson, and T.G. Connell., Raven Press, New York. (In press).
9. Caviness, V.S., Jr., and Rakic, P. (1978): Ann. Rev. Neurosci., 1:297-326.
10. Clendeninn, N.J., Petraitis, M., and Simon, E.J. (1976): Brain Res., 118:157-160.
11. Evrard, P., Caviness, V.S., Jr., Prats-Vinas, J., and Lyon, G. (1978): Acta Neuropath., 41:109-117.

12. Friede, R.L. (1975): Developmental Neuropathology, Springer-Verlag, New York.
13. Guillery, R.W.(1974): Sci. Am., 230:44-54.
14. Holmes, L.B., and Trelstad, R.L. (1977): Develop. Biol., 59: 164-173.
15. Holmes, L.B., and Trelstad, R.L. (1979): Develop. Biol., 72:41-49.
16. Honnebier, W.J., and Swaab, D.F. (1973): J. Obstet. Gynecol., 80:577-588.
17. Jacobson, M. (1978): Developmental Neurobiology, Plenum Press, New York.
18. Johnson, L.M. and Sidman, R.L. (1979): Biol. Reprod., 20: 552-559.
19. Kasamatsu, T., Pettigrew, J.D., and Ary, M.L. (1979): J. Comp. Neurol., 185:163-182.
20. King, J.W.B. (1955): J. Genet., 53:487-497.
21. King, L.S. (1936): J. Comp. Neurol., 64:337-363.
22. Krieger, D.T., and Liotta, A.S. (1979): Science, 205: 366-372.
23. Landis, D.M.D., and Sidman, R.L. (1978): J. Comp. Neurol., 179:831-864.
24. Landis, S., (1973): J. Cell Biol., 57:782-797.
25. Landis, S.C., and Mullen, R.J. (1978): J. Comp. Neurol., 177:125-143.
26. LaVail, M.M. and Mullen, R.J. (1976): Exp. Eye Res., 23: 227-245.
27. Levi-Montalcini, R., and Angeletti, P.U. (1968): Physiol. Rev., 48:534-569.
28. Lewis, U.J. (1967): Memoirs of the Society of Endocrinology, 15:179-191.
29. Lichtenstein, J.R., Martin, G.R., Kohn, L.D., Byers, P.H. and McKusick, V.A. (1973): Science, 182:298-299.
30. Liggins, G.C. (1974): In: Size at Birth, Ciba Foundation Symp. No. 27, edited by K. Elliott and J. Knight, pp. 165-183. Elsevier, Amsterdam.
31. Males, J.L., Townsend, J.L., and Schneider, R.A. (1973): Arch. Int. Med., 131:501-507.
32. Margolis, F.L. (1974): Science, 184:909-911.
33. Marx, J.L. (1979): Science, 205:886-889.
34. McGuire, J.C., and Greene, L.A. (1980): Neuroscience, 5: 179-189.
35. Meier, H., and MacPike, A.D. (1971): J. Hered., 62:297-302.
36. Menisini Chen, M.G., Chen, J.S., and Levi-Montalcini, R. (1978): Arch. Ital. Biol., 116:53-84.
37. Merrell, R., Pulliam, M.W., Randono, L., Boyd, L.F., Bradshaw, R.A., and Glaser, I. (1975): Proc. Natl.Acad. Sci. USA, 72:4270-4274.
38. Moscona, A.A. (1974): In: The Cell Surface in Development, edited by A.A. Moscona, pp. 67-99. Wiley, New York.
39. Mullen, R.J. (1977): In: Society for Neuroscience Symposia, Vol. 2, edited by W.M. Cowen and J.A. Ferrendelli, pp. 47-65. Society for Neuroscience, Bethesda.

40. Mullen, R.J., and LaVail, M.M. (1975): Nature, 258:528-530.
41. Mullen, R.J. and LaVail, M.M. (1976): Science, 192:799-801.
42. Olson, L., Ebendal, T., and Seiger, A. (1979): Dev. Neuro-sci., 2:160-176.
43. Pickel, V.M., Sumal, K.K., Miller, R.J., and Reis, D.J. (1979): Neurosci. Abstr., 5:174.
44. Pinto Lord, M.C., and Caviness, V.S., Jr., (1979): J. Comp. Neurol., 187:49-70.
45. Pledger, W.J., Stiles, C.D., Antoniades, H.N. and Scher, C.D. (1978): Proc. Natl. Acad. Sci. USA, 75:2839-2843.
46. Postlethwaite, A.E., Seyer, J.M., and Kang, A.H. (1978): Proc. Natl. Acad. Sci. USA, 75:871-875.
47. Purpura, D.P. and Suzuki, K. (1976): Brain Res., 116:1-21.
48. Rakic, P., (1974): Science, 183:425-427.
49. Rakic, P., and Sidman, R.L. (1973): J. Comp. Neurol., 152:103-132.
50. Rakic, P., and Sidman, R.L. (1973): J. Comp. Neurol., 152:133-162.
51. Rimoin, D.L. and Richmond, L. (1972): J. Clin. Endocrinol. Metab., 35:467-468.
52. Rimoin, D.L., and Schimke, R.N. (1971): In: Genetic Disorders of the Endocrine Glands, pp. 11-65. C.V. Mosby Co., St. Louis.
53. Sidman, R.L., and Rakic, P. (1973): Brain Res., 62:1-35.
54. Sokol, H.W., Zimmerman, E.A., Sawyer, W.H., and Robinson, A.G. (1976): Endocrinology, 98:1176-1188.
55. Sotelo, C. (1975): Brain Res., 94:19-44.
56. Sotelo, C. (1975): Adv. Neurol., 12:335-351.
57. Stiles, C.D., Capone, G.T., Scher, C.D., Antoniades, H.N., Van Wyk, J.J. and Pledger, W.J. (1979): Proc. Natl. Acad. Sci. USA, 76:1279-1283.
58. Swaab, D.F., Boer, G.J., Boer, K., Dogterom, J., Van Leeuwen, F.W., and Visser, M. (1978): Progr. Brain Res., 48:277-290.
59. Swisher, D.A., and Wilson, D.B. (1977): J. Comp. Neurol., 173:205-218.
60. Szutowicz, A., Frazier, W.A., and Bradshaw, R.A. (1976): J. Biol. Chem., 251:1516-1523.
61. Szutowicz, A., Frazier, W.A., and Bradshaw, R.A. (1976): J. Biol. Chem., 251:1524-1528.
62. Thoenen, H., Angeletti, P.V., Levi-Montalcini, R., and Kettler, R. (1971): Proc. Natl. Acad. Sci. USA, 68:1598-1602.
63. Williams, R.S. (1979): Ann. Neurol., 6:173.
64. Williams, R.S., Lott, I.T., Ferrante, R.J. and Caviness, V.S., Jr. (1977): Arch. Neurol., 34:298-305.
65. Wolfe, L.S., Kin, N.M.K., Baker, R.R., Carpenter, S., and Andermann, F. (1976): Science, 195:1360-1362.
66. Zeuchner, J., Rosengren, L., and Ronnback, L. (1979): J. Neurosci. Res., 4:177-182.

Neurosecretion and Brain Peptides,
edited by J. B. Martin, S. Reichlin, and K. L. Bick.
Raven Press, New York © 1981.

Problems of Peptide Analysis in Human Post-Mortem Brain

Edward D. Bird

Harvard Medical School, Boston, Massachusetts 02115; and McLean Hospital, Belmont, Massachusetts 02178

Much of the data on the distribution and concentration of neuropeptides in brain has been obtained using tissues removed from the skull immediately after death. Since this is not possible with human tissues, the validity of peptide studies of post-mortem brain must be established if clinical studies are to be meaningful. In this chapter, I review some of the methods and problems of obtaining neurochemical data from human brain.

Ehringer and Hornykiewicz's (20) original observation of decreased dopamine concentration in post-mortem brain from Parkinson's disease patients encouraged investigators to examine the neurochemistry of the brain in other neurological disorders. The inhibitory neurotransmitter, γ-aminobutyric acid (GABA), and its biosynthetic enzyme, glutamic acid decarboxylase (GAD), were found to be stable in post-mortem brain (Fig. 1). These markers of GABAergic neurons were found to be markedly decreased in the basal ganglia of brains obtained from subjects with Huntington's disease (HD) when compared to normals (38,10). Choline acetyltransferase (CAT), the biosynthetic enzyme for acetylcholine, is very stable in post-mortem tissue, and its activity has been found to be markedly reduced in the cerebral cortex of Alzheimer's disease patients (12,19,36).

Investigators have published data on the distribution and concentration of a number of neurotransmitters, as well as their biosynthetic enzymes and their receptors, in both HD and normal brain (8). The close relationship that many of the neuropeptides have with many of the classical neurotransmitters in nerve terminals within the brain (28) has stimulated an interest in measuring the concentration and distribution of these substances in the post-mortem brain in various neurological disorders.

NEUROCHEMICAL STABILITY

The post-mortem stability of neurochemical substances is of prime consideration in the evaluation of data on human brain.

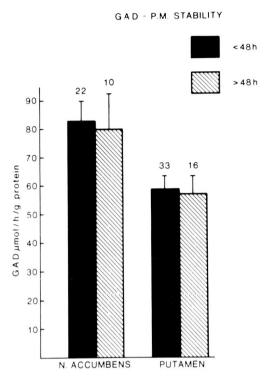

GAD - P.M. STABILITY

FIG. 1. GAD activity in the nucleus accumbens and putamen from human brain obtained from subjects who died without a neurological disorder. A comparison is made between those cases where the brain was removed within 48 hours of death and those that were removed between 48 and 72 hours after death. All corpses were placed in a 4°C refrigerator between 2 and 4 hours after death. (Reprinted with permission from ref. 30.)

Some substances are very unstable but there are many important neurochemical substances that are very stable, and neuropeptides appear to share this property. The activity of receptors in brain also appears to be stable in human post-mortem brain.

There are always a number of administrative details that need to be attended to before an autopsy can be carried out. Since these can alter the time that elapses between death and removal of the brain, it is important to determine the conditions necessary to maintain neurochemical stability in human brain. A review of the usual time and temperature changes that the human brain is subject to after death will provide a basis for establishing some guidelines in studying clinical neurological disorders.

After death, a corpse begins to cool to the environmental temperature. Human corpses, on arrival at mortuaries, are usually placed in a $4^{O}C$ refrigerator between 2 and 4 hours after death. Human tissues not subjected to these usual conditions should not be used. Control tissues obtained from cases dying without a disorder of the central nervous system and subjected to these same post-mortem conditions are essential for comparison. Since there are a number of other known and many unknown variables that exist in a heterogeneous group of human subjects, it is wise to use large groups for comparison.

In order to determine stability of a brain substance, many investigators have subjected the rat or mouse brain to time and temperatures that simulate the post-mortem human brain. However, the difference in the size of the animal brain would very likely result in temperature equilibration in a shorter time period than for human brain. Direct studies on this point have not been done. The center of a human brain takes 18 hours to reach $4^{O}C$ when the corpse is in a $4^{O}C$ refrigerator (50). Spokes and Koch (50) sacrificed mice by a blow to the head, and then placed the whole mouse in a controlled temperature chamber that duplicated the decrease in temperature that occurs in the human brain. They found that the activity of GAD decreased by about 25% in the mouse brain when subjected to human post-mortem conditions, whereas there was very little decrease in activity of the enzyme CAT.

An important question to address is how to handle the brain upon removal from the skull. Immunocytochemical studies require that the brain be placed in 4% buffered paraformaldehyde. Subcellular fractionation and uptake studies must be done on the fresh, unfrozen and unfixed brain. Since the time at which an autopsy is performed can vary, it is often inconvenient to carry out studies that require fresh brain. The majority of neurochemical measurements in man have been on frozen brain.

There have been a few studies on neuropeptide stability in human post-mortem material. We have dissected the human hypothalamus in the fresh state, chopped and divided the tissue into aliquots, and then placed the samples in a refrigerator for varying periods up to 7 days. Dr. George Fink at Oxford University found no loss of gonadotropic releasing hormone (GnRH) concentration in these aliquots as measured by radioimmunoassay (RIA) (unpublished data). Similarly, Okon and Koch kept fresh rat hypothalamus at $4^{O}C$ for 48 hours and were unable to find any loss of GnRH or thyrotropin releasing hormone (TRH) as measured by RIA (35). Whether other neuropeptides will remain stable under these circumstances is not known. For larger peptides, it will also be necessary to confirm stability by a method that will measure the whole peptide, since RIA may measure only a smaller peptide which could be a degradation product.

Neurochemical stability can be maintained when the whole brain is frozen immediately after removal from the skull. Brain tissues that are left at room temperature or homogenized and then frozen may show breakdown of a number of chemical substances. A frozen whole brain can be sliced, dissected while frozen, and the various neuroanatomical regions stored at -70°C until time of assay. Cooper et al. (17) found very little variation in somatostatin or Substance P concentrations in 10 normal human brains that were removed at varying time periods up to 24 hours after death, immediately dissected, and stored in 2N acetic acid until RIA. Peptide concentrations within these 10 brains did not correlate with the length of time stored at -70°C where brains had been kept from 5 to 9 months (17). Although this limited data would suggest that neuropeptides are very stable in post-mortem brain, what is not known is whether there is an immediate degradation of brain neuropeptides from a labile pool after death. This type of data could be obtained by sacrificing primates, immediately removing the brain, and comparing the data with brains kept under the usual post-mortem conditions. Another method might be to study human tissue removed at surgery.

CLINICAL STATUS BEFORE DEATH

The agonal effects associated with death of the patient may influence the concentrations of certain chemical substances in brain. Pneumonia and coma prior to death have been found to affect, particularly, the activity of the enzyme GAD. Bowen et al. (12) noted low post-mortem brain GAD values from patients dying with conditions likely to lead to cerebral hypoxia. Perry et al. (36) also pointed out the influence of pre-mortem status, and found that GAD activity was significantly lower in several brain regions from a group of chronically hospitalized patients than in a group of normal people who had suffered sudden death. This suggests that spuriously low GAD activities may be easily recorded in various pathological conditions in which post-mortem brain samples are obtained from elderly patients dying after chronic hospitalization, in whom the immediate cause of death is often bronchopneumonia or a related condition likely to lead to terminal hypoxia. GAD activity has been extensively studied in HD since marked reductions in the activity of this enzyme and of its product GABA have been found in the basal ganglia of patients dying with this disorder (38,10).

Iversen et al. (30) compared the GAD activities in post-mortem brain from both controls and choreics that suffered either an acute death or a prolonged death with bronchopneumonia. Although GAD activities in brain from controls that die from bronchopneumonia were lower than those dying suddenly, there was still a significant decrease in GAD activity in both the acute and prolonged death choreic groups when compared to respective control groups (Fig. 2).

GABA itself does not appear to be influenced by the agonal state since this neurotransmitter is fairly stable in post-mortem tissues from patients dying of pneumonia (37,49). Similarly, CAT activity and dopamine concentrations are not influenced by prior pneumonia (48).

The effects of the agonal state on neuropeptides have not been studied in detail. However, since many choreic patients die from

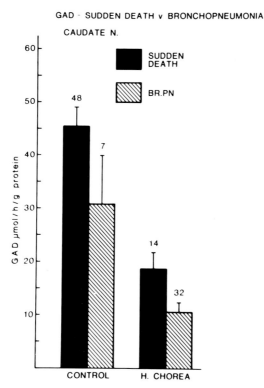

FIG. 2. A comparison of the GAD activity in the caudate nucleus of the post-mortem brain from control and Huntington's disease patients who died from either bronchopneumonia (Br. Pn.) or who had a sudden death. (Reprinted with permission from ref. 30).

bronchopneumonia, some suppositions could be drawn from the previous studies on HD brain tissues. Vasoactive intestinal polypeptide (VIP) concentrations were found to be the same in HD as in control brain where the majority of HD cases had died from pneumonia and most of the controls had not (22). Although regional differences for Substance P and somatostatin have been found in HD brain when compared to controls, there are a number of regions where there were no significant differences between control and HD brain (25). Thus far it would appear that preterminal pneumonia has little effect on neuropeptide concentrations in post-mortem brain.

It is conceivable that metabolic conditions such as diabetes mellitus, thryotoxicosis and steroid administration might affect brain neuropeptide concentrations but these have not been studied.

EFFECTS OF DRUGS

Consideration needs to be given to the effects of any medications that might have been administered prior to death. Particularly important would be knowledge of neuroleptic agents since these agents interfere with the effects of dopamine, a neurotransmitter that may affect the production and release of certain neuropeptides in the brain. A number of the endocrine changes that occur after the administration of neuroleptic agents are probably associated with neuropeptide changes in the brain. Conditions requiring long-term steroid treatment also might alter certain brain peptides.

With all the factors involved in the examination of human postmortem tissue, as discussed above, in addition to the possibility of sex and age differences, there is likely to be a wider variation in the neuropeptide concentrations measured in humans, as compared to inbred small animals. Therefore, it needs to be stressed again that to obtain representative mean values in humans, brain measurements should be carried out on as large a number of brains as possible.

NEUROPATHOLOGY

When interpreting the data on neuropeptide concentrations in human brain, some knowledge about the cellular pathology of the region under examination is needed. This applies particularly when investigating a degenerative disorder such as HD. The atrophy that occurs in the brain in this disease is reflected in the substantia nigra (SN), where there is marked atrophy of the zona reticulata region with a moderate degree of atrophy in the zona compacta (Fig. 3). The overall decrease in volume of the SN amounts to about 50% of normal (unpublished data). There is a twofold increase in the dopamine concentration in the SN in choreic brain as measured both in units per mg of tissue and as units per mg of protein (6). However, because there is atrophy of other neurons in this region, the measurements represent a relative increase in dopamine concentration in the choreic SN without a change in the total amount of dopamine in this nucleus. Ideally, the amount of neurotransmitter present needs to be related to the number of neurons producing the chemical substance. This has not been done, and indeed would be very difficult to do given the different conditions needed to count cells and carry out chemical assays on the same portion of tissue. Hopefully, immunocytochemical techniques can be developed to overcome this problem.

NEUROANATOMY

As has been previously noted with the measurement of catecholamines in post-mortem human brain, delineation of the

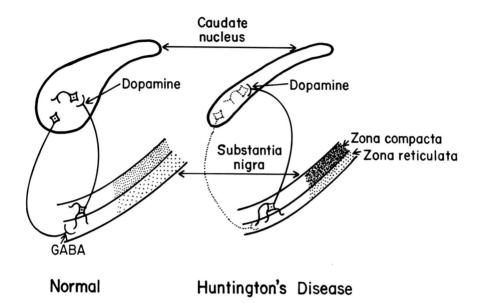

FIG. 3. Schematic representation of the striatal-nigral connections showing the dopaminergic pathways from the zona compacta
(ZONA C) of the substantia nigra to the striatum, and the GABAergic pathways from the striatum to the zona reticulata (ZONA R) in
both the normal and choreic brain. (Reprinted with permission
from ref. 8).

neuroanatomical boundaries of a nuclear area for subsequent
dissection is important (11). A great deal of the measurement
and distribution of neurochemical substances in the brain has
been carried out on small animals. Now that a number of neurochemical methods can be applied to the study of neurological disorders, there is a need to standardize the dissection of the
human brain in order that inter-investigator comparisons can be
made. Two good human brain atlases are those of Riley (45) and
Schaltenbrand (47).

Some precautions in dissecting the brain will be reviewed. When a specific area of the brain is dissected while frozen, it can be finely chopped, minced and weighed. If this tissue is to be stored, it should not be homogenized as this appears to lead to neuropeptide degradation. Similarly, any condition that might lead to the breakdown of cell membranes such as repeated thawing and freezing appears to result in degradation of neuropeptides. The weight of each area dissected ought to compare favorably with the weights for that area obtained by other investigators. If the total content of a neurochemical substance within a specific nucleus is required, ideally the whole sample should be used for the assay, unless previous measurements on a large number of aliquots from the area agree favorably with one another. Where there is an interest in comparing a presynaptic substance with the receptor density for that substance within a region, this should be carried out on aliquots from a homogenate consisting of the whole region. This may create problems where the homogenizing medium is different for different assays, and preliminary experimentation may be required to find a suitable, common homogenizing medium.

With sensitive RIA methods to measure neuropeptides, there is interest in measuring their concentrations in more minutely defined regions. Studies in the past have been reported on the whole SN, but the neurons more medially placed in the SN have their terminals in different areas of the brain than those originating in the lateral SN; so, there may be a need to further subdivide the zona compacta and zona reticulata into both medial and caudal regions and also, probably, rostral and caudal regions. Substance P has been shown to vary within regions of the SN and globus pallidus (31). A useful way to document areas taken for analysis is to photograph the slice of brain after a specific region has been removed.

HUNTINGTON'S DISEASE

Since most of the neuropeptide studies in man have been reported on brain tissues from HD as compared to controls, a brief description of this clinical disorder will be given.

Huntington's disease is an autosomal dominantly inherited disorder that can be followed through many generations. Vessie (53) traced the ancestors of many choreics in the Eastern United States to three immigrants who left Bures, England, in 1630 with the John Winthrop fleet that sailed from Yarmouth to Salem, Massachusetts. Abnormal movements usually begin about the age of 42 years in both males and females, although the disorder may, on occasion, be exhibited either in childhood or late life. No predictive tests are available to diagnose those "at risk" for HD. Genetic linkage studies have failed to define a marker for this disorder, and electroencephalogram (15) and computerized axial tomography cannot detect abnormalities in those at risk.

Although the onset of abnormal movements occurs in the majority of patients between the ages of 37 and 47 years, many patients manifest behavioral changes before the onset of movements. A woman married to a choreic will often suspect which of her children is affected long before choreiform movements occur. In the majority of cases, the abnormal movements begin in the extremities along with slight twitching of the face, occurring rather infrequently at first, and then progressively involving more muscle groups throughout the whole body until the patient is unable to stand. As a result of the involvement of the muscles used for deglutition, a number of choreics have died by choking on food. Aspiration pneumonia often occurs in the terminal phases of the disease, usually some 15 years after the onset of the choreiform movements. Dementia is a feature of HD that occurs in most cases, and the earlier the onset of chorea, the more progressive and severe the dementia.

In approximately 10% of HD cases, the onset may occur under the age of 15 years. The course of the illness is more rapid than in the adult; muscle rigidity occurs early and is usually present throughout the terminal stages of the disease. The onset of choreiform movement in some families may occur in the sixth or seventh decade and, when this happens, the progress of the disease is slower than usual. The neurologic examination reveals abnormalities mainly in the motor system, with the sensory system remaining intact. Many patients with HD have generalized rigidity in the terminal stage of their disease. Reflexes are always exaggerated, even in the early stages of the illness, and ataxia is usually present.

Increased appetite is a common phenomenon in choreic patients up to the time of death, but despite increased caloric intake, patients show progressive weight loss and, in the final stages of the disease, appear cachectic.

Alzheimer (1) was the first to associate the loss of neurons in the caudate nucleus and putamen with this clinical disorder. There is a general atrophy of neurons throughout the brain, but the greatest atrophy occurs in the basal ganglia. Campbell et al. (13) noted that the brains from children with chorea had a greater degree of basal ganglia atrophy than adults. Whole brain weight is usually reduced by 20%, whereas basal ganglia weight is usually less than 50% of normal. At the level of the anterior commissure, the striatal complex is less atrophic when compared to the more posterior regions of the caudate and putamen. This anterior striatal complex contains the nucleus accumbens, part of the limbic system. Microscopically the loss of tissue is most marked in the caudate, putamen, globus pallidus, and SN. There appears to be an increased number of glial cells in the basal ganglia. However, Lange et al. (33) counted glial cells in choreic and normal brain and found that the total glial cell number in each nucleus was the same as in normal brain; thus the concentration of glial cells is increased because of the loss of neuronal cells.

MEASUREMENTS OF SPECIFIC NEUROPEPTIDES IN HUMAN BRAIN

Substance P

The distribution of Substance P in four regions of the normal human brain is shown in Table 1. More detailed dissection of the SN and globus pallidus by Kanazawa (31) has shown that Substance P concentration is twofold greater in the zona reticulata compared to the zona compacta. This suggests that Substance P is related to the numerous synapses on dendrites in the zona reticulata. Also, the internal segment of the globus pallidus was found to have a much higher concentration of Substance P than the external segment.

TABLE 1.   Neuropeptides in human brain

	Hypothalamus	Amygdala	Substantia Nigra	Cortex	Ref.
GnRH[1]	27	---	--	<4	9
TRH[2]	54	---	--	1	35
Substance P[3]	5	3	47	2	25
Neurotensin[4]	32	5	23	1	17
Somatostatin[4]	278	339	24	53	17
Gastrin/CCK[2]	6	---	--	199	52
VIP[5]	23	21	2	17	22

1 = $pg/mg^{-1}/tissue$
2 = $pg/mg/tissue$
3 = $pmol/g/protein$
4 = $pg/g/tissue$
5 = $pmol/g/tissue$

Since these are regions that are atrophic in HD, it was of interest to measure this neuropeptide in HD tissues. Substance P concentration was markedly reduced in both the globus pallidus and the SN of HD brain (25). However, within the globus pallidus, Kanazawa (31) has now shown that Substance P is reduced significantly only in the inner segment.

What is more difficult to interpret is the failure to find a significant decrease in Substance P in the atrophic HD caudate nucleus where many of the Substance P cell bodies are located. Studies in animals show that lesions placed between the caudate and SN cause a decrease in Substance P in the SN suggesting a striato-nigral Substance P pathway (32). Therefore, the most logical explanation of the loss of Substance P in the HD globus pallidus and SN would be a degeneration of Substance P cell bodies in the striatum. However, recent immunocytochemical studies have shown that Substance P probably co-exists with serotonin neurons (28). Serotonergic neurons do not appear to degenerate in HD (18), so it is possible that the Substance P within serotonergic neurons in the striatum is also maintained in HD. There are very few serotonergic neurons in the SN, and therefore there is presumably another reason for decreased Substance P in choreic SN. This indicates the importance of combining immunocytochemistry with quantitative measurements of neuropeptides. This has not been done as yet in HD brain tissue.

Gonadotropin Releasing Hormone (GnRH)

There is evidence which suggests that the release of GnRH is modulated by dopaminergic neurons (40,24), and since dopamine is increased in choreic brain, it is conceivable that there may be an alteration in concentration of this peptide in HD brain. In human brain, GnRH has been found only within the hypothalamus and pituitary/infundibulum stem (see Table 1). In the post-mortem tissues from HD patients, the concentration of GnRH was found to be significantly increased in the median eminence of the female choreic brain, but not in that of the male (9). Whether these alterations have anything to do with the increased fertility reported in female choreics (42) or the increased libido that has sometimes been noted (29) will need to wait further confirmation and until kinetic studies on this neuropeptide in HD can be made.

Somatostatin

Patients with HD often have a ravenous appetite and, in spite of this, show a progressive weight loss. Thyroid function appears to be normal, although HD patients do have an increased basal metabolic rate (BMR) (unpublished observations). The increased BMR indicates increased fat metabolism, and plasma free fatty acids (FFA) are increased in HD patients (39). Since FFA can be increased by growth hormone (GH), its measurement in HD is of interest. There appears to be a hypersensitivity of GH-producing cells to stimulation by dopamine agonists in HD (for review see ref. 7). This in turn raises the question of brain somatostatin concentrations in HD brain. Cooper et al. (17) have measured somatostatin in a number of regions of human brain, and have found the highest concentrations to be in hypothalamus and amygdala (Table 1). Reichlin has found a twofold increase of somatostatin concentrations in the choreic hypothalamus when compared to normals (see Table 2, unpublished data). This is difficult to interpret at this time, and will have to await further data on somatostatin in HD.

TABLE 2. <u>Human brain peptides in Huntington's disease vs controls</u>

	Control	Huntington's Disease	Ref.
GnRH[1]			9
Median Eminence	314.0 ± 84.0 (11)	1231.0 ± 410.0 (9)*	
SOMATOSTATIN[2]			17
Hypothalamus	1.9 ± 0.2 (19)	3.9 ± 0.4 (20)**	
SUBSTANCE P[3]			25
Substantia Nigra	47.2 ± 4.8 (13)	22.2 ± 2.3 (9)**	
Globus Pallidus	18.0 ± 3.3 (18)	9.7 ± 1.9 (15)**	
Caudate	3.7 ± 0.8 (18)	3.5 ± 0.6 (19)	
Frontal Cortex	2.4 ± 0.3 (9)	1.7 ± 0.3 (7)	
VIP[4]			22
Frontal Cortex	17.3 ± 2.3 (21)	14.2 ± 1.5 (11)	
Caudate	4.6 ± 1.3 (12)	4.2 ± 2.0 (12)	
CCK-8[4]			21
Substantia Nigra	65.0 ± 10.0 (10)	25.0 ± 3.0 (10)**	
Globus Pallidus	21.0 ± 4.0 (10)	10.0 ± 2.0 (10)**	
Caudate	79.0 ± 23.0 (10)	71.0 ± 6.0 (10)	
Frontal Cortex	137.0 ± 14.0 (10)	138.0 ± 16.0 (10)	

Two-Tailed t Test   *$p < 0.05$ **$p < 0.001$

1 = $pg/mg^{-1}/tissue$
2 = $ng/mg/protein$
3 = $pmol/mg/protein$
4 = $pmol/g/tissue$
( ) = number of brains assayed.

Thyrotropin Releasing Hormone (TRH)

Okon and Koch (35) measured both TRH and GnRH in post-mortem brain from 15 normal subjects (Table 1). TRH was found in high concentrations in the infundibulum/pituitary stalk, the posterior hypothalamic nucleus and the premammillary area of the hypothalamus. TRH was found in certain other areas of normal brain, including thalamus, pineal body and cerebral cortex.

Angiotensin

Angiotensin II may possibly be a neurotransmitter in the brain (4). Specific angiotensin II receptor binding has been demonstrated in brain tissue (5). Angiotensin converting enzyme (ACE), a dipeptidyl peptidase which converts the inactive decapeptide angiotensin I to the active octapeptide angiotensin II, is ubiquitous throughout brain tissue, although the highest activities in the rat (54) and man (4) are in the corpus striatum.

In the post-mortem tissue from HD cases, a marked depletion in ACE activity has been observed in the corpus striatum (2), and in the SN (3), suggesting the possible involvement of a striato-nigral angiotensin-containing pathway.

Neurotensin

This tridecapeptide was originally purified and characterized by Carraway and Leeman (14). Neurotensin distribution has been determined in human brain where the highest concentrations were found in the hypothalamus and the SN (17).
Although the concentrations of this peptide are low in cerebral cortex, it has been shown by Uhl et al. (51) that specific binding of neurotensin is relatively high in cortex.

Gastrin and Cholecystokinin-8

Vanderhaeghen et al. (52) reported a peptide present in human brain that reacted with antigastrin antibodies. This brain gastrin peptide (BGP) was found in much higher concentrations in cerebral cortex than in basal ganglia and brainstem. Subsequent studies (46,43) indicated that the antigastrin antibody was also measuring a similar four-peptide sequence that appears at the COOH-terminal of cholecystokinin (CCK). With the recent production of specific antibodies to other portions of the CCK peptide (44), it would appear that most of what was measured earlier as gastrin in the cortex was CCK (Table 1). The biological activity of CCK lies within the terminal eight amino acids.
Since this peptide appears to affect satiety (26), it is conceivable that the continual hunger that so many choreics exhibit may be due to alterations of this peptide in the HD brain.
Measurements of CCK-8 in HD brain reveal that the only areas that show significant differences are decreased concentrations in the globus pallidus and SN (21). There are no significant differences found in the cortex, or caudate and putamen of HD brain. CCK-8 is similar to Substance P in this selective loss in HD, since Substance P is also decreased in globus pallidus and SN but not decreased in cerebral cortex, or caudate and putamen (25).
Both of these peptides have been associated with dopaminergic neurons, CCK-8 being found within the terminals (27) and Substance P found on dopamine dendrites in SN (34). In HD, the activity of the dopaminergic neurons in the striatum is maintained, and therefore this is a possible explanation for the maintenance of CCK-8 concentrations in HD striatum.

Vasoactive Intestinal Polypeptide (VIP)

VIP is also one of the intestinal peptides that is found in high concentrations in cerebral cortex. This peptide was measured in the cerebral cortex and caudate of control and choreic brain, and no differences were noted (22).

CONCLUSION

All of the measurements of neuropeptide concentration in human brain have been by RIA, using antibodies directed against peptide chains that have been either isolated and purified from mammalian brain or from a synthetic peptide chain that has been produced based on the characteristic sequence of amino acids in the natural neuropeptide. It is conceivable that there may be slight differences in the amino acid sequence for some neuropeptides in human versus subprimate brain. This would require the isolation and sequential analysis of the peptide obtained from human brain followed by the production of antibodies to the specific human neuropeptide. It appears that the use of antibodies to nonhuman mammalian neuropeptides has not been a problem in the analysis of human brain, but as larger peptide chains are studied, this conceivably may be a problem in the future.

Some peptides share a similar sequence of amino acids, e.g., CCK and gastrin (44), and this could introduce errors in interpretation in use of an antibody directed to this common sequence.

One other problem that might occur in human tissue is that RIA methods may measure either the larger inactive precursor, or one of the degradation products of the active peptide. This may require that the biological activity of the isolated neuropeptide be measured. For example, the biological activity of Substance P can be measured by effects on the submaxillary gland (16). However, for many of the peptides that are now found in the brain, there is no reliable method of measuring biological activity. Moreover, the biological activity for a peptide such as Substance P in the submaxillary gland may well reside on a different portion of the peptide chain than the portion of the peptide that is active in the brain. It is also conceivable that within different regions of the CNS, different portions of the peptide chain may be active.

The development of biochemical techniques such as high pressure liquid chromatography to separate peptides that differ by only single amino acids gives promise of being useful for identification of peptides and their degradation products in human brain (23).

Although a number of factors need to be considered when studying neuropeptides in human brain, the fact that they appear to be fairly stable in post-mortem tissue, and the availability of sensitive methods to measure them, should encourage further investigation of these substances in various neuropsychiatric disorders.

REFERENCES

1. Alzheimer, A. (1911): Z. Gesamte Neurol. Psychiatri., 3: 891-892.
2. Arregui, A., Bennett, J.P. Jr., Bird, E.D. Yamamura, H.T. Iversen, L.L., and Snyder, S.H. (1977): Ann. Neurol., 2: 294-298.

3. Arregui, A., Emson, P., Iverson, L.L.,and Spokes, E.G.S. (1979): Adv. Neurol., 23:517-525.
4. Barker, J.L. (1976): Physiol. Rev., 56:435-452.
5. Bennett, J.P. Jr., and Snyder, S.H. (1976): J. Biol. Chem., 251:7423-7430.
6. Bird, E.D. (1976): In: Biochemistry and Neurology, edited by H.F. Bradford, and C.D. Marsden, pp. 83-91. Academic Press, London.
7. Bird, E.D. (1978): In: Advances in Neurology, edited by T.N. Chase, N.S. Wexler and A. Barbeau, Vol. 23, pp. 291-297. Raven Press, New York.
8. Bird, E.D. (1980): In: Annu. Rev. Pharmacol. Toxicol., edited by R. Geroge, R. Okun, and A.K. Cho, 20:533-551.
9. Bird, E.D., Chiappa, S.A. and Fink, G. (1976): Nature, 260: 536-538.
10. Bird, E.D., and Iversen, L.L. (1974): Brain, 97:457-472.
11. Bird, E.D., Spokes, E.G.S., and Iversen, L.L. (1979): Science, 204:93-94.
12. Bowen, D.M., Smith, C.B., White, P., and Davison, A.N. (1976): Brain, 99:459-496.
13. Campbell, A.M.G., Corner, B., Norman, R.M., and Urich, H. (1961): J. Neurol. Neurosurg. Psych., 24:71-77.
14. Carraway, R., and Leeman, S.E. (1975): J. Biol. Chem., 250: 1907-1911.
15. Chandler, J.H. (1969): In: Progress in Neuro-Genetics, edited by A. Barbeau, and J.R. Brunette, pp. 564-565. Excerpta Medica Foundation, Amsterdam.
16. Chang, M.M., and Leeman, S.E. (1970): J. Biol. Chem., 245: 4784-4790.
17. Cooper, P.E., Fernstrom, M.H., Leeman, S.E., and Martin, J.B. (1980): Neurology, (in press).
18. Curzon, G: (personal communication).
19. Davies, P., and Maloney, A.J.R. (1976): Lancet, ii:1403.
20. Ehringer, H., and Hornykiewicz, O. (1960): Klin. Wschr., 38: 1236-1239.
21. Emson, P. (1980): Transmitter Biochemistry of Human Brain Tissue Symposium, abstract, Goteborg, Sweden.
22. Emson, P.C., Fahrenkrug, J., and Spokes, E.G.S. (1979): Brain Res., 173:174-178.
23. Feldman, J.A., Cohn, M.L., and Blair, D. (1978): J. Liq. Chromatograph, 1:833-848.
24. Fuxe, K. Hokfelt, T., and Nilsson, O. (1972): Acta Endocrinol. Copenhagen, 69:625-639.
25. Gale, J.S., Bird, E.D., Spokes, E.G.S., Iversen, L.L., and Jessell, T. (1978): J. Neurochem., 30:633-634.
26. Gibbs, Y., Youg, R.C., and Smith, G.P. (1973): Nature, 245: 323-325.
27. Hokfelt, T.: Eur. J. Pharmacol., (submitted).
28. Hokfelt, T., Ljungdahl, A., Steinbusch, H., Verhofstad, A., Nilsson, G., Brodin, E., Pernow, B., and Goldstein, M. (1978): Neuroscience, 3:517-538.
29. Huntington, G.S. (1872): Med. Surg. Reporter, 26:317-321.

30. Iversen, L.L., Bird, E., Spokes, E., Nicholson, S.H., and Suckling, C.J. (1978): GABA-Neurotransmitters, Alfred Benzon Symposium, 12:179-190. Munksgaard, Copenhagen.

31. Kanazawa, I., Bird, E.D., Gale, J.S., Iversen, L.L. Jessell, T.M., Muramoto, O., Spokes, E.G.S., and Sutoo, D.T. (1979): In: Advances in Neurology, edited by T.N. Chase, N.S. Wexler, and A. Barbeau, Vol. 23, pp. 495-504. Raven Press, New York.

32. Kanazawa, I., Emson, P.C., and Cuello, A.C. (1977): Brain Res., 119:447-453.

33. Lange, H. Thorner, G., Hopf, A., and Schroder, K.F. (1976): J. Neurol. Sci., 28:401-425.

34. Ljungdahl, A., Hokflet, T., Nilsson, G., and Goldstein, M. (1978): Neuroscience, 3:945-976.

35. Okon, E., and Koch, Y. (1976): Nature, 264:345-347.

36. Perry, E.K., Gibson, P.N., Blessed, G., Perry, R.H., and Tomlinson, B.E. (1977): J. Neurol. Sci., 34:247-265.

37. Perry, T.L., Buchanan, J., Kish, S.J., and Hansen, S. (1979): Lancet, i:237-239.

38. Perry, T.L., Hansen, S., and Kloster, M. (1973): N. Eng. J. Med., 288:337-342.

39. Phillipson, O.T., and Bird, E.D. (1977): Clin. Sci. Molec. Med., 52:311-318.

40. Porter, J.C., Kamberi, I.A., Goldma, B.D., Mical, R.S., and Grazia, Y.R. (1970): J. Reprod. Fertil. Suppl., 10:39-47.

41. Poth, N.M., Heath, R.G., and Ward, M. (1975): J. Neurochem., 25:83-85.

42. Reed, S.C., and Palm, J.D. (1951): Science, 113:294-296.

43. Rehfeld, J.F. (1978): J. Biol. Chem., 253:4022-4030.

44. Rehfeld, J.F. (1980): Trends in Neurosci., 3:65-67.

45. Riley, H.A. (1943): An Atlas of the Basal Ganglia, Brain Stem and Spinal Cord. Williams and Wilkens Co, Baltimore.

46. Robberecht, P., Deschodt-Lanchman, M., and Vanderhaeghen, J.J. (1978): Proc. Natl. Acad. Sci. USA, 75:524-528.

47. Schaltenbrand, G., and Wahren, W. (1977): Atlas for Stereotaxy of the Human Brain. Second edition. Georg Thienne Publishers, Stuttgart.

48. Spokes, E.G.S. (1979): Brain, 102:333-346.

49. Spokes, E.G.S., Garrett, N.J., and Iversen, L.L. (1979): J. Neurochem., 33:773-778.

50. Spokes, E.G.S., and Koch, D.J. (1978): J. Neurochem., 31: 381-383.

51. Uhl, G.R., Bennett, J.P. Jr., and Snyder, S.H. (1977): Brain Res., 130:299-313.

52. Vanderhaeghen, J.J., Signeau, J.C., and Gepts, W. (1975): Nature, 257:604-605.

53. Vessie, P.R. (1932): J. Nerv. & Ment. Dis., 76:553-573.

54. Yang, H.-Y. T., and Neff, N.H. (1972): J. Neurochem., 19: 2443-2450.

Neurosecretion and Brain Peptides,
edited by J. B. Martin, S. Reichlin, and K. L. Bick.
Raven Press, New York © 1981.

Potential Implications of Brain Peptides in Neurological Disease

Joseph B. Martin and Dennis M. D. Landis

Department of Neurology, Massachusetts General Hospital and Harvard Medical School, Boston, Massachusetts 02114

The preceding chapters of this volume have presented detailed accounts of recent advances in our understanding of peptide functions in the nervous system. More than 30 peptides have been found in neural tissue, and all of these appear to be localized in neurons (Table 1). Of these peptides, vasopressin and oxytocin (and their associated neurophysins), thyrotropin releasing hormone (TRH), gonadotropin releasing hormone, somatostatin, Substance P, neurotensin and the opioid peptides were first isolated from brain extracts. The remainder have been identified in brain primarily by the immunologic techniques of radioimmunoassay and immunocytochemistry. Despite the recent explosion of knowledge, information about peptide biosynthesis, transport, sites of cellular storage, mechanisms of release, molecular locus of action and processes of degradation or termination of biological effect remain fragmentary.

It is difficult to predict what roles peptides will be found to have in the pathogenesis of neurological diseases. In this chapter we shall speculate freely about the potential relationships of peptides to neuron dysfunction and premature death, about trophic effects of peptides, about the significance of peptides in the cerebrospinal fluid, about peptide effects on cerebral blood vessels, and about possible influences of peptides in pain and perception.

BRAIN PEPTIDES AS NEURONAL MARKERS

Many degenerative neurological disorders, particularly those with a genetic basis, are characterized by premature cell death in discrete neuronal populations. The charting of neuronal systems by peptidergic labelling may be of considerable importance in the further delineation of the cellular basis of such neuro-

TABLE 1. Peptides found in the central nervous system

A. HYPOPHYSIOTROPHIC HORMONES

    1. Thyrotropin releasing hormone (TRH)
    2. Luteinizing hormone/follicle stimulating hormone releasing
       hormone or gonadotropin releasing hormone (GnRH)
    3. Growth hormone release inhibiting hormone (somatostatin)

B. ADENOHYPOPHYSIAL HORMONES (including OPIOCORTINS)

    1. Growth hormone
    2. Prolactin
    3. Thyroid stimulating hormone
    4. Luteinzing hormone
    5. Follicle stimulating hormone
    6. Opiocortins
        a. Adrenocorticotropic hormone (ACTH)
        b. Beta-lipotropic hormone (β-LPH)
        c. Endorphins
        d. Enkephalins
        e. Alpha-melanocyte stimulating hormone (α-MSH)
        f. Beta-melanocyte stimulating hormone (β-MSH)

C. NEUROHYPOPHYSIAL HORMONES

    1. Vasopressin
    2. Oxytocin
    3. Neurophysins

D. BRAIN-GUT HORMONES

    1. Cholecystokinin (CCK)
    2. Gastrin
    3. Insulin
    4. Vasoactive intestinal polypeptide (VIP)
    5. Motilin
    6. Bombesin
    7. Substance P
    8. Neurotensin
    9. Glucagon

E. OTHERS

    1. Bradykinin
    2. Angiotensin II
    3. Carnosine
    4. Calcitonin
    5. Sleep peptide(s)

logical disorders. As described elsewhere in these proceedings, techniques of immunocytochemistry have now improved to the pointwhere description of the neuroanatomy of peptidergic pathways in the nervous system is feasible. Immunocytochemical characterization of the cell types commonly affected in the degenerative disease process together with quantitative measurement of regional peptide concentrations may provide clues to the neuronal specificity, or lack of it, involved in the disorder. Such information could provide fundamental insights into the pathogenesis of the degenerative process. One can imagine that the metabolic processes subserving synthesis, processing, or degradation of a specific peptide might be the locus of a pathological process reflected in cell death. Or, neuronal deterioration from processes unrelated to peptide content might eventually result in disordered release of peptide and subsequent loss of trans-synaptic or trophic effects on other neuronal populations.

The concept that a specific neurotransmitter deficiency may be important in the pathophysiology and clinical manifestations of a neuronal system degeneration has been developed most completely in Parkinson's disease and to a lesser extent in Hungtington's disease (Table 2). Both disorders have highly specific neuro-pathologic features. Parkinson's disease is sporadic, except in rare instances where it is associated with a more complex clinical syndrome, whereas Huntington's disease is an autosomal dominant genetic disorder with a high degree of penetrance and clinical predictability.

TABLE 2.   Postulated neurotransmitter deficits in selected central nervous system degenerative diseases

Clinical Disorder	Proposed Neurotransmitter Defects
Parkinson's Disease	Dopamine
Huntington's Disease	GABA
	Acetylcholine
	Substance P
Alzheimer's Disease	Acetylcholine
Spinocerebellar Degeneration	Unknown
	? TRH
Motor Neuron Disease	Unknown
Congenital Insensitivity to Pain	? Substance P
Familial Dysautonomia (Riley-Day Syndrome)	? Substance P
	?Other dorsal root ganglia peptides

The anatomical, physiological and pharmacological characterization of the dopaminergic cell loss in the substantia nigra in Parkinson's disease has led to remarkable therapeutic advances (13). Many of the clinical manifestations of Parkinson's disease, which include slowness of movement, rigidity, and tremor, appear to be related to imbalance of neurotransmitter function caused by the loss of dopaminergic neurons. One of the premises underlying the study of peptide physiology in human neurological disease is the presumption that peptides exert specific effects on particular neuronal populations in the central nervous system. It is possible that the lack of a peptide, like the lack of a conventional neurotransmitter, would be reflected by clinical manifestations. It is thus reasonable to expect that one may identify clinical symptoms associated with the loss of a specific peptide function, and thereby have a potential avenue for treatment of the disorder.

Although the dopaminergic deficiency hypothesis accounts for certain of the clinical features of Parkinson's disease, it has not provided any clues to the fundamental nature of the degenerative process. Hypothalamic dopaminergic neurons are largely preserved in this disorder, as is neuroendocrine function, so that a generalized dopaminergic cellular disorder does not appear to occur (see chapter by Thorner). Undefined additional factors must contribute to the selectivity of the degenerative process. It is probable that metabolic differences exist between cells that produce the same neurotransmitter in different brain regions. An example of this is the difference of drug effects on dopamine turnover in the striatum compared to the median eminence (11,18). Immunocytochemical studies may contribute further to a definition of these differences as they succeed in unravelling the as yet perplexing observation that more than one transmitter may cohabit in the same cell (19).

Huntington's disease is a noteworthy example of an hereditary degenerative disorder which affects specific neuronal populations and which is accompanied by alterations in neurotransmitter and peptide content (4,5,14,15,23). The brains of patients dying from this disorder invariably manifest cellular loss in the putamen and caudate nuclei, and to a lesser extent in the lower laminae of the cerebral cortex. The globus pallidus is also involved (26). There is no immediately obvious reason for this pattern of cell death. The neurons in these regions originate from different areas of brain during development, differ markedly in cytological features and connectivity, and probably have dissimilar functions. The cellular loss in the putamen and caudate is accompanied by a loss of the neurotransmitter gamma aminobutyric acid (GABA), but it is clear that the content of this neurotransmitter is not in itself the cause of cell death because GABA-containing neurons elsewhere in the brain survive (4,5).

The neuronal loss in Huntington's disease is also accompanied by changes in the brain content of certain peptides. The concentration of Substance P is decreased in the globus pallidus and in the substantia nigra, brain regions which receive efferent projections from the caudate and putamen (24,25). Levels of cholecystokinin are also decreased in the basal ganglia of patients with Huntington's disease (33), while vasoactive intestinal polypeptide (VIP) levels are reported to be normal (12).

There are currently three experimental techniques which can be applied to the assessment of brain peptides in neuropathologically defined disorders. The first of these is to determine the concentrations of peptides in brain tissues obtained by biopsy or at autopsy. This analysis requires homogenization and acid extraction of the tissue and measurement of the peptide fraction by radioimmunoassay. We have recently applied such techniques to the measurement of somatostatin, Substance P and neurotensin in human brains obtained at autopsy (7,10). Peptide levels were found to be remarkably constant from brain to brain. No significant differences in peptide concentration within a given region were found to occur in brains from men or women, aged 34-80 years, obtained at intervals up to 24 hours after death. From these observations we conclude that peptide concentrations in postmortem human brain can be reliably and reproducibly measured by radioimmunoassay, after acid extraction and boiling. It is possible now to compare peptide levels in normal brains with those obtained under comparable conditions from patients dying of neurological disease.

Several limitations to this approach require emphasis, some of which have been discussed by Bird (see chapter, this volume). One limitation is the lack of an appropriate method for expression of the data in order to correct for neuronal cellular loss and the proportional increase in glial tissue that occurs. Such changes complicate expression of the data when given either in units per wet weight of tissue or per protein content. The analysis of regional concentration of peptides will assume more significance when it is possible to correlate data obtained with reliable quantitation of cell types.

A second technique which appears to show promise is immunocytochemical study of postmortem human brain. With the preliminary evidence that peptides can be demonstrated in postmortem brain by radioimmunoassay, it is reasonable to attempt to localize those peptides to specific neuronal populations by immunocytochemical techniques (Fig. 1,2). Current techniques of peptide mapping by immunocytochemistry remain plagued by inconsistency (both false positives and false negatives) and by problems of quantitation. However, it can be anticipated that technical advances, such as the use of monoclonal antibodies, will improve the specificity of the method. Immunocytochemical localization data are most meaningful when they can be correlated with assay of the tissue content of peptides obtained by other methodologies.

A third general approach to the analysis of peptide distribution is postmortem human brain is an autoradiographic technique for demonstration of peptide receptors described by Pert and colleagues (20; see also chapter, this volume). The technique consists of application of radio-labelled peptide or receptor ligand to tissue sections followed by exposure to photographic emulsion to localize the sites of binding (39). Preliminary studies in postmortem tissue, including brain and spinal cord, indicate that receptors are preserved, although precise quanti-

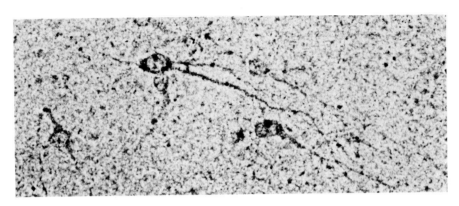

FIG. 1. Neuronal cell bodies containing leu-enkephalin-like immunoreactivity in the putamen of human brain (PAP technique, antibody courtesy Dr. A. Liotta, 515x).

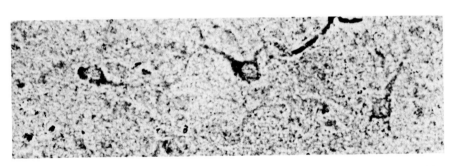

FIG. 2. Neuronal cell bodies containing Substance P-like immunoreactivity in the putamen of human brain (PAP technique, antibody courtesy Dr. S. Leeman, 515x).

tation has not yet been achieved (Pert, personal communication).

Together, these three techniques provide detailed and complementary information about peptide metabolism. Changes in tissue levels of peptide as assessed by radioimmunoassay can be correlated with cellular localization of peptides shown by immunocytochemistry, and both can be correlated with fluctuations in peptide receptor density. Such analyses are of great promise in unraveling the various neurological disorders which involve fairly consistent patterns of cellular death, including Alzheimer's disease, spinocerebellar degeneration and motor neuron diseases (Table 2).

Alzheimer's disease, which is the most common cause of dementia in the United States, is characterized by a loss of neurons in specific areas of the cerebral cortex (2). The disorder is usually sporadic, but occasionally familial, and occurs with increasing incidence in the seventh decade and beyond. Cellular loss is prominent in the association areas of the orbital-frontal, parietal and temporal cerebral cortices, and is particularly severe in the hippocampus. The nature of the degenerative process is entirely unknown. The earliest cellular lesion, the neurofibrillary tangle, includes specific, ultrastructurally identifiable changes in neurofilaments and neurotubules. Several recent studies have documented a decrease in acetylcholine content and of choline acetyltransferase activity in the cortical regions affected. There is, as yet, no information about peptide content and localization.

The spinocerebellar degenerations are a group of disorders that includes Friedreich's ataxia, olivopontocerebellar degeneration and primary cerebellar cortical atrophy. Each is characterized by loss of neurons in specific regions of the cerebellum, brainstem, and spinal cord. The precise distribution of cellular demise in affected areas is somewhat variable in each clinical category, and may even vary within a single kindred with a specific autosomal dominant disorder; as yet there is no explanation for the cell death. Interestingly, a recent report described rapid improvement of the ataxia and nystagmus associated with spinocerebellar degeneration after intravenous administration of 500 µg of TRH (36). The mechanism of this effect is entirely unknown, but it emphasizes the importance of assessing peptide physiology in the pathogenesis of a system disorder.

In motor neuron disease (amyotrophic lateral sclerosis), which is usually sporadic and occasionally familial, degeneration of motor neurons occurs in specific regions of cerebral cortex, brainstem, and the ventral horn of the spinal cord. TRH, neurotensin, and Substance P terminals are known to occur in the vicinity of motor neurons in the ventral horn of the spinal cord. The contribution of peptides to the degenerative process is unexplored.

There is a group of poorly defined sensory disorders that may be of great interest in terms of the analysis of peptidergic neurons. One of these is congenital insensitivity or indifference to pain, a condition in which the patient seems totally unreactive to pain stimuli. These patients also manifest anhidrosis, poor temperature regulation, and orthostatic hypotension, features that point to autonomic nervous system dysfunction. The cellular defect in this disorder includes loss of small neurons in the dorsal root ganglia, degeneration in the tract of Lissauer, and diminished size of the spinal tract of the trigeminal nerve. Interestingly, the neuropathological changes recognized thus far give no explanation for the commonly associated mental subnormality. Another disorder of potential importance for students of peptide function is familial dysautonomia (Riley-Day syndrome). This disorder, which is inherited as an autosomal recessive, is recognized by failure to thrive early in infancy, episodes of autonomic dysfunction, hyporeflexia, and impairment of pain and temperature sensation. Peripheral nerve biopsy reveals a loss of small myelinated and unmyelinated fibers. The disorder is presently thought to be a result of failure in neural crest cellular migration or differentiation, giving rise to a combination of dorsal root ganglion cell loss and autonomic disturbance. Associated peptide dysfunction remains to be determined.

While further assessment of the potential significance of peptide dysfunctions in human degenerative disorders is important, a note of caution should be made. As discussed above in the case of Parkinson's disease, the well characterized hypothalamic peptidergic neurosecretory cells appear to remain intact in the various degenerative neurological conditions where they have been assessed. Subtle defects in stimulated or inhibited growth hormone and/or prolaction secretion have been reported in Parkinson's disease and in Huntington's disease (see chapters by Thorner and Muller, this volume), but the results are contradictory and interpretations inconclusive. Indeed, peptidergic and biogenic amine systems of the hypothalamus appear to remain remarkably preserved in central nervous system degenerations. Thus neuroendocrine functions of the hypothalamus will not, it appears, provide a useful "window" to the study of human degenerative disease.

Perhaps the most intriguing development concerning peptide functions in relation to premature neuronal cell death will be the evolution of molecular biological probes that take advantage of peptide cellular products for tagging of cell dysfunctions. Purification of messenger RNA with preserved translation capabilities has been achieved using postmortem brain tissue (16). Attempts to determine abnormalities of cell products in single neurons obtained from Alzheimer's and Huntington's disease brain may soon be feasible.

TROPHIC FUNCTIONS OF PEPTIDES

Speculations concerning the potential importance of peptides as trophic substances within the central nervous system can be developed by analogy to the known effects of peptide hormones in peripheral tissues. In general terms, increasing molecular size of a chemical messenger in the blood is associated with greater specificity of function and with regulation of cellular events that include maintenance of target tissue mass. Circulating biogenic amines, such as norepinephrine and epinephrine, have widespread, rapid effects mediated by specific receptors that result in generalization of cardiovascular and other autonomic responses. Trophic effects are of minor importance. Vasopressin and oxytocin circulate in blood in extremely low concentrations and their effects are mediated by specific membrane receptors in responsive tissues such as kidney or uterus. These effects are transient and evidence does not support any effect on maintenance of tissue mass. ACTH, which contains 39 amino acid residues, not only acts on specific adrenal cortex receptors to stimulate immediate cortisol secretion, but also has profound trophic effects to maintain tissue integrity and mass. Hypophysectomy, with accompanying ACTH deficiency, results in adrenal atrophy. Larger polypeptides, such as thyroid stimulating hormone, growth hormone and the gonadotrophins, exert rather specific target effects via cell surface receptors, but are also of crucial importance for maintenance of cellular metabolism and target gland (or tissue) size, i.e., have trophic effects.

The peripheral effects of intermediate-sized peptides, such as gastrin, insulin, somatostatin and VIP, appear predominantly "endocrine" in the sense of immediate cellular regulatory effects, but trophic functions are clearly recognizable as well. Of particular interest is the function of the somatomedins, the formation of which is stimulated in the periphery by growth hormone. These substances consist of a family of polypeptides which are essential for somatic growth and which have close structural and functional relationships to insulin. As emphasized by Flier (see chapter, this volume), insulin acts in some tissues both on specific cell receptors to regulate glucose metabolism and also on other receptors to mediate trophic effects for maintenance of cellular growth. Interestingly, nerve growth factor, insulin and relaxin (obtained from muscle cells) have close structural homologies and similar metabolic effects.

The trophic functions of peptidergic substances within the central nervous system are not yet clear. In general terms, small molecular weight substances, such as acetylcholine, GABA, glycine and, in most cases, serotonin and norepinephrine, exert rather specific effects via receptors identifiable on either pre- or post-synaptic membranes. Many, but not all, of these effects involve alterations in ionic channels which secondarily influence transmembrane potentials and alter cellular electrical excitability. Some, such as dopamine, exert effects by activation of membrane bound adenylate cyclases. The effects of small peptides on neurons have been much more difficult to elucidate. They almost certainly include effects mediated by specific receptors (e.g., opioid peptides) or by adenylate cyclase (e.g., VIP), (see chapter by Nathanson this volume). Other biochemical effects at the cellular level can be presumed to occur within neuronal systems that respond to the administration of various peptides and cause changes in behavior, memory, alertness, specific drives (thirst, food intake), or temperature regulation. In addition, peptides may have chemotropic functions important for intercellular connectivity during development and, potentially, for nervous system regeneration. The central functions of nerve growth factor, the potential cellular effects of released precursor substances from peptidergic neurons and unidentified cellular products important for cellular interactions, as emphasized by Bunge et al. (see chapter), provide additional examples for such speculation. The extent to which peptidergic substances of low molecular weight may contribute to maintenance of developmental structure-function relationships and to plasticity within the brain is an important area for future investigation.

PEPTIDES IN CEREBROSPINAL FLUID

The assessment of in vivo neurotransmitter functions in the central nervous system in man has always been difficult and necessarily indirect. For the most part, measurement of plasma levels or of urinary excretion of brain neurotransmitter metabolites has not been shown to provide a reliable or interpretable correlation with brain or spinal cord metabolism. Both biogenic amines and peptides are found in several tissues outside the CNS, notably the gut and peripheral nerves, and blood or urine concentrations of these substances or their metabolic products are unlikely to reflect accurately changes in brain neurotransmitter function.

A variety of peptides have been detected by radioimmmunoassay in low concentration in the cerebrospinal fluid (CSF) of man (see chapter by Jackson, this volume). The immunoreactivity of peptides in CSF appears to be stable, even when fluid is kept at room temperature for several hours, suggesting that degradation is minimal. Full characterization of the molecules that demonstrate peptide immunoreactivity in the CSF has not been accomplished and further investigations are essential. The origin and fate of peptides found in the CSF are entirely unknown. It is possible, but not yet proven, that peptides in the CSF have a physiological role.

The extracellular space of the brain parenchyma is continuous with the CSF. The CSF is generally considered to be an important route for the removal of substances from brain extracellular space, but it is equally possible that material may be transported from the CSF to the brain extracellular space, and that secretion of a peptide into CSF represents one mechanism of disseminating the peptide to the brain parenchyma. While active secretion of peptides into the CSF is an appealing mechanism, it has not yet been documented. It is nonetheless fascinating that behavioral effects, more or less specific for individual peptides, can be demonstrated after introduction of peptides into the CSF either via the ventricular or cisternal route. Although the concentrations given generally exceed those normally found in CSF under physiological conditions, the specificity of the effects is striking and implies that the peptide is capable of reaching target receptor sites to evoke selective responses.

Because CSF is accessible to study, it is reasonable to pursue analysis of peptide content in a variety of neurological and psychiatric disorders which are currently unexplained. A few reports have appeared which document changes in CSF concentrations of peptides in neurological disorders or after experimental manipulation. CSF levels of Substance P are reported to be lowered in certain peripheral neuropathies and in the Shy-Drager syndrome and to be elevated in spinal arachnoiditis (see chapter by

Jackson, this volume). Studies in humans are now feasible and might be pursued in association with efforts to selectively stimulate or inhibit release of peptides by use of peripheral nerve stimulation or administration of neuropharmacologically-active drugs. It is possible that assessment of peptide concentrations in chronic pain conditions, narcolepsy, dystonia musculorum deformans, schizophrenia and manic-depressive illness may contribute useful clinical information. A limiting factor, as shown to be the case in measurement of CSF concentrations of the metabolic products of conventional neurotransmitters, is the fact that single lumbar CSF samples may reflect principally the release of spinal cord peptides.

A number of fundamental questions need to be addressed before a full understanding of peptides in CSF is achieved. These include sequential CSF sampling from lumbar, cisternal, and ventricular sites in man and experimental primate models to establish the dynamic characteristics of peptide secretion into CSF, further studies to document transport (active or passive) from blood to CSF and from CSF to blood, and anatomical studies of peptide penetration into brain extracellular space after CSF instillation. The methodologies for these and other studies are now available and should provide important information.

PEPTIDES AND CEREBRAL CIRCULATION

Autoregulation of cerebral circulation is extraordinarily complicated and only partially understood. Blood flow within brain parenchyma is regulated to a narrow range despite large fluctuations in systemic blood pressure. Neuronal activity increases local blood flow. A variety of physiological stimuli alter cerebral blood flow. for example, increased pCO_2 is accompanied by decreased cerebral vascular resistance and increased cerebral blood flow. As yet there is no evidence to suggest that peptides are involved in the normal physiology of cerebral circulation. However, it is clear that peptides with known vasoactive properties are found within the central nervous system and that axonal processes that terminate on or near blood vessels contain certain peptides.

Pathologic changes in cerebral circulation may reflect abnormalities or disorders of peptide regulatory function. Subarachnoid hemorrhage, caused by the rupture of a congenital aneurysm located on a major cerebral vessel, is often followed in seven to ten days by the occurrence of severe vasospasm in the blood vessels immediately adjacent to the aneurysm and in areas distant from it. The vasospasm may lead to severe ischemia and even infarction in the territories supplied by the narrowed vessels. Efforts to define the mechanisms of vasospasm following subarachnoid hemorrhage by examination of the effects of known neurotransmitters and their metabolites and of prostaglandins on normal and vasospastic vessels have been discouraging. It is appealing to speculate that the study of peptide physiology in relation to the cerebral circulation will yield fresh insights into this central problem in clinical neurology.

While it is not yet possible to relate peptides to cerebral circulation either in normal or pathological circumstances, it is pertinent to list the vasoactive peptides found in brain parenchyma and to review some of their properties (8,9). The peptides in the mammalian central nervous system able to produce changes in blood vessel tone or flow, either by direct or indirect effects, are listed in Table 3; of these, oxytocin, vasopressin, Substance P and vasoactive intestinal polypeptides (VIP) have been most thoroughly studied.

TABLE 3.   Vasoactive peptides in the central nervous system

1. Vasopressin or antidiuretic hormone (ADH)

2. Oxytocin

3. Thyrotropin releasing hormone (TRH)

4. Somatostatin

5. Angiotensin II

6. Substance P

7. Neurotensin

8. Bradykinin

9. Vasoactive intestinal polypeptide (VIP)

Vasopressin and Oxytocin

Vasopressin is known to play a primary role in salt and water metabolism, while oxytocin affects milk letdown and uterine contractions. In large doses both of these nonapeptides exert a direct effect on vascular smooth muscle: vasopressin causing contraction and oxytocin relaxation. In addition, as outlined by Raichle (see chapter, this volume), there is some evidence that vasopressin may act within the central nervous system to regulate water transport into the brain. Both substances and their precursor neurophysins have been observed by immunocytochemistry in extrahypothalamic areas of brain, including terminations in thalamus, brainstem, and spinal cord. Recent studies by Zimmerman and colleagues (personal communication) show that oxytocin-immunoreactive fibers also surround blood vessels at the base of the brain and have been seen to enter the midline pons in close apposition to the paramedian vessels in that region of the brainstem. Both peptides have been measured in the CSF of monkey and man. It is conceivable that release of these peptides into the

CSF or effects exerted via direct nerve terminations on blood vessels may produce changes in vascular reactivity.

Substance P

Intravenous injection of Substance P produces a transient fall in systemic blood pressure due to direct vasodilatory effects on vascular smooth muscle. The effect of intravenous Substance P on cerebral blood vessels is uncertain. Thulin et al. (37) found an increase in internal carotid artery blood flow after intravenous injection whereas Samnegard et al. (34) noted no significant changes after infusion of Substance P in patients undergoing carotid endarterectomy. Recently it was reported that high concentrations of Substance P produced contractions in the rabbit aorta, a response which was reversed by prior exposure to met-enkephalin (31). Anatomic findings suggest that Substance P-positive fibers course in the dorsal horn of the rat spinal cord in proximity to perivascular astrocytic processes (3), but it is not clear whether these fibers exert any vasoactive effect.

Vasoactive intestinal polypeptide (VIP)

VIP is a 28 amino acid peptide which causes vasodilation in a variety of vascular beds by a direct relaxation of smooth muscle. VIP has been shown to be present in nerve fibers distributed over cerebral blood vessels by Larsson et al. (27,28). Fibers containing VIP-immunoreactivity were found in the adventitia-media border of many pial vessels. There is a well developed plexus of fibers around the anterior cerebral, proximal middle cerebral and posterior communicating arteries. The adventitia of the basilar arteries also contains a few VIP-immunoreactive fibers arrayed in longitudinal course, but no fibers appeared to reach the muscle layer. VIP-containing fibers are also seen in the neuropil of the cerebral cortex and other areas without obvious relationship to vessels. The morphology and arrangement of nerve fibers containing VIP closely resemble the patterns formed by sympathetic and parasympathetic nerves. Larsson et al. (28) investigated the vasomotor effect of VIP on a segment of cat middle cerebral artery mounted in an organ bath. VIP had no effect on resting middle cerebral artery tone; however, in the presence of active tonic contraction induced by serotonin it produced a dose-related dilation, an effect that the authors interpreted to be neuromodulatory.

Currently available evidence does not permit any definite conclusions with respect to the physiological functions of peptides in cerebral vasoregulation. However, preliminary anatomic and pharmacologic observations noted above support the contention that further investigations of such properties may be informative.

PAIN AND MIGRAINE

One of mankind's first pharmacological discoveries was the recognition that the juice of the poppy could alleviate certain pains. It is presently clear that opium and its cogeners exert their effects at specific peptide receptors which are utilized in the afferent pathways that subserve pain perception (19).

Peptides appear to function in pain pathways at two principal sites: the midbrain and the spinal cord (see chapter by Fields, this volume). In the early 1970s it was shown that electrical stimulation of discrete brainstem areas, especially the ventro-lateral periaqueductal grey (PAG) of the midbrain, produced analgesia in animals (30) and man (1). Neuropharmacological studies demonstrated the presence of receptors in similar brain regions which bound opiate compounds stereospecifically (17). Considerable overlap was found between the sites for stimulation produced analgesia (SPA) and analgesia induced by intracerebral opiate injection, the PAG being the most effective site for both (29). Naloxone, a specific opiate antagonist, was found to partially block SPA (21). These data suggested that there was an endogenous substance with opiate activity which might be active in this system. Hughes et al. (22) found this to be the case when they isolated met- and leu- enkephalin from brain. Immuno-histochemically, rich concentrations of these substances were found in the PAG (35). Further evidence that opioid peptides are involved in the pain pathway at this level is the finding that analgesia produced by PAG stimulation in humans which is par-tially blocked by naloxone (21) is also associated with a rise in β-endorphin immunoreactivity in the CSF.

The small myelinated and unmyelinated peripheral nerve fibers that are considered to subserve primary pain perception are now known to contain several peptides. The evidence that Substance P may be a primary neurotransmitter for relay of pain has been reviewed by Jessell and Macdonald (see chapter, this volume). Other peptides localized to neurons of the dorsal root ganglia include somatostatin, angiotensin II, cholecystokinin (tetra-peptide) and VIP. It is likely that these same substances are present in cells of the Gasserian ganglion of the trigeminal nerve. Moreover, local neurons in laminae I-III of the dorsal horn of the spinal cord contain neurotensin, enkephalin and Substance P.

Animal studies support that feasibility of examining spinal cord release of peptides by measurement of their concentration in lumbar CSF (38). These techniques could be extended to humans. Selective peripheral nerve stimulation accompanied by analysis of peptide release from spinal cord into CSF should be undertaken. Pharmacologic and electrophysiologic studies in patients with chronic pain syndromes, peripheral neuropathy and radiculopathy may contribute to an understanding of spinal cord sensory mechanisms and to elucidation of the pathophysiology of pain, both acute and chronic.

The clinical syndrome of migraine remains a major enigma for the clinical neurologist. The disorder is a major health problem, as such headaches disturb 4-15 percent of the American population. The fundamental pathogenesis of migraine is not understood, but a prevailing hypothesis is that the neurological accompaniments (visual, sensory or motor) and the pain are caused by changes in diameter of cranial and cerebral blood vessels (6). Vasoconstriction is thought to coincide with neurological symptoms and vasodilation with the head pain. It has not, however, proven feasible to define the mechanisms which cause the blood vessels to first increase and subsequently decrease in caliber, and furthermore the distribution of the affected blood vessels and their relationship to the neurological symptoms remain poorly defined. Attempts to relate migraine to biogenic amine metabolism and vascular innervation patterns have led to unsatisfactory explanations. A fall in serum serotonin concentration has been described in the prodrome phase of classic migraine headaches and histamine release has been described in "cluster" headaches, but these observations have been inconsistent, and the suggestion for therapy based upon them have been only partially efficacious. As previously noted, peptides may play a role in the modulation of cerebral circulation and it is reasonable to pursue the possibility that abnormality of peptide function is associated with the clinical syndrome of migraine. Indeed, Moskowitz and coworkers (32) have recently speculated about the role of the trigeminal ganglia, Substance P, and other substances in the production of painful hemicranial syndromes. This area deserves further attention, if only because of its implications for the alleviation of human suffering. It is possible that an understanding of the physiology of peptides in the processing of pain information in the central nervous system will yield potential therapeutic avenues in the management of patients with migraine syndromes.

CONCLUSION

As more information is garnered concerning peptide effects in neural tissue, it can be expected that new important hypotheses will be developed to explain normal and disordered central nervous system functions. The field of endeavor is still young, but the potential for future investigations of neurologic disease is already evident. An attempt has been made in this chapter to highlight some of the areas of potential interest to the clinical neurologist, neurosurgeon and psychiatrist.

ACKNOWLEDGEMENTS

The authors thank Martha Conant, Kathy Sullivan and Rebecca Frost for secretarial assistance. Supported by USPHS grants AM26252, NS16367 and NS15573.

REFERENCES

1. Akil, H., Richardson, D.E., Barchas, J.D., and Li, C.H. (1978): Proc. Natl. Acad. Sci. USA, 75:5170-5172.
2. Alzheimer, A. (1911): Z. Gesamte Neurol. Psychiatr., 3:891-892.
3. Barber, R.P., Vaughn, J.E.,Slemmon, J.R., Salvaterra, P.M., Roberts, E., and Leeman, S.E. (1979): J. Comp. Neurol., 184:331-352.
4. Bird, E.D. (1978): In: Advances in Neurology, edited by T.N. Chase, N.S. Wexler, and A. Barbeau, 23:291-297. Raven press, New York.
5. Bird, E.D. (1980): In: Annual Review of Pharmacology and Toxicology, edited by R. George, R. Okun, and A.K. Cho, 20:533-551.
6. Caviness, V.S., Jr., and O'Brien, P. (1980): N. Engl. J. Med., 302:446-450.
7. Cooper, P.E., Fernstrom,M.H., Leeman, S.E., and Martin, J.B. (1980): Neurology, 30:374.
8. Cooper, P.E., and Martin, J.B. (1980): Annu. Neurol., (in press).
9. Cooper, P.E., and Martin, J.B. (1980): In: Cerebral Circulation and Neurotransmitters, edited by A. Bes and G. Gerard, pp. 247-253. Excerpta Medica, Amsterdam.
10. Cooper, P.E., Fernstrom, M.H., Rorstad, O.P., Leeman, S.E., and Martin, J.B. (1980): Submitted to Brain Research.
11. Demarest, K.T., and Moore, K.E. (1979): J. Neurol. Trans., 46:263-277.
12. Emson, P.C., Fahrenkrug, J., and Spokes, E.G.S. (1979): Brain Res., 173:174-178.
13. Erhinger, H., and Hornykiewicz, O. (1960): Klin. Wochensch., 38:1236-1239.
14. Finch, C.E. (1979): Ann. Neurol., 7:406-411.
15. Gale, J.S., Bird, E.D., Spokes, E.G.S., Iversen, L.L., and Jessell, T. (1978): J. Neurochem., 30:633-634.
16. Gilbert, J.M., Brown, B.A., Strocchi, P., Bird, E.D., and Marotta, C.A.: J. Neurochem., (in press).
17. Goldstein, A., Lowney, L.I., and Pal, B.K. (1971): Proc. Natl. Acad. Sci. USA, 68:1742-1747.
18. Gudelsky, G. Al, and Moore, K.E. (1976): J. Neurol. Trans., 38:95-105.
19. Hokfelt, T., Johansson, O., Ljungdahl, A., Lundberg, J.M., and Schultzberg, M. (1980): Nature, 284:515-521.
20. Herkenham, M., and Pert, C.B. (1980): Proc. Natl. Acad. Sci. USA, (in press).
21. Hosobuchi, Y., Adams, J.E., and Linchitz, R. (1977): Science, 197:183-186.
22. Hughes, J., Smith, T.W., Kosterlitz, H.W., Fothergill, L.A., Morgan, B.A., and Morris, H.R. (1975): Nature (London), 258:577-579.

23. Iversen, L.L., Bird, E., Suckling, C.J., Spokes, E., and Nicholson, S.H. (1978): GABA-Neurotransmitters, Alfred Benson Symp., 12:179-190.

24. Kanazawa, I., Bird, E.D., Gale, J.S., Iversen, L.L., Jessell, T.M., Muramoto, O., Spokes, E.G.S., and Sutoo, D.T. (1979): In: Advances in Neurology, Vol. 23, edited by T.N. Chase, N.S. Wexler, and A. Barbeau, pp. 495-504.

25. Kanazawa, I., Emson, P.C., and Cuello, A.C. (1977): Brain Res., 119:447-453.

26. Lange, H., Thorner, G., Hopf, A., and Schroder, K.F. (1976): J. Neurol. Sci., 28:401-425.

27. Larsson, L-I., Fahrenkrug, J., Schaffalitzky de Muckadell, O., Sundler, F., Hakanson, R., and Rehfeld, J.F. (1976): Proc. Natl. Acad. Sci. USA, 73:3197-3200.

28. Larsson, L-I., kEdvinsson, L., Fahrenkrug, J., Hakanson, R., Owman, C.H., Schaffalitzky de Muckadell, and Sundler, F. Brain Res., 113:400-404.

29. Lewis, V.A., and Gebhart, G.E. (1977): Brain Res., 124:283-303.

30. Mayer, D.J., kWolfe, T.L., Akil, H., Carder, B., and Liebeskind, J.C. (1971): Science, 174:1351-1354.

31. Moore, A.F. (1979): Res. Comm. Chem. Path. Pharm., 23:233-242.

32. Moskowitz, M.A., Reinhard, J.F., Jr., Romero, J., Melamed, E., and Pettibone, D.J. (1979): Lancet, 2:883-885.

33. Rehfeld, J.F. (1980): Trends in Neurosci., 3:65-67.

34. Samnegard, H., Thulin, L., Tyden, G., Johansson, kC., Muhrbeck, O., and Bjorklund, C. (1978): Acta Physiol. Scand., 104:491-495.

35. Simatov, R., Kuhar, M.J., Uhl, G.R., and Snyder, S.H., (1977): Proc. Natl. Acad. Sci. USA, 74:2167-2171.

36. Sobue, I., Yamamoto, H., Konagaya, M., Iida, M., and Takayanagi, T. (1980): Lancet, 1:418-419.

37. Thulin, L., Samnegard, H., and Holm, I. (1977): In: Substance P, Nobel Symposium 37, edited by U.S. von Euler, and B. Pernow, P. 295. Raven Press, New York.

38. Yaksh, T.L., Farb, D.H., Leeman, S.E., and Jessell, T.M. (1979): Science, 206:481-483.

39. Young, W.S., and Kuhar, M.J. (1979): Brain Res., 179:255-270.

Neurosecretion and Brain Peptides,
edited by J. B. Martin, S. Reichlin, and K. L. Bick.
Raven Press, New York © 1981.

Epilogue

Donald B. Tower

National Institute of Neurological and Communicative Disorders and Stroke,
National Institutes of Health, Bethesda, Maryland 20205

In a very real sense this entire volume has focused on the new frontiers for neurosecretion and brain peptides. The papers of the last section have highlighted several aspects of these frontiers. A report by Jessell and Kelly (1) has already addressed some of the important issues that arose from the conference.

One important aspect is the understanding of mechanisms of biosynthesis, delivery, receptor activation and inactivation of various peptide transmitters, hormones and modulators. Solutions are needed to the problems of precursors and how they are converted to active peptides; to the problems of cellular transport and turnover; and to problems of immunospecificity in their identification. Immunobiological techniques demand special care in application and in interpretation, since antigenicity is a major variable that can be a source of artifact resulting from tracers of unidentified but highly antigenic contaminants. This issue assumes practical importance if such molecules are to be exploited effectively in studies of neuropeptides by positron emission transverse tomography (PETT).

At the heart of these issues is specificity, which is demanded at the receptor, where evidently there may be subtle differences exemplified by structural analogs of the endogenous peptides. Specificity is also important to evaluate peptide-to-peptide interactions and to elucidate transitions from receptor activation to trophic or growth factor functions. And specificity must surely extend to functional correlations. Demonstration of receptors or peptidergic tracts in themselves is not enough. Such "peptide circuitry" is often too diffuse or too general; the element of "compartmentation" or functionally specific localization is needed. Thus, the significance of peptides in cerebrospinal fluid, either for distribution or as an index of turnover in the CNS, probably requires more "compartmental" information than is now available. However, Martin has pointed out that the specificity of the chemical messenger increases with increasing molecular size, and pari passu there is a functional shift from widespread generalized effects of a transmitter like norepinephrine, through more specific and restricted actions of a small peptide like vasopressin, to functions of cellular maintenance and tissue integrity as exemplified by ACTH and by growth hormone, respectively.

This spectrum of effects must be appreciated if we are to understand the roles of peptides in the central processing functions of the nervous system and if we are to elucidate many of the neurological and psychiatric disorders. In these contexts the sensory input systems and the autonomic systems assume particular importance. There is much to be learned from studies on peripheral nerves and neuropathies, as described by Reichlin. Hopefully the application of knowledge derived from such studies may bring new dimensions to the problems of central regeneration after spinal transection or other central lesions. Pain is another example of a clinical problem that is clearly a major part of the peptide research frontier. Vasoactive peptides and problems of vasospasm, hypertension and cerebrovascular disease comprise another important area for further investigation. A third example encompasses the problems of fluid and electrolyte distribution and regulation in the CNS.

Some of the most difficult and most prevalent disorders of the human nervous system are the dementias (e.g., Alzheimer's disease) and the psychoses (e.g., schizophrenia). The studies on urinary peptides reported here by the Scandinavian group (Reichelt et al.) appear to address these issues. However, the resort to investigation of urine constituents and the "overflow" hypothesis has a long but rather discouraging history in research on psychiatric disorders. The problems of non-neuronal sources and/or artifacts remain to be addressed.

Certainly a core issue in advances along the new frontiers in peptides is methodology. A number of relevant considerations is addressed in the papers by Vale, Bird and Caviness. Relatively little attention has been devoted to comparative and ontogenetic studies in a variety of species. Data on rodents must be supplemented with data from species like cat or monkey where the maturation of the central nervous system is expanded from a week or so to several months. Moreover, selected neurological mutations often provide invaluable insights into these various themes. Undoubtedly some of the most informative advances will come from probing the characteristics of peptide systems with structural analogs (both agonists and antagonists of peptides) and by pharmacological manipulation of peptide secretion and inactivation. Ultimately investigations on human specimens and on human subjects and patients will be required. Such studies place maximal demands on methodology, not only in the interests of definitive clinical results, but also within the ethical constraints demanded by human experimentation.

Endocrinology is evolving increasingly into neuro-endocrinology. Neurotransmitters like dopamine mimic peptide hormones or releasing factors. Hormones such as the gonadal hormones may assume transmitter-like properties at specific hypothalamic neurons. Many other peptides, like the enkephalins, clearly engage in transmitter functions. And most of the transmitters, hormones and peptides exert modulatory actions on central neuronal activity. Clearly there is important and extensive feedback by hormones from peripheral endocrine organs to the central nervous system, as well as direct input from pituitary and hypothalamus to other CNS areas. Thus, endocrinological emphasis is shifting from the traditional centrifugal focus on the peripheral target organs to a centripetal neural focus. This workshop has marked a significant milestone in that shift.

REFERENCES

1. Jessell, T.M.,and Kelly, J.S (1980): Nature, 285:131.

Subject Index

A

Acetylcholine
  in cortical cell cultures, 150
  in postnatal sympathetic neurons, 289
  receptors, 115
  in sensory neurons, 180
  trophic functions, 681
ACTH
  in Addison's disease, 551
  in Cushing's disease, 551–554
  in hypertension, 368
  immunocytochemistry, 77–84
  in Nelson's syndrome, 551–554
  nonpituitary, 43–44
  normal levels, 546–551
  peptide secretion, 42
  precursor, 24, 35–42, 50, 77
  synthesis and secretion, 35–44, 541–555
Acupuncture
  in heroin addicts, 348
  naloxone reversal, 207
Adenosine monophosphate, cyclic
  (cAMP)
  control of cell proliferation, 255–256
  interactions with amines and peptides,
    599–607
  role in brain ion and volume
    homeostasis, 331–334
Adenylate cyclase
  interactions with amines and peptides,
    599–607
  role in peptide hormone synthesis,
    28, 30
Aging cells, and extracellular matrix, 257
Alytesin, 401
Alzheimer's disease, 678–679
Amines, biogenic, interactions with
    peptides and cyclic nucleotides,
    599–607
γ-Aminobutyric acid (GABA)
  inhibition of cortical neurons, 150
  inhibition of Substance P, 182–187
  trophic functions, 681
Amygdala
  neurotensin, 102

  opiate receptors, 128
  Substance P, 142
Amyotrophic lateral sclerosis, 679
Analgesia
  endorphin-mediated systems, 199–208,
    213–216, 348
  naloxone inhibition, 213–216
  neurotensin effects, 95, 103
  pain and migraine, 686–687
  role of opioid distribution, 202–204
  of stress, 215–216
  thermal, and Substance P, 193–196
Angiotensin; see also Renin-angiotensin
    system of brain
  and blood pressure elevation, 365–369
  in cerebrospinal fluid, 349, 363
  competition with Substance P and
    prostaglandins, 382–383
  immunocytochemistry, 64–65
  induction of thirst and sodium
    appetite, 347, 373–385
  neurological mechanisms of induced
    thirst, 375–382
  in postmortem brain, 668
  receptors sites, 377–382
  role in brain ion and volume
    homeostasis, 332–334
  synthesis of, 359
Angiotensinases, 362
Angiotensin converting enzyme (ACE)
  in Huntington's disease, 508
  properties of, 362
  purification of, 55–59
Angiotensinogen, 359, 362–365
Antibodies
  ACTH, 546–549
  brain receptors, 110–115
  nerve growth factor, 264, 265, 290
  renin, 362
  somatostatin, 150
  Substance P, 175–177
Antidiuretic hormone. See Vasopressin
Antidromic firing, 9, 10
Apomorphine
  and growth hormone secretion,
    531–534